W9-CRB-385

INTERNATIONAL SERIES
OF
MONOGRAPHS ON PHYSICS

SERIES EDITORS

INTERNATIONAL SERIES OF MONOGRAPHS ON PHYSICS

Physics of Strongly Coupled Plasma

V.E. FORTOV

Institute for High Energy Densities, Russian Academy of Sciences

I.T. IAKUBOV

Institute of Theoretical and Applied Electrodynamics,
Russian Academy of Sciences

A.G. KHRAPAK

Institute for High Energy Densities, Russian Academy of Sciences

CLARENDON PRESS · OXFORD
2006

OXFORD
UNIVERSITY PRESS

Great Clarendon Street, Oxford OX2 6DP

Oxford University Press is a department of the University of Oxford.
It furthers the University's objective of excellence in research, scholarship,
and education by publishing worldwide in

Oxford New York

Auckland Cape Town Dar es Salaam Hong Kong Karachi
Kuala Lumpur Madrid Melbourne Mexico City Nairobi
New Delhi Shanghai Taipei Toronto

With offices in

Argentina Austria Brazil Chile Czech Republic France Greece
Guatemala Hungary Italy Japan Poland Portugal Singapore
South Korea Switzerland Thailand Turkey Ukraine Vietnam

Oxford is a registered trade mark of Oxford University Press
in the UK and in certain other countries

Published in the United States
by Oxford University Press Inc., New York

British Library Cataloguing in Publication Data

Data available

Library of Congress Cataloging in Publication Data

Data available

Printed in Great Britain
on acid-free paper by Biddles Ltd, King's Lynn, Norfolk

ISBN 0-19-929980-3 978-0-19-929980-5 (Hbk)

1 3 5 7 9 10 8 6 4 2

PREFACE

This book is dedicated to the physical properties of dense plasmas compressed so strongly that the effects of interparticle interaction are substantial, that is, nonideal or strongly coupled plasmas. Interest in this area of plasma physics has grown considerably over the last 20–25 years when states with high energy densities, which form the basis of modern technical projects and energy applications, became accessible to experiments.

Strongly coupled plasmas are essential from the standpoint of the operation of pulsed thermonuclear reactors with inertial confinement of hot plasma, powerful magnetic-flux and magnetohydrodynamic generators, power-generating plants and rocket engines with gas-phase nuclear reactors, plasmochemical and microwave reactors, plasma generators and powerful sources of optical and X–ray radiation. In the foreseeable future, strongly compressed and heated metallized plasmas will be used as a working body similar to the water vapor in modern thermoelectric power stations. Nonideal plasmas occur when matter is affected by strong shocks, detonation and electric-explosion waves, concentrated laser radiation, electron and ion fluxes, under conditions of powerful chemical and nuclear explosions, upon pulsed evaporation of the liners of pinches and magnetocumulative generators. Nonideal plasmas occur during hypersonic motion of bodies in dense planetary atmospheres, as a result of high–velocity impact, and in numerous situations characterized by extreme pressures and temperatures. The physics of electrode, contact and electric-explosion processes under conditions of vacuum breakdown are closely related to nonideal plasma, which is essential to the operation of powerful plasma accelerators, microwave generators and plasma switches. Modern progress in the understanding of the structure and evolution of giant planets in the solar system, as well as astrophysical objects, is largely based on the ideas and results from the field of highly compressed plasmas.

Along with pragmatic interest in high–pressure plasmas, purely fundamental interest is gaining momentum, because it is in this exotic state that the major part of matter in the universe finds itself. In fact, estimations show that about 95% of matter (without taking dark matter into account) are the plasmas of stars, pulsars, black holes, and giant planets of the solar system. Plasma nonideality defines the behavior of matter in a wide range of the phase diagram, from solid and liquid to neutral gas, the phase boundaries of melting and boiling, and the metal–dielectric transition region. The last problem is now at an advanced stage of consideration in experiments on the multiple shock compression of dielectrics and their metallization in the megabar pressure range, as well as in experiments on dielectrization of strongly compressed metals.

Investigation of strongly compressed Coulomb systems is now one of the hottest and most intensively developed fundamental branches of physics, which lies at the interfaces between different fields: plasma physics, physics of the condensed state, atomic and molecular physics. To the most impressive results of the last few years one can ascribe the pressure ionization of dielectrics and experimental observation of ordering in Coulomb systems ("plasma liquid" and "plasma crystal") including strongly coupled plasmas of ions cooled by laser radiation in electrostatic traps and cyclotrons; the condensation of the optically excited excitons in semiconductors; the two–dimensional crystallization of electrons at the surface of liquid helium and hydrogen; the Coulomb "freezing" of the colloid plasma, as well as laboratory and microgravity experiments with complex (dusty) plasmas. In spite of the wide variety of objects and experimental situations, they are all united by the dominant role of the strong collective interaction.

These facts provide a permanent stable stimulus to intense theoretical and experimental studies, which have recently produced a number of interesting and, more importantly, reliable data on the thermodynamic, optical, electrophysical and transport properties of dense plasmas. This special information is contained in a wide flow of original publications. This takes place against the background of an increasing number of specialists, both researchers and engineers, who make use of strongly coupled plasmas to solve diverse fundamental and applied problems.

We have attempted to systematize, generalize, and present from a single viewpoint, the theoretical and experimental results related to this relatively new field of science. The table of contents gives a good idea of the scope of this book. We have tried to expand the discussion as much as possible to cover the cases when nonideality shows most clearly in the plasma state of matter. For this reason the interesting problems related to dense plasmas of condensed metals and semiconductors, electrolyte and colloid plasmas, as well as a detailed discussion of plasma applications, have been omitted.

The physics of strongly coupled plasmas appears to present a very difficult subject for pure theory, because the strong interparticle interaction impedes the use of conventional methods of theoretical physics. Therefore, the recent progress in understanding the properties of compressed plasmas was only made possible by the emergence of experimental data obtained through nonconventional generation and diagnostic techniques. In this case, the experimental results provide a basis for model theories, as well as for defining the range of applicability of asymptotic approximations. We have tried to maintain the natural balance between theory and experiment while giving primary consideration to physical results. In our opinion, this is what distinguishes our work from the available (and rather few) review publications in the field.

The physics of strongly coupled plasmas is developing very rapidly, with more and more applications coming to light. Naturally, the material contained in this book will likewise be expanded and complemented. We would like to thank the readers in advance for their critical comments and suggestions.

We hope that this book will prove useful to broad sections of specialists by giving them access to original works and helping them get their bearings amid the present-day problems of dense plasmas. Knowledge of standard university courses is sufficient for productive reading.

V. E. Fortov

I. T. Iakubov

A. G. Khrapak

ACKNOWLEDGEMENTS

The authors are deeply grateful to all their colleagues who helped us to perform the numerous experiments and calculations which form the basis of this book. Of great value were stimulating discussions and creative contacts with the late A. M. Prokhorov, Ya. B. Zel'dovich, L. M. Biberman, and V. M. Ievlev. The authors are also grateful to A. Ivlev and S. Khrapak who assisted with the English translation of this book.

CONTENTS

1

NONIDEAL PLASMA. BASIC CONCEPTS

1.1 Interparticle interactions. Criteria of nonideality

1.1.1 *Interparticle interactions*

At low densities, a low–temperature, partly ionized plasma can be regarded as a mixture of ideal gases of electrons, atoms, and ions. The particles move with thermal velocities mainly along straight lines, and collide with each other only occasionally. In other words, the free path times are much greater than those of interparticle interaction. With an increase in density, the mean distances between the particles decrease and the particles start spending more time interacting with each other.

Under these conditions, the mean energy of interparticle interaction increases. When this energy gets to be comparable with the mean kinetic energy of thermal motion, the plasma becomes nonideal. The properties of such a plasma become very unusual, and cannot be described by simple relationships of the theory of ideal gases.

If the plasma is fully ionized, its state is defined by Coulomb interactions, whose specific feature is their long–range character. Therefore, in a rarefied plasma the particles move in weak but self–consistent fields developed by all of the particles. The interaction energy increases with compression, however, the contribution from strong short–range interactions becomes more and more important. Finally, under conditions of strong nonideality, the role of these interactions becomes predominant.

Discussed below are the peculiarities of interparticle interactions; parameters of nonidealliy are introduced and estimated. This enables one to perform the classification of states of a nonideal plasma. The conditions of strong nonideality correspond to a high energy density in the plasma. Under natural conditions the plasma nonideality plays a substantial role in a variety of phenomena which have always attracted researchers' attention. In recent years, investigations of nonideal plasmas assumed a practical significance in technological applications. Therefore, the last section of this chapter deals with nonideal plasma in natural phenomena and in technology.

1.1.2 *Coulomb interaction. Nonideality parameter*

The criterion of ideality of a plasma with respect to the interaction between charged particles may be provided by the smallness of the ratio between the average potential energy of Coulomb interaction and the mean thermal energy characterized by the temperature T. For nondegenerate singly ionized plasma this condition has the form

$$\gamma = e^2/kTr_{\rm s} \sim e^2 n_{\rm e}^{1/3}\beta \ll 1, \tag{1.1}$$

where $n_{\rm e}$ is the electron number density, $\beta = 1/kT$, k is the Boltzmann constant, and $r_{\rm s}$ is the mean interparticle distance. The value of $r_{\rm s}$ is usually defined as the radius of the Wigner–Seitz cell and is related to the plasma density by the simple relation

$$(4\pi/3)n_{\rm e}r_{\rm s}^3 = 1. \tag{1.2}$$

Given high temperatures, multiple ionization can be attained. In this case the degrees of correlation between different species are different and, hence, different nonideality parameters for ion–ion, ion–electron, and electron–electron interactions should be used. For example, in a fully ionized plasma with ions having charge number Z we have

$$\gamma_{ZZ} = Z^{5/3}e^2\beta/r_{\rm s} = Z^{5/3}\gamma_{\rm ee},$$
$$\gamma_{Ze} = Ze^2\beta/r_{\rm s} = Z^{2/3}\gamma_{\rm ee},$$
$$\gamma_{\rm ee} = e^2\beta/r_{\rm s}.$$

Note that γ is the parameter of nonideality of a classical Coulomb system. The electrons and ions in a plasma form a classical system if they are seldom found to be at distances comparable with the thermal electron wavelength $\lambda_{\rm e} = \hbar/\sqrt{2mkT}$. Since the characteristic radius of the ion–electron interaction is Ze^2/kT, the condition of classicality can be written as

$$\lambda_{\rm e} \ll Ze^2/kT. \tag{1.3}$$

Therefore, a plasma is classical at relatively low temperatures, $kT \lesssim {\rm Ry} = e^4m/\hbar^2$. In a more heated plasma one must allow for interference quantum effects due to the uncertainty principle.

Another manifestation of quantum effects is degeneracy and, primarily, the degeneracy of electrons having the greatest thermal wavelength. The possibility of employing classical statistics is defined by the smallness of the number of electrons in an elementary volume with a linear size of $\lambda_{\rm e}$. In other words, the inequality $n_{\rm e}\lambda_{\rm e}^3 \ll 1$ should be met. The condition of applicability of classical statistics corresponds to the smallness of Fermi energy $\varepsilon_{\rm F}$ as compared to temperature:

$$\xi = \varepsilon_{\rm F}/T \ll 1,$$
$$\varepsilon_{\rm F} = (3\pi^2 n_{\rm e})^{2/3}\hbar^2/2m, \tag{1.4}$$

where ξ is the degeneration parameter.

Therefore, the isothermal compression of low–temperature plasma ($T \lesssim {\rm Ry}/k = 1.58 \cdot 10^5$ K) leads to an increase of the Coulomb interaction energy which, after the parameter γ becomes larger than unity, exceeds the kinetic energy of particle motion. This complicates considerably the theoretical description of nonideal plasma while making impossible the use of perturbation theory and forcing one to employ qualitative physical models. Further compression of the plasma causes an increase of nonideality, although to a certain limit only. The point is that with increasing density degeneracy of electrons occurs at $n_e\lambda_e^3 \approx 1$. For example, in metals $n_e \sim 10^{23}$ cm^{-3}, and electrons are degenerate at $T \lesssim 10^5$ K, i.e., almost always. The Fermi energy, which increases with the plasma density, becomes the kinetic energy scale. The quantum criterion of ideality has the form

$$\gamma_q = e^2 n_e^{1/3}/\varepsilon_F \ll 1. \tag{1.5}$$

Since $\varepsilon_F \propto n_e^{2/3}$, the parameter γ_q decreases with increasing electron density. Therefore, a degenerate electron plasma becomes even more ideal with compression.

It should be emphasized that, at higher densities, only electrons represent an ideal Fermi gas. The ion component is nonideal. Depending on the degree of its nonideality, one may talk about ionic liquids, cellular or crystalline structures, and other model representations of the ion subsystem.

It is well known that even in a rarefied plasma, when $\gamma \ll 1$, one cannot directly employ the formulas of ideal gas theory for determining the thermodynamic and transport properties of the plasma. Quantities such as the second virial coefficient or the mean free path are diverging. This is due to the specific character of the Coulomb potential, i.e., its slow decrease at large distances and infinite increase at small distances. The divergence at small distances is eliminated when quantum effects are taken into account, while the effect of charge screening by the surrounding plasma eliminates divergences of the mean free path and the second virial coefficient at large distances.

Let us consider charge screening in plasmas. As a result of the long–range character of the potential Ze^2/r, multiparticle interactions at large distances $r \gg r_s$ prove to be substantial. Consider now the charge density distribution in the neighborhood of an arbitrary test particle with charge number Z. Such a particle repels like charges and attracts charges of the opposite sign. The resulting self–consistent potential created by the selected test particle and its plasma environment is known as the Debye–Hückel potential:

$$\varphi = (Ze/r)\exp(-r/r_D), \tag{1.6}$$

where r is the distance from the test particle, and r_D is the Debye screening distance or Debye radius

$$r_D = \left(4\pi e^2\beta \sum_k Z_k^2 n_k\right)^{-1/2}, \tag{1.7}$$

where subscripts k correspond to different charged plasma species. Therefore, the Coulomb field of the test particle is screened over a distance of the order of r_D. This, in fact, leads to the convergence of the basic quantities such as the mean free path and virial coefficient. The presence in the Debye sphere of a sufficiently large number of charged particles is the essential condition of validity of expressions (1.6) and (1.7).

Let us now estimate the intensity of interparticle interaction in a Debye plasma. Note that Eq. (1.6) represents a superposition of the Coulomb potential created by the test particle and the potential created by all of the remaining particles in the plasma. Deducting the test particle potential Ze/r from Eq. (1.6) and assuming $r \rightarrow 0$ we obtain the potential created by charged particles of the screening cloud at the location of the test particle, $\varphi = Ze/r_{\mathrm{D}}$. Then, the criterion of ideality for singly charged plasma can be written as

$$\Gamma = e^2 \beta / r_{\mathrm{D}} \ll 1. \tag{1.8}$$

The parameter Γ is referred to as the plasma parameter or the Debye parameter of nonideality.

It is readily seen that the criterion (1.8) can be expressed in terms of $N_{\mathrm{D}} = (4\pi/3)n_e r_{\mathrm{D}}^3$, i.e., in terms of the number of electrons in the Debye sphere. We have $\Gamma = (2\gamma^3)^{1/2} = (3N_{\mathrm{D}})^{-1}$ and, therefore, the plasma ideality criterion, i.e., the smallness of the energy of the Coulomb interaction as compared with kinetic energy, coincides with the condition of applicability of the Debye approximation: In both cases the number of charged particles in the Debye sphere must be large, $N_{\mathrm{D}} \gg 1$.

Under conditions when the electron component is degenerate, $n_e \lambda_e^3 \gg 1$, the screening distance by degenerate electrons is defined by the Thomas–Fermi length,

$$r_{\mathrm{TF}} = (\pi/3n_e)\sqrt{\hbar^2/4me^2}.$$

In a two–component electron–ion system, in which the electrons are degenerate while ions remain classical, the screening distance of the test charge is defined by the expression

$$r_{\mathrm{scr}}^{-2} = (r_{\mathrm{TF}}^{\mathrm{e}})^{-2} + (r_{\mathrm{D}}^{\mathrm{i}})^{-2} \approx (r_{\mathrm{D}}^{\mathrm{i}})^{-2}.$$

Hence the screening is mostly due to the ion component.

As the density increases, the Debye screening distance decreases and may become smaller than the interparticle distance r_{s}. Under these conditions, r_{D} loses the meaning of screening distance. For example, such is the situation in a liquid metal, where the charge is screened over distances of the order of the radius of the Wigner–Seitz cell.

Another characteristic singularity of the Coulomb potential is its behavior at small distances which makes a purely classical Coulomb system unstable. However, it is at small distances (of the order of the wavelength λ_e) that quantum effects leading to the formation of atoms and molecules become appreciable. One can speak of short–range quantum repulsion which is connected to the effect of the uncertainty principle and does not allow close approach of two particles with a preset relative momentum. This eliminates the divergencies and leads to quantum corrections at low temperatures (at $kT \lesssim \mathrm{Ry}$ in a plasma with singly

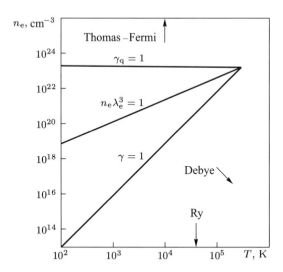

FIG. 1.1.   Ranges of existence of nonideal classical and degenerate low–temperature plasma.

ionized ions). Such a plasma is referred to as a "classical plasma". In the opposite case the plasma is called a "quantum plasma".

The ranges of existence of nonideal classical and degenerate plasma are given in Fig. 1.1. They are defined by the conditions $\gamma = 1$, $n_e \lambda_e^3 = 1$, and $\gamma_q = 1$. Also shown in the diagram are the ranges of applicability of limiting approximations describing the states of a weakly nonideal plasma, namely, the Debye and Thomas–Fermi approximations.

As we can see, only the states with extremely high pressures and temperatures located at the periphery of the phase diagram of matter are accessible to consistent theoretical analysis. The range of existence of nonideal plasma appears in Fig. 1.1 to be located within the triangle defined by the condition $\varepsilon_F = e^2 n_e^{1/3}$ and $\gamma = 1$, with the upper portion of this range relating to a degenerate plasma and the lower portion to a Boltzmann plasma, while the maximum possible values of the nonideality parameter are finite and do not exceed several units. At higher densities the plasma remains strongly nonideal due to ion–ion interaction as long as the ions are not degenerate.

1.1.3   *Electron–atom and ion–atom interactions*

Given the low ionization fraction in a low–temperature plasma, the interactions between charged particles can be ignored. Here, the interactions between charged and neutral particles can be dominant in view of the high density of the neutrals. In this case, the electron and ion components of a weakly ionized plasma cannot be described in an ideal–gas approximation. The nonideality due to charge–neutral interaction is appreciable primarily for the properties caused by the presence of charged particles, such as the electrical conductivity, thermal conductivity, and thermoelectromotive force.

In order to estimate the importance of charge–neutral interaction, let us calculate the potential produced by atoms (molecules) at the ion (electron) location. Contributions to the ion–atom (electron–atom) interaction are made by the exchange, electric, and polarization forces. Because of their long–range character, it is primarily the polarization force that is appreciable at low densities. This is especially true for the ions. An atom polarized by an ion sets up at the ion location a potential $\varphi(r) = -\alpha e/2r^4$, where r is the distance between them and α is the atomic polarizability. It is also assumed that $r > r_a$, where r_a is the atomic radius. The total potential set up by all atoms at the ion location is

$$\varphi = 4\pi \int_{r_a}^{\infty} \varphi(r)n_a(r)r^2 dr, \qquad (1.9)$$

where $n_a(r)$ is the atomic concentration which depends on the distance from the ion. If the interaction is not yet strong, this dependence (ion–atom correlations) can be neglected. Then,

$$\varphi = -2\pi e n_a \alpha/r_a.$$

The ideality criterion can be written as

$$\gamma_{ai} = 2\pi\alpha e^2 n_a\beta/r_a \ll 1, \qquad (1.10)$$

where r_a is the radius of the polarization interaction "cut-off", which is close to the atomic size. The nonideality caused by charge–neutral interaction can occur in highly polarizable gases such as metal vapors. For cesium, $\alpha = 400a_0^3$, and $r_a = 4a_0$, where a_0 is the Bohr radius. In a plasma of cesium vapors at $T = 2000$ K, $\gamma_{ai} \lesssim 0.1$ as long as $n_a \lesssim 10^{19}$ cm^{-3}.

The electron–atom interaction potential has the same polarization asymptote $\varphi(r)$; however, this potential may not always be determined unambiguously at small distances. Full information on the electron–atom interaction is contained in the scattering phases. A set of those phases enables one to calculate the mean interaction energy using the Beth–Uhlenbeck expressions (Landau and Lifshitz 1980). This leads to rather bulky expressions. However, the electron–atom interaction at low temperatures can be described by a single parameter, namely, the electron–atom scattering length L (as long as $n_a|L|^3 \ll 1$). Then, in solving a number of problems, the real potential $V(r)$ can be replaced with a δ–like potential

$$V(\mathbf{r}) = 2\pi\hbar^2 L\delta(\mathbf{r})/m. \qquad (1.11)$$

In this approximation, one can readily calculate the electron–atom interaction energy:

$$U = \int n_a(\mathbf{r}')V(\mathbf{r} - \mathbf{r}')|\Psi(\mathbf{r})|^2 d\mathbf{r}d\mathbf{r}',$$

where $\Psi(\mathbf{r})$ is the electron wavefunction. If the interaction is weak, one may neglect the electron–atom correlations and, using Eq. (1.11), obtain

$$U = 2\pi\hbar^2 L n_{\mathrm{a}}/m. \qquad (1.12)$$

The criterion of plasma ideality relative to the electron–atom interaction takes the form

$$\gamma_{\mathrm{ae}} = 2\pi|L|\hbar^2 n_{\mathrm{a}}\beta/m \ll 1. \qquad (1.13)$$

The approximation (1.11) is valid if $\lambda_{\mathrm{e}} \gg |L|$, i.e., at low temperatures.

Due to their high intensity and long–range character, the plasma interactions (Coulomb and polarization) can be strong under conditions when the interaction between neutral particles is still minor. Simple inequalities can be written using the van der Waals type equation of state

$$(p + n_{\mathrm{a}}^2 a) = \frac{n_{\mathrm{a}}kT}{1 - n_{\mathrm{a}}b}.$$

Then, the ideality criteria take the form

$$n_{\mathrm{a}}b \ll 1, \qquad n_{\mathrm{a}}a\beta \ll 1, \qquad (1.14)$$

where a and b are the parameters in the van der Waals equation, which can be expressed in terms of the critical temperature and density

$$T_{\mathrm{c}} = 8a/27b, \qquad n_{\mathrm{c}} = 1/3b. \qquad (1.15)$$

Interatomic interactions are substantial at densities close to and higher than the critical values and, in vapors, in the range of loss of thermodynamic stability.

A discussion of plasma nonideality criteria can also be found in Vedenov (1965); Kikoin et al. (1966); Kudrin (1974); Khrapak and Iakubov (1981); Iosilevskii (2000); and Iosilevskii et al. (2000).

1.1.4   Compound particles in plasma

Consider first how atoms are formed in a plasma. We shall proceed from a two–component classical plasma consisting of electrons and singly charged ions only. The inequality (1.3) can be rewritten as

$$\beta\mathrm{Ry} \gg 1, \qquad \mathrm{Ry} = me^4/2\hbar^2. \qquad (1.16)$$

It follows from this inequality that the energy of the bound state of an electron and ion, i.e., the bond energy of the atom, considerably exceeds the energy of thermal motion of free particles, thus pointing to the thermodynamic efficiency of the presence of atoms in the plasma. Such partly ionized plasma is often referred to as a three–component plasma, in view of the atoms and of the electrons and ions that remain unbound.

The concentrations of atoms, electrons, and ions are related by Saha equation

$$\frac{n_e n_i}{n_a} = \frac{2\Sigma^+}{\Sigma} \left(\frac{m}{2\pi\hbar^2\beta}\right)^{3/2} \exp(-\beta I), \qquad (1.17)$$

where I is the ionization energy of the atom, and Σ^+ and Σ are the internal statistical sums of ion and atom, respectively. The quantity $x = n_e(n_a + n_i)^{-1}$ is referred to as the degree of ionization.

It is important that the electron and ion bound in the atom are separated from each other by a distance of the order of the Bohr radius a_0, which is another characteristic dimension in the system. Therefore, the electron and ion making up the atom are screened by each other such that, in a first approximation, the interaction of such pairs with the surrounding particles may be generally ignored. Indeed, note that the bond energy of the hydrogen atom, $H = p + e$, is equal to $Ry = 13.6$ eV whereas the bond energy of the ion, $H^- = p + 2e$, is only 0.8 eV, while the possibility of the stability of $H^{-2} = p + 3e$ is not confirmed and only discussed in the literature.

The residual quantum effects in a classical plasma can be accounted for by corrections. While proportional to λ_e/r_D, they are small if

$$\lambda_e/r_D \sim \gamma^{3/2}(\beta Ry)^{-1/2} \ll 1.$$

Given a high density or sufficiently low temperatures, a plasma may become multicomponent. The A_2 molecules, A_2^+ molecular ions, and A^- negative ions are formed in the plasma. These are the result of interatomic interaction and charged–neutral particle interactions. The concentrations of these composite particles are calculated using equations of chemical equilibrium, Saha's equation being one of those. For instance, for the reaction of negative ion formation, $A^- \rightleftarrows A + e$, we obtain the equation

$$\frac{n^-}{n_e n_a} = \frac{\Sigma^-}{2\Sigma} \left(\frac{2\pi\hbar^2\beta}{m}\right)^{3/2} \exp(\beta E), \qquad (1.18)$$

where n^- is the density of negative ions, Σ^- and Σ are the internal statistical sums of the negative ion and atom, respectively, and E is the energy of the electron–atomic affinity.

1.2 The range of existence of nonideal plasma. The classification of states

1.2.1 *Two–component plasma*

If a plasma is fully ionized, it is two–component, with the electron concentration n_e being equal to the ion concentration n_i (for singly charged ions) due to quasineutrality. Such a plasma is characterized by two independent parameters, namely, the degeneracy parameter ξ (Eq. 1.4) and the Coulomb coupling (nonideality) parameter γ (Eq. 1.1) if $\xi \ll 1$ and γ_q (Eq. 1.5) if $\xi \gg 1$.

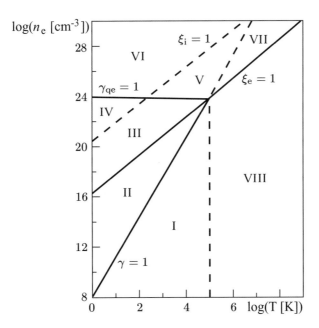

FIG. 1.2.   Regions of parameters of a two–component plasma

It is convenient to show on the (T, n_e) plane in logarithmic coordinates the lines corresponding to the following conditions: $\gamma = 1$, $\gamma_q = 1$, and $\xi = 1$. These lines will divide the (T, n_e) plane into a number of characteristic regions. This is shown in Fig. 1.2 for the case of hydrogen plasma (Klyuchnikov and Triger 1967). The degeneracy parameter for electrons is $\xi_e \sim \hbar^2 n_e^{2/3} \beta / m$, for ions $\xi_i \sim \hbar^2 n_i^{2/3} \beta / M$, where M is the ion mass. The quantum coupling parameter for electrons is $\gamma_{qe} = r_s / a_0$, where $a_0 = \hbar^2 / m e^2$, and for ions $\gamma_{qi} = (r_s / a_0)(M/m)$.

Let us now analyze the parameter regions emerging in Fig. 1.2:

Region I: $\xi_e < 1$, $\xi_i < 1$, $\gamma < 1$: a classical plasma with weak interaction of the electrons and ions.

Region II: $\xi_e < 1$, $\xi_i < 1$, $\gamma > 1$: a classical plasma with strong interaction of the electrons and ions.

Region III: $\xi_e > 1$, $\xi_i < 1$, $\gamma_{qe} > 1$, $\gamma > 1$: the electrons form a degenerate system with strong interaction while the ions form a classical system with strong interaction.

Region IV: $\xi_e > 1$, $\xi_i > 1$, $\gamma_{qe} > 1$, $\gamma_{qi} > 1$: a quantum plasma with strong interaction of the electrons and ions.

Region V: $\xi_e > 1$, $\xi_i < 1$, $\gamma_{qe} < 1$, $\gamma > 1$: the electrons form a degenerate system with weak interaction while the ions form a classical system with strong interaction.

Region VI: $\xi_e > 1$, $\xi_i > 1$, $\gamma_{qe} < 1$, $\gamma_{qi} > 1$: the electrons are degenerate and interact weakly, the ions are degenerate and interact strongly.

Region VII: $\xi_e > 1$, $\xi_i < 1$, $\gamma_{qe} < 1$, $\gamma < 1$: an electron/ion plasma with weak interactions, in which the electron component is degenerate.

Region VIII: $\xi_e < 1$, $\xi_i < 1$, $\gamma < 1$: a classical plasma with weak interaction of the electrons and ions.
It is seen from this analysis that regions I, VII, and VIII represent gaseous plasmas at various temperatures and densities; regions V and VI correspond to a solid in which the electrons form a degenerate gas with weak interaction; and in regions III and IV, the states corresponding to moderate values of γ for electrons (a typical value of γ in metals is in the range 2–5) and rather high values of γ for ions are also condensed.

Region I represents a weakly nonideal low–temperature plasma well known in the physics of gas discharge and realized in numerous natural phenomena. Region VIII corresponds to high–temperature almost ideal plasma. The plasma parameters observed in the various systems are shown schematically in Fig. 1.3; see also Ebeling *et al.* (1976) and Smirnov (1982).

In the regions of $\gamma < 1$, perturbation theory can be employed to calculate the thermodynamic quantities. In those regions the interparticle interactions are weak and the system may be regarded as a mixture of almost ideal gases. In the regions where $\gamma > 1$, perturbation theory is inapplicable; the particle interaction is appreciable and the system is similar to a liquid. Such is region III where the liquid–metal states are located. The region of classical strongly nonideal plasma, region II, is restricted from below with respect to charge concentrations by the condition of strong interaction, $\gamma = 1$, and from above by the electron degeneracy. Naturally, the regions of nonideality are restricted from above with respect to temperature as well, because at high temperatures the kinetic energy prevails over the interaction energy.

The information provided by the (T, n_e) diagram is incomplete. The regions of real existence of a stable two–component plasma are not indicated in Fig. 1.2. With a temperature decrease electrons and ions recombine, the plasma becomes partly ionized, may become molecular, and, finally, the matter condenses and crystallizes.

1.2.2   *Metal plasma*

Figure 1.4 shows in what range of temperature and density one can experimentally realize the nonideal plasma of mercury and cesium, which have the lowest values of critical temperatures for metals.

Curve 5 corresponds to the degree of ionization of $x = 0.5$. This curve conventionally separates the region of two–component fully ionized plasma from the region where the ionization is only partial. A rise in temperature causes an increase of the configurational weight of free states of electrons and ions. As a result, thermal ionization occurs. In accordance with this, the high–temperature branch of curve 5 is constructed using Saha equation (1.17). It is known, however, that, with a strong interparticle interaction, the conventional Saha equation does not describe real ionization equilibrium. One of the most important effects

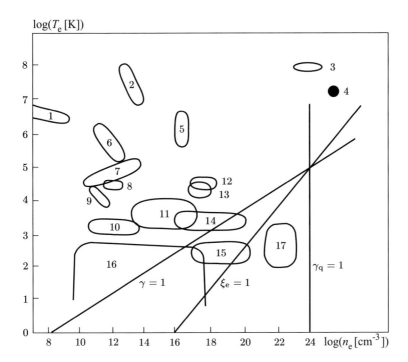

FIG. 1.3.   Plasma parameters realized in nature and in various technical devices (Ebeling *et al.* 1976): 1, solar corona; 2, tokamak; 3, laser–induced fusion; 4, core of Sun; 5, Z–pinch; 6, stellarator; 7, gas lasers; 8, plasmotron; 9, chromosphere of Sun ; 10, plasma of hydrocarbon fuel combustion products; 11, electric arcs; 12, cathode spot; 13, spark; 14, MHD generator utilizing nonideal plasma; 15, semiconductor plasma; 16, metal–ammonia solutions; 18, metals.

is that, in becoming strong, the interaction leads to a decrease in the bond energy of atoms or, in other words, reduces the ionization potential. Given a very strong interaction, the bound states of the electron and ion completely disappear. It is sometimes said that pressure ionization takes place. These situations correspond to the high–density branch of curve 5.

The region occupied by a nonideal electron plasma on the low–temperature side is limited by the curves of the existence of vapor and liquid. On the high–density side, the region of classical nonideality is limited by the electron degeneracy, $n_e \lambda_e^3 = 1$. At $\rho > \rho_c$, the metallization of the plasma occurs. The states above curve 4 can be regarded as liquid–metal ones. It should be noted, however, that the direct measurements in mercury by Kikoin *et al.* (1966) have shown that a truly metal state exists only at $\rho = 11$ g cm^{-3}, i.e., at $n_a = 3.3 \cdot 10^{22}$ cm^{-3}.

On the low–density side, the region of nonideality is defined by the condition of equality between the interaction energy and thermal energy. The charge–neutral interaction becomes strong at $n_a \gtrsim 10^{20}$ cm^{-3} in cesium and at $n_a \gtrsim 10^{21}$

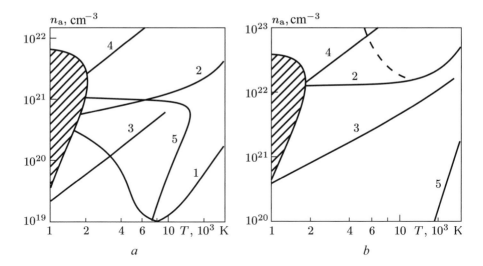

FIG. 1.4. Density–temperature diagram for cesium (a) and mercury (b). The numbered curves are calculated from the following conditions: $\Gamma = 1$ (1); $\gamma = 1$ (2 or dashed line if n_e is determined according to Saha equation); $\gamma_{ai} = 1$ (3); electron degeneracy (4); $x = 0.5$ (5). The regions of liquid–vapor coexistence are shaded.

cm^{-3} in mercury (the large difference in these values of n_a is mainly due to the difference in polarizability). In the plasma of metal vapors it is this interaction which becomes pronounced with an increase of n_a at low temperatures. This interaction is responsible for a whole number of qualitatively new effects observed in metal vapors at temperatures in the neighborhood of the critical temperature.

The charge–neutral interaction shifts the ionization equilibrium toward the increase of n_e. At high densities the degree of ionization does not drop down with an increase of n_a according to Eq. (1.17), but increases. As a result, the ionization observed in the neighborhood of the critical point of cesium is almost full while in the neighborhood of the critical point of mercury the degree of ionization x amounts to several tens of percent.

In view of this, in Fig. 1.4 are plotted the curves $\Gamma = 1$ and $\gamma = 1$ indicative of nonideality in interaction between charges. In order to illustrate the extent of the discrepancy between the ionization equilibrium in nonideal plasma and ideal plasma equlibrium (Eq. 1.17), further shown in Fig. 1.4(b) with a dashed line is a $\gamma = 1$ curve calculated using Saha equation (1.17).

It is evident from Fig. 1.4 that the region occupied by a non–Debye plasma is very wide. In the last few years it was subjected to very intensive studies. Much less data are available on a strongly nonideal plasma in which $\gamma \gtrsim 1$, i.e., $\Gamma \gtrsim 5$.

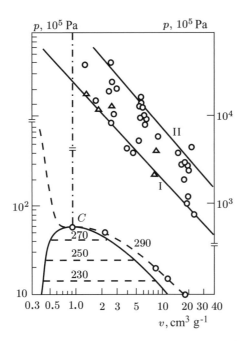

$\textsc{Fig.}$ 1.5.   Phase diagram of xenon (Iosilevskii *et al.* 2000). Lines I and II correspond to the limits of single and double ionization, respectively; dashed line show isotherms; *C* is the critical point; symbols correspond to the available measurements of thermodynamic parameters.

1.2.3 *Plasma of hydrogen and inert gases*

As distinct from alkali metals and mercury, in the case of an inert gas plasma one manages to realize an extremely wide range of parameters and continuously follow the variation of the physical properties of such a plasma from gas ($n_e \sim 10^{14}$ cm^{-3}) to solid state ($n_e \sim 10^{23}$ cm^{-3}) densities. One deals with extremely high pressures, i.e., from hundreds to a million atmospheres at maximum temperatures up to 10^5 K. Under these strongly supercritical conditions, a plasma with developed ionization is realized.

Of particular interest is the investigation of properties of hydrogen (the most widely–distributed substance in the universe) plasma. At high pressures and densities the electrical conductivity of hydrogen sharply increases up to the values which are characteristic of metals, mostly due to pressure ionization. This affects the properties of giant planets, for instance, the value of their magnetic field. The intriguing possibility of the existence of a metastable metallic and even superconductive phase of hydrogen at zero pressure is being investigated.

In the phase diagram of a xenon plasma, shown as an example in Fig. 1.5, solid curves indicate the regions of single and double ionization, while symbols correspond to thermodynamic measurements. In view of the high temperatures and peculiarities of electron shells, molecular and cluster formations are absent in such plasma, while the Coulomb interaction is the prevailing type of interparticle interaction because of the small number of neutrals. The nonideality parameter reaches, in this case, maximum values of about $\gamma = 10$. With the maximum plasma densities realized in experiment, $\rho = 4.5\text{–}9.7$ g cm^{-3} (which exceeds by several times the solid state density of xenon of $\rho = 3$ g cm^{-3}), the electron component of such a plasma is partly degenerate, the parameter $n_e \lambda_e^3$ reaching values close to 0.5.

A characteristic feature in the description of inert gas plasma when compressed to supercritical densities is the need to allow for atom–atom and ion–ion repulsion. Indeed, at $\rho > \rho_c = 1.1$ g cm^{-3}, the characteristic interparticle distance in the plasma is comparable to the size of atoms and ions in the ground state ($\sigma_{Ar} = 3.4 \cdot 10^{-8}$ cm, $\sigma_{Xe} = 4 \cdot 10^{-8}$ cm) and, the more so, in excited states. The dimensionless parameter $n_a \sigma^3$ characterizing this interaction reaches in xenon a value of 0.2–0.3.

Finally, in a high–temperature plasma ($T > 3 \cdot 10^4$ K) the amplitude of the Coulomb scattering for the electron Ze^2/T appears comparable to the characteristic ion size. This leads to non–Coulomb behavior of electron scattering. These conditions are attained during compression of a gas in strong shock waves and the conditions with more moderate parameters, during adiabatic compression. The principles of dynamical plasma generation are discussed in Chapter 3, while the properties of strongly compressed hydrogen and inert gases have discussed in Chapter 9.

1.2.4 *Plasma with multiply charged ions*

Powerful pulsed energy contributions to condensed matter give rise to a nonideal superdense plasma with multiply charged ions. The density of such a plasma is close to that of the condensed state, the pressure reaches several terapascals, and the temperature is of the order of tens of electron volts. The plasma consists of electrons and highly charged ions, $Z \gtrsim 10$. In expanding, it passes through a whole gamut of peculiar states, in which the degree of degeneracy of the electron component also varies.

Relatively uniform volumes of such plasma are produced under laboratory conditions during compression of porous metals by powerful shock waves ($n_e \lambda_e^3 \approx 0.5\text{–}2$, $\gamma_q = 10$, $\Gamma = 2$).

During adiabatic expansion of shock–compressed metals, a wide range of plasma states is realized from a highly compressed metallic liquid to a weakly nonideal classical plasma.

If the electrons are nondegenerate, interaction of three different types is observed, namely, ion–ion, ion–electron, and electron–electron. The nonideality parameters for these interactions are

$$\gamma_{ZZ} \sim Z^2 e^2 n_i^{1/3} \beta, \qquad \gamma_{Ze} \sim Z e^2 n_e^{1/3} \beta, \qquad \gamma_{ee} \sim e^2 n_e^{1/3} \beta. \qquad (1.19)$$

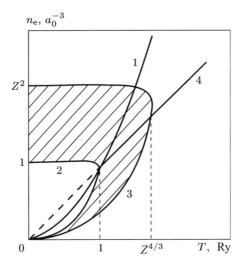

FIG. 1.6. Diagram of the parameters of highly compressed matter (Kirzhnits *et al.* 1975).
Curve 1 corresponds to electron degeneracy, curve 2 indicates the condition $\gamma_{ee} = 1$, curve
3 indicates the condition $\gamma_{Ze} = 1$, and curve 4 corresponds to the equality of the exchange
and quantum–mechanical effects to the correlation effects.

If $Z \gg 1$, then, in view of the quasineutrality condition, one can write the
inequalities

$$\gamma_{ZZ} \gg \gamma_{Ze} \gg \gamma_{ee}. \tag{1.20}$$

If the electrons are degenerate, the values of their interaction energy must be
compared with that of the Fermi energy (1.5). The ions still remain very far from
degeneracy. Therefore,

$$\gamma_{ZZ} \sim Z^2 e^2 n_i^{1/3}\beta, \qquad \gamma_{Ze} \sim Z e^2 n_e^{1/3}/\varepsilon_F, \qquad \gamma_{ee} \sim e^2 n_e^{1/3}/\varepsilon_F.$$

and the hierarchy (1.20) is maintained. This implies the possibility of the existence of a number of interesting physical systems. These include a plasma formed
by a quasicrystal system of multiply charged ions ($\gamma_{ZZ} \gg 1$) and an ideal electron gas ($\gamma_{ee} \ll 1$) with different intensity of interaction with ions.

In the diagram of the parameters of highly compressed matter (Fig. 1.6), a
region is shaded in which the energy of interaction of a pair of electrons is low as
compared with their kinetic energy. However, the electrons interact with nuclei,
as well as with their numerous partners that are present in the sphere effectively
covered by the Coulomb forces. At $Z \gg 1$, these interactions are far from being
small. In the shaded region, the plasma is divided into Wigner cells and may
be described by the Thomas–Fermi method with corrections or by the method
of functional of the energy density (thermodynamic potential) (Kirzhnits *et al.*
1975).

1.2.5 *Dusty plasmas*

Over the last decade much interest has been attracted to investigations of properties of nonideal dusty plasmas (Yakubov and Khrapak 1989; Fortov *et al.* 1999; Bouchoule 1999; Shukla and Mamun 2002; Tsytovich *et al.* 2002; Piel and Melzer 2002; Merlino and Goree 2004; Vladimirov and Ostrikov 2004; Fortov *et al.* 2004; Ignatov 2005; Fortov *et al.* 2005), which consist of electrons, singly charged ions, and highly charged particles of the condensed disperse phase.

In a thermal plasma, the particles emitting electrons and acquiring a positive charge, can substantially increase the electron concentration in the plasma. In a nonequilibrium plasma of a gas discharge, the cold dust particles are charged negatively due to higher mobility of electrons in comparison with ions. The particle charge grows with size in both types of plasma and can achieve values of the order of $Q \sim (10^3 – 10^4) e$ for micron–size particles. The nonideality parameter describing dust–dust interactions γ_d is proportional to $Q^2 n_d^{1/3}$. For this reason, nonideality of the dust subsystem can be achieved much more easily than in the electron–ion subsystem even for relatively low dust densities. This allowed the realization of all possible states in dusty plasmas: ideal gas fully disordered state, liquid–like with short–range order in dust particle positions, and crystal–like with the clear long–range order.

At first glance dusty plasmas may seem to be a full analogue of multicomponent plasmas with multiply charged ion components. However, this is far from being true. One of the most important differences is the variability of the particle charge. The point is that that the particle charge depends on the surrounding plasma parameters such as temperature, electron and ion concentration, dust particle concentration, etc. Moreover, charge fluctuations are always present due to the stochastic nature of the charging process. This leads to a variety of unexpected and interesting effects, which can be observed in dusty plasmas. A detailed discussion of dusty plasma properties is given in Chapter 11.

1.2.6 *Nonneutral plasmas*

Usually plasmas are assumed to be neutral or quasineutral systems with equal numbers of charges of opposite signs. However, plasmas consisting exclusively of particles with a single sign of charge can also exist. Examples of plasmas that have been realized in recent experiments include pure electron plasmas, positive–ion plasmas of one or more species, positron plasmas, and electron–antiproton plasmas. Obviously, because of the strong Coulomb repulsion, nonneutral plasmas must be confined. This can be achieved by using various types of electric and magnetic traps, e.g., Penning traps, r.f. or Paul traps, Kingdon traps, storage rings, and accelerators (Davidson 1990; Dubin and O'Neil 1999). Two–dimensional electron plasmas can be also realized above the surface of liquid helium (Cole 1974; Shikin and Monarkha 1989; Leiderer 1995) or on the inner surface of multielectron bubbles in the bulk of liquid helium (Volodin *et al.* 1977; Artem'ev *et al.* 1985). Pure electron and pure ion plasmas can be confined for hours and even days in a state of thermal equilibrium. They can be cooled down to cryogenic temperatures, which allows one to observe liquid–like as well as crystal–like states. In Chapter 10 the properties of nonneutral plasmas, which can substantially differ from those of the usual quasineutral plasmas, are discussed.

1.3 Nonideal plasma in nature. Scientific and technical applications

A high–density plasma with appreciable nonideality effects is realized in numerous natural phenomena and technical devices. This includes the electron plasma in solid and liquid metals, semiconductors and electrolytes, superdense plasma of the matter of white dwarfs, the Sun, deep layers of the giant planets in the solar system, and astrophysical objects whose structure and evolution are defined by plasma properties. In studying the giant planets of the solar system, one will run, both in the literal and figurative sense, into a nonideal plasma formed as a result of hypersonic travel of space vehicles in the dense atmospheres of those planets.

Recently, growing purely pragmatic interest has been observed in high–pressure plasma studies in view of the realization of a number of major energy–related projects and devices which depend for their action on the pulsed local concentration of energy in dense media. A nonideal plasma appears to provide an advanced working medium in powerful continuous-operation and pulsed MHD generators (Biberman *et al.* 1982), power-generating facilities and rocket engines with gas–phase reactors (Gryaznov *et al.* 1980; Thom and Schneider 1971), in commercial–scale plasmochemical apparatus. A nonideal plasma is formed as a result of nuclear explosions (Ragan *et al.* 1977), explosive evaporation of liners of pinches and magnetocumulative generators, the effect of powerful shock waves, laser radiation or electron and ion beams on condensed matter, and in a number of other cases.

A special need for information on the physical characteristics of nonideal plasma arises when realizing the idea of pulsed fusion accomplished by way of laser, electron, ion or explosion compression of spherical targets (Prokhorov *et al.* 1976), as well as when solving the most important problems related to high–velocity shocks.

Of decisive importance from the standpoint of physical analysis and calculation of the hydrodynamic consequences of such effects are the data on the physical characteristics of plasmas in a wide range of phase diagram of matter from highly compressed condensed states to ideal degenerate and Boltzmann gases, including the high–temperature boiling curve and the neighborhood of the critical point.

In this section we shall mention very briefly the most representative practical situations in which a nonideal plasma is formed and utilized.

The most natural example of a nonideal plasma is provided by the plasma of conduction electrons in solid and liquid metals. We refer to a degenerate plasma ($\varepsilon_F \gg T$) with an electron concentration of $n_e = 10^{22}$–$2.5 \cdot 10^{23}$ cm^{-3}. Because

$2 \leq r_s \leq 5.6$ (here r_s is in atomic units), a strong Coulomb interaction between ions leading to crystallization is realized in such system. At the same time the interaction between electrons is weak (due to degeneracy), which allows us to consider them within the framework of an ideal degenerate gas. The effect of powerful shock waves on metals (see Chapter 3) allows us to compress such a plasma by a factor of three–four to bring the maximum values of n_e to about $6 \cdot 10^{23}$ cm^{-3} and the maximum temperatures to $5 \cdot 10^5$ K, thereby approaching the limit of degeneracy.

Liquid electrolytes , in particular, ammoniacal solutions of alkali metals (Lepouter 1965) represent a strongly nonideal plasma in a very wide range of variation of the degeneracy and interaction parameters. This is attained by varying the fraction of metal dissolved in ammonia. Under these conditions, a powerful charge–neutral interaction is realized in the system along with the strong Coulomb interaction. These interactions have for their result unusual phase transitions and anomalously high electrical conductivities attained at moderate temperatures even with small fractions of metal in the solution.

In intrinsic and impurity semiconductors the number of electrons and holes is varied over a wide range by varying the temperature and concentration of impurities. Under conditions of intense light irradiation, electrons optically excited to the conduction band form a plasma. In a number of cases, the interparticle interaction in this plasma is so strong as to lead to a phase transition, i.e., formation of an exciton liquid (Jeffries and Keldysh 1983).

According to Shatzman (1977), the parameters of the iron plasma in the center of the Sun are extremely high: $\rho \approx 120$ g cm^{-3}, $T \approx 13 \cdot 10^6$ K. The nonideality parameter $\gamma \approx 40$ realized under these conditions is apparently insufficient for plasma crystallization. Calculations using the method of molecular dynamics give the crystallization limit of $\gamma \approx 170$. However, even at lower values of γ, partial crystallization of the plasma is possible, thus causing, in particular, a variation of its optical properties determining the internal structure of the Sun.

A plasma with extreme parameters is realized at late stages of the evolution of stars in the so-called white dwarfs (Zel'dovich and Novikov 1971), decaying stars with a mass of less than 1–1.2 times the mass of the Sun. In this case, the matter is in the state of equilibrium when the pressure of a quasiuniform degenerate gas balances out the gravity forces compressing the star. At high densities, pyknonuclear reactions (quantum tunnelling of nuclei through the Coulomb barrier upon "cold" compression above the density of $\sim 10^{34}$ cm^{-3}) occur in such a plasma and, at sufficiently high temperatures, thermonuclear reactions occur. The effects of nonideality increase the rates of nuclear reactions and define the structure, stability, and evolution of these exotic objects (Ichimaru 1982; Slattery *et al.* 1980).

Experimental investigations of the giant planets of the solar system using unmanned spacecraft provide rich information about their physical properties, which stimulates construction of modern models, involving the theory of nonideal plasmas. The point is that the strong gravitational field of those planets forms a very dense atmosphere. Under conditions of hypersonic travel of space craft in such a dense atmosphere, a shock wave is formed in front of the space-craft. Within the shock wave, there occur compression and irreversible heating of the plasma to tens of thousands of degrees at pressures of hundreds and thou-sands of atmospheres. In order to provide effective protection of the spacecraft against the effects of such a plasma, as well as to ensure reliable radio communi-cation, one needs reliable data on the thermodynamic, transport, and radiation characteristics of strongly nonideal shock–compressed plasmas.

Of most importance among the numerous technical applications of nonideal plasmas are energy–related applications because the development and realization of a whole series of advanced energy-related projects are associated with ionized high–density plasmas. Along with thermonuclear systems of magnetic confine-ment of hot plasma, inertial controlled thermonuclear fusion is being developed as an alternative approach (Kadomtsev 1973; Brakner and Djorna 1973). With this approach, the thermonuclear reaction is accomplished as a "microexplosion" over a short time (several nanoseconds) defined by the time of inertial expan-sion of the hot plasma. The energy threshold of initiation in systems of inertial fusion is achieved by compressing the thermonuclear fuel of the target to a den-sity approximately 1000 times higher than that of a solid. Diverse possibilities are considered as regards the compression and heating of a deuterium/tritium mixture in spherical microtargets, namely, powerful laser or "soft" X–ray radia-tion, streams of relativistic electrons and light and heavy ions, and macroscopic liner shock. In so doing, an energy of the order of 10^6 J must be delivered to a composite layered target of about 0.1 cm over a time of about 10^{-9} s. This results in the emergence of a complicated unsteady state flow of dense nonideal plasma with a pressure of up to 10 TPa. In order to calculate the hydrodynamics of such flow and develop the optimum structures of thermonuclear microtargets, one needs extensive data on the thermodynamic, optical, and transport proper-ties of nonideal plasmas of various composition in an extremely wide range of pressures and temperatures.

A strongly nonideal plasma characterized by a wide range of parameters emerges also in the case of interaction between powerful pulsed sources of radia-tion and matter. When the surface of a solid material is subjected to a high–power laser pulse, a region of inhomogeneous nonideal plasma occurs, in which a wide range of states is realized. A relatively rarefied heated plasma ($n_e < 10^{21}$ cm^{-3}, $kT \sim 1$ keV) moves toward the laser beam and absorbs its energy. This energy is transferred to the region of strongly nonideal plasma ($n_e \approx 10^{21}$–10^{23} cm^{-3}), where at "critical" density (at which the frequency of laser radiation becomes equal to the characteristic plasma frequency) unsteady–state hydrodynamic mo-tion emerges. The strong recoil reaction causes the target material to compress, and a region of superdense matter forms in the target ($n_e > 10^{23}$ cm^{-3}), see Fig. 1.7 (More 1983).

Also directed toward the utilization of nonideal plasma is another futuris-tic energy project, that of the gas–phase nuclear reactor (Ievlev 1977). This is a

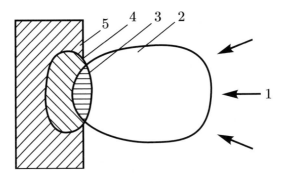

FIG. 1.7. Results of interaction between radiation and matter (More 1983). 1, laser pulse; 2, corona flame ($\rho \leq 10^{-2}$ g cm^{-3}, $kT \approx 1$ keV); 3, nonideal plasma of ablation products ($\rho \approx 0.01$–1 g cm^{-3}, $kT \approx 100$ eV); 4, superdense matter ($\rho = 10$–50 g cm^{-3}, $kT \approx 10$–100 eV); 5, cold matter of normal density.

cavity-type reactor (see Fig. 1.8) with a high–pressure uranium plasma in the center. Flowing between the uranium and the walls is a working medium heated by the thermal radiation of the uranium plasma. The mixing of the working medium and uranium is suppressed by a stabilizing magnetic field, profiling of the velocity field and by other means. Such an apparatus can provide a basis for the development of nuclear power plants, compact space-borne power-generating facilities, rocket engines, etc. (Ievlev 1977). The high temperatures and requirements of criticality result in the need to have a pressure of hundreds of atmospheres in the gas–phase reactor cavity. Under these conditions, at temperatures of tens of thousands degrees, uranium and the working media (mixtures of alkali metals and hydrogen) are in the state of nonideal plasma.

A high level of electrical conductivity with appreciable compressibility, characteristic of a nonideal plasma, renders such a plasma a suitable working medium for magnetohydrodynamic generators (Iakubov and Vorob'ev 1974; Nedospasov 1977). The operating principle of the magnetohydrodynamic (MHD) generator provides for the travel of a conducting medium in a transverse magnetic field. The power which can be extracted per unit active volume is, other things being equal, proportional to the electrical conductivity of the working medium. The electrical conductivity of a nonideal plasma may exceed the ideal plasma conductivity by several orders of magnitude at relatively moderate temperatures. A closed-cycle MHD scheme utilizing a nonideal cesium and sodium plasma is treated by Iakubov and Vorob'ev (1974).

Biberman et al. (1982) proposed a MHD power plant scheme wherein the MHD generator operates in a closed cycle utilizing a nonideal plasma of saturated cesium or potassium vapors. For the case of cesium, the thermodynamic parameters of the working medium are at the level of $T \leq 1800$ K, $p \leq 7$ MPa. The facility operates on the Rankine cycle. The cycle performed by the alkali metal (see Fig. 1.9) includes heating and evaporation in the steam generator, ex-

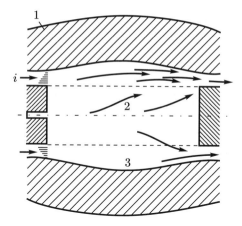

FIG. 1.8. Schematic of a gas-phase nuclear reactor (Ievlev 1977). 1, neutron reflector; 2, fissile uranium plasma; 3, flow of the working medium.

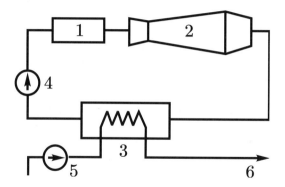

FIG. 1.9. Schematic of a MHD facility utilizing alkali metal vapors (Biberman *et al.* 1982). 1, steam generator; 2, MHD unit (nozzle, MHD channel, diffuser); 3, condensing heat exchanger; 4, liquid metal pump; 5, water pump; 6, to the turbogenerator.

pansion in the MHD unit, condensation in the condensing heat exchanger with heat transfer to the steam–water loop, and an increase of pressure in the pump.

A nonideal plasma of cesium can also be regarded as the working medium for plasma thermoelements. This is favored by the high values of the thermoelectric coefficient of cesium ($S = 10^3 \, \mu V \, K^{-1}$) at moderate pressures and temperatures ($p \leq 10$ MPa, $T \leq 2000$ K) (Alekseev *et al.* 1970). The thermal conversion element is a sleeve of niobium in which there is inserted a beryllium oxide tube inside which cesium is poured. On top, the tube is closed with a metal plug serving as one of the electrodes. The temperature difference is developed by a resistance furnace placed in a protective casing under the desired pressure of an inert gas. The thermoelement may have a power of up to 10 W cm^{-2} and an efficiency of up to 20%. Such converters are advantageous over semiconductor ones because they can greatly vary their characteristics upon variation of the applied pressure.

Another example of strongly nonideal plasmas is provided by dusty plasmas, which have various applications. These include electrophysics of rocket-fuel combustion products, production of microelectronic devices, plasma deposition and etching technologies, technologies of surface modification, etc. Dust particles are not only deliberately introduced in a plasma, but can also form and grow due to different physical processes, e.g., volume condensation upon the efflux of plasma from nozzles, upon the plasma expansion in the channel of an MHD generator, upon the expansion of aggregates of matter into a vacuum, in the products of ablation of a solid surface subjected to the effect of high energy fluxes, etc. Dust and dusty plasmas are quite natural in space. They are present in planetary rings, comet tails, and interplanetary and interstellar clouds (Goertz 1989; Northrop 1992; Tsytovich 1997; Bliokh et al. 1995). Dusty plasmas are found in the vicinity of artificial satellites and space stations (Whipple 1981; Robinson and Coakley 1992) and in thermonuclear facilities with magnetic confinement (Tsytovich and Winter 1998; Winter and Gebauer 1999; Winter 2000).

A nonideal plasma results from powerful electric discharges in water and solids and its properties define the dynamics of motion of generated shock waves. Pulsed electric discharges in a liquid have found extensive industrial application, e.g., for intensifying mechanical and chemical production processes (Ivanov et al. 1982). The operation of such facilities is based on the use of high–voltage discharge in the liquid as the process of rapid conversion of the energy of a capacitor bank to mechanical work. The discharge duration is 10^{-5}–10^{-4} s, with an energy density of 10^{14}–10^{15} J m^{-3}, a temperature of 10^4–10^5 K and the pressure in the discharge channel up to 1 TPa. Under these conditions, the properties of a nonideal plasma, especially its electrical conductivity, affect the process of both the formation and expansion of the current–conducting channel, as well as shock wave generation.

References

Alekseev, V. A., Vedenov, A. A., Krasitskaya, L. S., and Starostin, A. N. (1970). Thermoelectric power of cesium near critical temperatures and pressures. *JETP Lett.*, **12**, 351–354.

Artem'ev, A. A., Khrapak, A. G., and Yakubov, I. T. (1985). Multi–electron states in small dielectric particles. *Sov. J. Low Temp. Phys.*, **11**, 555–563.

Biberman, L. M., Likal'ter, A. A., and Yakubov, I. T. (1982). MHD generator on nonideal plasma of saturated alkali vapors. *High Temp.*, **20**, 565–572.

Bliokh, P., Sinitsin, V., and Yaroshenko, V. (1995). *Dusty and self–gravitational plasmas in space*. Kluwer Academic, Dordrecht.

Bouchoule, A. (1999). Technological impacts of dusty plasmas. In *Dusty plasmas: Physics, chemistry and technological impacts in plasma processing*, Bouchoule, A. (ed.), pp. 305–396. Wiley, Chichester.

Brakner, K. and Jorna, S. (1973). *Laser driven fusion*. Fusion Inc., New York.

Cole, M. W. (1974). Electronic surface states of liquid helium. *Rev. Mod. Phys.*, **46**, 451–464.

Davidson, R. C. (1990). *Physics of nonneutral plasmas*. Addison–Wesley, Redwood City.

Dubin, D. H. E. and O'Neil, T. M. (1999). Trapped nonneutral plasmas, liquids, and crystals (the thermal equilibrium states). *Rev. Mod. Phys.*, **71**, 87–172.

Ebeling, W., Kreft, W. D., and Kremp, D. (1976). *Theory of bound states and ionization equilibrium in plasmas and solids*. Akademie-Verlag, Berlin.

Fortov, V. E., Molotkov, V. I., Nefedov, A. P., and Petrov, O. F. (1999). Liquid– and crystallike structures in strongly coupled dusty plasmas. *Phys. Plasmas*, **6**, 1759–1768.

Fortov, V. E., Khrapak, A. G., Khrapak, S. A., Molotkov, V. I., and Petrov, O. F. (2004). Dusty plasmas. *Phys. Usp.*, **47**, 447–492.

Fortov, V. E., Ivlev, A. V., Khrapak, S. A., Khrapak, A. G., Morfill, G. E. (2005). Complex (dusty) plasmas: Current status, open issues, perspectives. *Phys. Reports*, **421**, 1–103.

Goertz, C. K. (1989). Dusty plasmas in the solar system. *Rev. Geophys.*, **27**, 271–292.

Gryaznov, V. K., Iosilevskii, I. L., Krasnikov, Y. G., Kuznetsova, N. I., Kucherenko, V. I., Lappo, G. B., Lomakin, B. N., Pavlov, G. A., Son, E. E., Fortov, V. E. (1980). *Thermophysical properties of gas core nuclear engine*. Atomizdat, Moscow.

Iakubov, I. T. and Vorob'ev, V. S. (1974). MHD generator on nonideal plasma. *Astronautics Acta*, **18**, 79–83.

Ichimaru, S. (1982). Strongly coupled plasmas: high–density classical plasmas and degenerate electron liquids. *Rev. Mod. Phys.*, **54**, 1017–1059.

Ievlev, V. M. (1977). Some results of a study of a cavity–type gas–phase nuclear reactor. *Energetika i Transport*, No. 6, 24–31.

Ignatov, A. M. (2005). Basics of dusty plasma. *Plasma Phys. Reports*, **31**, 46–56.

Iosilevskii, I. L. (2000). General characteristic of the thermodynamic description of low temperature plasma (LTP). In *Encyclopedia of low temperature plasma. Introductory volume 1,* Fortov, V. E. (ed.), pp. 275–293. Nauka, Moscow.

Iosilevskii I. L., Krasnikov, Y. G., Son, E. E., and Fortov, V. E. (2000). *Thermodynamics and transport in nonideal plasma*. MFTI, Moscow.

Ivanov, V. V., Shvets, I. S., and Ivanov, A. V. (1982). *Underwater spark discharges*. Naukova Dumka, Kiev.

Jeffries, C. D. and Keldysh, L. V. (eds). (1983). *Electron–hole droplets in semiconductors*. North–Holland, Amsterdam.

Kadomtsev, B. B. (1973). *Lasers and thermonuclear problem*. Atomizdat, Moscow.

Khrapak, A. G. and Iakubov, I. T. (1981). *Electrons in dense gases and plasma*. Nauka, Moscow.

Kikoin, I. K., Senchenkov, A. P., Gelman, E. V., Korsunskii, M. M., and Naurza-kov S. P. (1966). Electrical conductivity and density of a metal vapor. *JETP*, **22**, 89–91.

Kirzhnits, D. A., Lozovik, Y. E., and Shpatakovskaya, G. V. (1975). Statistical model of matter. *Sov. Phys. Uspekhi*, **18**, 3–48.

Klyuchnikov, N. I. and Triger, S. A. (1967). Thermodynamics of a system of strongly interacting charged particles. *JETP*, **25**, 178–181.

Kudrin, L. P. (1974). *Statistical physics of plasma*. Nauka, Moscow.

Landau, L., and Lifshitz, E. (1980). *Statistical physics*. Pergamon Press, Oxford.

Leiderer, P. (1995). Ions at helium interface. *Z. Phys. B*, **98**, 303–308.

Lepouter, M. (1965). *Metal ammoniac solutions*. Academic Press, New York.

Merlino, R. L. and Goree, J. (2004). Dusty plasmas in the laboratory, industry, and space. *Phys. Today*, **57**, 32–38.

More, R. M. (1983). Atomic processes in high–density plasmas. In *Atomic and molecular physics of controlled thermonuclear fusion*, Joachain, C. J. and Post, D. E. (eds), pp. 399–440. Plenum Press, New York.

Nedospasov, A. B. (1977). The physics of MHD generators. *Sov. Phys. Uspechi*, **20**, 861–869.

Northrop, T. G. (1992). Dusty plasmas. *Phys. Scr.*, **45**, 475–490.

Piel, A. and Melzer, A. (2002). Dynamical processes in complex plasmas. *Plasma Phys. Control. Fusion*, **44**, R1–R26.

Prokhorov, A. M., Anisimov, S. I., and Pashinin, P. P. (1976). Laser thermonu-clear fusion. *Sov. Phys. Uspekhi*, **19**, 547–560.

Ragan. C. E. , Silbert, M. G., and Diven, B. C. (1977). Shock compression of molybdenum to 2.0 TPa by means of a nuclear explosion. *J. Appl. Phys.*, **48**, 2860–2870.

Robinson, P. A. and Coakley, P. (1992). Spacecraft charging–progress in the study of dielectrics and plasmas. *IEEE Trans. Electr. Insul.*, **27**, 944-960.

Shatzman, E. (1977). (private communication).

Shikin, V. B. and Monarkha, Y. P. (1989). *Two–dimensional charged systems in helium (in Russian)*. Nauka, Moscow

Shukla, P. K. and Mamun, A. A. (2002). *Introduction to dusty plasma physics*. IOP, Bristol.

Smirnov, B. M. (1982). *Introduction to plasma physics*. Nauka, Moscow.

Slattery, W. L., Doolen, G. D., and DeWitt, H. E. (1980). Improved equation of state for the classical one–component plasma. *Phys. Rev. A*, **21**, 2087–2095.

Thom, K. and Schneider, R. T. (eds). (1971). *Research of uranium plasmas and their technological application*. NASA, Washington.

Tsytovich, V. N. (1997). Dust plasma crystals, drops, and clouds. *Phys. Uspekhi*, **40**, 53–94.

Tsytovich, V. N. and Winter, J. (1998). On the role of dust in fusion devices. *Phys. Uspekhi*, **41**, 815–822.

Tsytovich, V. N., Morfill, G. E., and Thomas H. (2002). Complex plasmas: I. Complex plasmas as unusual state of matter. *Plasma Phys. Rep.*, **28**, 623–651.

Vedenov A. A. (1965). Thermodynamics of plasma. In *Reviews of plasma physics. Vol. 1*, Leontovich M. A. (ed.), pp. 312–326. Consultants Bureau, New York.

Vladimirov, S. V. and Ostrikov, K. (2004). Dynamic self–organization phenomena in complex ionized gas systems: New paradigms and technological aspects. *Phys. Reports*, **393**, 175–380.

Volodin, A. P., Khaykin, M. S., and Edel'man, V. S. (1977). Development of instability and bubblon production on a charged surface of liquid helium. *JETP Lett.*, **26**, 543–546.

Whipple, E. C. (1981). Potentials of surfaces in space. *Rep. Prog. Phys.*, **44**, 1197–1250.

Winter, J. (2000). Dust: A new challenge in nuclear fusion research? *Phys. Plasmas*, **7**, 3862–3866.

Winter, J. and Gebauer, G. (1999). Dust in magnetic confinement fusion devices and its impact on plasma operation. *J. Nucl. Mater.*, **266**, 228–233.

Yakubov, I. T. and Khrapak, A. G. (1989). Thermophysical and electrophysical properties of low temperature plasma with condensed disperse phase. *Sov. Tech. Rev. B. Therm. Phys.*, **2**, 269–337.

Zel'dovich, Y. B. and Novikov, I. D. (1971). *Relativistic astrophysics*. University of Chicago Press, Chicago.

2

ELECTRICAL METHODS OF NONIDEAL PLASMA GENERATION

The electrical methods of nonideal plasma generation include two principally different techniques, namely, the heating in resistance furnaces of ampoules containing the material under investigation and Joule heating of samples of material by an electric current passed through the latter. A resistance furnace can be used to obtain homogeneous volumes of plasma without passing an electric current through it and can record the plasma parameters to a relatively high accuracy. On the other hand, the static character of the method causes the restriction of its capabilities to temperatures of up to 3000 K. This is due to great difficulties associated with the choice of structural materials. The techniques utilizing Joule heating include high–pressure gas discharges in thick–walled capillaries, the heating of materials with current in an atmosphere of inert gas, discharges in liquids, and some other techniques. These techniques, which are mainly pulsed, permit of plasma generation at considerably higher temperatures of up to 10^5 K. The principal difficulties of these methods are caused by the difficulty of attaining the homogeneity of plasma volumes and by various plasma instabilities.

2.1 Plasma heating in furnaces

A detailed description of stationary methods of plasma generation can be found in the reviews by Alekseev and Iakubov (1983, 1986). These methods are based on heating an ampoule containing the material under investigation (measuring cell) in electric furnaces of various designs. The entire unit is placed in a chamber under a high pressure of inert gas. Shown by way of example in Fig. 2.1 is the schematic of a facility which was used (Alekseev 1968) to investigate alkali metals at high temperatures and pressures. Argon from cylinder (1) was fed via a system of valves to chamber (3) for cleaning. Cleaned argon was delivered to a nitrogen thermocompressor (4) with the pressure in the latter rising to 600 bar, and from the thermocompressor to measuring chamber (5) accommodating a heater and a measuring cell. The pressure was monitored by gauges (2). The temperature was determined using a thermocouple with its cold junction placed in a melting ice bath. If the required pressure level exceeded 600 bar, the facility was modified: the gas was compressed to 5000 bar using a booster (Kikoin and Senchenkov 1967).

High temperature values in the region of nonideality present the principal difficulty of static methods which is due to the difficulty of selection of the structural materials for measuring cells. Especially serious is the extremely narrow

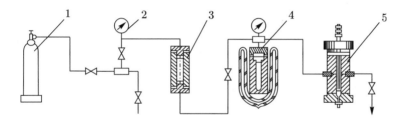

FIG. 2.1.   Schematic of a facility for the generation and investigation of nonideal plasma of
metals in the region of high temperatures and pressures (Alekseev 1968): 1, cylinder with
argon; 2, pressure gauges; 3, argon cleaning system; 4, thermocompressor; 5, high–pressure
chamber.

choice of insulating materials. The best insulating properties are observed in insulators of pure aluminum, zirconium, beryllium, and thorium oxides. Speaking
of metals, the best materials include plastic tungsten, tungsten and molybdenum alloys, and molybdenum. The construction of a reliable cell calls for a high
degree of experimental craftsmanship. As regards the development of the static
methods of nonideal plasma research, note should be made of the importance of
the pioneering work by Kikoin *et al.* (1966) and Hensel and Franck (1968). In
the 1970s, new designs of measuring cells were developed which helped considerably to improve the accuracy of the experimental data. The results of previous
studies were often considerably refined and, in some cases, revised. At present,
the experimental data obtained in different laboratories are usually found to be
in adequate agreement with each other.

The most convenient object for investigations of the nonideality effect is cesium which has a low ionization potential ($I = 3.89$ eV), low critical pressure
(about 110 bar), and a critical temperature that is quite accessible to static experiments (about 2000 K). A fairly high saturated vapor pressure at relatively
low temperatures permits the investigation a nonideal cesium plasma at different levels of interparticle interaction (Renkert*et al.* 1969; Alekseev *et al.* 1970a,
1970b, 1975, 1976; Renkert *et al.* 1971; Barol'skii *et al.* 1972; Pfeifer *et al.* 1973;
Sechenov *et al.* 1977; Dikhter and Zeigarnik 1977; Isakov and Lomakin 1979;
Kulik *et al.* 1984; Borzhievsky *et al.* 1987).

2.1.1   *Measurement of electrical conductivity and thermoelectromotive force*

The requirements of measuring cells include high corrosion resistance of their
walls upon contact with the plasma, air–tightness of contacts between the insulating ceramics and metal electrodes, and the protection of material under investigation against contact with an inert gas. Figure 2.2 illustrates a measuring
cell of beryllium oxide with an H–shaped measuring volume in which the metal
under investigation does not come in contact with the pressure-transmitting gas
(Alekseev *et al.* 1976). Two through holes with a jumper in the center were made
in the cell body (2) of beryllium oxide. Current (solid) and potential (tubular)

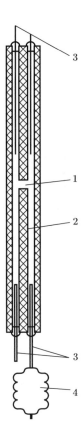

FIG. 2.2.   Cell (closed) for measuring the electrical conductivity (Alekseev *et al.* 1976): 1, H–shaped measuring volume; 2, beryllium oxide ceramics; 3, electrodes; 4, expanders of the material under investigation.

electrodes (3) were sealed at the ends. Using such a structure, one can measure the resistance in a volume whose uniformity is reliably guaranteed. An elastic expander, bellows (4), was attached to one of the ends (in a closed ampoule, the material expands upon heating) while through the other end liquid cesium was delivered in vacuum. After filling the cell and bellows with cesium, the measuring cell was sealed off.

The measuring current was passed through the bottom electrodes while the top electrodes were used to measure the voltage drop. Given the jumper dimensions, one can calculate the resistance. The error in measuring the cesium resistance at a temperature of up to 1300 K amounts to 3% and in the region of metallization ("metal–dielectric" transition) to 10% while in dense cesium vapors at a temperature above 2000 K this error may exceed 50%. The temperature is measured by W/Rh thermocouples. Special precautions may be taken

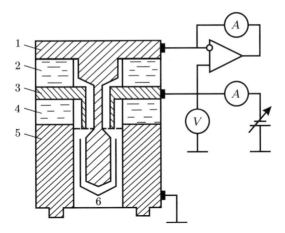

F<small>IG.</small> 2.3.   Cell for measuring the coefficient of electrical conductivity, equipped with a
guard-ring electrode (Borzhievsky *et al.* 1987): 1 and 5, measuring electrodes; 2 and 4,
insulation ring; 3, guard ring; 6, measuring volume.

to eliminate the effect of leakage currents on the electrode surface. Borzhievsky
et al. (1987) measured the coefficient of electrical conductivity by the two–probe
technique using a guard-ring electrode 3 made of sapphire (Fig. 2.3). Equal potentials of the electrode and guard-ring electrode 3 are maintained. Therefore,
the current being measured (between the electrodes 1 and 5) could only flow in
the radial direction in the plasma volume 6. It was the objective of Borzhievsky *et
al.* (1987) to measure the coefficient of electrical conductivity of low–temperature
cesium vapors, $T \leq 1200$ K.

A thermoelectric effect shows up in matter in the presence of a temperature
gradient,

$$\mathbf{F} = \mathbf{j}/\sigma + S\nabla T, \qquad (2.1)$$

where S is the thermoelectric coefficient usually referred to as the thermoelectromotive force, and F is the gradient of the electrochemical potential. In static
experiments, the potential difference U is measured, which appears at the ends
of an open circuit consisting of dissimilar conductors. If the circuit contains two
contacts so that the end conductors are of like metal (tungsten), with a temperature gradient developed along the middle conductor (cesium), a potential
difference appears at the circuit ends,

$$U = \int_{T_1}^{T_2} (S_{\mathrm{Cs}} - S_{\mathrm{W}})\, dT, \qquad (2.2)$$

where T_1, T_2 define the temperature difference.

Since the temperatures are high, during the experiment it is very difficult
to maintain with adequate accuracy small temperature differences between the

junctions. One should bear in mind that under these conditions the thermal e.m.f. may strongly depend on temperature. Averagings may also produce an appreciable error even in small temperature intervals. Therefore, the thermo-electromotive force is measured by a so–called integral technique. The "cold" junction is maintained at some known low temperature T_1 while the "hot" junction temperature T_2 varies. As a result, the quantity $U(T_2)$ is measured. By differentiating the relationship $U(T_2)$, the thermal e.m.f. is obtained.

Consider now the structure of a measuring cell used to measure the thermoelectric properties of cesium and mercury plasma. The cell illustrated in Fig. 2.4 was employed for measurements using the integral technique (Alekseev *et al.* 1975, 1976). This cell was manufactured from a tungsten tube (3) terminating in an expander 8. Arranged coaxially inside the tube was an insulator (4) comprising a top electrode (2) in the form of a thin–walled tungsten tube with a thin tungsten bottom welded on its lower end and a W/Rh thermocouple (1) welded in the bottom. The thermocouple ends insulated by beryllium oxide ceramics passed out through the top end of the tube.

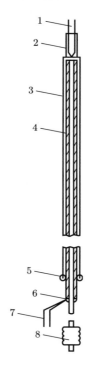

FIG. 2.4.   Cell for measuring the thermal e.m.f. (Alekseev *et al.* 1975, 1976): 1, thermocouple; 2, tungsten electrodes; 3, tungsten casing; 4, beryllium oxide insulator; 5, junction between the ceramics and the casing; 6, junction between the bottom electrode and the insulator; 7, thermocouple; 8, bellows.

The measuring element was filled with mercury or cesium. The working volume was essentially a cylinder with a height of 0.5–2 mm and diameter of 3–5 mm defined by the insulator, the lower end of the top electrode and the bottom of the outer tungsten ampoule serving the function of the second electrode 6. The thermal e.m.f. being measured occurs owing to the temperature gradient developed along the cell.

Let us now turn to the discussion of the results of measurements of electrical conductivity and thermal e.m.f. The electrophysical properties of cesium appear to be the most studied at present. Shown in Fig. 2.5 is a region in which the electrical conductivity and thermoelectromotive force of cesium were studied using both static and dynamic methods by Alekseev (1968), Barolskii *et al.* (1972), and Borzhievsky *et al.* (1987).

The pioneering study of cesium was performed by Alekseev (1968). It was shown that the electrical conductivity in the region of high temperatures on the isobars had dropped upon heating by up to four orders of magnitude.

Figure 2.6 gives the results of more recent measurements of the electrical conductivity of rubidium (Franz *et al.* 1981), which demonstrate clearly how the conductivity decreases upon heating and transition to the nonideal plasma state. Analogous data are cited by Pfeifer *et al.* (1979) and Freyland and Hensel (1972) for rubidium and potassium. These values are characterized by an error of 3–4% in liquid and 50–60% in the transition region.

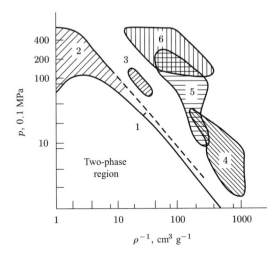

FIG. 2.5.   Experimentally studied regions of $p - V$ diagram of cesium. Static experiments: 1, phase coexistence curve (Renkert *et al.* 1971); 2, measurements of σ (Alekseev 1968; Alekseev *et al.* 1975; Renkert *et al.* 1969, 1971; Kulik *et al.* 1984) and measurements of S (Alekseev *et al.* 1970b, 1975; Pfeifer *et al.* 1973); dynamic measurements of σ: 3, adiabatic tube (Isakov and Lomakin 1979); 4, 5, shock tube (Sechenov *et al.* 1977); 6, ohmic heating in an inert atmosphere (Dikhter and Zeigarnik 1977).

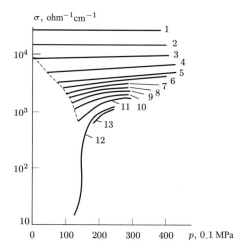

FIG. 2.6. The electrical conductivity of rubidium in the "metal–dielectric" transition region (Franz *et al.* 1981): 1, $T = 300$ K; 2, 500 K; 3, 900 K; 4, 1200 K; 5, 1400 K; 6, 1500 K; 7, 1550 K; 8, 1600 K; 9, 1650 K; 10, 1700 K; 11, 1750 K; 12, 1815 K; 13, 1830 K.

Let us now turn to the data on the thermoelectromotive force. The absolute magnitude of the thermal e.m.f. of cesium had grown rapidly with temperature in the region of the transition from the metal to plasma state (Fig. 2.7). One can see that the solid and broken curves, while featuring the same behavior of the variation of thermal e.m.f. with temperature, diverge somewhat in the "metal–nonideal plasma" transition zone.

Consider in more detail the results of measuring the electrical conductivity of

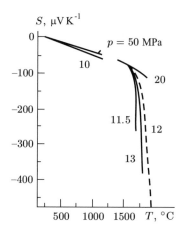

FIG. 2.7. Experimental values of the thermal e.m.f. of cesium: broken line, Alekseev *et al.* 1975; solid lines, Pfeifer *et al.* (1973).

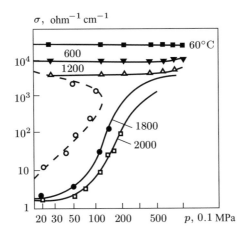

FIG. 2.8. Isotherms of cesium electric conductivity (solid) and electric conductivity on phase coexistence curve (dashed) (Renkert *et al.* 1971).

cesium in a nonideal plasma of metal vapors. The results obtained in this region are far from being numerous. Here, however, qualitative results were also obtained which had proved to be of importance in selecting the theoretical models of a nonideal plasma. Figure 2.8 gives the isotherms of the electrical conductivity of cesium, as well as the electrical conductivity of cesium on the phase coexistence curve (Renkert *et al.* 1971). Figure 2.9 represents a chart combining the results obtained in different laboratories (Alekseev *et al.* 1975; Renkert *et al.* 1971; Lomakin and Lopatin 1983) for the 2 MPa isobar and for saturated vapors.

Attention is drawn to the anomalously high level of the electrical conductivity of saturated vapor. It exceeds by orders of magnitude the level of standard estimates for an ideal plasma. Further, on isobaric heating, the electrical conductivity starting from these high values does not rise but, on the contrary, drops sharply. It decreases on the 2 MPa isobar to the minimum value of about $2 \cdot 10^{-2}$ $\text{ohm}^{-1}\text{cm}^{-1}$ and, on further heating, starts increasing in the usual manner as in a weakly nonideal plasma.

The region of weakly nonideal cesium plasma corresponds to low pressures of $p \lesssim 0.1$ MPa at which the measurements are performed in vacuum chambers 1 (Fig. 2.10) accommodating heaters 2. Inserted in the heaters are two insulated electrodes 3 with applied potential used to measure the current–voltage characteristics. From the latter, the resistance of the plasma gap between the electrodes is determined. In order to weaken the end effects, an insulated protective ring 4 is set on one of the electrodes.

The plasma of desired pressure is provided with the aid of a cesium delivery system 5. For this purpose, a furnace 6 maintains a preset temperature of the liquid cesium surface. An optical pyrometer or thermocouple was used to determine the plasma temperature.

Figure 2.11 shows the results of measuring the electrical conductivity of cesium at low pressures. These values of electrical conductivity are characterized by an error of from 10 to 25%. At $p \lesssim 0.1$ MPa, the plasma of cesium vapors is already nonideal. Therefore, the electrical conductivity increases as the pressure decreases, as in the case of an ideal plasma.

Mercury, like cesium, has a low critical temperature. Therefore mercury was

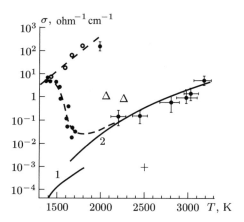

FIG. 2.9.   The electrical conductivity of nonideal plasma of cesium vapors. Measurement results: saturated vapor, ○ (Renkert *et al.* 1971); on the 2 MPa isobar, ● (Alekseev *et al.* 1975), △ (Renkert *et al.* 1971), + (Lomakin and Lopatin 1983). Calculated results: 1, saturated vapor by ideal gas approach; 2, isobar 2 MPa with account for molecular ions (Gogoleva *et al.* 1984).

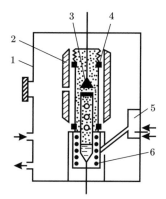

FIG. 2.10.   Schematic of the facility for generation and studies of a low–pressure cesium plasma ($p \lesssim 0.1$ MPa) in the high–temperature region (Ermokhin *et al.* 1971): 1, vacuum chamber body; 2, heaters; 3, electrodes; 4, protective ring; 5, cesium delivery system; 6, liquid cesium heating furnace.

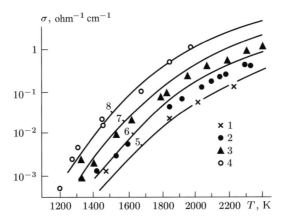

FIG. 2.11. The electrical conductivity of cesium at low pressures. Measurements: 1, 0.1 MPa; 2, 10^{-2} MPa; 3, 10^{-3} MPa (Ermokhin *et al.* 1971); 4, $0.66 \cdot 10^{-2}$ MPa (Rochling 1963). Calculations: 5, 0.1 MPa; 6, 10^{-2} MPa; 7, 10^{-3} MPa ; 8, $0.66 \cdot 10^{-2}$ MPa (Gogoleva *et al.* 1984).

the first substance whose properties were investigated in the "metal–nonmetal" transition region by Kikoin *et al.* (1966) and Hensel and Franck (1968). Isochores of electrical conductivity of dense mercury vapors are given in Fig. 2.12. For this purpose, the density was determined by extrapolating the experimental $p - \rho - T$ data obtained at higher densities. It is seen from Fig. 2.12 that the electrical conductivity of mercury plasma at constant density grows with temperature if $\rho \lesssim \rho_c$. The temperature dependence of electrical conductivity in this region exhibits a semiconductor behavior, $\sigma \sim \exp(-\Delta E/T)$. As the density increases, the slope of the curves decreases indicating that the "energy gap" ΔE decreases.

Figure 2.13 shows the temperature coefficient of conductivity of mercury as a function of density, $\Delta E = -2d\ln\sigma/d\beta$, where $\beta = 1/kT$. The energy gap ΔE is closed at $\rho \cong \rho_c$. An interesting, though still not properly understood, peculiarity in the $\Delta E(\rho)$ dependence was observed by Kikoin and Senchenkov (1967) at $\rho > \rho_c$. In the liquid state region, the gap opens on further compression and finally closes again only at $\rho \gtrsim 9$ g cm^{-3}. Kikoin and Senchenkov concluded that mercury had metallized only at 9 g cm^{-3}.

The great uncertainty in this region is demonstrated by the data on the thermal e.m.f. of mercury. There is no doubt, however, that an interesting phenomenon of a sharp variation of thermal e.m.f. near the critical point has been observed in four laboratories by Alekseev *et al.* (1972, 1976), Dukers and Ross (1972), Tsuji *et al.* (1977), and Neale and Cusack (1979), which so far has no theoretical interpretation.

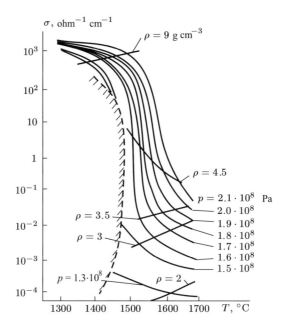

FIG. 2.12.   Isochores of the electrical conductivity of mercury (Hensel and Franck 1968).

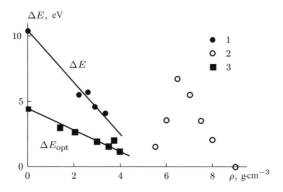

FIG. 2.13.   The transport, ΔE, and optical, ΔE_{opt}, gaps in the energy spectrum of mercury, $T = 1820$ K. Experiments: 1 – ΔE, Kulik *et al.* (1984); 2 – ΔE, Kikoin and Senchenkov (1967); 3 – ΔE_{opt}, Uchtmann and Hensel (1975).

2.1.2   *Optical absorption measurements.*

The optical properties of strongly nonideal mercury plasma were measured by Hensel (1970, 1971, 1977) and Ikezi *et al.* (1978). Optical cells in the 1.3–4.9 eV energy range were developed. The cell design, enabling one to perform the measurements of light transmission in mercury at temperature of up to 1700 °C and pressures of up to 0.2 GPa, is illustrated in Fig. 2.14.

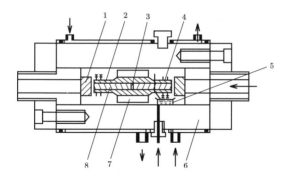

FIG. 2.14. Optical cell for determining the absorption in a dense mercury plasma (Hensel 1971): 1, high–pressure windows; 2, niobium cylinder; 3, measuring volume; 4, Teflon seals; 5, container with mercury; 6, autoclave; 7, furnace; 8, optical sapphires.

The cell consisted of a niobium cylinder (2) closed on both ends by windows of polished sapphires (8) whose outer ends stayed cold. They are sealed by Teflon spacers (4). Provided between the ends of the sapphires was a clearance whose width could be varied from $5 \cdot 10^4$ to 0.1 cm. Mercury was delivered to that clearance from a special container (5). The clearance served a measuring element for determining the absorption index of mercury plasma. The cell was heated by a furnace (7) inside an autoclave (6). Argon provided the pressure-maintaining medium. Light from the autoclave passed out through windows arranged in ends (1).

The absorption index k could be determined from the measured radiation transmission $T = I/I_0$

$$T = (1 - R)^2 \exp(-kd), \qquad (2.3)$$

where d is the sample thickness, I_0 and I are the incident and emitted radiation intensities, and R is the reflection coefficient. The values of k were determined by Hensel (1971) without using the values of R, but as a result of measurement of T with two values of d.

The sources of errors such as the uncertainty in the optical path, the nonuniformity of the temperature distribution and the inaccuracy of intensity recording led to errors in the absorption coefficient not exceeding 20%. The main source of uncertainty is provided, as in the case of measuring the electrophysical quantities, by errors in recording the temperature and density. The stimulated radiation correction to the absorption index, made by the factor $[1 - \exp(-\beta\hbar\omega)]$ in the conditions under investigation, is small.

Figure 2.15 shows the frequency dependence of the absorption index of dense mercury vapors obtained in two different laboratories. The absorption index $k(\omega)$ varies exponentially upon variation of the incident light wavelength. The first thing to strike one's eye is the abruptness of the variation of $k(\omega)$. A variation of $k(\omega)$ by an order of magnitude corresponds to a variation of $\hbar\omega$ by only several

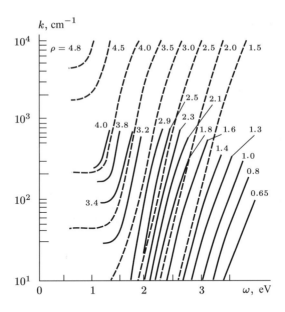

FIG. 2.15.   Absorption index of mercury as a function of the incident light frequency, $k(\omega)$, at $T = 1600$ °C and at different densities ρ, g cm^{-3}: solid curves, Ikezi *et al.* (1978); broken curves, Uchtmann and Hensel (1975).

tenths of an electron volt. The data of the two experiments are in good agreement with each other.

The mercury reflection coefficient R was measured by Ikezi *et al.* (1978) and Hefner *et al.* (1980). Figure 2.16 shows the results of measurements of the reflection coefficient of incident radiation, which is almost normal to the surface of the plasma layer (Ikezi *et al.* 1978). These results cover the region in which mercury loses its metallic properties at temperatures somewhat exceeding the critical point and changes to the state of strongly nonideal plasma. As in the case of the results of electrical conductivity measurements, the density of 9 g cm^{-3} separates the metal and dielectric regions. The measured absorption and reflection coefficients were used by Hefner *et al.* (1980), Cheshnovsky *et al.* (1981), and Hefner and Hensel (1982) to obtain the frequency dependencies of dielectric permeability and electrical conductivity (cf. Chapter 8).

The emissivity of mercury was measured by Uchtmann *et al.* (1981) in an apparatus analogous to that shown in Fig. 2.14. The radiation intensity of a plane layer of plasma was determined as the difference between its values recorded in an empty cell and in a mercury–filled cell. The emissivity ϵ was determined as the ratio between the intensity being measured (normal to the surface of plasma layer) and the intensity of the perfect radiator. The isochores of emissivity are presented in Fig.2.17.

A characteristic feature of these curves resides in the occurrence (at densities

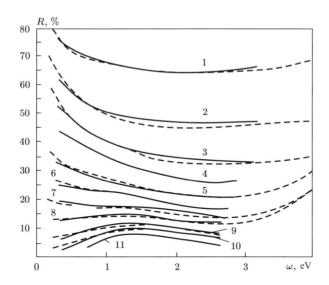

FIG. 2.16.   Reflection coefficient of mercury as a function of frequency (Ikezi *et al.* 1978):
solid curves, experiment; broken lines, calculated curves (Chapter 8); 1 – ρ = 13.6 g cm^{-3}
/ p = 120 MPa / T = 20 °C; 2 – 12 / 186.5 / 650; 3 – 11 / 177.5 / 1090; 4 – 10.6 / 169 /
1220; 5 – 10 / 174 / 1325; 6 – 9.6 / 177 / 1390; 7 – 9 / 179 / 1440; 8 – 8.3 / 181 / 1475; 9
– 6.5 / 185 / 1510; 10 – 5 / 187.5 / 1575; 11 – 4 / 175 / 1550.

exceeding 3 g cm^{-3}) of a continuum in the infrared region, the intensity of which
grows sharply with further increase in density. This is in good agreement with
the results of measuring the absorption index at the same densities and will be
discussed in Chapter 8.

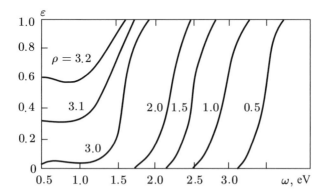

FIG. 2.17.   The emissivity of mercury plasma at 1600 °C and various densities ρ (Uchtmann
et al. 1981).

2.1.3 *Density measurements.*

For the purpose of measuring the density of nonideal plasma of metals, a technique was developed based on measuring the intensity of γ–radiation of metal under investigation. The apparatus for this is described by Alekseev (1968) and Korshunov *et al.* (1970). A thick–walled tungsten tube is defined by a thin tungsten bottom at one end and by an elastic bellows at the other. This element is filled with cesium under vacuum and sealed off. Located beneath the cell is a point measuring volume. In order to avoid convection, the vacant space is filled with a tungsten insert. The metal under investigation was pre–activated in the reactor by slow neutrons. The density was measured using the method of recording the γ–radiation from the measuring volume, led out via a special port in a high–pressure chamber and collimated with the aid of a lead collimator. For γ-quantum recording, use was made of a CsI scintillator and a photomultiplier. The number of pulses from the detector (proportional to density) and the temperature were recorded in digital form for subsequent computer processing. This method was used to measure the cesium density at temperatures of up to 2500 °C and pressures of up to 60 MPa. The method suffers from poor accuracy (ca. 10%) in measuring low vapor densities. This is because the absolute error in measuring the cesium density amounted to ± 0.005 $\mathrm{g\,cm^{-3}}$.

More accurate data were obtained using a constant-volume piezometer where the cesium vapor pressure was measured by the compensation method. Stone *et al.* (1966) and Novikov and Roshchupkin (1967) used an elastic membrane introduced in the piezometer as a pressure cell. This reduced the range of temperatures under investigation because of the adverse effect of high temperatures on the elastic properties of the membrane. Volyak and Chelebayev (1976) removed the membrane from the heated volume. This permitted $p - \rho - T$ measurements of cesium vapors at 5.2 MPa and 1940 K.

A schematic of the apparatus is shown in Fig. 2.18. Essentially, it comprises a piezometer consisting of a cylindrical chamber (1) with a volume of 50–100 $\mathrm{cm^{-3}}$ placed in a heater (2). On heating the piezometer, the pressure of cesium vapors in the chamber is transmitted through a column of liquid cesium in capillary (3) to membrane (4) of an induction null pressure cell (5). Consequently, the membrane deflects. By applying an argon pressure to its other side, the membrane is brought into "zero" position.

The specific volume of cesium vapor was determined as the ratio of the piezometer chamber volume (with due regard for its thermal expansion) to the mass of cesium vapor in the chamber. The accuracy of experiment ($\pm 0.65\%$) was estimated from the error of the compressibility coefficient.

The results of density, temperature, and pressure measurements enable one to construct the $p - \rho - T$ equation of state (Alekseev *et al.* 1970a; Kikoin *et al.* 1973; Chelebaev 1978). The thermodynamic equation of state for mercury and cesium are shown in Figs. 2.19 and 2.20, respectively.

The most exact measurements of curves of phase equilibrium for cesium and rubidium were performed by Jungst *et al.* (1985) (Fig. 2.21). The revealed asym-

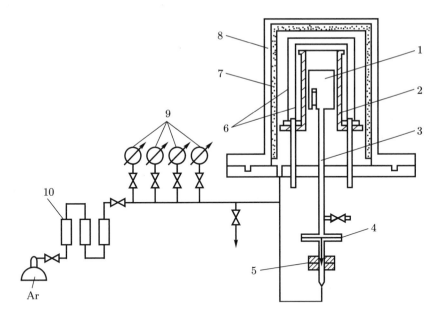

FIG. 2.18. Element for measuring the cesium vapor density (Volyak and Chelebaev 1976):
1, piezometer; 2, graphite heater; 3, capillary with liquid cesium; 4, membrane; 5, "zero"
manometer pressure cell; 6, heat shields; 7, heat insulation; 8, apparatus body; 9, pressure
gauges; 10, argon–cleaning system.

metry of branches of these curves is indicative of the invalidity of the law of the
rectilinear diameter in a relatively wide temperature range.

Important information on the structure of the material is provided by neutron
diffraction studies of the structure factor. The static structure factor $S(q)$ is
directly related to the binary correlation function $g(r)$:

$$S(q) = 1 + n \int d\mathbf{r} \, \exp(-i\,\mathbf{q}\cdot\mathbf{r})[g(\mathbf{r}) - 1], \qquad (2.4)$$

where n is the particle density of the medium. The static structure factor of
rubidium was measured by Noll $et\ al.$ (1988) and Winter $et\ al.$ (1988) (Fig.
2.22).

The basic error is due to the inaccuracy of discrimination of the background
intensity whose quantitative evaluation is very difficult to accomplish. However, it can be evaluated from the requirements of satisfying the rules of the
sum $\int_0^\infty q^2 dq[S(q) - 1] = 2\pi^2 n$, the asymptotics $S(\infty) = 1$, and the value at zero
$S(0) = nkT\chi_T$, where χ_T is the isothermal compressibility. Proceeding from this
and the reproducibility of $S(q)$, which is measured using different measuring cells
and furnaces, it was concluded on the error of $S(q)$ amounts to $\pm 5\%$ for $q > 13$
nm^{-1} at low temperatures. On heating, the error increases to $\pm 15\%$.

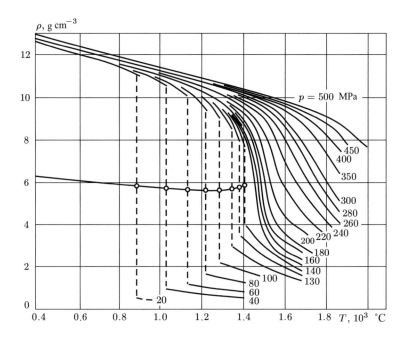

FIG. 2.19.   Thermodynamic equation of state for mercury (Kikoin *et al.* 1973).

It follows from Fig. 2.22 that, at $\rho < 0.66$ g cm^{-3} (which is twice as high as than the critical density ρ_c), the behavior of $S(q)$ with high values of q is fully smoothed while the behavior at low values of q points to sharply increasing scatter, which is analogous to the behavior of $S(q)$ of argon. From here, it follows that the critical fluctuations are not a factor defining the beginning of the "metal–nonmetal" transition.

2.1.4   *Sound velocity measurements*

In the acoustic method of "fixed distance" (Vasil'ev and Trelin 1969; Winter *et al.* 1988) the sound velocity is determined by the time, taken by a sound pulse to pass the known speed base in the medium under investigation, and the attenuation of sound is determined by the decrease in its amplitude. The test assembly is shown diagrammatically in Fig. 2.23. Longitudinal sound oscillation is excited in the medium being investigated by means of piezoelectric transducers (1). Two sound ducts (3) are separated by a niobium ring (5). The ring is filled with the test substance, and the ring width is the speed base. A niobium ampoule (2) is heated to the desired temperature. As the pressure varied, mercury introduced into the chamber (6) was forced out into an expander (which is not shown in Fig. 2.23).

The measuring scheme enables one to register, separately, the pulse which passed through both sound ducts (3), and the pulse which was singly reflected from the internal interface between the medium and receiving duct line before

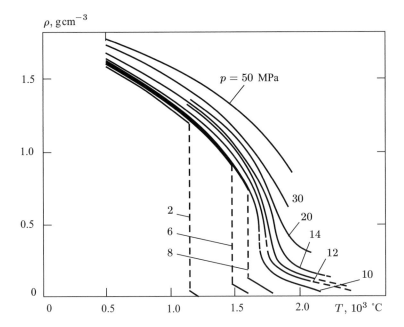

FIG. 2.20. Thermodynamic equation of state for cesium (Alekseev *et al.* 1970a).

returning to the radiating piezoelectric transducer. The difference between the in–phase time coordinates of both pulses gives rise to the time taken by the sound wave to pass the speed base. This enables one to calculate the sound velocity. This procedure is described in more detail by Kulik *et al.* (1984).

The measurement results are given in Fig. 2.24. One can see the rise in the slope of isobars, especially in the liquid phase, as the critical point is approached. The behavior of supercritical isobars becomes smooth. Apparently, the smoother it is, the higher the value of p/p_c. The $v_S(\rho)$ relationship, derived during the rearrangement of these data, clearly reveals a kink in the case of a density of 9 $g\,cm^{-3}$. It is known that at this density mercury becomes metallized.

2.2 Isobaric Joule heating

Isobaric Joule heating is realized in a capillary or in an atmosphere of inert gas. These methods help to attain a high concentration of charged particles (up to 10^{20} cm^{-3}) and high temperatures (up to 10^5 K) at pressures of up to 0.1 GPa. However, the plasma generated proves to be substantially inhomogeneous. More-over, in this case, one fails to realize direct reliable measurements of temperature, density and pressure.

2.2.1 *Isobaric heating in a capillary*

This technique was developed to investigate the properties of a nonideal plasma of cesium and, later, of sodium, potassium and lithium (Kulik *et al.* 1984). A

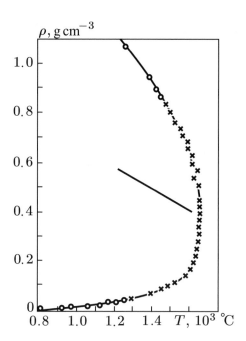

FIG. 2.21.   Curve of phase equilibrium for cesium (Jungst *et al.* 1985).

schematic of the apparatus is shown in Fig. 2.25. A transparent quartz (or glass) capillary 0.7 mm in diameter and 20 mm long is filled with liquid cesium (or it can accommodate a wire of solid potassium, sodium, or lithium). Cesium is subjected to constant argon pressure maintained throughout the experiment. Filled with argon simultaneously, the volume serves as an expander for cesium, part of which is forced out of the capillary upon heating. The cesium in the capillary is heated by a current pulse shaped by an electric circuit comprising a capacitor bank, inductor and a control thyristor. The apparatus was greatly improved in the course of subsequent studies.

In the experiment, the current–voltage characteristic is measured (from the oscillogram). In the oscillogram, the region corresponding to steady–state conditions is recorded. It is obvious that in doing so, the capillary is completely filled with plasma. By processing the data of the oscillogram, one can obtain the $F(i)$ dependence, where F is the electric field intensity and i the discharge current. This dependence provides the initial experimental data used to obtain, by calculations, the isobars of electrical conductivity, $\sigma(T)$, and thermal conductivity, $\chi(T)$.

The basic equations include the heat balance equation, which is known in the physics of gas discharge as the Elenbaas–Heller equation, and ohm's law. For a sufficiently long capillary, these equations take the form

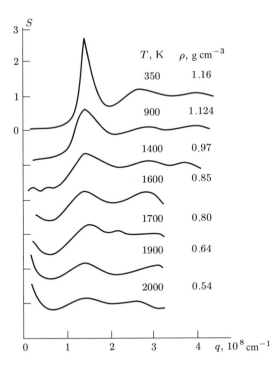

FIG. 2.22.   Curve of phase equilibrium for cesium (Jungst *et al.* 1985).

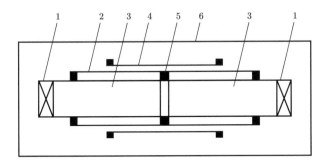

FIG. 2.23.   Cell for measuring the sound velocity in mercury (Suzuki *et al.* 1977): 1, quartz piezoelectric transducers; 2, niobium ampoule; 3, sound ducts; 4, heater; 5, niobium ring; 6, high–pressure chamber.

$$\frac{1}{r}\frac{d}{dr}\left(r\kappa\frac{dT}{dr}\right) + \sigma F^2 = 0, \tag{2.5}$$

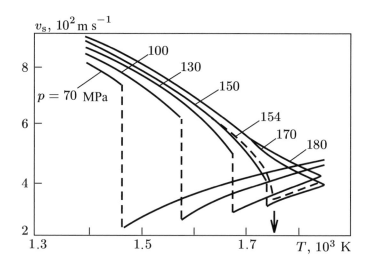

FIG. 2.24.   Isobars of the sound velocity in mercury (Suzuki *et al.* 1977).

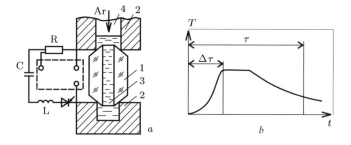

FIG. 2.25.   Schematic of an apparatus for isobaric heating in a transparent capillary (Barol-skii *et al.* 1972). (a): 1, capillary; 2, electrodes; 3, liquid cesium; 4, argon purge. (b) The behavior of the time dependence of plasma temperature.

$$i = 2\pi F \int_0^R \sigma r \, dr, \tag{2.6}$$

where r is the distance from the capillary axis and R is its radius. The boundary conditions are: $dT/dr = 0$ at $r = 0$; $T = T_s$ at $r = R$, where T_s is the pyrometrically measured temperature of the outside surface of the plasma.

In the plasma under investigation, the main mechanism for the extraction of heat and the shaping of the temperature distribution $T(r)$ is provided by radiation transfer. Since the mean radiation path length, that is, the Rosseland length, is $l_R \ll R$, the radiative thermal conductivity approximation is valid while $\kappa(T)$ is equal to the coefficient of radiative thermal conductivity $\kappa_R(T)$. The sought-for functions $\kappa_R(T)$ and $\sigma(T)$ were obtained as a result of substitution in the equations

$$T(r) = T_\mathrm{s} + F^2 \int\limits_r^R dr(r\kappa_\mathrm{R})^{-1} \int\limits_0^r \sigma r'\, dr', \qquad (2.7)$$

$$i = (\pi R^2 F)\frac{2}{R^2} \int\limits_0^R \sigma r\, dr \qquad (2.8)$$

of the trial functions $\sigma(T)$ and $\kappa_\mathrm{R}(T)$ followed by the iteration procedure. During the experiments, the maximum pressure of 0.111 GPa was attained and the temperature range of $(4\text{--}18)\cdot 10^3$ K was covered.

It is noted by Dikhter and Zeigarnik (1981) that the main drawback of the technique resides in the impossibility of measuring either the plasma density in the process of its expansion or the plasma temperature. The temperature gradient near the discharge column boundary is very great. Therefore, the calculation of temperature based on its boundary value undoubtedly leads, under such conditions, to great uncertainties. The situation is aggravated by the presence of a quartz wall also having a temperature gradient; there is the further possibility of a reversible loss of transparency by quartz. Moreover, one cannot rule out the possibility of fusion and evaporation of quartz at the boundary with the cesium plasma, as well as partial mixing of the cesium plasma with oxysilicon quartz plasma, that is, a violation of the purity of the material under investigation.

Vorobiov et al. (1981) and Yermokhin et al. (1981) obtained the $\kappa_\mathrm{R}(T)$ and $\sigma(T)$ values for lithium, sodium, potassium and cesium at pressures of 10–100 MPa and temperatures of $(4\text{--}14)\cdot 10^3$ K. Figure 2.26 shows the isobars of the coefficient of radiative thermal conductivity $\kappa_\mathrm{R}(T)$ for cesium.

At low temperatures, that is, at maximum nonidealities, the κ_R values obtained exceed considerably the values calculated in weak nonideality approximation by Kuznetsova and Lappo (1979). This means that a nonideal alkali plasma proves more transparent than one might assume from ideal–gas concepts. This is especially clear from Fig. 2.27, which shows the ratios between the Rosseland lengths measured by Vorobiov et al. (1981) and those calculated by Kuznetsova and Lappo (1979).

Figure 2.28 gives the isobars and one of the isentropes of the electrical conductivity of cesium plasma.

2.2.2 Exploding wire method

The exploding wire method was used for the first time by Lebedev (1957, 1966) (see, also, Lebedev and Savvatimski 1984) for the investigation of the liquid state of metals. Afterwards, this method was employed by Gathers et al. (1974, 1976) to measure the electrical conductivity and the equation of state for liquid uranium in the high–temperature region. Dikhter and Zeigarnik (1975, 1981) investigated the equation of state and electrical conductivity of nonideal cesium and lithium plasma. The apparatus is shown schematically in Fig. 2.29. A cesium

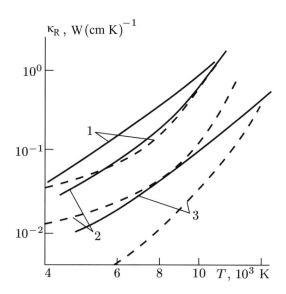

FIG. 2.26. Isobars of the coefficient of radiative thermal conductivity of cesium plasma (Vorobiov *et al.* 1981): 1, 10 MPa; 2, 30 MPa; 3, 100 MPa; broken line, calculation by Kuznetsova and Lappo (1979).

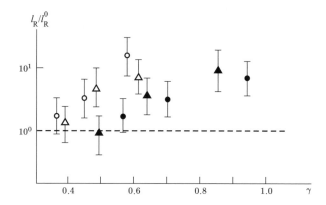

FIG. 2.27. Ratio between the values measured by Vorobiov et al. (1981) and those calculated by Kuznetsova and Lappo (1979) of Rosseland lengths depending on the nonideality parameter γ, $T = 6000$ K.

(or lithium) wire is placed in a high–pressure chamber 5, and a current with a density of $(1-5)\cdot10^6$ A cm^{-2} is passed through the wire. The plasma formation contained by a high–pressure inert gas expands upon heating. While doing so, the pressure in the column remains constant and equal to the inert gas pressure.

The experiment involves the measurement of pressure in the chamber, the

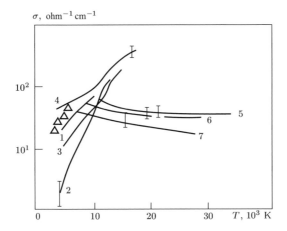

FIG. 2.28.   Coefficient of electrical conductivity of cesium plasma: isobars 1, $p = 13$ MPa; 2, $p = 27.5$ MPa; 3, $p = 60$ MPa; 4, $p = 110$ MPa (Kulik *et al.* 1984); 5, $p = 50$ MPa; 6, $p = 26$ MPa; 7, $p = 126$ MPa (Dikhter and Zeigarnik 1981); Δ, isentrope (Sechenov *et al.* 1977).

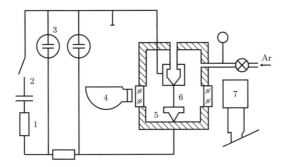

FIG. 2.29.   Schematic of the apparatus for isobaric heating in an argon atmosphere (Dikhter and Zeigarnik 1981): 1, ballast resistor; 2, capacitor bank; 3, oscillograph; 4, high–speed photorecorder; 5, high–pressure chamber; 6, heated wire; 7, spectrograph.

oscillography of current in the circuit and the voltage drop across the plasma column, and photorecording of the column expansion process with time. The latter enables one to measure the time dependence of the column diameter. Special estimates, as well as the variation of the experimental conditions, demonstrate that the mass of material in the discharge is constant (the mixing of argon and cesium can be ignored) while all the losses of energy, including the radiation losses, are small. Therefore, one can use the measured values to calculate the energy input per unit mass, that is, the enthalpy, density, and electrical conductivity.

According to the estimates by Dikhter and Zeigarnik (1981), the error in measuring the current and voltage does not exceed 7% and for the diameter,

3%. This leads to the following estimates of errors in measuring the enthalpy, density, and electrical conductivity: $\Delta H = 15\%$, $\Delta \rho = 10\%$ and $\Delta \sigma = 20\%$. The scatter of experimental thermodynamic data does not exceed 22%, that is, it is in good agreement with error estimates. However, the scatter of electrical conductivity data reaches 40%.

In Figure 2.28, the results of electrical conductivity measurements are compared with the data obtained by other authors. While leaving the discussion of the peculiarities in the behavior of electrical conductivity of nonideal plasma for Chapter 7, we shall only draw the reader's attention to one important detail. The electrical conductivity values on isobars, obtained by the method under discussion, do not increase with heating but, on the contrary, decrease. This can be attributed to the principal shortcoming of the method: the inhomogeneity of the plasma column caused by instability, which leads to superheating.

An interesting phenomenon was observed during the experiments: the plasma column separated into dark and light strata extending across the current. The strata are observed in the opening frames and are clearly visible during the first several hundred microseconds (Fig. 2.30). There is no doubt that, given such fairly slow heating of material, there is enough time for local thermodynamic equilibrium to set in. Therefore, these strata can be referred to as thermal, which is distinct from ionization strata well known in the physics of gas discharges. It was shown by Iakubov (1977) that the observed effect could be a result of the development of thermal instability of a nonideal plasma with current. We emphasize that the strata revealed by Dikhter and Zeigarnik (1977) still remain the most clearly defined manifestation of the instability of nonideal plasma in external fields.

In a series of experiments (Gathers *et al.* 1974, 1976, 1979, 1983; Shaner *et al.* 1986), the exploding wire method was used to study the thermophysical properties of metals in the liquid phase with subsequent evaluation, on this basis, of critical point parameters. As distinct from the techniques used by Dikhter and Zeigarnik, use was made of a slower introduction of energy into the sample under investigation, which had a length of about 25 mm and a diameter of about 1 mm, thereby improving the uniformity of parameters and allowing the avoidance of the instability of the moving metal–gas interface. At the same time, the introduction of energy was sufficiently "fast" to preclude the collapse of the plasma column in the gravitational field. During the experiments, the current–voltage characteristics, surface temperature, channel diameter and, in a number of cases, the velocity of sound were recorded. The results of those measurements provided a basis for subsequent determination of the electrical conductivity, density, enthalpy, and heat capacity of about ten metals at pressures of up to 0.5 GPa and temperatures of up to $9{\cdot}10^3$ K. Whilst doing so, Gathers, Shaner, and others, during their investigations, deliberately confined themselves to the single–phase liquid–metal region of parameters. This was because it had been established that the crossing of the boiling curve in the process of "slow" isobaric expansion causes a sharp development of nonuniformities in heated plasma parameters.

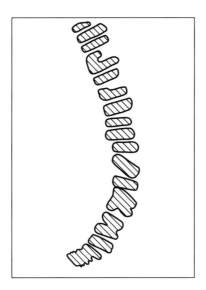

FIG. 2.30. Structure of plasma column under isobar heating of cesium in a rare gas atmosphere (Dikhter and Zeigarnik 1977).

In order to describe the thermodynamics of dense metal plasma, Gathers *et al.* (1974) and Shaner *et al.* (1986) employed the soft–sphere model, which helped in the evaluation of the parameters of critical points for copper and lead; these are in good agreement with the data of dynamic measurements of Altshuler *et al.* (1980) and are listed in Appendix A.

For measuring the sound velocity, Altshuler *et al.* (1980) subjected a column of matter to the effect of a focused laser pulse (0.1–0.5 J, 25 ns, 100 μm). An acoustic wave, which was excited in matter, was propagated across the column and caused a compression wave in the gas upon exit from the latter. This enables one to determine the time of the wave exit from the column and, given the column diameter, to calculate the sound velocity. It has been found that the empirical Birch law, which stipulates that the sound velocity depends linearly on the density of matter, is quite valid. However, the reasons for this are not quite clear.

Kloss *et al.* (1998) and Rakhel *et al.* (2002) used the exploding wire method for investigation of thermodynamic and electrical properties of liquid tungsten. The experiments were performed with tungsten wires placed in the Teflon covering. The set of measuring parameters was as follows: the current through the sample, the drop voltage, the diameter of the conductor, and the temperature at its surface. A fast framing photo camera was used for control of the sample geometry. For images, with an exposure time 10 ns each, it is important to ensure that no parallel discharges and no change of the wire length occur during the observation time of about 200–300 ns. Distribution of the measured quantities throughout the wire cross–section was investigated by means of computational modelling. Heating conditions which ensured inhomogeneity of the conductor down to (3–4)–fold expansion were determined. Estimation of the critical point parameters was the main result of these works: the critical temperature, $T_c = (1.5-1.6) \cdot 10^4$ K, critical pressure, $p_c = (1.1-1.3)$ GPa, and critical density, $\rho_c = (4-5)$ g cm^{-3}.

The exploding wire method was used for measuring the conductivity of metals in a wide density range (DeSilva and Kunze 1994; DeSilva and Katsouros 1998; Saleem et al. 2001). A rectilineal piece of wire was placed in condensed matter (water, glass capillary, etc.) and heated by the current impulse with density $(3-5) \cdot 10^7$ A cm^{-2}. The plasma conductivity was determined under the assumption of inhomogeneous distribution of temperature, pressure, and other quantities in the column. Shadow photographs of the column allowed workers to be sure that it is really inhomogeneous lengthways and axially symmetric. However, radial distribution of measuring quantities could not be controlled. Besides, it was not possible to exclude the influence of evaporation. In these works unexpected results on the conductivity of dense copper, aluminum, and tungsten plasma were obtained: a minimum at the isotherms of conductivity as a function of density at temperatures 1–2 eV was found. This fact testifies to the loss of metallic conductivity only under significant (in 5–7 times) expansion of condensed metals.

In the experiment by Savvatimski (1996) tungsten wires were heated for 10 μs in glass and quartz capillaries in air. The moment of capillary infilling by liquid metal was controlled by conductivity and luminescence. With decreasing density of liquid tungsten from 7.5 to 1.0 g cm^{-3}, the specific resistance increased from 0.5 to 5 μohm cm. The estimation of temperature gave values between $1.0 \cdot 10^4$ and $1.4 \cdot 10^4$ K. The radial distribution of the measuring quantities also was not controled by Savvatimski.

Benage et al. (1999) determined the conductivity of aluminum plasma in exploding wire experiments in thick capillaries made from lead glass. The resistance was determined by measured current and voltage, and the wire cross–section was calculated using the one–dimensional hydrodynamic model and the wide–range equation of state SESAME (Lyon and Johnson 1992). The initial phase of the process, when the metal is in a condensed state, was not modeled. Nevertheless, at the initial stage of the process an abrupt increasing of the resistance was observed induced, apparently, by evaporation. It relates to the fact that the increase of the resistance took place at the moment when the scattered in the wire heat did not amount the sublimation heat of aluminum.

The conductivity of aluminum and tungsten plasmas in conditions of isochoric heating were measured by Renaudin et al. (2002). Foil samples wrapped in spiral (around the axis in the current direction) were placed inside thick sapphire cylinders and heated by a current. The ratio of the inner cylinder volume to the initial volume amounts to about 30. At the initial stage the foil evaporated, vapors filled up the inner cylinder volume, and the homogeneous distributions of such values as temperature, density, etc. were established. The authors assume that the plasma became homogeneous after 100 μs. The beginning of the monotonic resistance drop is a sign of the establishment of the homogeneous state. After 400 μs from the beginning of the process, the resistance starts to increase. Renaudin *et al.* (2002) connect this with plasma cooling as result of the radiative heat transfer. Analyzing the method of the conductivity measurements it is necessary to draw attention to many uncertainties and, first of all, to the problem of the distribution of measurable values in the plasma volume.

Korobenko *et al.* (2002) developed and used a method that makes it possible to investigate the transition of a metal from a condensed to a gaseous phase while maintaining almost uniform temperature and pressure distributions in the sample. The method consists in the pulsed Joule heating of a sample in the form of a thin foil strip placed between two relatively thick glass plates. This method was used to measure the conductivity of tungsten in a process during which the pressure in the sample is maintained at a level of 46 GPa and the density of the sample decreases from the normal solid density to a density 20 to 30 times lower. Effects connected to the sample evaporation are absent because these pressures exceed in several times the critical pressure of the liquid–gas phase transition. In Fig. 2.31 resistivity ρ of tungsten as a function of relative volume v/v_0 (v_0 is a specific volume of solid sample at normal conditions) obtained in the pulsed Joule heating experiments with tungsten foil strips placed in the cells of glass or sapphire. One can see that this dependence has different character in solid, liquid and gaseous states. In liquid state the specific resistivity is approximately proportional to specific volume, that is the ratio of specific conductivity to density remains almost constant. Over the range of relative volumes v/v_0 from 9 to 11, the dependence changes its character (in experiments with the glass cell). For larger volumes, the resistivity approaches a constant value. Note that, for relative volumes larger than 5, the heating process was nearly isobaric. Thus, the dependence of resistivity on the specific volume changes its character along an isobar. As the pressure increases from 4 to 6 GPa, as it follows from the figure, the dependence becomes substantially flatter, in which case the range of relative volumes where the character of the dependence changes becomes larger by a factor of approximately 3. It should be noted that no such effect was detected in nonisobaric processes. The new data on supercritical state of Al were obtained by Korobenko *et al.* (2005).

2.3 High–pressure electric discharges

Because of a high density of material and high level of temperature, the charged particle concentration in a plasma of high–pressure electric arcs and discharges is as high as $(10^{18}-10^{21})$ cm^{-3}. In such a plasma, the Coulomb nonideality parameter Γ may reach substantial values. Whilst doing so, at pressures $p \gtrsim 1$ MPa, the plasma of high–pressure discharge is locally equilibrium. This section will briefly characterize the principal approaches to the investigation of high–pressure discharge plasma. A good review of these studies was made by Asinovskii and Zeigarnik (1974).

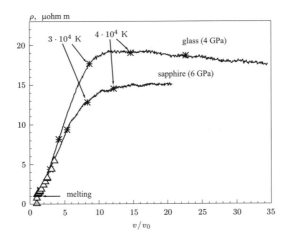

FIG. 2.31. Specific resistivity of tungsten as a function of relative volume (Korobenko *et al.* 2002). The bend marked by the arrow corresponds to the end of melting. Stars refer to temperature in the range from 10^4 to $5 \cdot 10^4$, with a step 10^4. Triangles are for the data of Kloss *et al.* (1998) and Rakhel *et al.* (2002).

Facilities with steady high–pressure arcs appeared as a result of improvement of a wall–stabilized Maecker arc. In the late 1960s, facilities operating at pressures of up to 2 MPa were developed in a number of laboratories in the USSR and FRG. It gave a possibility performing of a series of investigations on the properties of nonideal plasma. Batenin and Minayev (1970) measured the electrical conductivity and emissivity of argon plasma with the following parameters: $p = 1$ MPa, $T = (12-15) \cdot 10^3$ K, $n_e = 10^{18}$ cm^{-3}, nonideality parameter $\Gamma \lesssim 0.5$, and specific power input to the discharge was 4.5 kW cm^{-1}. In steady stabilized high–pressure arcs and high–pressure discharges, one manages, as a rule, to perform measurements of the electrical conductivity and emissivity of plasma. However, the possibilities of wall-stabilized arcs are greatly restricted from the standpoint of nonideal plasma generation. The relatively low level of specific power restricts the level of temperature and, consequently, the values of the degree of ionization.

High parameters of plasma were obtained by Peters (1953) who stabilized the arc with a liquid wall. The arc was burning in water vapors formed due to evaporation of liquid adjoining the discharge column. The liquid proper formed a rotating liquid wall. This was due to the rotation of the casing as a whole. The plasma parameters were as follows: $p = 0.1$ GPa, $T = 18 \cdot 10^3$ K, $n_e = 8 \cdot 10^{18}$ cm^{-3}, $\Gamma = 1.25$, and a specific power of 120 kW cm^{-1}.

Considerably more intensive studies were made into pulsed discharges, which featured a number of advantages. The problem of wall cooling is nonexistent. The specific power is readily increased to the 100 kW cm^{-1} level, thereby providing high degrees of ionization of the material. Quasisteady burning conditions

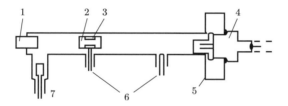

FIG. 2.32. Discharge tube (Radtke and Günter 1976): 1, quartz window; 2, auxiliary electrode; 3, steel ring; 4, pressure sensor; 5, cathode; 6, tungsten probes; 7, anode.

are maintained in a typical arc for a period of approximately one millisecond, which is sufficient to perform measurements and to establish local thermodynamic equilibrium. Because of high pressures and temperatures, the attainable concentrations of charged particles are two (and even three in the case of discharges in liquid) orders of magnitude higher than the characteristic electron concentrations in sustained arcs and reach $(10^{19}$–$10^{23})$ cm^{-3}. However, a possibility arises with the emergence of inhomogeneities in plasma: wave phenomena, and even manifestations of nonequilibrium, are possible.

Pulsed discharges in gases are generated upon discharging a tank of capacitors through an interelectrode gap. The discharge techniques are rather varied. Figure 2.32 illustrates one of the possible designs for a discharge tube. A detailed description of this design was given by Günter (1968) and Radtke and Günter (1976). Essentially, it consists of a quartz tube with four electric leads: anode, cathode (comprising a pressure cell), and two measuring probes. Inside the tube, a movable auxiliary electrode required for ignition is mounted. The tube has a length of 10 cm and a diameter of 1 cm. It is filled with an inert gas at an initial pressure of up to 0.1 MPa.

The basic element of the discharge circuit includes several parallel-connected LC elements shaping a square pulse. Figure 2.33 is a schematic of the apparatus employed by Popovic et al. (1974). Measurements of the current strength i and axial field intensity F help to determine the mean electrical conductivity $\bar{\sigma}$,

$$\bar{\sigma} = (2/R^2) \int_0^R \sigma(r) r \, dr, \qquad (2.9)$$

where R is the radius of the tube. Since the radiation presents the main mechanism of heat transfer at high pressures, the transverse inhomogeneity of the plasma is relatively low. Figure 2.34 gives the radial profiles of the electron temperature measured in flash lamps which have a length of 1 m and radius of 13 mm and are filled with xenon at a pressure of $4 \cdot 10^3$ Pa.

Mitin (1977) developed a procedure for producing superhigh pressures (about 0.1 GPa) in the plasma of a high–pressure pulsed free arc. As distinct from stationary discharges, the power of a pulsed arc may be high, up to 10 kW cm^{-1}. The discharge has a narrow central high–temperature zone (up to $7 \cdot 10^4$ K) and a

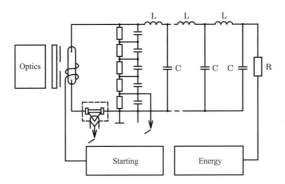

FIG. 2.33. Schematic of the experimental apparatus of Popovic *et al.* (1974).

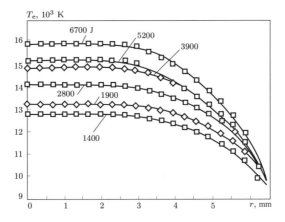

FIG. 2.34. Radial profiles of T_e in a xenon plasma in a flash-lamp for different values of discharge energy (numerals by the curves) (Vitel *et al.* 1990).

peripheral low–temperature zone (approximately 10^4 K). Under these conditions, conventional optical diagnostic methods provide information about the discharge surface and the registered radiating power characterizes the integral properties of inhomogeneous plasma. A pulsed cascade arc, unlike a free one, makes it possible to produce a plasma column that is homogeneous along the axis. However, the inhomogeneity over the column cross–section persists. The coefficient of electrical conductivity of such a plasma was measured for the nonideality parameters of $0.1 \leq \gamma \leq 0.3$. The density of material was measured by X–rays. An X–ray beam from a standard X–ray tube passed through a top hollow electrode (its end, turned into the chamber, had been sealed with a tungsten–tipped beryllium plug), axial region of the arc and a bottom electrode, and was recorded by a NaI photomultiplier.

In order to study the physical properties of the plasma at pressures above 10 MPa and temperatures of up to 20 000 K, special plasma generating sources must

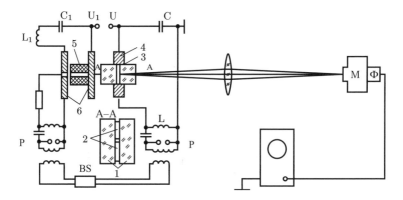

FIG. 2.35. Plasma source design and optical scheme of temperature measurements (Andreev and Gavrilova 1974). Plasma source : 1, quartz plates; 2, quartz spacers; 3, platinum electrodes; 4, tungsten cylinder. Transmission source: 5, Teflon capillary; 6, steel electrodes; M, monochromator; Φ, photomultiplier; O, oscillograph; P, ignition; BS, synchronizer; A-A, cross–section of the plasma source (see insertion).

be provided. Andreev and Gavrilova (1974, 1975a, 1975b) described a plasma source which was used to measure the electrical conductivity and emissivity of air. Its design is shown in Fig. 2.35. This source allows the derivation of an optically transparent plasma at pressures above 10 MPa with uniform and controlled distribution of temperature and density. The apparatus developed by Andreev and Gavrilova (1974, 1975a, 1975b) offers great possibilities for investigations in a high region of parameters.

The plasma volume of $1.0 \times 1.1 \times 10$ mm^3 is defined by quartz plates 1 and spacers 2. Air–tightness was ensured by the high quality of machining performed on the plates and spacers and by tightly pressing the electrodes against the quartz. A bank of capacitors ($C = 3.75$ μF) discharged through the plasma volume. The discharge mode was aperiodic ($L = 0.51$ μH) with a current amplitude reaching 300 A and a duration of about 50 μs. The radial fluctuations of plasma density cease after ignition in less than 5 μs.

Measurements of brightness of the transmission source and transparency of the plasma helped to determine the temperature. Spatial resolution of the optical scheme allowed the derivation of the temperature (and density) distribution in the plasma volume. The regress of wall and electrode materials into the plasma was monitored spectroscopically.

Andreev and Gavrilova (1974) used this source to measure the electrical conductivity of air at $13\,400$ K $\leq T \leq 18\,100$ K and $9.5 \leq p \leq 15$ MPa, thus corresponding to $N_e \cong 10^{18}$–10^{19} cm^{-3} and $\Gamma \cong 0.7$–1. The plasma gap resistance was

$$R(t) = \left(U_0 - \frac{1}{C} \int\limits_0^{t_{\mathrm{m}}} i\,dt - L\frac{di}{dt} \right) i^{-1}. \qquad (2.10)$$

At the moment of current passage through the maximum t_{m}, it is sufficient to know U_0, $i(t_{\mathrm{m}})$, $C^{-1} \int\limits_0^{t_{\mathrm{m}}} i\,dt$. The measured values of $R(t_{\mathrm{m}})$ were on the order of 1 ohm. The error in subsequent determination of the values of electrical conductivity did not exceed 15%, with the values of σ in the central region of the plasma volume differing from its average value by only 8–10%. The measured values proved to be approximately twice lower than those given by the Spitzer formula. These results are discussed in Chapter 7.

The highest values of charged particle concentrations, ca. 10^{21} cm^{-3}, were attained in pulsed discharges in liquids. In doing so, pressures of a tenth of a GPa are attained. The plasma is generated by discharging a bank of capacitors between two electrodes immersed in a liquid, usually water. The current-heated plasma tends to expand the discharge column. This is impeded by the mechanical inertia of the ambient water. Consequently, high pressures develop. For this purpose, the bank of capacitors must be discharged as soon as possible while the plasma column volume is still small. The discharge is shaped over a period of time of about 1 µs. The plasma generated upon discharge in water has the following parameters: $T = (1\text{–}5)\cdot10^4$ K, $p = 0.1\text{–}5$ GPa. Therefore, high pressures are combined with high temperatures, which may hardly be accomplished in other types of facilities. However, in view of the high temperature values, the parameter of nonideality is not very high, being $\Gamma \leq 1.5$. The discharge is usually initiated by electric explosion of a thin wire interconnecting the electrodes. Therefore, the plasma consists of a multicomponent mixture of the products of water vapor decomposition with a small admixture of metal from which the wire was made.

The first investigations of a plasma of pulsed discharge in water were performed by Robinson (1957) and Martin (1960). At present, such discharges are employed in a wide range of investigations related to the development of advanced technologies (cf. Ivanov et al. 1982; Naugol'nykh and Roy 1971). Naturally, such investigations may yield interesting physical results as well.

The capillary discharge with evaporating wall is a pulsed discharge of a special kind. The characteristic feature of this discharge resides in the purging of the discharge channel by vapors of the wall material evaporating due to the Joule heat of the discharge. Ogurtsova et al. (1967, 1974) obtained a plasma with a charge concentration of up to 10^{20}cm^{-3} at a temperature ranging from $3\cdot10^4$ to 10^5 K and at a pressure of 20–50 MPa. The chemical composition of the plasma depends solely on the composition of the wall material. The capillaries were manufactured from textolite. Therefore, the plasma had a complex elementary composition, namely, 47% hydrogen, 37% carbon, and 16% oxygen. The plasma was investigated under quasistationary conditions. To this end, (i) the discharge was maintained by fairly long current pulses of up to 10^{-3} ms; (ii) use was made of capillaries with open ends, through which the discharge current had been introduced, while the sink of the plasma generated under conditions of wall evaporation occurred in the form of plasma jets; (iii) the relative variation of the capillary diameter due to its burning out was small. Investigations have shown that the capillary discharge with evaporating wall allows the derivation of a quasistationary homogeneous plasma of desired chemical composition.

The capillary discharge with evaporating wall is characterized by a high level of temperatures at moderate pressures. Depending on the level of energy contribution, the nonideality parameter ranges from 0.5 to 1.25. The capillary discharge with evaporating wall was used to measure the spectral absorption factor of continuous radiation and the electrical conductivity.

References

Alekseev, V. A. (1968). Electrical conductivity of cesium at supercritical temperatures and pressures. *High Temp.*, **6**, 923–927.

Alekseev, V. A. and Iakubov, I. T. (1983). Nonideal plasmas of metal vapors. *Phys. Rep.*, **96**, 1–69.

Alekseev, V. A. and Iakubov, I. T. (1986). Electrical conductivity and thermoelectromotive force of alkali metals. In *Handbook on thermodynamic and transport properties of alkali metals* , Ohse, R. W. (ed.), pp. 703–734. Blackwell, Oxford.

Alekseev, V. A., Ovcharenko, V. G., Rizhkov, Y. F., and Senchenkov, A. P. (1970a). Equation of state of cesium at pressures 20–600 atm and temperatures 500–2500 C. *JETP Lett.*, **12**, 207–210.

Alekseev, V. A., Vedenov, A. A., Krasitskaya, L. S., and Starostin A. N. (1970b). Thermoelectric power of cesium near critical temperatures and pressures. *JETP Lett.*, **12**, 501–504.

Alekseev, V. A., Starostin, A. N., Vedenov, A. A., and Ovcharenko, V. G. (1972). Nature of thermoelectric power of mercury in transcritical state. *JETP Lett.*, **16**, 49–53.

Alekseev, V. A., Vedenov, A. A., Ovcharenko, V. G., Krasitskaya, L. S., Rizhkov, Y. F., and Starostin, A. N. (1975). The effect of saturation on the thermo-e.m.f. of caesium at high temperatures and pressures. *High Temp. High Press.*, **7**, 677–679.

Alekseev, V. A., Ovcharenko, V. G., and Rizhkov, Y. F. (1976). The metal–dielectric transition in liquid metals and semiconductors at high temperatures and pressures in the region of the critical point. *Sov. Phys. Uspechi*, **19**, 1027–1029.

Altshuler, L. V., Bushman, A. V., Zhernokletov, M. V., Zubarev, V. N., Leontiev, A. A., and Fortov, V. E. (1980). Release isentropes and the equation of state of metals at high energy densities. *JETP*, **51**, 373–383.

Andreev, S. I. and Gavrilova, T. V. (1974). Investigation of a pulsed stabilized discharge in air at a pressure above 100 atm. *High Temp.*, **12**, 1138–1142.

Andreev, S. I. and Gavrilova, T. V. (1975a). Measurement of electrical conductivity of air plasma at pressures above 100 atm. *High Temp.*, **13**, 151–153.

Andreev, S. I. and Gavrilova, T. V. (1975b). Measurement of spectral absorption coefficient and total radiation coefficient of an air plasma at pressures above 100 atm. *High Temp.*, **13**, 584–586.

Asinovskii, E. I. and Zheigarnik, V. A. (1974). High pressure discharges. *High Temp.*, **12**, 1120–1135.

Barolskii, S. G., Kulik, P. P., Ermokhin, N. V., and Melnikov, V. M. (1972). Measurement of electrical conductivity of dense highly nonideal cesium plasma. *JETP*, **62**, 176–182.

Batenin, V. M. and Minaev, P. V. (1970). High–temperature stationary plasma brightness source. *High Temp. High Press.*, **2**, 597–607.

Benage, J. F., Shanahan, W. R., and Murillo, M. S. (1999). Electrical resistivity measurements of hot dense aluminum. *Phys. Rev. Lett.*, **83**, 2953–2956.

Borzhievsky, A. A., Sechenov, V. A., and Horunzhenko, V. I. (1987). Experimental investigation of electrical conductivity of cesium vapour. In *Proceedings of the 18th International Conference on Ionization Phenomena in Gases*. Contributed Papers, Swansea. Vol. 1, pp. 250–251.

Chelebaev, A. K. (1978). Measurements of equation of state parameters for dense steams of cesium. D. Phil. thesis, MAI, Moscow.

Cheshnovsky, O., Even, U., and Jortner, J. (1981). The polarization catastrophe and the metal–nonmetal transition in disordered materials. *Phil. Mag. A*, **44**, 1–7.

DeSilva, A. W. and Katsouros, J. D. (1998). Electrical conductivity of dense copper and aluminum plasmas. *Phys. Rev. E*, **57**, 5945–5951.

DeSilva, A. W. and Kunze, H.–J. (1994). Experimental study of the electrical conductivity of strongly coupled copper plasmas. *Phys. Rev. E*, **49**, 4448–4454.

Dikhter, I. Y. and Zeigarnik, V. A. (1975). Electrical explosion of a cesium wire at pressures up to 500 atm. *High Temp.*, **13**, 447–454.

Dikhter, I. Y. and Zeigarnik, V. A. (1977). Equation of state and conductivity of a highly ionized cesium plasma. *High Temp.*, **15**, 196–198.

Dikhter, I. Y. and Zeigarnik, V. A. (1981). Equation of state and electrical conductivity dense strongly ionized plasma of alkali metals. In *Review on thermophysical properties of matter. No. 4*, pp. 59–102. IVTAN, Moscow.

Duckers, L. J. and Ross, R. G. (1972). Thermoelectric power of supercritical fluid mercury. *Phys. Lett. A*, **38**, 291–294.

Ermokhin, N. V., Ryabyi, V. A., Kovalev, B. M., and Kulik, P. P. (1971). Experimental investigation of coulomb interaction in dense plasma. *High Temp.*, **9**, 611–620.

Franz, G., Freyland, W., and Hensel, F. (1981). (unpublished).

Freyland, W. and Hensel, F. (1972). Electrical properties of metals in liquid–gas critical region. *Ber. Bunsenges. Phys. Chem.*, **76**, 348–349.

Gathers, G. R. (1983). Thermophysical properties of liquid copper and aluminum. *Int. J. Termophys.*, **4**, 209–226.

Gathers, G. R., Shaner, J. W., and Young, D. A. (1974). Experimental, very high temperature, liquid uranium equation of state. *Phys. Rev. Lett.*, **33**, 70–72.

Gathers, G. R., Shaner, J. W., and Bri, R. L. (1976). Improved apparatus for thermophysical measurements on liquid metals up to 8000 K. *Rev. Sci. Instruments*, **47**, 471–479.

Gathers, G. R., Shaner, J. W., and Hodson, W. M. (1979). Thermodynamic characterization of liquid metals at high temperature by isobaric expansion measurements. *High Temp. High Press.*, **11**, 529–538.

Gogoleva, V. V., Zitserman, V. Y., Polishchuk, A. Y., and Iakubov, I. T. (1984). Electrical conductivity of an alkali–metal–vapor plasma. *High Temp.*, **22**, 163–170.

Günter, K. (1968). Elektrische Eigenschaften von Xenon–Impulsplasmen. *Beitr. Plasma Phys.*, **8**, 383–401.

Hefner, W. and Hensel, F. (1982). Dielectric anomaly and the vapor–liquid phase transition in mercury. *Phys. Rev. Lett.*, **48**, 1026–1028.

Hefner, W., Schmutzler, R. W., and Hensel, F. (1980). Measurements of optical parameters of mercury in region of metal–dielectric transition. *J. de Phys.*, **41**, suppl., 8–11.

Hensel, F. (1970). Optical absorption measurements for gaseous mercury at supercritical temperatures and high densities. *Phys. Lett. A*, **31**, 88–89.

Hensel, F. (1971). Pressure dependence of optical absorption of gaseous mercury up to 1700 C and 2200 bar. *Ber. Bunsenges. Phys. Chem.*, **76**, 847–851.

Hensel, F. (1977). (unpublished).

Hensel, F. and Franck, E. U. (1968). Metal–nonmetal transition in dense mercury vapor. *Rev. Mod. Phys.*, **44**, 697–703.

Iakubov, I. T. (1977). Thermal instability of nonideal current–carrying plasmas of metal vapors. *Beitr. Plasma Phys.*, **17**, 221–227.

Ikezi, H., Schwarzenegger, K., Simons, A. L., Passner A. L., and McCall, S. L. (1978). Optical properties of expanded fluid mercury. *Phys. Rev. A*, **18**, 2494–2499.

Isakov, I. M. and Lomakin, B. N. (1979). Measurement of electric conductivity during adiabatic compression of cesium vapors. *High Temp.*, **17**, 222–225.

Ivanov, V. V., Shvets, I. S., and Ivanov, A. V. (1982). *Underwater spark discharges*. Naukova Dumka, Kiev.

Jungst S., S., Knuth, B., and Hensel, F. (1985). Observation of singular diameters in the coexistence curves of metals. *Phys. Rev. Lett.*, **55**, 2160–2163.

Kikoin, I. K. and Senchenkov, A. P. (1967). Electrical conductivity and equation of state of mercury in temperature range 0–2000 °C and pressure range of 200–5000 atm. *Phys. Metals and Metallography–USSR*, **24**, 74–87.

Kikoin, I. K., Senchenkov, A. P., Gelman, E. V., Korsunskii, M. M., and Naurzakov S. P. (1966). Electrical conductivity and density of a metal vapor. *JETP*, **22**, 89–91.

Kikoin, I. K., Senchenkov, A. P., Naurzakov, A. P. (1973). (unpublished).

Kloss, A., Rakhel, A. D., and Hess, H. (1998). Experimental results on tungsten wire explosions in air at atmospheric pressure – Comparison with one–dimensional numerical model. *Int. J. Thermophys.*, **19**, 983–991.

Korobenko, V. N., Rakhel, A. D., Savvatimskiy, A. I., and Fortov, V. E. (2002). Measurement of the electrical resistivity of tungsten in continuous liquid–to–gas transition. *Plasma Phys. Rep.*, **28**, 1008–1016.

Korobenko, V. N., Rakhel, A. D., Savvatimskiy, A. I., and Fortov, V. E. (2005). Measurement of the electrical resistivity of hot aluminum passing from the liquid to gaseous state at supercritical pressure. *Physical Review B*, **71**, 014208/1–10.

Korshunov, Y. S., Senchenkov, A. P., Asinovskii, E. I., and Kunavin, A. T. (1970). Measurement of P–V–T dependence for cesium at high temperatures and pressures, and estimation of parameters of critical point. *High Temp.*, **8**, 1207–1210.

Kulik, P. P., Ryabii, V. A., and Ermokhin, N. V. (1984). *Nonideal plasma.* Energoatomizdat, Moscow.

Kuznetsova, N. I. and Lappo, G. B. (1979). Optical properties of alkali–metal plasmas consisting of lithium, sodium, and potassium. *High Temp.*, **17**, 32–40.

Lebedev, S. V. (1957). Explosion of a metal by an electric current. *JETP*, **5**, 243–252.

Lebedev, S. V. (1966). Initial heating stage of exploding wires, *JETP*, **23**, 337–346.

Lebedev, S. V. and Savvatimskii A. I. (1984). Metals under fast heating by electric current of high density. *Phys. Uspechi*, **144**, 215–250.

Lomakin, B. N. and Lopatin, A. D. (1983). Electrical conductivity of compressed vapors of cesium and potassium. *High Temp.*, **21**, 190–193.

Lyon, S. P. and Johnson, J. D. (eds). (1992). SESAME: The Los Alamos national laboratory equation of state database. Report No. LA–UR–92–3407.

Martin, E. A. (1960). Experimental investigation of high–energy density, high–pressure arc plasma. *J. Appl. Phys.*, **31**, 255–267.

Mitin, P. V. (1977). Stationary and pulse arcs of high and ultrahigh pressure and methods of theirs diagnostics. In *Properties of low temperature plasmas and diagnostics methods*, Zhukov, M. F. (ed.), pp. 105–138. Nauka, Novosibirsk.

Naugol'nykh, K. A. and Roy, N. A. (1971). *Electrical discharges in water.* Nauka, Moscow.

Neale, F. E. and Cusack, N. E. (1979). Thermoelectric power near the critical point of expanded fluid mercury. *J. Phys. F*, **9**, 85–94.

Noll, F., Pilgrim, W. C., and Winter, R. (1988). Electrical–conductivity of Na-NH$_3$ and cesium in the critical region. *Z. Phys. Chemie.*, **156**, 303–307.

Novikov, I. I. and Roshchupkin, V. V. (1967). Experimental evaluation of PVT relationship of cesium vapors. *Measurement techniques–USSR*, **10**, 1193–1195.

Ogurtsova, N. N. and Podmoshenskii, I. V. (1967). (unpublished).

Ogurtsova, N. N., Podmoshenskii, I. V., and Smirnov, V. L. (1974). Measurement of the electrical conductivity of a nonideal plasma at 38 000 K and pressure (5-25)·10^7 N/m^2. *High Temp.*, **12**, 559–561.

Peters, T. H. (1953). Temperatur und Strahlungsmessungen am wasserstabilisieten Hochdruckbogen. *Z. Phys.*, **135**, 573–592.

Pfeifer, H. P., Freyland, W. F., and Hensel, F. (1973). Absolute thermoelectric power of fluid cesium in metal–nonmetal transition range. *Phys. Lett. A*, **43**, 111–112.

Pfeifer, H. P., Freyland, W., and Hensel, F. (1979). Equation of state and transport data on expanded liquid rubidium up to 1700 °C and 400 bar. *Ber. Bunsenges. Phys. Chem.*, **83**, 204–211.

Popovic, M. M., Popovic, S. S., and Vukovic, S. M. (1974). Distribution of electrical conductivity and density along section of plasma Coulomb of high pressure arc. *Fizika*, **6**, 29–35.

Radtke, R. and Günter, K. (1976). Electrical conductivity of highly ionized dense hydrogen plasma. 1. Electrical measurements and diagnostics. *J. Phys. D*, **9**, 1131–1138.

Rakhel, A. D., Kloss, A., and Hess, H. (2002). On the critical point of tungsten. *Int. J. Thermophys.*, **23**, 1369–1380.

Renaudin, P., Blancard, C., Faussurier, G., and Noiret, P. (2002). Combined pressure and electrical resistivity measurements of warm dense aluminum and titanum plasmas. *Phys. Rev. Lett.*, **88**, 215001/1–4.

Renkert, H., Hensel, P., and Franck, E.U. (1969). Metal–nonmetal transition in dense cesium vapor. *Phys. Lett. A.*, **30**, 494–495.

Renkert, H., Hensel, P., and Franck, E. U. (1971). Electrical conductivity of liquid and gaseous cesium up to 2000 °C and 1000 bar. *Ber. Bunsenges. Phys. Chem.*, **75**, 507–512.

Robinson, J. W. (1957). Measurements of plasma energy density and conductivity from 3 to 120 kbar. *J. Appl. Phys.*, **38**, 210–216.

Rochling, G. (1963). *Adv. Energy Conversion*, **3**, 69–75.

Saleem, S., Haun, J., and Kunze, H.–J. (2001). Electrical conductivity measurements of strongly coupled W plasmas. *Phys. Rev. E*, **64**, 056403/1–6.

Savvatimski, A. I. (1996). Experiments on expanded liquid metals at high temperatures. *Int. J. Thermophys.*, **17**, 495–505.

Sechenov, V. A., Son, E. E., and Shchekotov, O. E. (1977). Electrical conductivity of a cesium plasma. *High Temp.*, **15**, 346–349.

Shaner, J. W., Hixson, R. S., and Winkler, M. A. (1986). (unpublished).

Stone, J. P., Ewing, C. T., Spann, J. R., Steinkul, E. W., Williams, D. D., and Miller, R. R. (1966). High temperature *PVT* properties of sodium, potassium, and cesium. *J. Chem. Eng. Data*, **11**, 309–311.

Suzuki, K, Inutake, M., and Fujiwaka, S. (1977). (unpublished).

Themperly, G. M., Rowlinson, J., and Rushbrooke, G. (eds). (1968). *Physics of simple liquids*. North–Holland, Amsterdam.

Tsuji, K., Yao, M., and Endo, H. (1977). Electrical conductivity and thermoelectric power of expanded mercury and dilute amalgams. 1. Hg and Cd amalgams. *J. Phys. Soc. Jap.*, **42**, 1594–1600.

Uchtmann, H. and Hensel, F. (1975). Density dependence of optical gap of compressed mercury vapor. *Phys. Lett. A*, **53**, 239–240.

Uchtmann, H., Popielawski, J., and Hensel, F. (1981). Radiation emitted by a slightly ionized nonideal high–pressure plasma. *Ber. Bunsenges. Phys. Chem.*, **85**, 555–558.

Vasil'ev, I. N. and Trelin, Y. S. (1969). Fixed distance acoustic pulse measurement of speed of ultrasound in gaseous and vapor media at high temperatures. *High Temp.*, **7**, 1035–1041.

Vitel, Y., Mokhtari, A., and Skowronek, M. J. (1990). Electrical conductivity and pertinent collision frequencies in nonideal plasma with only a few particles in the Debye sphere. *J. Phys. B: At. Mol. Phys.*, **23**, 651–660.

Volyak, L. D. and Chelebaev, A. K. (1976). Measurement of the *PVT* parameters of cesium vapor. *High Temp.*, **14**, 913–921.

Vorobiov, V. V., Kulik, P. P., Pallo, A.V., Rozanov, E. K., and Riabyi, V. A. (1981). Experimental study of the non–ideality of radiative heat conductivity of dense alkali plasmas. In *Proceedings of the XV symposium on phenomena in ionized gases*. Contributed papers, Part 1. Minsk, pp. 361–362.

Winter, R., Noll, F., Bodensteiner, T, Glaser, W., and Hensel, F. (1988). Conductivity scattering and neutron–scattering experiments on expanded fluid cesium in the metal–nonmetal transition region. *Z. Phys. Chemie.*, **156**, 145–149.

Yermokhin, N. V., Kulik, P. P., Riabyi, V. A., and Semionov, V. K. (1981). Dense plasma electric conductivity measurements in the region of pressure ionization. In *Proceedings of the XV symposium on phenomena in ionized gases*. Contributed papers, Part 1. Minsk, pp. 363–364.

3

DYNAMIC METHODS IN THE PHYSICS OF NONIDEAL PLASMA

The highest plasma parameters have been obtained by dynamic techniques (Fortov 1982) which permitted the realization, under controlled conditions, of record–high local concentrations of energy. These techniques are based on the accumulation of energy in the material under investigation either as a result of viscous dissipation in the front of shock waves which propagate throughout the material to cause its compression, acceleration, and irreversible heating, or as a result of adiabatic variation of pressure in the material. Without going into a detailed comparison between the electric and dynamic techniques, we shall emphasize that the high purity and homogeneity of the volume under investigation, the absence of electric and magnetic fields (hampering the diagnostics and causing the development of various instabilities in the plasma), the high reproducibility of results, and the possibility of attaining extremely high parameters render the dynamic techniques a convenient means for attaining and investigating the physical properties of highly nonideal plasma under extreme conditions. In addition, the application of the general laws of conservation of mass, momentum, and energy enables one to reduce (see Section 3.2) the recording of the thermodynamic characteristics of plasma to the registration of the kinematic parameters of the motion of shock discontinuities and interfaces (i.e., to the measurement of times and distances), which presents an additional important advantage of the dynamic techniques.

The use of shock waves in high–pressure physics enabled one to attain pressures of condensed matter of hundreds of thousands megapascals and, while so doing, perform extensive thermodynamic, optical, and electrophysical investigations (Altshuler 1965; Kormer 1968; Minaev and Ivanov 1976; Duvall and Graham 1977; Davison and Graham 1979). The use of such techniques in the physics of nonideal plasma (Fortov 1982) offered possibility of considerably expanding the range of plasma pressures and temperatures to be investigated and made possible laboratory investigations of states with extremely high energy concentrations. Hence physical measurements in the regions of the phase diagram of matter, which used to be inaccessible to the traditional methods of plasma experiments, are possible.

In this chapter we shall consider three methods of dynamic generation of plasma, namely, shock and adiabatic compression and the method of adiabatic expansion of shock–compressed matter.

The shock compression technique appears most effective from the standpoint of studies into nondegenerate plasma of materials which, in their initial state, are gases. A combination of preheated pneumatic, electric discharge and explosion tubes provided information on the thermodynamic, electrophysical, and optical properties of plasma at pressures of up to 30 GPa, electron concentrations of about 10^{23} cm^{-3}, and densities of up to 4.5 g cm^{-3}(1.5 times higher than the crystallographic density of xenon). Under these conditions, the Coulomb energy is an order of magnitude higher than the kinetic energy of particle motion so that, from the physical viewpoint, such a plasma resembles a liquid while differing from the latter by a richer and complex spectrum of interparticle interactions. The plasma behavior at ultramegabar pressures is currently of special interest from the standpoint of ascertaining the role of band structure effects and determining the lower range of application of quasiclassical models of the equation of state. In order to reach this pressure range, nuclear explosions, laser radiation, and rapid electric explosion of metal foils have been used to advantage (see Section 3.5).

Shock and isentropic compression techniques make it possible to attain high pressures and temperatures in a medium of increased density and, because of thermodynamic restrictions, do not allow an investigation of the boiling curve and the near–critical states of metals. The most effective method, from the standpoint of obtaining plasma with a less–than–solid–state density, is isentropic expansion of metals (see Section 3.5) prestressed in a powerful shock front. This method helps in the investigation of an extremely wide region of the phase diagram of metals from a highly condensed metallic liquid to ideal gas, including the region of nonideal degenerate and Boltzmann plasma and the neighborhood of the critical point.

The results of registration of release isentropes served as a basis for the construction of modern wide-range equations of state and, in addition, helped form a more conclusive opinion on the general qualitative form of the phase diagram of matter at high pressures and temperatures. This problem is not trivial for a highly nonideal plasma, in view of the existence of numerous predictions of exotic phase transitions due to the "metal–dielectric" transformation, as well as to a strong interaction of charged particles with each other and with neutrals. We shall discuss the relevant experimental studies in Sections 3.4 and 5.9.

3.1   The principles of dynamic generation and diagnostics of plasma

The classification of plasma states in Chapter 1 shows (cf. Fig. 1.1) that only the plasma states with extremely high pressures and temperatures located at the periphery of the phase diagram of matter, where the interparticle interaction is low, are accessible to consistent theoretical analysis. The range of existence of nonideal plasma appears restricted and is located in Fig. 1.1 within the triangle defined by the $\gamma = 1$ and $\varepsilon_F \sim e^2 n_e^{1/3}$ curves, the upper part of this triangle relating to a degenerate plasma and the lower part to a Boltzmann plasma. The maximum attainable values of the nonideality parameter are finite and do not exceed several units. Figure 1.1 demonstrates that optimum conditions exist for the generation of highly nonideal plasma. These conditions are further complemented by their optima dictated by the processes of ionization under conditions of the method selected for the realization of those states (cf. Sections 3.2 and 3.3). Given the isochoric heating of a Boltzmann plasma in the initial stages of ionization, the electron concentration grows sharply with temperature, $n_e^2 \sim n_a T^{3/2} \exp(-I/T)$, thus leading to a faster growth of the Coulomb energy as compared with kT. A further temperature increase causes full ionization of the given electron shell ($n_e \sim$ const.), and a decrease in γ in view of the increase in kinetic energy. While putting off the qualitative assessment of this effect until Sections 3.2 and 3.3, we note that these optima formed the basis for the designs of most devices for dynamic compression of plasma.

Although it appears restricted in Fig. 1.1, the nondeal plasma takes an exceedingly wide region in the real $p - V$ diagram of the state of matter (Fig. 3.1). It adjoins directly and actually penetrates the region of the condensed state (Ebeling, Kreft, and Kremp 1976; Fortov 1982) where the physical description is extremely difficult because it depends on the specific electron spectra of atoms and molecules. Therefore, priority is given in this case to the experimental, primarily dynamic, methods (Altshuler 1965; Kormer 1968; Minaev and Ivanov 1976; Duvall and Graham 1977; Davison and Graham 1979) which enabled one to perform elegant measurements of the thermodynamic, optical, and electrophysical properties of condensed matter in the megabar pressure range. Modern high–temperature experimental techniques have facilitated the advancement into the region of the boiling curve and the neighborhood of the critical point of low–boiling metals, such as mercury, cesium, and rubidium, and the investigation of the liquid phase region of a number of other elements at $p \lesssim 0.4$ GPa, $T \lesssim 5 \cdot 10^3$ K (Shaner and Gathers 1979) (Chapter 2). However, not only have the critical parameters not been found for the great majority of other metals (amounting to about 80% of all elements in the periodic system), but the qualitative form of the phase diagram has also not been clarified. This is because the greater part of the latter is taken by the region of nonideal plasma, which is inaccessible to traditional experimental techniques.

In order to produce a plasma with strong interparticle interaction, a considerable amount of energy release is required in a higher–density medium. The concomitant pressures and temperatures, as a rule, appreciably exceed the thermal–strength limits of the structural materials of the respective facilities. Therefore, such a plasma can only be maintained for a short period of time, which is defined by its inertial expansion. This compels one to conduct the experiment under heavy–duty pulsed conditions at a high power level. The appropriate experimental facilities must provide a fast energy supply to the material under investigation, and should feature geometric dimensions that are sufficient for reliable diagnostics. The existing possibilities for plasma generation are listed in Table 3.1, which lists the characteristic (not necessarily maximum) parameters of energy sources used in the compression and heating of the material. Whilst doing so, because of the above restrictions due to plasma degeneracy and superheating, the extreme parameters listed in Table 3.1 do not correspond at all to

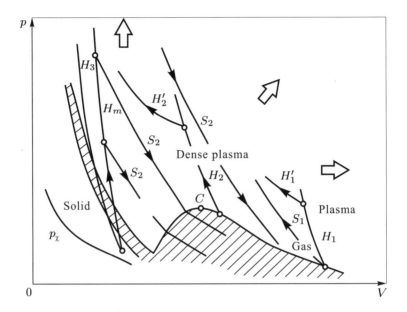

FIG. 3.1. Schematic illustrating the principles of dynamic generation of plasma. p_χ is the boundary of maximum compressions of the material, that is, the "cold" ($T = 0$ K) compression curve. The shaded areas show two–phase regions of melting and evaporation. C is the critical point. Circles indicate the initial states of the medium. H_1 and H_2 are the curves of cesium and inert gas compression by incident and reflected (prime) shock waves. H_3 and H_m are shock–wave compression of solid and porous metals. S_1 is the curve of adiabatic compression of cesium while S_2 is the unload adiabats for shock–compressed metals.

the maximum Coulomb nonideality effects.

The dynamic methods of plasma generation are based on hydrodynamic methods of heating a material as a result of viscous dissipation of energy in a shock front or adiabatic compression of the medium (Zel'dovich and Raizer 1966). By now, three techniques have come to be most widely used in plasma investigations, namely, isentropic and shock compression, as well as isentropic expansion of pre–shock–compressed material. The use of various energy sources to excite shock and compression waves, such as compressed and electric discharge–heated gas, condensed and nuclear explosives, and powerful laser and neutron fluxes, enabled one to produce a nonideal plasma of different elements in a wide and continuously varying range of densities: from solid state to gas at pressures of hundreds of thousands of megapascals and temperatures of tens of kiloelectron volts.

A considerable amount of these data were obtained by propagating shock waves over the material under investigation to cause its irreversible compression, heating, and acceleration. In doing so, the initial states of the material may be

Table 3.1 *Sources of energy and experimental facilities employed in the physics of high energy concentrations.*

Primary energy source	Final form	Energy density, MJ cm^{-3}	Tempe-rature, eV	Pressure, 10^5 Pa	Total energy, MJ	Dura-tion, s	Power, W
Chemical explosive		10^{-2}	0.5	$5 \cdot 10^5$	10^2	10^{-7}	10^{10}
	Metal plates	0.3	60	10^7	3	10^{-6}	10^{12}
	Magnetic field, 1 MOe	$4 \cdot 10^{-3}$	0.3	$5 \cdot 10^4$	5	10^{-6}	$5 \cdot 10^{12}$
	Magnetic field, 25 MOe	2.5	200	$2.5 \cdot 10^7$	1	10^{-7}	10^{13}
	Explosion plasma generators	10^{-2}	60	10^5	30	10^{-6}	10^{12}
Nuclear explosive		10^4	10^7	10^{10}	10^{11}	10^{-6}	10^{22}
	Neutron heating	10	50	$2 \cdot 10^7$	10^3	10^{-6}	10^{15}
	Shock waves in solid	5	50	$5 \cdot 10^7$	10^4	$3 \cdot 10^{-6}$	10^{15}
	Shock waves in gas	0.3	40	$2 \cdot 10^5$	10^7	10^{-5}	10^{18}
Compressed gas	Adiabatic compression	$2 \cdot 10^{-5}$	0.3	150	10^3	$6 \cdot 10^{-3}$	10^5
	Pneumatic shock tubes	10^{-4}	1	250	10^{-2}	10^{-4}	$3 \cdot 10^8$
	Combustion shock tubes	10^{-6}	2	10	$2 \cdot 10^{-2}$	$3 \cdot 10^{-4}$	10^8
	Electric discharge shock tubes	10^{-7}	2	1	10^{-2}	10^4	10^8
Capacitor		10^{-8}	-	-	40	10^{-5}	10^{12}
Rotor machine		10^{-3}	-	-	100	10^{-4}	10^{12}
Induction storage		10^{-4}	-	-	100	10^{-4}	10^{12}
Storage cell		$5 \cdot 10^{-4}$	-	-	1000	10^{-3}	10^{12}
	Rapid explosion of wires	$5 \cdot 10^{-2}$	4	10^5	10^{-3}	10^{-6}	10^9
	Slow explosion of wires	$2 \cdot 10^{-2}$	0.5	$5 \cdot 10^2$	10^{-3}	10^{-4}	10^7
	Pulsed discharges	10^{-3}	10	10^4	10^{-4}	10^{-3}	10^9
	Plasma focus	10^{-2}	10^3	10	10^{-4}	10^{-5}	10^{10}
	High–pressure arcs	10^{-5}	2	10^4	10^{-4}	Steady	10^{-5}
	Furnace	10^{-3}	0.3	$5 \cdot 10^3$	10^{-3}	Steady	10^3
Laser		10^{-6}	-	-	$5 \cdot 10^{-4}$	10^{-10}	10^{13}
	Target	10^4	$5 \cdot 10^3$	10^8	0.5	10^{-10}	10^{13}
Electron beam		10^{-6}	-	-	1	10^{-8}	10^{13}
	Target	50	$5 \cdot 10^3$	10^7	0.1	10^{-8}	10^{13}

either in the solid or gas phase. The latter case is especially convenient to obtain a highly nonideal Boltzmann plasma (Sections 3.2 and 3.3). It is generated as a result of shock compression of high–pressure gases whose initial states are (cf. Fig. 3.1) in the neighborhood of the saturation curve (cesium, argon) or under supercritical conditions (xenon). By recording the state of single or double compression, it is possible to obtain a plasma with supercritical parameters in a wide range of pressures (up to 11 GPa) and temperatures (up to 10^5 K) and to penetrate, from the "gas" phase side, the condensed state region. The maximum density of xenon plasma amounted to 4.5 g cm^{-3}. This is about 1.5 times greater than the density of solid xenon and the order of the density of aluminum. The compression of liquid xenon using light–gas "guns" in the pressure range of up to 140 GPa yields valuable information on the electron spectrum of superdense plasma in the range of 10^5 MPa. Adiabatic compression of saturated cesium vapors, S_1 (Fig. 3.1), makes it possible (Section 3.2) to attain lower degrees of plasma heating where the charge–neutral interaction prevails. The compression of metals by powerful shock waves (Altshuler 1965; Kormer 1968; Minaev and Ivanov 1976; Duvall and Graham 1977; Davison and Graham 1979), a result of the detonation of condensed explosive, transforms the metal into a state with a pressure of up to 1000 GPa and temperature of tens of thousands degrees when the metal is molten. As a matter of fact, a disordered electron–ion plasma, in which the electron component is degenerate or partly degenerate, is realized.

The use of powerful underground explosions (Altshuler *et al.* 1968; Trunin *et al.* 1969, 1972; Vladimirov *et al.* 1984; Simonenko *et al.* 1985; Avrorin *et al.* 1987), neutron radiation resulting from the blasting of nuclear charges (Thom and Schneider 1971), and concentrated laser radiation (Anisimov *et al.* 1984) increase the pressure up to 100 TPa. It also serves a basis for extrapolation checks of quasiclassical theories (Kirzhnits *et al.* 1975).

One characteristic feature of the shock–wave technique is that it helps the attainment of high pressures and temperatures in compressed media while the region of lower plasma densities (including the boiling curve and the neighborhood of the critical point) prove inaccessible for such techniques. The intermediate plasma states between solid and gas can be investigated using the isentropic expansion technique (Section 3.5). It is based on plasma generation under conditions of adiabatic expansion of condensed matter (S_2, Fig. 3.1) precompressed and irreversibly heated in a powerful shock front. In this way, it was possible to investigate the properties of metals in a wide range of phase diagram, from a highly compressed condensed state to an ideal gas, including the region of degenerate and Boltzmann low–temperature plasma, near–critical and two–phase states, as well as the "metal–dielectric" transition region.

We see that the dynamic methods, in their various combinations, enable one to realize and investigate a wide spectrum of plasma states with variegated and strong interparticle interaction. In doing so, it proves possible to both actually realize states with high energy concentrations and fully diagnose those exotic states.

The dynamic diagnostic methods are based on the utilization of the relationship between the thermodynamic properties of the medium under investigation and the experimentally observed hydrodynamic phenomena occurring upon accumulation of high–energy densities in the matter (Zel'dovich and Raizer 1966). In the general form, this relationship is expressed by a system of nonlinear (three–dimensional) differential equations of nonstationary gas dynamics. Its complete solution appears to be beyond the capabilities of most powerful modern computers. For this reason, one tends to use in dynamic investigations self–similar solutions of the type of stationary shock wave or centered Riemann expansion wave (Altshuler 1965; Zel'dovich and Raizer 1966), which express the conservation laws in a simple algebraic or integral form. While doing so, the conditions of self–similarity of flow must be observed in the experiment in order to use these simplified relationships.

Upon propagation of a stationary shock–wave discontinuity through the material, the laws of conservation of mass, momentum, and energy are satisfied in the front of the latter (Zel'dovich and Raizer 1966):

$$v/v_0 = (D - u)/D; \quad p = p_0 + Du/v_0, \\ E - E_0 = (1/2)(p + p_0)(v_0 - v), \tag{3.1}$$

which allows the derivation of the hydrodynamic and thermodynamic characteristics of the material from the recording of any two out of five parameters characterizing the shock wave discontinuity (E, p, v, D, u). The shock velocity D is measured most readily and accurately using baseline techniques. The choice of the second parameter to be measured depends on the actual experimental conditions.

An analysis of errors in the relationships included in Eq. (3.1) indicates (Lomakin and Fortov 1973; Kunavin et al. 1974) that in the case of readily compressible ("gas") media, it is practical to perform the recording of density, $\rho = v^{-1}$, of shock–compressed material. A procedure has been developed for such measurements based on the registration of absorption by cesium (Section 3.2), argon (Section 3.5), and the air (Section 3.6) plasma of "soft" X–ray radiation. In the case of lower compressibility of the system (condensed media), tolerable accuracies are ensured (Altshuler 1965) by recording the mass velocity of travel u.

In the experiments which involve the recording of the isentropic expansion curves for shock–compressed matter (Section 3.5), the states in a centered release wave are described by the Riemann integrals (Zel'dovich and Raizer 1966):

$$v = v_{\mathrm{H}} + \int_p^{p_{\mathrm{H}}} \left(\frac{du}{dp}\right)^2 dp; \quad E = E_{\mathrm{H}} - \int_p^{p_{\mathrm{H}}} p \left(\frac{du}{dp}\right)^2 dp, \tag{3.2}$$

which are calculated along the measured isentrope $p_S = p_S(u)$. By recording under different initial conditions and shock–wave intensities, one can determine the caloric equation of state $E = E(p, v)$ in the region of the $p - v$ diagram, which is overlapped by Hugoniot and/or Poisson adiabats. In the experiments involving dynamic stimulation of plasma performed today, the variation of shock wave intensity was effected by varying the power of excitation sources, such as the pressure of propelling gas (Section 3.2), types of explosives, projectile launchers, and targets (Section 3.5). In addition, use was made of the various ways to vary the parameters of the initial states, such as the initial temperature and pressure (a plasma of inert gas (Section 3.5), cesium (Section 3.4), liquid (Lysne and Hardesty 1973)), as well as the use of finely divided targets with a view to increasing the effects of irreversibility (Zel'dovich and Raizer 1966).

Therefore, the dynamic diagnostic techniques which are based on the general conservation laws enable one to reduce the problem of determining the caloric equation of state $E = E(p, v)$ to the measurement of the kinematic parameters of motion of shock waves and contact surfaces, that is, to the registration of distances and times which can be done with a high level of accuracy. The internal energy, however, is not the thermodynamic potential in relation to the p, v variables and, in order to construct closed thermodynamics of the system, one needs an additional temperature dependence $T = T(p, v)$. In optically transparent and isotropic media (gases, Section 3.5), the temperature is measured jointly with other shock–compression parameters. Condensed media and, primarily, metals are, as a rule, nontransparent. Therefore, the light radiation of a shock–compressed medium is inaccessible to registration.

Based on the first law of thermodynamics and the experimentally known relationship $E = E(p, v)$, a thermodynamically complete equation of state can be constructed directly from the results of dynamic measurements without introducing a priori considerations on the properties and nature of the material under investigation (Zel'dovich and Raizer 1966; Fortov and Krasnikov 1970; Fortov 1972a). This leads to the linear inhomogeneous differential equation for $T(p, v)$

$$\left[p + \left(\frac{\partial E}{\partial v} \right)_p \right] \frac{\partial T}{\partial p} - \left(\frac{\partial E}{\partial p} \right)_v \frac{\partial T}{\partial v} = T. \tag{3.3}$$

The solution is constructed by the method of characteristics

$$\frac{\partial p}{\partial v} = -\frac{p + (\partial E/\partial v)_p}{(\partial E/\partial p)_v}; \quad \frac{\partial T}{\partial v} = -\frac{T}{(\partial E/\partial p)_v} \tag{3.4}$$

or in the integral form

$$E = E_0 \exp\{- \int_{v_0}^{v} \gamma(v, E) \, d \ln v\};$$

$$T = T_0 \frac{pv}{p_0 v_0} \exp \left\{ - \int_{v_0}^{v} \left(\frac{\partial \ln \gamma(E, v)}{\partial \ln v} \right) d \ln v \right\}. \tag{3.5}$$

Equations (3.3)–(3.5) are complemented by boundary conditions, namely, the temperature is assigned in the low–density region where its reliable theoretical calculation is possible (cesium plasma, Section 3.2) or the temperature is known from experiment (Fortov 1972a).

The $E(p,v)$ or (p,v) relationships required in the calculation of the right–hand sides of Eqs. (3.4) and (3.5) were determined from the experimental data in the form of power polynomials and rational functions. The accuracy of the obtained solution, as a function of the experimental errors and inaccuracy of initial data, was determined using the Monte Carlo method by computer modelling of the probability structure of the measuring process (Fortov and Krasnikov 1970).

This method is free of restricting assumptions on the properties, nature, and phase composition of the medium under investigation. This is because it makes use of the first principles of mechanics and the fundamental thermodynamic identity (3.3). This thermodynamic universality enables one to construct, using a unified procedure, the equation of state for a wide spectrum of condensed media and to use it to describe phase transformations (Fortov 1972b). The method proved to be especially effective in the study of the thermodynamics of nonideal cesium plasma on the basis of experiments in shock (Lomakin and Fortov 1973; Bushman *et al.* 1975) and adiabatic (Kunavin *et al.* 1974) compression of saturated vapors.

3.2 Dynamic compression of the cesium plasma

Cesium has the lowest ionization potential, ca. 3.89 eV, of the practically accessible elements, which helps attain a high charge concentration n_e at moderate temperatures, thereby ensuring a substantial value of the nonideality parameter at relatively low energy contributions. Therefore, this element is the most popular choice for use in nonideal plasma experiments. Its phase diagram indicating regions of parameters accessible for various techniques is shown in Fig. 3.2. The dynamic generation of cesium plasma was accomplished by way of adiabatic (state 4) and shock compression of saturated vapors in the incident (6) and reflected (7) shock fronts.

Experiments in dynamic compression of cesium vapors were performed in a pneumatic, diaphragm shock tube (Lomakin and Fortov 1973; Bushman *et al.* 1975), as illustrated schematically in Fig. 3.3. In order to attain initial high pressures of saturated vapors, the experimental apparatus having a length of 4 m and an inside diameter of 4.5 cm was heated to 700 °C. Along with the marked aggressive properties and chemical activity of cesium, these high temperatures presented the main reason for the difficulties involved in such an experiment. An ionizing shock wave emerged upon expansion to saturated cesium vapor of helium, argon or their mixtures precompressed to about 0.1 GPa. An additional increase of shock–compressed plasma parameters was attained with the aid of shock waves reflected from the tube end.

In accordance with the dynamic approach, two mechanical parameters characterizing shock compression were independently registered in each experiment. The optical and X–ray (Lomakin and Fortov 1973) baseline techniques were employed for determining the shock front velocity (to an accuracy of about 1%) and

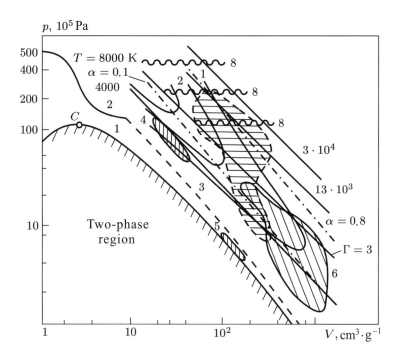

FIG. 3.2.   Phase diagram for cesium: saturation curve, 1 (Chapter 2); static experiments,
2 and 3; region of isentropic compression, 4 (Pfeifer *et al.* 1973; Kunavin *et al.* 1974),
from the initial states, 5; compression by incident, 6, and reflected, 7, shock waves; electric
explosion, 8; α, degree of ionization; C, critical point; Γ, nonideality parameter.

for monitoring the steadiness of flow of shock–compressed plasma. The plasma
density was recorded using the flash radiography technique based on measuring
the plasma attenuation of "soft" X–ray radiation. The wavelength of the latter
radiation was selected for reasons of maximum sensitivity and minimum statistical error of the measuring scheme. It amounted to $(0.2–0.5)\cdot 10^{-8}$ cm. The X–ray
photographs obtained point to the absence of marked condensation of shock–
compressed cesium and permit the determination of the plasma density with
an error of 5–10%. This is supported by the results of radiography of reliably
calculated shock waves in xenon (Bushman *et al.* 1975).

Experiments in shock compression of cesium were conducted in the optimum
(as regards nonideality effects) region of parameters at initial pressures of $(0.04$–
$0.54)\cdot 10^5$ Pa and shock velocities of $(0.8–3)\cdot 10^5$ cm s^{-1}. These correspond to a
sizable region of the phase diagram (Fig. 3.2), namely, $p \sim (1.4–200)\cdot 10^5$ Pa,
$T \sim 2600 - 20\,000$ K and $n_{\rm e} \sim 5\cdot 10^{15}$–$5\cdot 10^{19}$ cm^{-3}, with the maximum Coulomb
nonideality ($\Gamma = 0.2$–2.2). It is significant that in the phase diagram (Fig. 3.2),
the regions of parameters behind the incident and reflected waves partially overlap and, in their lower part, correspond to a quasi–ideal plasma. This enables

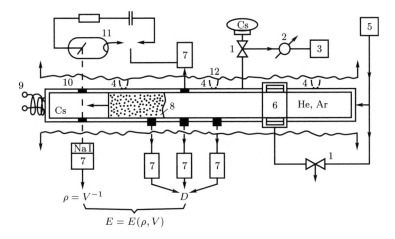

FIG. 3.3. Schematic of a heated cesium shock tube (Lomakin and Fortov 1973; Bushman
et al. 1975): 1, air–operated valves; 2, system for measuring the initial cesium pressure; 3,
liquid cesium vessel; 4, thermocouples; 5, propelling gas; 6, diaphragm unit; 7, photomultipliers; 8, shock–compressed plasma; 9, electrical conductivity measuring coil; 10, beryllium
windows; 11, X–ray tube; 12, electric heater.

one to make full use of the temperature calculation technique according to Eq.
(3.3). This is because the characteristics (3.4) and (3.5) of this equation (isentropes in Fig. 3.1) lie fully in the region overlapped by experiment, including
the region of weakly ionized plasma, $V = 1600 \text{ cm}^3\text{g}^{-1}$, where the initial data
for Eqs. (3.3)–(3.5) are assigned. A reliable calculation of derivatives entering
into Eqs.(3.3)–(3.5) is possible in case the array of experimental points is sufficiently wide in the p and V variables. This caused one to exceed the limits of
the optimum conditions during the experiments (Fig. 3.2).

The static electrical conductivity of shock–compressed cesium plasma was
measured using the induction technique (to an accuracy of 20–40%) in a parallel oscillatory circuit (Sechenov et al. 1977) at frequencies of 0.2–4 MHz. An
inductance coil placed as a flat coil in a shock tube end (Fig. 3.3) changed its
inductance under the effect of plasma formed behind a reflected shock front.
The experiments were performed at $1 \lesssim p \lesssim 15$ MPa and $4000 \lesssim T \lesssim 25\,000$
K, with Coulomb interaction prevailing ($0.3 < \Gamma < 2$). Under these conditions,
the effects of unpairing and correlation of atoms, as well as cluster effects, are
small whereby the Coulomb component of electrical resistance can be isolated
and compared with plasma theories (Fig. 7.5).

In the case when a weakly ionized plasma with strong charge–neutral interaction is of most interest, the adiabatic compression technique proves effective.
This technique is free from irreversibility effects. Therefore, it is possible to attain
lower (as compared to the shock wave technique) temperatures and appreciable
densities of material (Ryabinin 1959; Kunavin et al. 1974; Pfeifer et al. 1973; Lo-

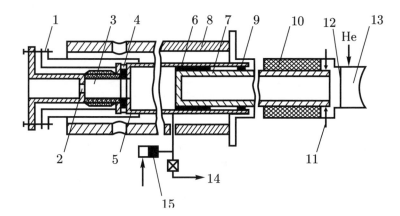

FIG. 3.4. Schematic of a heated cesium shock tube (Pfeifer *et al.* 1973): 1, seal adjustment;
2, foil; 3, conductivity sensor; 4, beryllium windows; 5, barrel; 6, copper attachment; 7,
piston; 8, electric heater; 9, Teflon; 10, displacement sensor; 11, contacts for triggering the
instruments; 12, diaphragm; 13, propelling gas; 14, evacuation; 15, liquid gas.

makin and Lopatin 1983). In doing so, compression must be rather "smooth" so
as to preclude shock wave generation and, at the same time, sufficiently "rapid"
to eliminate heat transfer effects. Suitable apparatus was developed in the 1940s
(Ryabinin 1959) and permitted unique measurements of the characteristics of
overdense gases at temperatures of up to $9 \cdot 10^3$ K and pressures of up to 0.1
GPa. The use of such techniques in the physics of nonideal plasma called for
substantial changes in the basic arrangement of adiabatic compression facilities.

In a two–stage adiabatic tube, compression of cesium is effected (Kunavin *et
al.* 1974) by means of an unloaded lightweight piston (7) moving in a chamber (5)
preheated to 1000 °C. The fact is that in this apparatus, the leakage of cesium
through the lightweight piston renders the realization of the principal advantage
of the adiabatic technique impossible, that is, the high compression ratios which
in this case did not exceed the value of six, this being close to the values characteristic of shock waves. This shortcoming was eliminated in a single–stage tube
(Pfeifer *et al.* 1973) using a heavy piston (2) with an extended seal surface (Fig.
3.4). The compressed plasma density was measured using the "soft" X–ray technique. Electrical conductivity was measured, using the induction method. The
use of the crusher technique to record the pressure was also propose. During the
first experiments in this facility, the electrical conductivity of plasma was measured at a compression ratio of about 25. This corresponds to rated pressures of
up to 33 MPa and temperatures of 4900–6100 K.

3.3   Compression of inert gases by powerful shock waves

An analysis of cesium experiments (Section 3.2) revealed a sizable contribution
of bound states in the dense plasma. This emphasized the need for extension of
the region of parameters under investigation and for a transition to experiments

with other chemical elements. This problem called for a substantial increase in shock wave intensity and the use of new high–enthalpy working media for shock wave generation.

Tkachenko *et al.* (1976) and Volkov *et al.* (1978) used a shock tube to generate argon, xenon, and air plasma. An increase in the propelling gas enthalpy had been attained by means of a discharge of a tank of capacitors with an energy of about 60 kJ. This facility was employed to perform precision measurements of the thermodynamic (registration of the reflected shock velocity), optical (measurement of the growth of radiation intensity), and electrophysical (induction method) properties of a plasma compressed in a reflected shock wave. Since the selected method of generation produces a weakly nonideal plasma of $\Gamma \leq 0.2$ at a moderate pressure of several megapascals, these experiments register only initial manifestations of plasma nonideality and a slight shift of the ionization potential and photoionization continuum, as well as of minor deviations of the electrical conductivity values from those derived by Spitzer.

Substantially higher parameters can be attained by using powerful condensed explosives in view of their high specific energy content and rate (ca. 10^{-7} s) of detonation transformation, which enabled the development of high–power facilities ($10^{10} - 10^{12}$ W). Christian and Yarger (1955) were the first to employ directly the explosion technique for the registration of shock adiabat of gaseous argon. A similar technique was thereafter used by Deal (1957) to register shock adiabats of air at atmospheric pressure with subsequent determination, on that basis, of the nitrogen dissociation energy. In subsequent studies, explosion shock waves in inert gases and air were used as a source of intense optical radiation for high–speed photography, laser pumping, excitation of detonation, the study of the effect of radiation on matter, spectroscopic investigations, and so forth (Tsukulin and Popov 1977). Since the initial gas pressures in the respective facilities did not exceed 10^5 Pa, there was not enough time for the nonideality effects in such plasma to show up vividly. Hence the nonideality effects did not serve the subject of investigation.

Special experiments aimed at studying the effect of nonideality upon the physical properties of explosion plasma were staged in the early 1970s. Inert gases were chosen as the subject of investigation in view of the possibility of their efficient heating in the shock front. This is due to the elimination of dissociation losses of energy and high molecular weight, as well as the much more simple interpretation of the measurement results. The latter is attributed to the absence of complex molecular and ion–molecular clusters and to prior detailed studies of elementary processes.

In order to assess the optimum experimental conditions and the choice of plasma generator designs, machine computations were performed (Gryaznov, Iosilevskii, and Fortov 1973) on the thermophysical and gas–dynamic properties of shock waves in dense inert gases. It has been revealed that the optimum values of the nonideality parameter are attained at shock velocities D of about $9 \cdot 10^5$ cm s^{-1} in argon and $5\,10^5$ cm s^{-1} in xenon. A further increase of D leads to overheating and an increase in the ionization multiplicity of the plasma, while a rise of initial pressure leads to plasma degeneracy. It is important that the optimum conditions can be realized by using a simple linear scheme of shock wave excitation, while to produce a highly heated, multiply ionized plasma one must employ reflected shock waves or geometric cumulation effects.

In a linear explosion shock tube (Fig. 3.5) (Fortov *et al.* 1975a; Bespalov *et al.* 1975; Ivanov *et al.* 1976; Mintsev and Fortov 1979), an ionizing shock wave is generated upon expansion into the gas under investigation of the hexogen detonation products (hexogen is a condensed explosive with a specific energy content of approximately 10^4 J cm^{-3}), which feature, after the completion of the detonation transformation reaction, high dynamic characteristics of $p \sim 37$ GPa, $T \sim 5 \cdot 10^3$ K, $\varrho \sim 2.3$ g cm^{-3}. An additional (as compared with conventional shock tubes) increase of the propelling gas enthalpy is caused by the fact that the disintegration of pressure discontinuity occurs in a coordinate system moving at a rate of $2.1 \cdot 10^5$ cm s^{-1}. The use of a specially shaped detonation lens and proper choice of the size of active explosive charge accounted for the one–dimensional nature and stationary position of the detonation front upon its exit from the explosive and into the test gas. The total energy release in each experiment amounted to approximately $3 \cdot 10^6$ J with a power of about 10^{11} W. Naturally, this leads to the destruction of the entire apparatus and the need to use premises with special protection and to take the appropriate safety measures. The results of photographic, electrophysical, and X–ray measurements revealed the one–dimensional nature and quasi–steadiness of plasma flow, which are accomplished through the inertial confinement of shock–compressed plasma by the massive walls of the shock tube channel.

The explosion shock tube (Fig. 3.5) was fully equipped for independent thermodynamic and electrophysical measurements. The shock front velocity was measured (Bespalov *et al.* 1975) by optical and electrocontact baseline techniques with the aid of high–speed cameras and ionization sensors to an accuracy of 1–1.5%. Within this error, the ionization front coincided with the shock wave luminescence front and the position of hydrodynamic shock.

The density of shock–compressed argon plasma was registered at an accuracy of about 8% by flash radiography (Section 3.2), which is characterized by high time (ca. 10^{-7} s) and space (ca. 2 mm) resolution and the absence of perturbation in the plasma flow.

In view of the transparency of the plasma before the shock front and the small dimensions of the viscous shock, thermal radiation is capable of leaving the plasma volume without any obstacles and yielding experimental data on equilibrium temperature and absorption factors of shock–compressed plasma (Model' 1957; Tsukulin and Popov 1977). The intensity of this radiation, in the case of using the brightness method (at an accuracy of 5–10%), for temperature recording was determined (Bespalov *et al.* 1975, 1979a) by photometric comparison of time scans of the luminescence of shock–compressed plasma and reference light sources, such as a flash lamp with brightness temperature of 8600 ± 200 K,

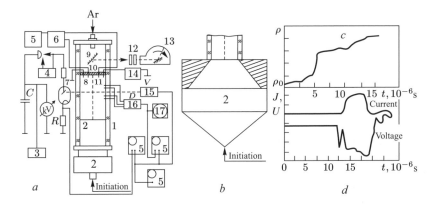

FIG. 3.5. Scheme of measurements in an explosion generator of nonideal plasma: a, diagnostics; b, explosive charge; c, roentgenogram; d, oscillogram; 1, generator channel; 2, explosive charge; 3, supply to X–ray tube; 4, X–ray control unit; 5, oscillographs; 6, differential amplifier; 7, X–ray tube; 8, potential and current probes for registration of electrical conductivity; 9, mirror; 10, Plexiglas obstacle; 11, probes for measuring the shock front velocity; 12, filter and attenuator; 13, high–speed camera; 14, d.c. source; 15, photomultiplier with NaI scintillator; 16, supply to electric contacts; 17, oscillograph for velocity recording.

pulsed capillary light source with $T = 39700 \pm 700$ K, and a shock wave in air with $T = 11800 \pm 600$ K.

As demonstrated in a special series of test experiments, as well as by estimates, the self–shielding effect is absent in view of the heating of gas by ultraviolet plasma radiation. According to the data by Tsikulin and Popov (1977), this effect becomes conspicuous for argon ($p_0 = 10^5$ Pa) beginning at $D \geq 15$ km s^{-1} and from $D > 8$ km $^{-1}$ for xenon, and decreases with a rise in the initial gas pressure.

The registration of the shock front velocity and the density yields, in accordance with the conservation laws of Eq. (3.1), the caloric equation of state for nonideal plasma, $E = E(p, v)$, which, together with the temperature measured, $T = T(p, V)$, provides one with thermodynamically complete information on argon plasma at $p \sim (1.6–5.8){\cdot}10^8$ Pa, $T \sim (15.5–23){\cdot}10^3$ K, $\Gamma \sim 1.3$–2.2 and $n_e \sim (1.5–3.5){\cdot}10^{20}$ cm^{-3}.

Electrical conductivity is an important characteristic of plasma which offers valuable information on its structure and elementary processes. In view of the high level of electrical conductivity, its measurements (Ivanov et al. 1976; Mintsev and Fortov 1979; Mintsev, Fortov, and Gryaznov 1980; Ebeling et al. 1991) were performed using the probe technique (Fig. 3.5). This is characterized by high space resolution and relative simplicity of realization under conditions of a single dynamic experiment. A characteristic oscillogram of transport current and voltage is given in Fig. 3.5 where one can clearly see a "plug" of shock–compressed plasma.

Extremely high densities of plasma were attained by two–stage compression of near–critical states of xenon in the incident and reflected shock waves (Mintsev and Fortov 1979). To this end, a Plexiglas obstacle was placed at a distance of 7 cm from the charge end face (Fig. 3.5). The interaction with the obstacle causes a reflected shock wave in the plasma cluster. This leads to additional compression and heating of the xenon plasma. While doing so, a shock wave emerged in the Plexiglas obstacle. Its velocity enabled one to determine the pressure, density, and enthalpy of xenon following its double compression (cf. point 10 in Fig. 3.6). A consistent description of these thermodynamic data was accomplished using a "chemical" model of the plasma. In this model, the Coulomb interaction is described by the modified Debye approximation and the repulsion at small distances, by virial corrections. This thermodynamic model was then used to interpret (Section 3.5) the results of measurements of the electrical conductivity of the plasma under supercritical conditions.

These experiments have yielded data on the electrical conductivity of xenon (Mintsev and Fortov 1979) under the highly supercritical conditions of $\varrho \sim 1$–4 $\mathrm{g\,cm^{-3}}$ ($\rho_c \sim 1.1 \mathrm{\ g\,cm^{-3}}$) at high pressures and temperatures ($p \sim 2$–11 GPa and $T \sim (1$–$2)\cdot 10^4$ K), where a wide spectrum of strong interparticle interactions involving the participation of neutral and charged particles is realized. Note that in this manner, one manages to generate plasma under unusual conditions, that is, the plasma density is 1.5 times higher than that of solid xenon and is comparable to the density of solid metals. The investigated parameters in region I in Fig. 3.6 extend from the lower–density states II, where the thermodynamic and electrophysical properties of the material are described by plasma models, to adjoin directly region III of solid–state densities described by the band theory of solids. The latter is formed as a result of dynamic compression of liquid xenon.

The urge to obtain a plasma with high nonideality parameters compels one to perform experiments at relatively low temperatures of the order of 20 000 K, at which the plasma is, as a rule, not fully ionized (Model' 1957; Kunavin et al. 1974; Bushman et al. 1975; Bespalov et al. 1975). At the same time, the study of a multiply ionized nonideal plasma is of considerable interest and enables one to investigate purely Coulomb effects in a highly heated medium.

The use of reflected shock waves made it possible to advance, at reduced initial gas pressures, into the region of multiple ionization and to produce plasma at temperatures above 20 000 K (Mintsev et al. 1980; Ebeling et al. 1991). The same purpose is served by cumulative explosion tubes (Mintsev et al. 1980; Ebeling et al. 1991), which depend for their action on the increase in the parameters of propelling gas under conditions of centripetal motion (Zaporozhets, Mintsev, and Fortov 1987) of the latter in a conical chamber (Fig. 3.5) with apex angles of 60-120°. In this manner, shock velocities in xenon of 8–15 $\mathrm{km\,s^{-1}}$ were attained, with initial pressures of up to 1 MPa. This corresponds to extremely high, $(3$–$10)\cdot 10^4$ K, heating of nonideal ($\Gamma \sim 2$) plasma. While leaving the interpretation of these data until Chapter 5, we shall point out that it is the thermal electron scattering by the inner shells of ions which is important under the given

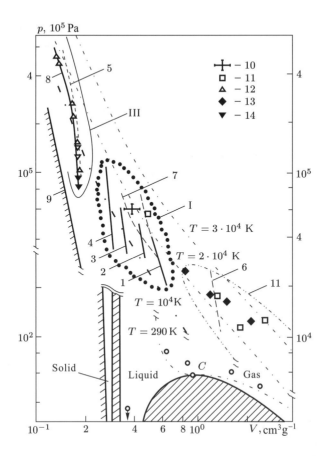

FIG. 3.6.   Phase diagram for xenon. I, experiments with reflected shock waves; II, experiments with incident shock waves; III, shock compression of liquid xenon. Calculated adiabats: 1, $p_0 = 5 \cdot 10^6$ Pa; 2, $p_0 = 6 \cdot 10^6$ Pa; 3, $p_0 = 7 \cdot 10^6$ Pa; 4, $p_0 = 8 \cdot 10^6$ Pa; 5, $p_0 = 2.8 \cdot 10^5$ Pa; $T_0 = 165$ K; 6, $p_0 = 2 \cdot 10^5$ Pa; 7, shock adiabat of double compression (Mintsev and Fortov 1979); 8, calculation under the band model (Keeler *et al.* 1965); 9, $T = 0$ K isotherm. Thermodynamic measurement data: 10, "reflection" method (Mintsev and Fortov 1979); 11, Fortov *et al.* (1976); 12, Keeler *et al.* (1965); Nellis *et al.* (1982). Electrical conductivity measurement data: 13, Ivanov *et al.* (1976); 14, Nellis *et al.* (1982). C, critical point.

conditions.

In order to measure the thermodynamic and optical characteristics of plasma at higher (as compared with shock tubes) pressures, use was made of linear explosion generators of step shock waves (Fig. 3.7) of varying intensity (Fortov *et al.* 1976; Gryaznov *et al.* 1980) and duration, as well as of light–gas propulsion devices (Keeler *et al.* 1965; Nellis *et al.* 1982). In such apparatus, the ionizing shock wave emerged upon expansion of metal or polymer targets, precompressed

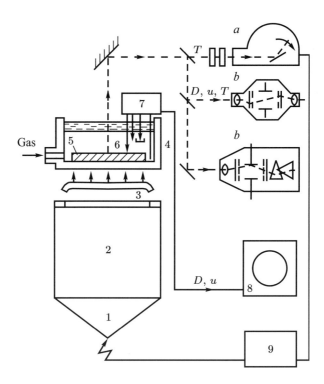

FIG. 3.7.   Schematic illustrating the experiments in explosive compression of inert gas plasma: 1, detonation lens; 2, explosive charge; 3, metal projectile; 4, test assembly; 5, target; 6, test gas; 7, electric contacts and pulse-shaper unit; 8, oscillographs; 9, synchronization and detonation unit; a, high–speed photorecorder; b, image converter.

to about 0.1 TPa, into the gases under investigation (argon, xenon). Powerful shock waves in the targets were excited by linear explosion propulsion devices (Altshuler 1965; Fortov 1982) operated by acceleration of flat metal projectiles to 2–6 km s^{-1} by the detonation products. The energy release in these experiments amounted characteristically to $(2$–$30)\cdot10^6$ J at a power of the order of 10^{11} W.

In accordance with the dynamic approach, electrocontact and optical baseline techniques were employed (Fortov *et al.* 1976; Gryaznov *et al.* 1980) in the experiments in these generators for simultaneous independent registration of two kinematic parameters: the front velocity D and the plasma mass velocity u. Open electrocontact pins recorded the front velocity D at an accuracy of the order of 1% with the aid of fast oscillographs. The mass velocity u was registered (at an accuracy of 1.5–2%) by enclosed pins of special design which are unresponsive to the shock wave in the plasma. They also operate upon the arrival of the plasma/target heavy contact interface. The system of sensor pins was arranged to allow for possible shock wave tilt and to monitor the parameters of the driving pressure pulse from the projectile accelerated by the detonation products. In the case of the optical registration technique, a Plexiglas obstacle (Fig. 3.7) was placed at a preset distance from the target; the emission of shock–compressed plasma was registered through the obstacle with the aid of high–speed cameras or image converters. The escape of the shock wave from the target into the test gas leads to the appearance of luminescence which intensifies the moment the shock wave is reflected from the obstacle (determination of D), and disappears when the heavy contact interface arrives at the target (registration of u). In addition to shock compression parameters, the temperature of shock–compressed plasma was also recorded in a number of experiments using the brightness method.

The results of recording the shock adiabats of argon and xenon in kinematic variables are illustrated in Fig. 3.8, where one can clearly see good agreement between the data obtained by the optical, electrocontact, and X–ray techniques in explosion shock tubes and step–wave generators using metals, polymers and condensed explosives as targets. The results obtained in determining the equation of state for nonideal argon plasma cover a wide range of parameters (Fig. 3.9), namely, $p \sim (0.3\text{--}40) \cdot 10^8$ Pa, $T \sim (5.2\text{--}60) \cdot 10^3$ K and $n_e \sim 10^{14}\text{--}3 \cdot 10^{21}$ cm^{-3}, which is characterized by developed ionization of $\alpha \sim 3$ and strong Coulomb interaction of $\Gamma \sim 0.01\text{--}5.2$. It is interesting to note that in the experiments described above, the equation of state for a plasma with $\rho \sim 1.3$ g cm^{-3}, was studied. It exceeded the critical density of xenon. Note that under these conditions, the interaction proves so strong that it causes a perceptible shift and deformation of the energy levels in the plasma.

Investigations of the electron spectrum and equation of state for the plasma of argon and xenon of solid–state density, using a lightweight ballistic facility (Nellis et al. 1983) for shock–wave compression, were performed by Keeler, van Thiel, and Alder (1965), Ross, Nellis, and Mitchell (1979), and Nellis, van Thiel, and Mitchell (1982). In a two–stage light–gas "gun" (Fig. 3.10), powder gases (1) resulting from the charge combustion accelerate a heavy piston (2) which moves in a chamber (3), having a length of 10 m and inside diameter of 90 mm on its inside, to compress hydrogen (4). Following the rupture of the diaphragm, expanding hydrogen acts to accelerate a light impactor (6), which measures 28 mm in diameter, in a 9 m long evacuated acceleration channel (5). In this manner, a highly symmetric "smooth" and heating–free acceleration of the impactor is effected with a mass of about 20 g to velocities of 7–8 km s^{-1}, which are somewhat higher than those attained in the case of plane propulsion in explosion devices such as those found in Fig. 3.7. The impact of the accelerated impactors leads to the generation in liquid argon, as well as in aluminum shields, of plane stationary shock waves whose velocities were registered by a system of electrocontact sensors with a time resolution of hundreds of picoseconds. These measurements were used by Keeler, van Thiel, and Alder (1965), Ross, Nellis, and Mitchell (1979), and Nellis, van Thiel, and Mitchell (1982), along with the regularities of disintegration of discontinuity on the "shield–plasma" interface and with the conservation laws (3.1), to obtain the equations of state for a highly compressed argon and xenon plasma at pressures of up to 0.13 TPa and maximum temperatures of up to

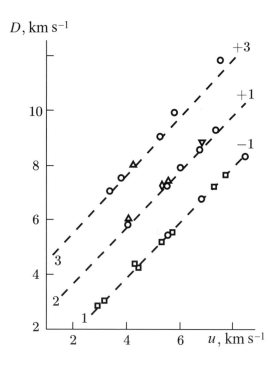

FIG. 3.8.   Shock adiabats of argon and xenon (1, $p_0 = 0.78 \cdot 10^5$ Pa; 2, $p_0 = 5 \cdot 10^5$ Pa; 3, $p_0 = 10^6$ Pa) registered using different techniques: 4, optical (Christian and Yarger 1955); 5, electrocontact (Fortov *et al.* 1976); 6, X–ray (Bespalov *et al.* 1975); 7, optical (Ebeling *et al.* 1991); 8, electrocontact and optical (Gryaznov *et al.* 1980); broken line, calculation by Gryaznov, Iosilevskii, and Fortov (1973). The numerals on the right indicate the values of shifts of each plot with respect to the front velocity D (km s^{-1}); u, mass velocity.

30 000 K ($n_\mathrm{e} \sim 5.7 \cdot 10^{22}$ cm^{-3}). Under such conditions of high compression (maximum nuclear density of the order of $5.7 \cdot 10^{22}$ cm^{-3}), one observes a marked deformation of the electron energy spectrum of the plasma which causes a triple variation of pressure. In doing so, a considerable number of electrons are in a thermally excited state and, on the whole, a highly compressed plasma possesses "semiconductor properties" (Nellis *et al.* 1982).

Interesting results in the thermal physics of nonideal argon and xenon plasma were obtained in the facility for adiabatic compression. These were described by Hess (1989). At an initial pressure of about 2000 Pa, the test gas was compressed by a heavy (4.9 kg in mass) piston in a steel channel measuring 7.5 m long and 150 mm in diameter. The pressure of adiabatic compression was registered by a piezoelectric transducer and the compression ratio was defined by the position of the piston. In addition, an optical method was used to record the thermal radiation of the plasma. This provides information about the emission spectra and equilibrium temperature. Probe techniques were used to register the electrical

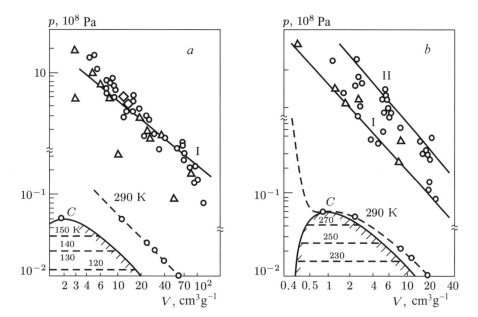

FIG. 3.9.   Phase diagram for argon (a) and xenon (b). The two–phase region boundaries and the critical point C are shown. Broken lines, isotherms of initial states. I and II, single and double ionization limits, respectively. Dots, results of dynamic experiments.

conductivity of the dense plasma. This unique facility helped to produce lower (as compared with shock wave generators) values of the plasma heating temperature $T \sim 14\,000$ K at a high pressure p of up to 14.5 GPa and electron number density n_e of up to 10^{20} cm^{-3}. This resulted in a considerable expansion of the experimentally accessible region of the phase diagram of the nonideal plasma. It also provided the possibility of checking on the validity of asymptotic theories.

Optical properties are of considerable interest from the standpoint of plasma physics. That is because they enable one to follow the effect of nonideality on the dynamics and energy spectrum of electrons in a dense disordered medium. The plasma absorption coefficient κ_ν was measured by registering the rise in time of the intensity of optical radiation, which emanates from a plane plasma layer enclosed between the shock front and contact surface (Model' 1957; Bespalov et al. 1979b). In view of the small time of photon relaxation as compared with the characteristic gas–dynamic time, the spectral radiation intensity of such a layer, $J_\nu(t)$, in a quasistationary approximation (Kirzhnits, Lozovik, and Shpatakovskaya 1975) takes the form

$$J_\nu(t) = J_\nu^0 (1 - \exp\{-\kappa_\nu'(D - u)\,t\}),$$

where $J_\nu^0 = J_\nu(\infty)$ is the Planck radiation intensity, $\kappa_\nu' = \kappa_\nu(1 - e^{-h\nu/kT})$ is the absorption coefficient corrected for induced radiation, and D and u are the rates

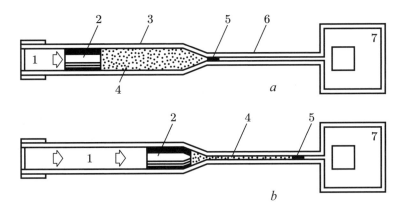

FIG. 3.10. Schematic of a two–stage ballistic facility (Nellis *et al.* 1983): a, prior to experiment; b, during experiment; 1, powder gases; 2, heavy piston; 3, high–pressure chamber;
4, hydrogen; 5, impactor; 6, acceleration channel; 7, sample.

of travel of the front and rear boundaries of the radiating layer, respectively.

An important requirement of this technique is the provision of conditions of
one–dimensional and steady flow of plasma in the zone of recording. This enables
one, when interpreting the measurement results, to use the simple relationship
(3.6) instead of solving the integral–differential equation of radiation transfer.
During experiments by Bespalov *et al.* (1979b), these conditions were monitored
by electrocontact and optical measurements, as well as by gas–dynamic calculations.

A facility utilizing explosion propulsion devices is shown schematically in Fig.
3.11. Light radiation emanating from a plasma is diverted with the aid of mirrors
and optical system to polarization filters and photomultipliers. The respective
signals are recorded by high–speed oscillographs. The kinetic parameters of the
plasma, which are essential in determining the absorption coefficients, were determined by electrocontact and optical methods. The exit of the shock wave from
the target and into argon led to a smooth rise (Fig. 3.11b) of light radiation until
its saturation in accordance with Eq. (3.6). A flash–up of brightness marks the
time (and, consequently, the velocity D) of the front arrival to a transparent
obstacle. The subsequent radiation cutoff is caused by the destruction of the
obstacle by a massive target moving at a velocity u.

Note that such an organization of experiments enables one to obtain diversified physical information in a single experiment. The rise of intensity $J_\nu(t)$ in
the initial portion of flow determines, from Eq. (3.6), the light absorption coefficient, and the level of radiation at the stage of saturation, $J_\nu(\infty)$, determines
the brightness temperature. The registration of shock wave reflection from the
obstacle helps measure D and u which, with due regard for the conservation
laws (3.1), is equivalent to direct measurements of pressure, density, and internal energy. Thus the data obtained are consistent and complement the results of

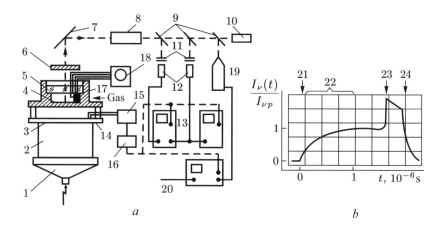

FIG. 3.11. Schematic illustrating the experiment in registering the absorption coefficient of shock–compressed plasma (Bespalov et al. 1979b): a, diagnostics; b, oscillogram; 1, detonation lens; 2, explosive charge; 3, impactor; 4, target; 5, transparent obstacle; 6, diaphragm; 7, mirror; 8, optical system; 9, semitransparent mirrors; 10, adjustment laser; 11, interference light filters; 12, photomultipliers; 13, oscillographs; 14, electrocontacts for measuring the shock velocity; 15, pulse-shaping circuit; 16, delay line; 17, electrocontacts for measuring the shock velocity; 18, oscillograph for measuring the shock velocity; 19, fast spectrometer; 20, system of digital spectrum processing; 21, shock wave entry into the gas; 22, zone of radiation rise; 23, shock wave collision with the obstacle; 4, beginning of the destruction of obstacle by the target material.

special thermodynamic measurements (Bespalov et al. 1975; Fortov et al. 1976; Gryaznov et al. 1980).

In order to reveal the compression effects and to facilitate the interpretation of the results, the gas–dynamic characteristics of explosion generators were selected to provide for approximate constancy of the temperature of shock–compressed plasma, $T \sim 2 \cdot 10^4$ K. The measured coefficients of light absorption (an error of the order of 30–40%), with a frequency of $(5.17 \pm 0.05) \cdot 10^{14}\,\mathrm{s}^{-1}$, are presented in Fig. 8.13. Every experimental point is obtained by the averaging of five to ten experiments with two to three independent frequency recordings in every experiment. One can see that the measurement results cover a wide range of plasma compressions and are in reasonable agreement with the estimates of κ_ν by Tsikulin and Popov (1977). Spectral measurements were made by Bespalov, Kulish and Fortov (1986) in argon plasma at $T \sim 1.74 \cdot 10^4$ K and electron densities up to $1.2 \cdot 10^{19}$ cm^{-3}. Higher n_e broadening does not allow a separation of the spectral lines.

The foregoing method of registration of the absorption coefficient turns out to be restricted on the side of low plasma densities by the finite time of existence of the steady flow of plasma. Therefore, when registering low absorption coefficients, Bespalov et al. (1979a) performed a somewhat different version of such experiments by involving the use of explosion shock tubes as the source of air plasma. A shock wave decay in such systems led to the emergence of a maximum for the relationship $J_\nu(t)$ being registered. The position of the latter had helped to find the value of κ'_ν in every experiment and to evaluate the other parameters of shock compression on the basis of gas–dynamic data. The results obtained provided a basis for evaluating the photodisintegration cross–section of the negative ion of nitrogen.

The reflection of laser radiation from nonideal plasma yields interesting information on the physical properties of a shock–compressed medium, whereby one can study the high–frequency conductivity of this medium in the field of an electromagnetic wave. This method is fairly widely used in the physics of metals and semiconductors to determine the energy spectrum, concentration, and effective mass of charge carriers. Of special interest is the region of resonance interaction, in which the incident electromagnetic wave of frequency ω_λ is comparable to the plasma frequency $\omega_\mathrm{p} \sim \sqrt{\frac{4\pi\, n_e\, e^2}{m}}$. In this case, the resonance condition $\omega = \omega_\mathrm{p}$ corresponds to a strong reflection of incident radiation and defines the electron number density, and the dispersion of the reflection curve characterizes the losses of electron energy, thereby enabling one to find the electron collision frequency.

The experiments in measuring the reflection of laser radiation from nonideal plasma were performed by Zaporozhets, Mintsev and Fortov (1987) using explosion generators of square shock waves in the scheme of Fig. 3.11. The electron number density of xenon plasma was varied by varying the initial parameters of xenon, namely, $p_0 \sim 1$–5.7 MPa, $\rho_0 = 0.06$–0.8 g cm^{-3}, and $T_0 = 270$ K. In view of considerable proper thermal radiation of shock–compressed plasma with $T \sim 3 \cdot 10^4$ K, the probing pulse was developed by a pulsed (with $\tau \sim 10^{-8}$ s) aluminum-garnet laser producing a high $(T \sim 6 \cdot 10^7$ K) spectral temperature and a small angular divergence of the light flux. The laser system was equipped with an electronically controlled Pockels shutter which enabled one to synchronize the explosion generation and laser probing with a difference in time of not worse than 10^{-8} s. The sounding radiation was introduced into the explosion chamber by means of a special optical system and, after being reflected from shock–compressed matter, it was applied to fast photomultipliers equipped with interference filters with $\Delta\lambda \sim 0.02$ µm. For the focal spot diameter of about 0.6 mm, the specific intensity of laser radiation in the shock wave front amounted to $(1\text{-}7)\cdot 10^5$ W cm^{-2}, which is insufficient for substantial heating of matter.

The optical reflectivity of plasma R was determined by comparing the intensities of incident radiation with the radiation reflected to the aperture of the receiving ring–shaped lens. In order to perform an overall check on the procedure in the dynamic mode, experiments were carried out to register the coefficient R for shock–wave compressed carbon tetrachloride and silicon. The values obtained for R are in good agreement with the results of previous measurements in the region of the metal–dielectric transition. In each experiment, in order to determine the thermodynamic parameters of shock–compressed plasma, the electrocontact baseline method was used to measure the velocity of shock wave propagation in plasma and the impactor approach velocity.

The measurements cover a wide range of values of plasma pressure, $p \sim 1.6$–17 GPa and density $\varrho \sim 0.5$–4 g cm^{-3} that exceed considerably the values of the density of xenon at the critical point. Under these conditions, a highly heated $(T \sim 10^4$ K) disordered plasma with considerable Coulomb interaction ($\Gamma \sim 2 - 7$) is realized. In doing so, the electron number density, calculated within the ring Debye approximation, exceeds by an order of magnitude the critical (for $\lambda \sim 1.06$ μm) value of $n_c \sim 10^{21}$ cm^{-1}. The reflectivity also reaches the value of 50%, which is characteristic of metals. The estimates of the space structure of the shock front, caused by the ionization kinetics and radiant heat transfer, have shown that it is the resonant properties of plasma electrons that govern the reflection of radiation. These measurements can be used to analyze the physical properties of strongly nonideal plasma.

3.4   Isentropic expansion of shock–compressed metals

The shock wave technique described in Sections 3.2 and 3.3 enables one to attain high pressures and temperatures in compressed media whereas the region of reduced pressures (including the saturation curve and the neighborhood of the critical point for metals) proves inaccessible to such investigation techniques. Steady–state experiments under conditions of normal pressure and temperatures below 2500 K allow the determination of heat capacity and isothermal and adiabatic compressibility, as well as the entropy and density jumps occurring upon melting. By now, the melting curves for metals and the general form of the phase diagram have been defined at pressures of up to 5 GPa, and the isothermal compressibility at pressures of up to 30 GPa. In the pressure range of several hundred gigapascals, the properties of metals are determined on the basis of the absolute method of registration of the shock and isentropic compressibility of solid and porous samples (Altshuler 1965; Davison and Graham 1979) and, at pressures of up to hundreds/thousands of megapascals, by way of comparative measurements (Altshuler et al. 1968; Trunin et al. 1969, 1972; Thom and Schneider 1971; Vladimirov et al. 1984; Simonenko et al. 1985; Avrorin et al. 1987).

Therefore, what remains uninvestigated is an extensive and practically important part of the phase diagram (Fig. 3.1) characterized by great diversity and utter complexity of the description of the physical processes occurring there. The following are realized in this region: a plasma that is nonideal with respect to a wide spectrum of interparticle interactions; a dense heated metallic liquid whereby during its expansion, a "release" of degeneracy of the electron component, its recombination, takes place; "metal–dielectric" transition; and high–temperature evaporation of metal to the gas or plasma phase. The current information on these processes is very limited and exists in the form of semi–empirical estimates and a few measurement results. Suffice it to mention that, out of more than 80 metals in the periodic system, the critical point parameters have only been determined for three lowest–boiling ones (Likal'ter 2000), to say nothing of more detailed information on the form of the phase diagram for metals at high pressures and temperatures.

One can move into the region of lower densities and high pressures of plasma using the isentropic expansion technique (Altshuler *et al.* 1977, 1980; Fortov *et al.* 1974; Leont'ev *et al.* 1977). It is based on the generation of dense plasma under conditions of isentropic expansion of condensed material precompressed and irreversibly heated in a shock front. To assess the possibilities of dynamic methods for isentropic generation of nonideal plasma, Leontiev and Fortov (1974) performed calculations of energy release leading, upon its disintegration, to melting and evaporation in the release wave. For such an assessment, one needs data on the critical parameters of metals, reference points on the thermodynamic surface of the material. In view of the collectivization of valence electrons, their bond energies and, consequently, critical point parameters are extremely high (Likal'ter 2000) and, in most cases, inaccessible to conventional thermophysical experimental techniques. An exception is made for alkali metals and mercury, for which the registration of high–temperature boiling parameters makes it possible to verify the applicability, in their case, of the principle of thermodynamic similarity and, on this basis, to assess (Fortov *et al.* 1975b) the critical parameters for 80 metals that are still uninvestigated (cf. Appendix A). It follows from these estimates that the critical temperatures of metals in many cases prove to be comparable with their ionization potentials, which are considerably reduced in the plasma due to strong interaction of charges with each other and with neutrals. Therefore, the metal vapors on the right side of the binodal are, apparently, in a thermally ionized state. Meanwhile, the high–temperature evaporation of metals corresponds to the transition, directly, to a highly nonideal plasma state while bypassing, as distinct from the already investigated elements, the ionized gas region. This fact may affect the kinetics of high–temperature phase transitions (Fortov and Leont'ev 1976), as well as drastically altering the usual form of the phase diagram of matter by causing the emergence of additional regions of phase separation and new exotic phase transitions (cf. Section 5.8).

The region of densities intermediate between solid and ideal gas, that is, characterized by the maximum uncertainty of theoretical predictions, prevents one from performing direct calculations of release isentropes under these conditions. Therefore, in order to assess the energy release leading to phase transitions under conditions of adiabatic expansion of metals, use was made of the entropy criterion (Zel'dovich and Raizer 1966) which allows for the condition of isentropic behavior of flow in the release wave. This criterion is based on the comparison of the critical point entropy and tabular values of entropy of phase transformations with the value of entropy of the condensed phase of metals, in which the semi–empirical equations of state are valid (Bushman and Fortov 1983) at high pressures and temperatures. The most detailed calculations have been performed by Leontiev and Fortov (1974) for aluminum, nickel, copper and lead, whereby perfect equations of state (Kormer *et al.* 1962) with variable heat capacity are available to describe the shock compression of solid and porous samples and fea-

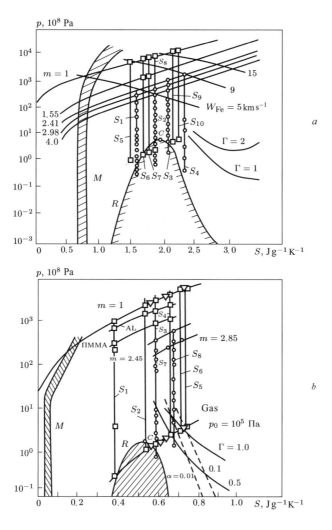

FIG. 3.12. Entropy diagrams for (a) copper and (b) bismuth: M and R, boundaries of melting and boiling, respectively; m, shock adiabats of different porosities; S, expansion isentropes; W_{Fe}, deceleration curves for iron impactors. The diagram in (b) gives the adiabats for materials in which bismuth was released. The points indicate the experimental data by Altshuler $et\ al.$ (1977); Anisimov $et\ al.$ (1984); Bazanov $et\ al.$ (1985); Bushman $et\ al.$ (1986a); Glushak $et\ al.$ (1989).

ture high–temperature asymptotics of ideal gas. The results of estimations for copper, and bismuth are given in Fig. 3.12 and those for other metals (using the equation of state by Kormer, Urlin and Popova 1961) in Fig. 3.13. The states which occur upon impact against the target of an iron impactor accelerated to a velocity of the order of 5–15 km s^{-1} are marked in Fig. 3.12 and Table 3.2.

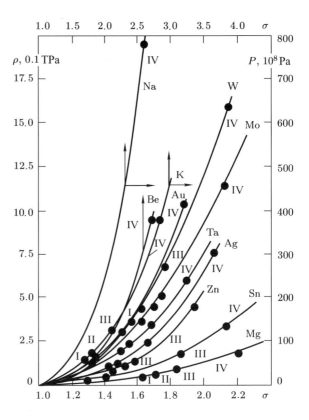

FIG. 3.13. Shock wave intensities causing melting (I, onset; II, end), boiling (III) and arrival at the critical point (IV) in an isentropic release wave: $\sigma = \rho/\rho_0$, compression ratio of the material.

Table 3.2

	Ni		Al		Cu		Pb	
Porosity, m	1	3	1	3	1	4	1	2
ρ_{ev}	3.8	0.3	1.7	0.2	2.7	0.2	0.7	0.1
ρ_c	7.3	0.5	4.4	0.5	7.5	0.7	1.9	0.4
ρ_{cond}	20	1.9	10	2.7	20	3.9	5.0	1.3

The results of these calculations show that the dynamic methods of plasma generation enable one to attain high–energy states of metals in a wide range of parameters near the "liquid–vapor" equilibrium curve. However, in order to attain supercritical conditions in the release wave, shock waves of extremely high intensity need to be generated, which is just within the energy capabilities of chemical explosives (Altshuler 1965). The required shock wave intensities may be substantially lowered by the use of porous targets offering a more effective increase of shock compression entropy (Zel'dovich and Raizer 1966). It is apparent that an increase of the initial porosity of targets at W =const. leads to a decrease of shock pressure. However, this is accompanied by a rise in entropy, thereby enabling one to considerably expand the region of parameters accessible to dynamic experiment.

The few experiments involving the registration of plasma expansion under the effect of shock waves can be divided into two groups based on the nature of the physical information they yielded. The experiments of the first group (Altshuler et al. 1977; de Beaumont and Leygenie 1970; Hornung and Michel 1972; Kerley and Wise 1987; Asay, Trucano and Chhabildas 1987) are based on Zeldovich's idea to determine the entropy of shock compression of solids by registering only the final parameters of the expanded material. The experiments of the second group (Fortov et al. 1974; Leont'ev et al. 1977; Altshuler et al. 1980; Bushman et al. 1986a) involve a detailed registration of release isentropes in the entire region of parameters from solid to plasma or gas.

It is known (Zel'dovich and Raizer 1966; Fortov 1982) that only the dynamic methods make it possible to determine the caloric equation of state for a shock–compressed medium. This equation lacks important characteristics of material such as the temperature or entropy, which are extremely difficult to measure directly in shock–compressed material. Following the passage through the material of waves of extreme intensities, the expanded material changes over to the state of an ideal gas or plasma (Zel'dovich and Raizer 1966) whose entropy, coinciding with that of shock–compressed condensed material, can be reliably calculated from the measured values of temperature, pressure, or density. The technique under discussion still cannot be realized in this form in a practical way. This is due to considerable difficulties involved in attaining, under conditions of shock compression, high–energy states (cf. Figs. 3.12 and 3.13, Table 3.2) which are essential in reaching, upon release, the regions of ideality (Bazanov et al. 1985; Bushman et al. 1986a; Glushak et al. 1989; Ageev et al. 1988).

In the case of shock waves of lower intensity, the final states prove to be in the solid or liquid phase, which enabled Zhernokletov et al. (1984) to use the photoelectric technique of residual temperature registration. It made it possible for Fortov and Dremin (1973) to find, on this basis, the entropy and temperature of copper at pressures of up to 190 GPa. Scidmore and Morris (1962) derived the entropy of sodium, strontium, barium, and uranium by means of optical adsorption measurements of the evaporated metal fraction under the effect of short shock waves with $p \sim$ 20–300 GPa excited by thin (0.1–1 mm) impactors moving at velocities of 2–6 $\mathrm{km\,s}^{-1}$. It follows from the entropy analysis (Leont'ev and Fortov 1974) that shock waves with such moderate intensities may cause appreciable evaporation of metals, only as a result of their expansion to extremely low pressures, which compelled experiments in a vacuum of the order of 10^{-5} mm Hg (Scidmore and Morris 1962) to be carried out. Adsorption measurements performed under conditions of substantially non–one–dimensional flow enable one to derive the condensate fraction by relating it, on the basis of quality considerations on the kinetics of the nonequilibrium process of evaporation and condensation, to the entropy of shock–compressed metal plasma. In a series of subsequent studies, the general knowledge of evaporation and ionization in the release wave was used to analyze the high–frequency collision of solids and to develop fluorescent barium clouds in outer space for the purpose of studying the Earth's magnetosphere (Fortov and Dremin 1973). The latter case involved explosions in space of metallic barium–lined cumulative charges.

Altshuler *et al.* (1977) used the results of measurements of the expansion rate W for lead after the passage of shock waves with amplitudes of 40–300 GPa to analyze the vaporization of lead in release waves. The deviation from the rule of "velocity doubling" is extreme evidence of the vaporization effect in release waves.

Zhernokletov *et al.* (1984) made an attempt to complement the traditional kinematic measurements with a registration of the temperature of isentropic expansion of lead in the neighborhood of its high–temperature boiling. The experiments involved adiabatic expansion of metal, shock–compressed to a pressure of about 600 GPa, into helium at different values of initial pressure, which remained transparent in the shock–compressed state and therefore transmitted the thermal radiation of expanded lead. This radiation was delivered, via optical–fiber communication lines, to fast silicon photodiodes and two–channel pyrometers, whereby the temperature could be determined using the brightness technique. The results of these measurements are given in Fig. 3.14. The transition of isentrope from the single–phase state (power dependence of T and p) to the region of phase mixture (exponential dependence) manifests itself in a sharp variation of the slope of the S and R curves. It is worth noting that the special series of experiments to measure the temperature of states in the two–phase region, which is produced by expansion from the initial pressure of different values $p \sim 160$ GPa and $p \sim 240$ GPa, resulted, in accordance with Gibbs' phase rule, in identical temperature values. These serve as proof of equilibrium of the process of adiabatic release in the two–phase region. This equilibrium is apparently disturbed at a pressure below $3 \cdot 10^6$ Pa, where the recorded values of temperature exceed those on the saturation curve. This result corresponds to calculations of the kinetics of evaporation and condensation of metals (Ikezi *et al.* 1978). Therefore, these experiments enabled one to register, with sufficient certainty, the beginning of evaporation of metal upon entry of the isentrope into the two–phase region, and to determine the curve of high–temperature boiling of lead in a wide range of pressure values, $p \sim (50–1000) \cdot 10^5$ Pa, which is more than 100 times higher than those in the pressure range accessible to static methods.

The results of experiments by Zhernokletov *et al.* (1984), as well as earlier recordings of kinematic parameters (Fortov *et al.* 1974; Leont'ev *et al.* 1977; Altshuler *et al.* 1977, 1980; Bushman *et al.* 1986a), point to the absence of thermodynamic anomalies that could be associated with phase transitions in a dense plasma (for more details, see Section 5.8).

The same opinion on the plasma phase transition is held by Kerley and Wise

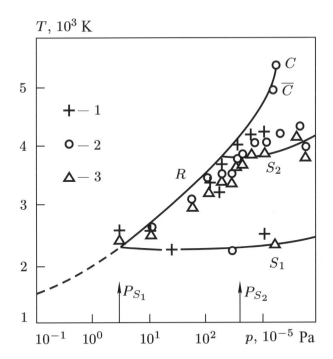

FIG. 3.14.   Temperature measurements under conditions of adiabatic expansion of lead:
the solid curves indicate the results of calculation, by the semiempirical equation of state
(Ikezi *et al.* 1978), of the boiling curve R, critical point C, and release isentropes S; the
broken curve indicates the static data on the boiling curve; $\overline{C}$, estimates of the critical
point parameters from (Leont'ev and Fortov 1974). The symbols indicate the results of
recording of temperature by photodiodes (1) and two–channel pyrometer: $\lambda_1 = 449$ nm
(2); $\lambda_2 = 560$ nm (3). $p_{S_1} = 3 \cdot 10^5$ Pa and $p_{S_2} = 4.2 \cdot 10^7$ Pa, pressure on the isentropes
upon their entry into the two–phase region.

(1987) and Asay *et al.* (1987), who performed, analogously with Altshuler *et al.*
(1980), experiments in adiabatic expansion into the vacuum of porous aluminum,
copper and polymethyl methacrylate. They were preloaded by a tantalum impactor accelerated by a two–stage light–gas gun to a velocity of 6–7 km s^{-1}. In
these experiments, laser interferometry techniques were used to register the rate
of motion of a condensed obstacle caused by the impact of evaporated metal
against the latter. This rate of motion was compared against the results of calculation by a one–dimensional gas–dynamic computer code, in which various
equations of state were entered.

The use of hydrodynamic recordings in the isentropic expansion method to
obtain thermodynamic information assumes that local thermodynamic equilibrium in the system is present. It corresponds to the smallness of characteristic
times of phase transformations as compared to the typical time the of experi-

ment of the order of 10^{-7} s. It is interesting to note that in the case of isentropic expansion, one observes a unique (and unrealizable in most other cases) possibility (Fortov $et~al.$ 1975b) of penetrating the region of absolute instability of a single–phase system $[(\partial p/\partial V)_T > 0]$ by means of nonstatic variation of pressure. This is because at the critical point where $(\partial p/\partial V)_T = 0$, the quantity $(\partial p/\partial V)_S$ remains negative, which is the case inside the spinodal. The kinetics of removing the nonequilibrium state in this case are nonactivative; they depend mainly on heat transfer and permit, under the experimental conditions as shown by estimates, negligible supercooling of not more than several degrees (Fortov and Leont'ev 1976).

An analysis of the kinetics of disintegration of metallic metastable liquid has shown (Fortov and Leont'ev 1976) that high values of critical temperatures (comparable with corresponding ionization potentials) lead to a marked thermal ionization of metal vapor on the binodal. In this case, the metastable phase contains a considerable number of charges which, while being effective condensation centers, lead to a phase relaxation time of the order of 10^{-9} s under experimental conditions. This is in agreement with the results of kinetic measurements upon evaporation of metals into vacuum (Hornung and Michel 1972; Kerley and Wise 1987; Asay $et~al.$ 1987) and expansion of shock–compressed lead into air (Leont'ev and Fortov 1974) and helium (Zhernokletov $et~al.$ 1984). According to the estimates by Fortov and Leont'ev (1976), marked nonequilibrium can only be expected at pressures lowered to several megapascals. Apparently, this defines the lower limit of application of the isentropic expansion technique in thermodynamics. The measurement data support this viewpoint (Fig. 3.14) while pointing to the quasi–equilibrium behavior of high–temperature condensation of metals. To further clarify this point, a series of experiments was staged (Altshuler $et~al.$ 1977) involving the registration of the rate of expansion of metals into low–pressure air and analysis of lead vapor impact against a copper shield. In the equilibrium case, a shock wave with $p = 73$ GPa was excited in copper whereas an impact by superheated lead produced a pressure which is two times higher.

Mention should be made of a number of other studies on the expansion of material following the passage of powerful shock waves (Scidmore and Morris 1962; de Beaumont and Leygenie 1970; Fortov and Dremin 1973). Using the converging conic wave technique, de Beaumont and Leygenie (1970) obtained two points for the expansion of uranium and copper plasma into air at atmospheric pressure. Based on these data, the semiempirical equation of state has been corrected. The results relating to the reflection from soft obstacles of shock waves in porous copper are presented by Alekseev $et~al.$ (1971), who were mainly interested in the problems of two–phase gas dynamics.

Another group of experiments (Scidmore and Morris 1962; Fortov $et~al.$ 1974; Altshuler $et~al.$ 1980; Zhernokletov $et~al.$ 1984; Ageev $et~al.$ 1988; Glushak $et~al.$ 1989), which involved the release adiabats, is based on registering both the final and intermediate states occurring upon isentropic expansion of the plasma (Fig. 3.15). A shock wave propagates through the test substance M and causes the

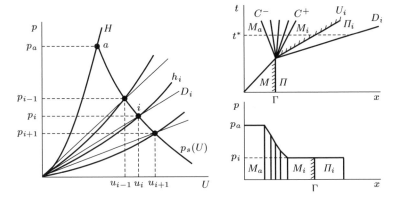

FIG. 3.15. Schematic gas–dynamic diagram of experiments using the adiabatic expansion technique.

compression of the latter and its irreversible heating to an a state. The shock wave arrives at an interface Γ with a dynamically softer obstacle Π. It also generates a centered Riemann wave C^+C^- in which an adiabatic expansion of shock–compressed plasma from the a state to an i state takes place. Such expansion of metal generates a shock wave propagating at a velocity D_i in the obstacle. The registration of D_i enables one to find, from the known shock adiabat h_i, the pressure and mass velocity of the obstacle which, because of the condition of continuity (Zel'dovich and Raizer 1966) at the contact interface Γ, coincide with the respective characteristics of the expanding plasma.

By using obstacles of different dynamic rigidities and recording the respective values of p and u, one can continuously trace the variation of the expansion isentrope $p = p_s(U)$ from states on the Hugoniot adiabat to lower pressures and temperatures. The use of propulsion systems of different power allows one to vary the increment of entropy in the shock wave and, thereby, to use different isentropes overlapping the selected phase diagram region. The transition from the hydrodynamic variables $p - u$ to thermodynamic variables $p - v - E$ may be accomplished by calculating the Riemann integrals (3.2) which express the conservation laws for the given type of self–similar flow.

Scidmore and Morris (1962) employed this technique, while using linear and spherical explosion systems, to register three expansion isentropes for shock–compressed uranium. The experiments were performed mainly in the liquid phase while the most interesting neighborhoods of the critical point, namely, the two–phase and plasma regions, remained unstudied. This caused Scidmore and Morris (1962) to restrict themselves to qualitative estimates in describing the isentrope behavior in the region of lower pressures.

A number of authors (Altshuler *et al.* 1980; Bazanov *et al.* 1985; Bushman *et al.* 1986a; Ageev *et al.* 1988; Glushak *et al.* 1989) describe systematic investigations of adiabatic expansion of lead and copper used in the form of finely divided powder to enhance the dissipation effects. In doing so, special attention was given to the plasma region of parameters attained by using gases compressed to $(1\text{--}50) \cdot 10^5$ Pa as obstacles, while an increase in the density and pressure of metals was attained through the use of "soft" condensed obstacles such as light metals (aluminum, magnesium), Plexiglas, polyethylene, fluoroplastic, and various foam plastics. The experimental scheme using linear explosion systems and gas obstacles is analogous to that shown in Fig. 3.11.

The velocity of propulsion of aluminum and steel impactors was $5\text{--}6.5$ km s^{-1}. The approach velocity of impactors and the velocity of shock waves in shields, targets and "light" obstacles were measured by the electromagnetic and optical baseline methods. The error in electrocontact registration of wave velocity in condensed media was about 1%. The velocity of shock waves in gas obstacles (error of 1.5–2%) was measured by the optical baseline method using a fast photochronograph by means of recording the beginning of plasma luminescence, when a shock wave leaves the sample for a gas, and a sharp increase of luminescence after the shock wave is reflected from a transparent obstacle installed at a preset distance.

In order to increase the propulsion velocity and, consequently, the amplitude shock–wave pressure, layered cumulative propulsion systems have been developed (Bushman et al. 1984, 1986a; Glushak et al. 1989) whose principle of operation is based on a mechanism analogous to the acceleration of a light body upon its elastic collision with a heavy one (Zababakhin 1970). In order to optimize multi–stage propulsion systems, entropy and gas–dynamic calculations were performed, which described the thermal and gas–dynamic processes with due regard for melting, evaporation, elastoplastic effects, and destruction.

As a result of these calculations and numerous experiments, two– and three–stage propulsion devices of various designs were developed. In one of these devices, a steel impactor with a thickness and diameter of 1 mm and 60 mm, respectively, was accelerated by the detonation products to $5\text{--}5.5$ km s^{-1} over a speed base of 30 mm and struck against an organic–glass obstacle with a thickness of 1 mm. Outside of the latter a molybdenum impactor was placed with a thickness of 0.1–0.2 mm and diameter of about 30 mm. As a result of wave interaction, this impactor would fly a distance of about 1.5 mm before striking against the target assembly, where plane shock waves 10–12 mm in diameter could be produced in the latter. A three–stage propulsion system differs from a two–stage one in that a steel impactor with a thickness of 2.5 mm was accelerated in the first stage to load a layer of plastic explosive, which is about 5 mm thick. A steel impactor with a thickness of 1 mm was placed on the outside of the latter. The design of the outer stage and target assembly was the same as that found in a two–stage generator.

In doing so, the velocity range of molybdenum impactors is extended to 13 km s^{-1} as compared with 6 km s^{-1} for conventional propulsion. The characteristic time intervals measured in these experiments are approximately 10^{-8} s, which is an order of magnitude less than the characteristic registration time in typical explosion experiments (Altshuler 1965).

Because of this, in experiments involving layered systems, the shock compressibility and adiabatic expansion of solid samples of aluminum, copper, and bismuth were determined by the optical baseline method using step targets. For this purpose, recesses were made in the target on the side of the approaching impactor and, on the outside, a transparent Plexiglas observation window was provided at a fixed distance from that surface. Agate–SF high–speed, high–sensitivity image converters operating in the photochronograph mode were used for registration, which made it possible to reduce the error of measurement of time intervals to 1% in the scanning range of approximately 2 ns mm^{-1}. Owing to this procedure, continuous and independent information can be obtained in each experiment at about the time of arrival of the impactor and the flow structure in the target, while monitoring the quasi–one–dimensionality and stationarity of the flow and allowing for a substantial degree of duplication of the kinematic information obtained.

With this experimental procedure, the approach velocity W of the molybdenum impactor and the rate of shock wave propagation in the target over different speed bases are registered, as well as the rate of adiabatic expansion of metal into air at atmospheric pressure. Given these parameters, the "deceleration" method (Fortov 1982) enables one to find (using the shock adiabat for molybdenum) the pressure p and mass velocity u of a shock–compressed target, as well as the states on the release isentrope. In doing so, every experimental point is obtained by averaging out the results of four to six experiments.

Along with experiments involving linear cumulative devices, experiments were performed using conical explosion generators of Mach shock waves (Glushak et al. 1989), which utilize the effect of geometric cumulation under conditions of irregular (Mach) reflection of conically convergent shock waves (Bazanov et al. 1985). As compared to the plane case, these devices are characterized by an additional concentration of energy during the shock wave convergence to the axis of symmetry and, at the same time, by an increased stability of flow as compared to the spherical convergence of shock waves (Bushman et al. 1986b).

A typical generator of conical shock waves (Fig. 3.16) consisted of a solid metal cone with an angle of 45–60°(Bazanov et al. 1985; Glushak et al. 1989). The shock wave was excited either by the direct effect of the detonation products or by conical copper and aluminum liners with a thickness of about 3 mm, and accelerated by the detonation products to $3–4 \text{ km s}^{-1}$. The asynchronism of shock wave motion on the cone generator was at least 50 ns. The Mach shock wave, at a distance of 40–50 mm from the apex of the copper cone, had a diameter $\sim 7–8$ mm, a velocity $\sim 12.1 \text{ km s}^{-1}$ and a pressure $\sim 6 \cdot 10^5$ Pa.

Similar to layered systems, the kinematic parameters were registered using step targets. The optical signals were recorded by Imacon-640 (scanning of 50–100 ns mm^{-1}) and Kadr–2 (scanning of 10–20 ns mm^{-1}) high–speed image converters. In some experiments, the radiation was delivered to the photocathode of image converter from the basal planes of the targets via quartz light guides

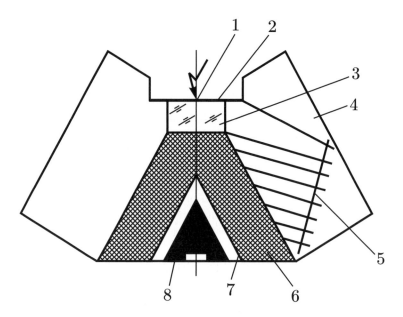

FIG. 3.16. Schematic of explosion generator of Mach shock waves: 1, detonator; 2 and 5, lines of detonation arrangement; 3, absorbing washer; 4, textolite plate of detonation arrangement; 6, conical explosive charge; 7, conical liner projectile; 8, conical target.

with diameter of about 60 μm.

With this organization of experiments, one could register, in every experiment, the shock wave velocity in a copper cone (shield) over different bases in a target of test material, as well as the rate of its adiabatic expansion. In addition, the resultant photochronograms offer the possibility of monitoring the curvature and attenuation of shock waves. The subsequent use of the reflection method (using copper as the standard) helps find the remaining thermodynamic parameters. When registering the release isentrope, shock–compressed bismuth and copper were expanded into aluminum, Plexiglas, and air at atmospheric pressure. Each experimental point was obtained by averaging out the results of four to ten independent experiments with two to four recordings in every experiment. The error of individual measurement is estimated at 2–3%, and the averaged error, at 1%.

A number of experiments were performed in generators utilizing, simultaneously, the effects of geometric and gradient cumulation. In these experiments, a conical Mach shock wave exert influence, via a spacer of condensed explosive, a molybdenum (about 0.1 mm thick) or tungsten (about 0.25 mm thick) impactor, which accelerated over a base of 100–500 μm to the velocity of 16 km s^{-1} and collided with a step copper target. As a result of these experiments, a shock compression pressure of approximately 1400 Pa ($T \sim 5$ eV) was attained, and isentropic scattering of copper into air at atmospheric pressure was registered. Note that these parameters of scattering are in reasonable agreement with the

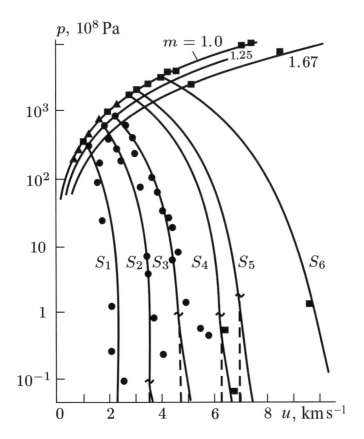

FIG. 3.17. $p - u$ diagram for lead (Altshuler *et al.* 1977): m, shock adiabats of different porosities; S, expansion isentropes. Dashed lines, metastable branches; dots, experiment.

results of measurements by de Beaumont and Leygenie (1970).

Figure 3.12 gives the $p - S$ diagram for copper and bismuth, and Figs. 3.17 and 3.18 give the $p - u$ diagram for lead and copper. In the case of bismuth, the experimental results cover a wide range of states from highly compressed matter on the shock adiabat with $p \sim 670$ GPa and $\rho \sim 2.6\rho_0$ to rarefied metal vapor with $\rho \sim 10^{-2}\rho_0$.

The experimental data presented in Figs. 3.17 and 3.18 demonstrate that the process of adiabatic expansion covers a very wide range of parameters (four orders of magnitude with respect to pressure and two orders of magnitude with respect to density): from a highly compressed metal plasma, where ions are disordered and electrons degenerate, to a quasi–nonideal Boltzmann plasma and rarefied metallic vapor. As expansion proceeds, diverse little–studied physical processes occur in the system: the electron degeneracy is removed; the electron energy spectrum is radically restructured; partial recombination of dense plasma

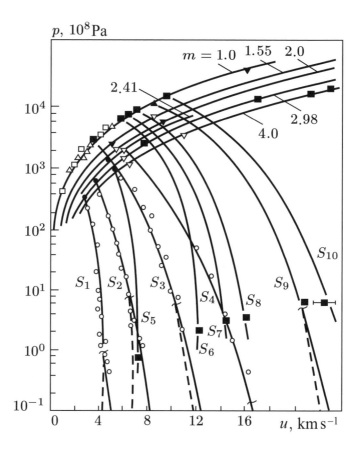

FIG. 3.18. $p - u$ diagram for copper (Glushak *et al.* 1989). Designations are the same as
    Fig. 3.17.

is effected; the "metal–dielectric" transition in a disordered electronic structure
is accomplished; and a nonideal (with respect to various types of interparticle
interaction) plasma emerges.

    The presence of strong collective interaction makes it difficult to provide a
consistent theoretical description of material in this range of parameters. A number of heuristic models have been proposed (cf. Chapter 5) to describe individual
effects in relatively narrow parts of the phase diagram. The main qualitative result yielded by most models resides in the possibility of the loss of thermodynamic
stability and separation of highly nonideal plasma into new exotic phases, which
would distort considerably the usual form of the phase diagram for metals. The
experiments performed have demonstrated the absence of marked discontinuities of thermodynamic functions or any hydrodynamic anomalies which could
be interpreted as specific plasma phase transformations. Note that it is in the
investigated range of parameters that the plasma phase transitions discussed in

the literature are most likely to occur. This is because an increase in temperatures and a decrease in density of a Boltzmann plasma, as well as an increase in the degenerate plasma pressure, lead to a relative decline of nonideality effects (Section 3.1). The same conclusion on plasma phase transitions was made by US researchers (Kerley and Wise 1987; Asay *et al.* 1987).

The highest stages of metal plasma expansion in experiments correspond to the realization of near–critical states. The entry of isentropes into the two–phase "liquid–vapor" region on the liquid phase side is accompanied by evaporation and on the gas side, by condensation. This leads to a variation of their slope and to an additional increment of velocity in the release wave. The presence of kinks on the experimental curves in Figs 3.17 and 3.18 and their agreement with the *a priori* estimates of the evaporation effects (Fortov *et al.* 1975b; Fortov and Leont'ev 1976) provide further evidence of the equilibrium nature of the process of two–phase expansion. The nature of thermodynamic information obtained is also in agreement with the estimates of the parameters of the critical point and phase boundary of boiling. It follows from the experiments performed that the phase diagrams for copper and lead have the usual form with a single critical point of the "liquid–vapor" phase transition.

Of special interest is the region of strongest expansion of metals, where a metal is under highly supercritical conditions. Figure 3.12 gives the values of the nonideality parameter γ and of the degree of ionization $\alpha = n_e/(n_e + n_i)$ calculated from the chemical model of the plasma (Fortov 1982). The results of an analogous treatment for copper demonstrate that the adiabatic expansion from the states on the shock adiabat with $p \sim 1410$ GPa, $T \sim 5.25 \cdot 10^4$ K, and $v \sim 0.0521$ cm^3 g^{-1} leads to a weakly nonideal plasma with parameters $p \sim 0.73$ GPa, $T \sim 9200$ K, $v \sim 1.3$ cm^3 g^{-1}, $\Gamma \sim 0.1$ and $\alpha \sim 0.003$. This means that the entropy of a slightly ionized metallic vapor in these states may be reliably calculated by a quasi–ideal–gas approximation and, by virtue of the condition of isentropicity, compared with the values determined by the semiempirical equations of state for highly compressed metal. A comparison with the equation of state by Altshuler *et al.* (1980) produced close results.

One can see that the results, produced with modern explosion generators of powerful shock waves, made it possible, for the first time, to combine the regions of the phase diagram that correspond to radically different physical states (Bushman and Fortov 1988).

The expansion isentropes relate the states with similar entropy of a superdense degenerate plasma on the shock adiabat with the near– or supercritical states of slightly ionized vapor. In addition, during the entry into the liquid–vapor two–phase region, they have the values of energy and volume that are in agreement with the parameters of the equilibrium line. This made possible the experimental realization of Zel'dovich's idea (Zel'dovich and Raizer 1966) of deriving a thermodynamically complete equation of state from the results of mechanical measurements, namely, the entropy found for a highly expanded ideal vapor proves, by virtue of adiabaticity, to be equal to the entropy of shock compression. The equilibrium temperature of shock–compressed matter is then calculated by the thermodynamic identity of the second law of thermodynamics (Fortov and Dremin 1973).

The experimental data on isentropic expansion served as a basis for the construction of semiempirical equations of state (Bushman and Fortov 1983, 1988; Bushman *et al.* 1988) describing the entire available totality of static and dynamic data in the solid, liquid, and plasma phases. They reproduce the melting and evaporation effects and feature, at superhigh pressures and temperature, the correct asymptotic behavior with respect to Thomas–Fermi and Debye–Hückel theories. We shall discuss these problems in more detail in Chapter 5.

3.5   Generation of superdense plasma in shock waves

According to the existing concepts (Kirzhnits *et al.* 1975; Fortov 1982), the description of plasma properties is radically simplified at extremely high pressures and densities when the inner electron shells of atoms and ions turn out to be "crushed" and a quasi–uniform distribution of electron density within Wigner–Seitz cells is realized. In this case, a quasiclassical approximation to the self–consistent field method is valid (cf. Section 5.5) when the description is made in terms of the mean density of the electron component of the plasma rather than in the quantum–mechanical language of wavefunctions and eigenvalues. These concepts form a basis for the Thomas–Fermi model. Its range of application depends on the smallness of exchange, correlation, and shell effects (Kirzhnits *et al.* 1975). The smallness of the corresponding latter criteria leads to a general estimate of the lower limit of application of the Thomas–Fermi model of $p_{lim} \sim e^2/a_0^4$. This corresponds to extremely high pressures which are approximately equal to or higher than 30 TPa $(T = 0$ K). These exceed considerably the capabilities of experimental facilities based on the use of powerful condensed explosives. Therefore, an active search is currently underway for alternative methods of generation of superhigh plasma parameters. This would enable one to approach the region of a quasiclassical description of matter with a view to estimating the real range of its application. Such methods include the use of powerful shock waves emerging in the short–range band of strong explosions (Altshuler *et al.* 1968; Trunin *et al.* 1969, 1972; Thom and Schneider 1971; Volkov *et al.* 1980a, 1980b; Vladimirov *et al.* 1984; Simonenko *et al.* 1985; Model' *et al.* 1985; Avrorin *et al.* 1987; Bushman *et al.* 1988; Avrorin *et al.* 1990), the use of coherent laser radiation (Anisimov, Prokhorov, and Fortov 1984; Duderstadt 1982), and of relativistic electron and ion beams, electric explosion of thin metal foils, and the use of rail–gun electrodynamic accelerators. The last method is still being developed while quantitative information on the thermodynamics of superdense plasma is provided by measurements in strong–explosion experiments. For obvious reasons, the amount of experimental information in this case is rather small and, in the foreseeable future, will be hardly comparable with the scope of laboratory investigations. Therefore, it is practical to use the unique possibilities offered by such experiments for the solution to key problems of the physics of high–energy processes. In recent years, the most representative example of such problems is provided by the study of the effect of the electron shell structure of atoms on the thermodynamic properties of dense substances (Avrorin *et al.* 1990). The shell effects show up in the oscillating behavior of thermodynamic functions (such as adiabats or isochores). This, in turn, may cause the existence of regions with $(\partial^2 p/\partial v^2)_S < 0$, which are responsible for anomalous behavior (Zel'dovich and Raizer 1966) of substances in dynamic processes (the release occurs in a shock wave, and compression is smooth). The inclusion of such properties of real substances in gas–dynamic calculations would call for a modification of the algorithms employed in modern mathematical computer codes (Bushman *et al.* 1988). This fact further rouses interest in shell effects.

At present, the vast majority of experiments with powerful shock waves are performed by a comparison method (the method of "reflections"; Fortov 1982). It records the values of the velocity of shock waves passing successively through layers of the materials being investigated, one of which is the standard. Clearly, such a method involves an uncertainty associated with the extrapolation procedure of constructing the shock adiabat for the standard in that very region of parameters, in which direct measurements must be performed. This uncertainty is absent from experiments in recording the absolute shock compressibility of molybdenum (Thom and Schneider 1971) and aluminum (Simonenko *et al.* 1985) using penetrating physical fields (neutron and γ–radiation) for diagnostics.

Thom and Schneider (1971) suggested a scheme to measure absolute shock compressibility in the high–pressure region. It is based on shifting the resonances of the interaction of neutrons with nuclei of moving matter with respect to their position in the case of nuclei at rest (Doppler shift). The energy of fission of uranium nuclei by neutrons, formed during a nuclear explosion, is used to produce high pressures. In the experiments performed by Ragan *et al.* (1977), a block of ^{235}U, placed at a distance of 1.1 m from a nuclear charge behind a B$_4$C slow–neutron absorber, supported the molybdenum sample being investigated with light guides. The guides were provided therein for baseline registration of the shock front velocity (Fig. 3.19). The neutron flux generated upon detonation of the nuclear device A caused a rapid uniform heating of the uranium to about 50 eV. This led to an expansion of the uranium which generates a plane shock wave with an amplitude pressure of 2 TPa in molybdenum. The second kinetic parameter of the velocity of shock–compressed molybdenum was found from the Doppler shift of resonance lines of neutron absorption in the 0.3–0.8 keV energy range. It was registered by a time–of–flight neutron spectrometer.

From the standpoint of experimental support of the method, the neutron flux in the resonance range of energy 10-10^3 eV is essential. The number of such neutrons in the fission spectrum is very small. In order to increase their number, Prokhorov *et al.* (1976) placed a thin layer of hydrogen–containing material, such as organic glass, between the uranium and the sample.

The method discussed above on the measurement of mass velocity is not universal. For the materials under investigation, the cross–sections in the resonances

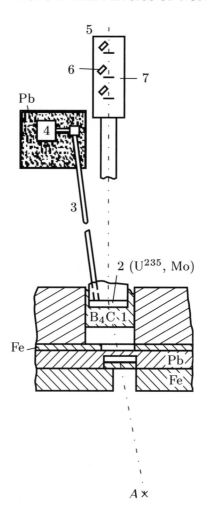

FIG. 3.19. Schematic illustrating the generation of strong shock waves by a nuclear explosive (Thom and Schneider 1971): A, nuclear charge; 1, B_4C slow–neutron absorber; 2, test assembly of ^{235}U and molybdenum; 3, light guides (12 m long); 4, optical radiation recorders; 5, time–of–flight neutron spectrometer; 6, solid–state detectors; 7, lithium and plutonium foils.

must support a well–recorded attenuation of neutron flux by both stationary and moving matter, that is, for thickness comparable to the measuring bases. Such properties are characteristic of molybdenum, iron, and copper. Besides, there are elements whose nuclei have anomalously large resonance cross–sections (tungsten, gold, cobalt). Later, molybdenum was used by Avrorin et $al.$ (1990) as a standard in experiments involving the measurement of relative compressibility of uranium at a pressure of the order of 6.7 TPa, which yielded pressures higher

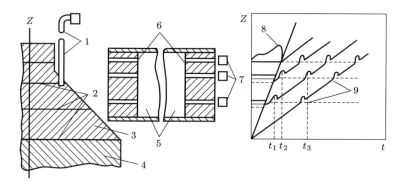

FIG. 3.20. Scheme of experiments by Simonenko *et al.* (1985) in recording the absolute
shock compressibility of aluminum: 1, optical channel; 2, reference layers; 3, material being
investigated; 4, channel where the shock wave is formed; 5, collimating system; 6, collimating slits; 7, γ–radiation detectors; 8 and 9, signals of radiation from stationary and moving
reference points.

than those predicted by the quasiclassical theory. Note that in processing these
data, Avrorin *et al.* (1990) had to resort to rather far–reaching (from 2 to 5 TPa)
extrapolation of the standard adiabat of molybdenum.

In the case of absolute measurements (Simonenko *et al.* 1985), the phase and
mass velocities of shock waves in aluminum were registered with the aid of reference layers and slit collimators L arranged parallel with each other, with the distance between them representing the gauge lengths (Fig. 3.20). γ–radiation was
recorded by scintillation detectors through neutron irradiation of reference samples of material with a large radiative–capture cross–section (europium) placed
in aluminum.

In the experiments, the "plane" geometry of the shock front, reference layers
and collimating slits, in which the planes of respective surfaces are parallel to one
another (Fig. 3.20), are the most easily realized. For this purpose, a cylindrical
channel was installed in the direction of wave propagation. It was manufactured
from a material (such as magnesium and organic matter) whose density is less
than the material in which the experimental facility is accommodated. It follows
from the results of two–dimensional gas–dynamic calculations that the provision
of such a channel ensures sufficient advance of the shock front relative to the
front in the surrounding medium. The measuring unit is mounted at the cylinder
end. The protection of the collimating system, which rules out the possibility of
damage before completion of the recording of the moments when the reference
layers pass the control positions, is accomplished by placing dense matter (lead
or steel) in the direction of wave propagation.

In the experiments of Simonenko *et al.* (1985), an intensive gamma–source
was obtained with pulsed irradiation by neutrons of a substance whose nuclei
had a radiative–capture cross–section exceeding those of the test material by a

factor of approximately 10^3. The existing pulsed sources usually give rise to fast neutrons ($E \sim 1$ MeV). The radiative–capture cross–sections proceed efficiently at lower values of energy. Therefore, a neutron pulse must be ahead of the gas–dynamic motion being recorded by the time interval required for deceleration of neutrons in the test material to optimum values of energy. In a number of cases, europium may be used in the reference layers, for which the cross–section of the (n, γ)–reaction at $E = 10$-100 eV is $q = (220-80) \cdot 10^{24}$ cm^2.

The deceleration of neutrons in the sample leads to the heating of the test material. In general, this heating affects the shock compressibility of the material. To render the interpretation of the experimental results easier, it is necessary that this effect be weak, which restricts the flux from above by the value of $\Phi \leq 10^{17}$ cm^{-2}.

The measuring unit (Fig. 3.20) was constructed as a set of plates manufactured from AD–1 aluminum, in which the reference layers were embedded in the form of tablets. This form of layer agrees with the cylindrical geometry of the channel in which the shock wave is formed and can, when required, accommodate additional collimating systems. The shape of shock wave is monitored with the aid of three optical channels located at the angles of an isosceles triangle on the diameter of 150 mm (one channel is shown in Fig. 3.20). The data of measurements by Simonenko *et al.* (1985) point to considerable (in aluminum) errors of the quantum–statistical model at $p \sim 1.1$ TPa. These will be treated in more detail in Chapter 5.

The principal experiments in the ultrahigh–pressure region have been performed using the comparison method. The standard in experiments in dynamic compression of solid materials was provided by lead (Altshuler *et al.* 1968; Trunin *et al.* 1972) and iron (Volkov *et al.* 1980a). Interpolation shock adiabats were constructed relating the ultrahigh–pressure ($p \geq 10$ TPa) region with a lower–pressure range of about 1 TPa, which is accessible to direct experiments (Fortov 1982). In such experiments, the error of measurement of time intervals was 0.7–1.0%. The wave decay is at the level of 1–2%. By and large, it is an accurate estimate of the values of the wave velocity on the contact boundary. These results may be used to verify the plasma models. However, they do not provide any answers to the question of the effect of the shell structure: the measurements were performed in the vicinity of the lower (with respect to pressure) boundary of the manifestation of shell effects. The transition from plane to spherical geometry of experiments in the measurements of Lepouter (1965) enables one to raise the experimentally attainable pressure level to the exotically high values of about 400 TPa. However, such a transition results in a stronger decay of shock wave in samples, and the indeterminacy of its inclusion once again made a major contribution to the total measuring error.

Unconventional experiments in studying the effect of the shell structure were performed by Model' *et al.* (1985): the experimentally recorded times, required for shock wave passage in samples of different materials, were compared with the calculated values. During the passage in samples, the pressure at the shock front dropped by a factor of three or four, that is, the simulation of the conditions of wave passage strongly calls for knowledge of both the shock adiabat region and the release isentropes, which makes it difficult to interpret the results of these measurements.

The contribution by the shell effects was recorded most explicitly by Shatzman (1977). Shock waves with a clearly defined front shape were excited in test assemblies, and measurements were performed using a unified procedure for several values of pressure at the shock front. The gauge lengths were selected such that the behavior of the wave velocity inside the layers being investigated should be, according to calculations, monotonic.

The moments of shock front arrival at the reference surfaces were recorded by the optical glow of the air layers adjoining those surfaces, which was transmitted to a detector, such as a coaxial photocell, via a light–guide channel with polished metal inner walls. In order to ensure the accuracy of measurements, three reference surfaces corresponding to the selected pressure were placed within the field of vision of a single photocell, namely, two surfaces of the standard and one surface of the sample. Two layers of the same material were provided for experimental detection of the wave decay. Every light–guide channel accommodated two detectors, with each recording a full signal. The complexity of the three–stage form of signal necessitated the use of oscillographic equipment.

Measurements were performed for the following pairs of materials: iron–aluminum, iron–lead, lead–iron, iron–water, and iron–quartzite. The first material in each pair was placed first in the direction of wave propagation. The results will be discussed in Chapter 5. However, we would like to note that Avrorin *et al.* (1987) have reliably registered the shell effects on the shock adiabats in lead and aluminum.

The plasma effects are reflected most vividly in the experiments of Trunin *et al.* (1969), who studied the compression of porous copper by powerful shock waves. These experiments involved the generation, behind the shock front, of a plasma with a specific internal energy of up to 0.75 MJ cm^{-3}, an electron concentration $n_e \sim 10^{23}$ cm^{-3}, and a pressure of up to 2 TPa (see Section 5.6). At the maximum temperature $T \sim (3\text{-}5) \cdot 10^5$ K, the electron degeneracy $n_e \lambda_e^3 \sim 1$ in a quintuply ionized plasma is removed while the Coulomb and short–range interaction remains strong, $\Gamma \sim 2$. Under these conditions, as shown in Section 5.6, the quasichemical plasma model, in view of the Coulomb interaction within the ring Debye approximation in the grand canonical ensemble of statistical mechanics, gives an acceptable description of the dynamic experiments of Trunin *et al.* (1969), whereas the differences from the quasiclassical model may be as great as 20-30% with respect to density, and several times with respect to pressure.

3.6   Nonideal plasma generation by powerful fluxes of energy

Along with the customary techniques of shock wave generation (chemical and nuclear explosives, light–gas, and powder guns), researchers are expressing an ever growing interest in unconventional methods based on the stimulation of matter by intense fluxes of directional energy. We refer to lasers, generators of pulsed electron and ion fluxes, electric–explosion, and electrodynamic propulsion devices. The modern pulsed generators of directional energy, developed for the purposes of inertial controlled thermonuclear fusion, remote stimulation and technological applications, are capable of delivering to the test material energy in the kilojoule/megajoule range at the megawatt power level. By focusing this energy into the submillimeter ranges of space, one is able to attain exotically high values of power density (10^{15}-10^{20} W cm^{-3}), leading to extremely high values of pressure and temperature of a highly compressed plasma.

A broad spectrum of plasma states emerges under conditions of pulsed stimulation of condensed targets by intense fluxes of directional energy. It is difficult to identify a specific range of parameters for the thermodynamic description of those states (Akkerman *et al.* 1986a; Bushman *et al.* 1987). It is necessary to have the data on the properties of matter in an extremely wide region of the phase diagram, ranging from the highly compressed condensed state to ideal gas, including the region taken up by strongly nonideal plasma. We will explain this using the example of stimulation of a condensed target by intense laser radiation (Fig. 3.21) (Anisimov *et al.* 1984). The laser radiation of frequency ω is absorbed in a heated plasma corona (at $\omega \sim \omega_{\mathrm{p}} \sim \sqrt{4\pi n_e e^2/m}$) taken up by a high–temperature ideal plasma. The thermal energy is transferred to the region of relatively cold shock–compressed high–pressure plasma by electron thermal conductivity (shown by the broken line in Fig. 3.21) via a nonideal Boltzmann plasma embraced by complex hydrodynamic motion. In problems of controlled thermonuclear fusion with inertial confinement (Duderstadt 1982), the target shown in Fig. 3.21 is an external ablator of a multi–layer spherical thermonuclear micro–target. On its inside, a density of $\rho \sim 10^3 \rho_0$, pressure $p \sim 10^{14}$ Pa and temperature $T \sim 10$ keV develop. Stimulation by intense fluxes of relativistic electrons and ions (Iakubov 1977; Gathers, Shaner, and Hodson 1979) reminds one, qualitatively, of the situation corresponding to Fig. 3.21. The difference in this case is that the main energy of the beam is released in the high–density region $\rho \sim \rho_0$ (Fig. 3.21b) and the overfly of the plasma corona is close to adiabatic.

The results of analysis demonstrate that the effect of concentrated fluxes of directional energy on materials is a promising tool for generation of powerful shock waves, which produce nonideal plasma of extreme states.

Most physical experiments involve the use of pulsed lasers (Anisimov *et al.* 1984), which offer a unique possibility of focusing coherent electromagnetic radiation onto small ($\sim 10^{-4}$ cm^{-2}) surfaces. This leads to extremely high values of the local concentration of energy. The specific powers currently attained and applied to targets are in the range of 10^{14}–10^{17} W cm^{-2} and may be brought to 10^{21} W cm^{-2} in the near future. The recoil impulse, which occurs as a result of the effect of such light fluxes, generates a powerful shock wave in the target. It can be used for compression and irreversible heating of a dense plasma of mate-

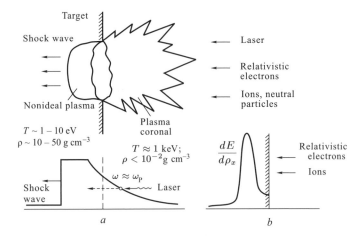

FIG. 3.21. Irradiation of a condensed target by fluxes of directional energy: a, laser stimulation; b, charged particles.

rials under investigation. An analysis by Trainor, Graboske, and Long (1978) of
hydrodynamic calculations of the intensity of shock waves, which emerge upon
stimulation of various materials by existing and projected laser systems, demonstrates (Fig. 3.22) that in this case there is a real possibility of advancing into the
ultramegabar range of pressures and investigating the properties of superdense
plasma.

In designing laser targets for such experiments, a number of specific requirements arise (Trainor, Graboske, and Long 1978; More 1981; Anisimov et al. 1984)
which are dictated by the physical singularities of the process and characteristics of the diagnostic equipment. The target thickness depends on the absence
of distorting release waves at the moment of laser pulse termination, as well
as the small effect of "nonthermal" electrons occurring in the laser radiation
resonance–absorption band. The target diameter is selected such that it attains
a fairly high radiation intensity and ensures the absence of side release waves. In
addition, the target size should be sufficiently small in order to reduce the effect
of surface currents from the heated plasma. Otherwise, a special shield should
be used.

The initial experiments in the excitation of shock waves in hydrogen and
Plexiglas involved the use of a low–power neodymium laser (Van Kessel and
Sigel 1974) with an energy of $E \sim 12$ J and pulse duration of $\tau \sim 5 \cdot 10^{-9}$ s.
Because of small size of the focal spot (40 μm), the shock waves decayed rapidly
and degenerated to spherical. In order to generate plane shock waves, Veeser and
Solem (1978) used a more powerful laser system ($E \sim 30$ J, $\tau \sim 0.3 \cdot 10^{-9}$ s).
Measurements of the time of shock wave passage through a step–like aluminum
sample enable one to register a shock front velocity of 13 km s^{-1} corresponding to
a pressure of the order of 0.2 TPa. These pressures were increased (Trainor et al.

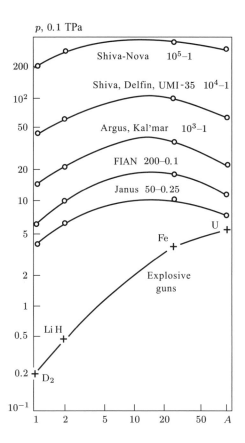

FIG. 3.22. Maximum plasma pressures (Trainor, Graboske, and Long 1978) generated by different laser systems (the first numeral gives the energy in joules, the second numeral the pulse duration in 10^{-9} s); the bottom curve indicates the parameters accessible to the chemical explosive and light gun techniques; A, atomic number of the target element.

1979) by an order of magnitude (Fig. 3.23) by using a laser of higher parameters ($E \sim 100$ J, $\tau \sim 0.3 \cdot 10^{-9}$ s). It produces a radiation intensity of $8 \cdot 10^{13}$– $3 \cdot 10^{14}$ W cm^{-2} on the target. In this case, use was made of a small diameter target reducing. In the authors' opinion, the surface current effect, and agreement between theory and experiment, were observed for the first time. Maximum pressures of 3.5 TPa were attained in the Shiva laser facility (Trainor, Holmes, and Anderson 1982) upon the irradiation of a composite aluminum/gold target by a light flux with a power density of the order of $3 \cdot 10^{15}$ W cm^{-2}. In subsequent studies (Rosen and Phillipson 1984), 10 beams of this facility (energy of about 1 kJ and pulse duration of the order of 3 ns) were used to accelerate a carbon foil 10 μm thick to approximately 100 km s^{-1}. Consequently, impact against a carbon target produced a shock wave with an amplitude pressure of approximately 2

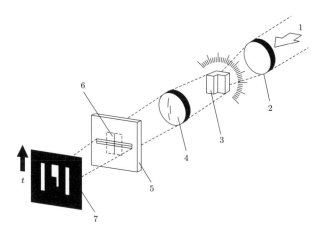

F‍IG. 3.23.   Schematic illustrating the experiments in laser generation of shock waves (Veeser
and Solem 1978): 1, laser radiation; 2, focusing system; 3, step–like target; 4, image converter optics; 5, image converter slit; 7, photographic film.

TPa. The same range of velocity of propulsion of plastic impactors was realized
in experiments involving the use of an iodine laser (Bondarenko *et al.* 1981), with
the characteristic values of power density ranging from 10^{12} to $6 \cdot 10^{15}$ W cm^{-2}.

To obtain information on the mass velocity of the plasma, Veeser and Solem
1978 used the "reflection" method in their experiments, which involve the shock
wave transition from aluminum ($p \sim 0.3$ TPa) to gold ($p \sim 0.6$ TPa). The use
of a CO_2 laser, with the nonthermal electrons playing a minor role, yielded a
shock wave intensity in aluminum of 5 TPa, the value of the power density J on
a laminar target ranging from $5 \cdot 10^{13}$ to $4 \cdot 10^{14}$ W cm^{-2}. A layer of gold, with
thickness of 10 μm, was included in the structure of this target to absorb the
nonthermal electrons emerging from the aluminum ablation plasma.

An interesting example of the use of nonthermal electrons for isochoric heating of plasma is provided by Burnett *et al.* (1981). According to them, nonthermal
electrons with an energy $E \sim 15$ keV can cause a rapid heating and explosion of
a 3–μm layer of aluminum. This leads, upon expansion, to the generation of a
shock wave with an amplitude value of pressure of about 1.3 TPa.

The treatment of the physical processes occurring in a laser plasma leads one
to conclude (Anisimov *et al.* 1984) that the main limitation in the generation
of shock waves in homogeneous targets is the heating of matter by high–energy
electrons emerging in the resonance absorption zone. Therefore, the use of short-
wave radiation (produced with the aid of nonlinear crystals or in KrF, Se, and
other laser systems) makes for a considerable extension of the range of attainable
pressure. For example, in the experiments of Trainor *et al.* (1982), the effect of the
second harmonic of a neodymium laser ($\lambda \sim 0.52$ μm) at $J \sim 1$–$1 \cdot 10^{14}$ W cm^{-2}
produced a shock wave in an aluminum target with an amplitude pressure of 1.0–
1.2 TPa. Meanwhile, the use of the first harmonic $\lambda \sim 1.06$ μm would produce

a pressure which is two times lower. For this reason, the experiments in laser shock waves are preferably performed with short–wave radiation (Ng et al. 1985, 1986; Roman et al. 1986; Hall et al. 1988). Hall et al. (1988) used the fourth harmonic of a neodymium laser ($\lambda \sim 0.26 \, \mu$m) to generate a shock–wave pressure of about 5 TPa in an aluminum target. This was doubled during the shock–wave transition from aluminum to gold. The maximum values of pressure in this series of experiments (about 15 TPa) were obtained as a result of the impact of aluminum foil with a thickness of 12 μm against the target.

Experiments with laser systems place severe demands on the diagnostic equipment: the time and space resolution should be as good as 10^{-11} s and 10^{-4} cm. Therefore, in the majority of laser experiments, it is only possible now to measure the shock velocity. The most diverse possibilities have been considered that would facilitate the registration of one more dynamic parameter while most diverse possibilities are considered for registering one more dynamic parameter: that of the plasma velocity. The possibilities include the "obstacle" technique (Veeser and Solem 1978; Roman et al. 1986), flash radiography (Rosen and Phillipson 1984; Hall et al. 1988), Doppler level shift, and the like (for more details, see the review by Anisimov et al. 1984).

In order to study the structural properties of nonideal plasma, Hall et al. (1988) employed the method of flash radiography of aluminum. The latter was heated to $T \sim 10^4$ K and triply compressed by colliding plane shock waves using the second harmonic of a neodymium laser ($J \sim 2 \cdot 10^{13}$ W cm^{-2}). The source of external X–radiation was provided by a uranium target irradiated by a special laser. It was recorded using a fast micro–crystal X–ray spectrometer. The experiments revealed an oscillation in the X–ray spectra. This was caused, in the opinion of Hall et al. (1988), by the emergence of short–range order in the dense plasma, and is attributed to a plasma phase transition.

Of considerable interest in experiments with laser shock waves is the study of the physical effects accompanying the arrival of shock waves to the free surface (Zel'dovich and Raizer 1966). Ng et al. (1985) used a fast image converter to analyze the emitted optical radiation of the rear side of aluminum target after the passage of a shock wave with a pressure of about 0.03–1.2 TPa. The intensity of this radiation was then compared with the temperature of shock compression within the framework of too simplified a model. Then, for similar physical conditions, Ng et al. (1986) measured the reflectivity of laser radiation with the wavelength $\lambda \sim 0.57 \, \mu$m from an adiabatically expanding plasma, which provided information about the high–frequency electrical conductivity of high density nonideal plasma $\rho \lesssim 1$ g cm^{-3}.

Powerful (of the order of 10^{14} W) pulse generators of relativistic electrons and ions (Bondarenko et al. 1981; Rosen and Phillipson 1984), developed for the purposes of controlled fusion and the solution of applied problems, permit us to focus high–intensity beams in targets which are several millimeters in diameter. Specific powers of the order of 10^{14}–10^{18} W cm^{-2}, applied in this manner, cause evaporation and expansion of the outer part of the target. This leads to the ablation generation of powerful shock waves.

Assuming the same (as in the case of laser stimulation) requirements (Vladimirov *et al.* 1984), that is, the plasma flow in the target should be one–dimensional and quasistationary, and taking into account the characteristic ranges of electrons with energy in the mega–electron–volt range in metals are of the order of 0.1–1 mm, one can estimate the amplitudes of shock waves in plasma at a level of terrapascals (Akkerman *et al.* 1986a). In the initial experiments performed to generate shock waves in metals with the aid of relativistic electron beams (Perry and Widner 1976; Demidov and Martynov 1981; Akkerman *et al.* 1986b), the maximum pressures do not exceed 300 GPa. The rate of propagation and time profile of the shock wave were registered: they provide information about the strength properties of materials and parameters of the equation of state.

A series of experiments with shock waves was staged in an effort to reveal the importance of anomalous deceleration ("magnetic stopping") of an electron beam in condensed targets (Perry and Widner 1976; Bogolyubskii *et al.* 1976; Akkerman *et al.* 1986b). This anomalous deceleration, which is characteristic of intense relativistic electron beams, is attributed to the elongation of the electron trajectory in the plasma as a result of the effect of the intrinsic magnetic field of the beam. It shows up very clearly during the passage of the beam through thin metal foils (Akkerman *et al.* 1986a). In order to clarify this effect, Akkerman *et al.* (1986b), in measuring the group and phase velocities of shock waves, used diverse techniques which include monomode quartz light guides, two–step targets, and targets with a complex configuration. Figure 3.24 gives the results of measurement of the decay of the mass velocity of a dense aluminum plasma as the shock wave moves away from the zone of energy release. A comparison of these results with those for one–dimensional (curve 2) and two-dimensional (curves 3 and 4) hydrodynamic calculations (Akkerman *et al.* 1986b) points to the validity of the classical (calculation by the Monte Carlo method is shown at 3) mechanism of energy release, as well as to the insignificance (curve 4) of "magnetic stopping" as regards thick targets.

Interesting possibilities are offered by the method of ablation acceleration of foils under the effect of X–radiation, which occurs upon deceleration of relativistic electron beam in the outer, metallic part of the target. In this manner, the rate of travel of the order of $5 \cdot 10^6$ cm s^{-1} of a polyethylene envelope was attained in a cone target. A neutron yield of the D–D reaction of about $3 \cdot 10^6$ neutrons per pulse was registered.

Along with the methods of pulsed controlled fusion excited by laser radiation or by charged particle fluxes, which have become traditional, ever increasing attention is paid to the scheme which utilizes the impact by macroscopic liners (with a mass of the order of 0.1 g) accelerated to velocities of 10^7–10^8 cm s^{-1} (Manzon 1981). Obviously, the high–velocity propulsion devices based on the electrodynamic acceleration methods, which are considered for such uses, can be used to generate superpower shock waves and to investigate the properties of plasma at extreme pressures and temperatures. In a linear magnetodynamic

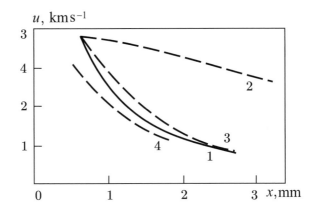

FIG. 3.24.   Mass velocity of a shock wave generated by a high–current relativistic electron
beam: 1, experiment of Demidov and Martynov (1981); 2, one–dimensional calculation; 3,
two–dimensional calculation with classical energy release (calculation by the Monte Carlo
method); 4, energy release with due regard for "magnetic stopping".

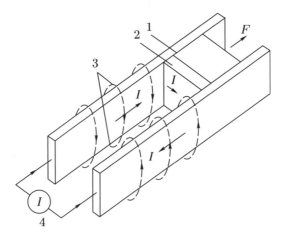

FIG. 3.25.   Schematic of rail–gun generator (Rashleigh and Marshall 1978; Kondratenko *et
al.* 1988): 1, impactor; 2, electric arc; 3, current–carrying rails; 4, power supply.

accelerator, a superconducting impactor is accelerated in a nonuniform magnetic
field. The latter is generated by coils whose switching is synchronized with the
impactor travel. In one of the projects, the impactor acceleration is accomplished
through a series of z–pinches collapsing toward the axis of symmetry (Manzon
1981).

    The rail–gun accelerator (Rashleigh and Marshall 1978; Kondratenko *et al.*
1988) appears to be the most well–developed among a great number of electrodynamic accelerators. In this device (Fig. 3.25), an impactor (1) is accelerated

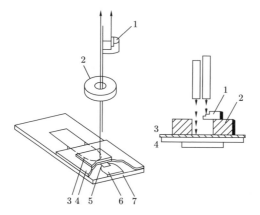

FIG. 3.26. Schematic of electric–explosion facility (Froshner and Lee 1984): 1, step–like target; 2, channel for impactor travel; 3, projectile plate; 4, plastic insulation spacer; 5, foil being exploded; 6, current–carrying buses; 7, massive substrate.

by a ponderomotive force $F = (1/2)I^2 L$ upon the passage of electric current I in electrodes (3) (linear inductance L). In doing so, the accelerating force F should not exceed the ultimate strength of the impactor material ($I \leq 1$ MA). The optimal electric contact with the rails is provided by an electric arc (2) burning in the rear portion of dielectric liner and urged against the latter by a magnetic field. The working capacity of such systems was demonstrated in experiments by Rayleigh and Marshall (1978), who supplied the rail gun from an inductive storage and homopolar generator (energy of 500 MJ) to attain velocities of the order of 6 $\mathrm{km\,s^{-1}}$ ($m \sim 3$ g) and by Kondratenko *et al.* (1988), who had brought the propulsion velocity to 10–11 km $\mathrm{s^{-1}}$ with the use of a bank of capacitors (energy of about 0.8 MJ) and a linear magnetocumulative generator.

An interesting means of advancing through the pressure scale is offered by an electric explosion of conductors occurring upon the discharge of a powerful bank of capacitors (Fig. 3.26) into the latter. According to Froshner and Lee (1984), a bank of capacitors with a capacitance of 15.6 mF and voltage of 100 kV imparts to a thin aluminum foil a specific energy exceeding by a factor of 10–100 the characteristic internal energy of condensed explosives. The foil expands to accelerate a plastic impactor, which has a thickness of 300 μm, to a velocity of 30 km $\mathrm{s^{-1}}$ and a tantalum impactor, which has a thickness of 30 μm, to 10 km $\mathrm{s^{-1}}$. In this experiment, the shock front velocity (step–like target) is registered and the variation of the rate of impactor flight planned (laser Fabry–Perot interferometer). In accordance with the reflection method, these can provide quantitative information on the dynamic compressibility of the plasma in the megabar range of pressures. The equations of state obtained for tantalum in the 190–780 GPa pressure range are in good agreement with data obtained in light–gas facilities.

References

Ageev, V. G., Bushman, A. V., Kulish, M. I., Lebedev, M. E., Leont'ev, A. A., Ternovoi, V. Y., Filimonov, A. S., and Fortov, V. E. (1988). Thermodynamics of a dense lead plasma near the high–temperature boiling curve. *JETP Lett.*, **48**, 659–663.

Akkerman, A. F., Demidov, B. A., Fortov, V. E., *et al.* (1986a). *Application of the heavy–current relativistic electron beams in dynamic physics of high temperatures and pressures.* Joint institute of chemical physics, Chernogolovka.

Akkerman, A. F., Bushman, A. V., Demidov, B. A., Zavgorodnii, S. F., Ivkin, M. V., Ni, A. L., Petrov, V. A., Rudakov, L. I., and Fortov, V. E. (1986b). Effect of size of energy absorption zone on the characteristics of shock waves exited by strong relativistic electron beam in a metallic target. *JETP*, **64**, 1043–1045.

Alekseev, Y. L., Ratnikov, V. P., and Rybakov, A. P. (1971). Shock wave adiabats of porous metals. *J. Appl. Mech. Techn.Phys.*, **12**, No. 2, 101–106.

Altshuler, L. V. (1965). Use of shock waves in high–pressure physics. *Sov. Phys. Uspekhi*, **8**, 55–91.

Altshuler, L. V., Moiseev, B. P., Simakov, G. V., and Trunin, R. F. (1968). Relative compressibility of iron and lead at pressures of 31 to 34 Mbar. *JETP*, **27**, 420–422.

Altshuler, L. V., Bakanova, A. A., Bushman, A. V., Dudoladov, I. P., and Zubarev, V. N. (1977). Evaporation of shock–compressed lead in release waves. *JETP*, **46**, 980–983.

Altshuler, L. V., Bushman, A. V., Zhernokletov, M. V., Zubarev, V. N., Leontiev, A. A., and Fortov, V. E. (1980). Release isentropes and the equation of state of metals at high energy densities. *JETP*, **51**, 373–383.

Anisimov, S. I., Prokhorov, A. M., and Fortov, V. E. (1984). Application of high–power lasers to study matter at ultrahigh pressures. *Sov. Phys. Uspekhi*, **27**, 181–205.

Asay, J. R., Chhabildas, L. C., Kerley, G. I., and Trucano, T. G. (1987). High pressure strength of shocked aluminum. In *Shock waves in condensed matter,* Shmidt, S. C. and Holmes, N. C. (eds), pp. 145–149. Elsevier, New York.

Avrorin, E. N., Vodolaga, B. K., Voloshin, N. P., Kovalenko, G. V., Kuropatenko, V. F., Simonenko, V. A., and Chernovolyuk, B. T. (1987). Experimental study of the influence of electron shell structure on shock adiabats of condensed materials. *JETP*, **66**, 347–354.

Avrorin, E. N., Vodolaga, B. K., Simonenko, V. A., *et al.* (1990). *Intense shock waves and extremal states of matter.* IVTAN, Moscow.

Bazanov, O. V., Bespalov, V. E., Zharkov, A. N., Rumyantsev, B. V., Fedotova, T. B., Fortov, V. E., and Misonochikov, A. L. (1985). Irregular reflection of conically converging shock waves in plexiglas and copper. *High Temp.*, **23**, 976–982.

Bespalov, V. E., Gryaznov, V. K., Dremin, A. N., and Fortov, V. E. (1975). Dynamic compression of nonideal argon plasma. *JETP*, **42**, 1046–1049.

Bespalov, V. E., D'yachkov, L. G., Kobzev, G. A., and Fortov, V. E. (1979a). Radiation from high–pressure air plasmas. *High Temp.*, **17**, 226–231.

Bespalov, V. E., Gryaznov, V. K., and Fortov, V. E. (1979b). Radiation emitted by a shock–compressed high–pressure argon plasma. *JETP*, **49**, 71–75.

Bespalov, V. E., Kulish, M. I., and Fortov, V. E. (1986). Shift of spectral lines of argon in nonideal plasma. *High Temp.*, **24**, 995–997.

Bogolyubskii, S. L., Gerasimov, B. P., Liksonov, V. I., Popov, Y. P., Rudakov, L. I., Samarskii, A. A., Smirnov, V. P., and Urutskoev, L. I. (1976). Heating of thin foils by a large–current electron beam. *JETP Lett.*, **24**, 178–181.

Bondarenko, Y. A., Burdonskii, I. N., Gavrilov, V. V., ZhuZhukalo, E. V., Koval'skii, A. N., Kondrashov, V. N., Mkhitar'yan, L. S., Pergament, M. I., and Yaroslavskii, A. I. (1981). Experimental study of the acceleration of thin foils by high–power laser pulses. *JETP*, **54**, 85–89.

Burnett, N. H., Josin, G., Ahlborn, B., and Evans, R. (1981). Generation of shock waves by hot electron explosions driven by a CO_2 laser. *Appl. Phys. Lett.*, **38**, 226–228.

Bushman, A. V. and Fortov, V. E. (1983). Model equations of state. *Phys. Uspekhi*, **26**, 465–496.

Bushman, A. V. and Fortov, V. E. (1988). Wide–range equation of state for matter under extreme conditions. *Sov. Tech. Rev. B. Therm. Phys.*, **1**, 162–181.

Bushman, A. V., Lomakin, B. N., Sechenov, V. A., and Sharipdzhanov, I. I. (1975). Thermodynamics of nonideal cesium plasma. *JETP*, **42**, 828–831.

Bushman, A. V., Krasyuk, I. K., Pashinin, P. P., Prokhorov, A. M., Ternovoi, V. Y., and Fortov, V. E. (1984). Dynamic compressibility and thermodynamics of a dense aluminum plasma at megabar pressures. *JETP Lett.*, **39**, 411–413.

Bushman, A. V., Glushak, B. L., Gryaznov, V. K., Zhernokletov, M. V., Krasyuk, I. K., Pashinin, P. P., Prokhorov, A. M., Ternovoi, V. Y., Filimonov, A. S., and Fortov, V. E. (1986a). Shock compression and adiabatic decompression of a dense bismuth plasma at extreme thermal energy densities. *JETP Lett.*, **44**, 480–483.

Bushman, A. V., Glushak, B. L., Gryaznov, V. K., Zhernokletov, M. V., Krasyuk, I. K., Pashinin, P. P., Prokhorov, A. M., Ternovoi, V. Y., Filimonov, A. S., and Fortov, V. E. (1986b). *Shock compression and adiabatic unloading of a dense plasma of bismuth at extremal concentrations of thermal energy.* Joint Institute of Chemical Physics, Chernogolovka.

Bushman, A. V., Vorob'ev, O. Y., and Fortov, V. E. *et al.* (1987). *Computational modelling of the action of a strong ion beam on metal targets.* Joint Institute of Chemical Physics, Chernogolovka.

Bushman, A. V., Kanel', A. V., Ni, A. L., *et al.* (1988). *Thermophysics and dynamics of intense pulse actions.* Joint Institute of Chemical Physics, Chernogolovka.

Christian, R. H. and Yarger, F. L. (1955). Equation of state of gases by shock wave experiments. 1. Experimental method and the Hugoniot of argon. *J.*

Chem. Phys., **33**, 2042–2044.

Davison, L. and Graham, R. A. (1979). Shock compression of solids. *Phys. Rep.*, **55**, 255–379.

de Beaumont, P. and Leygonie, L. J. (1970). Vaporization of uranium after shock loading. In *Proceedings of the 5th International Symposium on Detonation*. Pasadena, pp. 430–442.

Deal, W. E. (1957). Shock Hugoniot of air. *J. Appl. Phys.*, **28**, 782–794.

Demidov, B. A. and Martynov, A. I. (1981). Experimental investigation of shock waves excited in metals by an intense relativistic electron beam. *JETP*, **53**, 374–377.

Duderstadt, J. J. (1982). *Inertial confinement fusion*. Wiley, New York.

Duvall, G. E. and Graham, R. A. (1977). Phase transitions under shock–wave loading. *Rev. Mod. Phys.*, **49**, 523–579.

Ebeling, W., Kreft, W. D., and Kremp, D. (1976). *Theory of bound states and ionization equilibrium in plasmas and solids*. Akademie–Verlag, Berlin.

Ebeling, W., Förster, A., Fortov, V., Gryaznov, V., and Polishchuk, A. (1991). *Thermophysical properties of hot dense plasmas*. Teubner, Stuttgart–Leipzig.

Fortov, V. E. (1972a). Calorific equation of state of silicon oxide and silicone fluid. *Comb., Expl., Shock Waves*, **8**, 428–433.

Fortov, V. E. (1972b). Equations of state of condensed matters. *J. Appl. Mech. Techn. Phys.*, **13**, No. 6, 156–166.

Fortov, V. E. (1982). Dynamic methods in plasma physics. *Sov. Phys. Uspekhi*, **25**, 781–809.

Fortov, V. E. and Dremin, A. N. (1973). Determination of temperature of shock–compressed cooper by measuring of parameters in release wave. *Comb., Expl., Shock Waves*, **9**, 743–781.

Fortov, V. E. and Krasnikov, Yu. G. (1970). Construction of a thermodynamically complete equation of state of a nonideal plasma by means of dynamic experiments. *JETP*, **32**, 897–902.

Fortov, V. E. and Leont'ev, A. A. (1976). Kinetics of vaporization and condensation with isentropic expansion of metals. *High Temp.*, **14**, 634–639.

Fortov, V. E., Leont'ev, A. A., Dremin, A. N., and Pershin, S. V. (1974). Isentropic expansion of shock–compressed lead. *JETP Lett.*, **20**, 13–14.

Fortov, V. E., Dremin, A. N., and Leont'ev, A. A. (1975b). Evaluation of the parameters of the critical point. *High Temp.*, **13**, 984–992.

Fortov, V. E., Ivanov, Yu. V., Dremin, A. N., Gryaznov, V. K., and Bespalov, V. E. (1975a). Explosive generator of nonideal plasma. *DAN USSR*, **221**, 1307–1309.

Fortov, V. E., Leont'ev, A. A., Dremin, A. N., and Gryaznov, V. K. (1976). Shock–wave production of a nonideal plasma. *JETP*, **44**, 116–122.

Froshner, K. F. and Lee, R. S. (1984). (private communication).

Gathers, G. R., Shaner, J. W., and Hodson, W. M. (1979). Thermodynamic characterization of liquid metals at high temperature by isobaric expansion measurements. *High Temp.–High Press.*, **11**, 529–538.

Glushak, B. L., Zharkov, A. N., Zhernokletov, M. V., Ternovoi, V. Y., Filimonov, A. S., and Fortov, V. E. (1989). Experimental investigation of the thermodynamics of dense plasma formed from metals at high energy concentrations. *JETP*, **69**, 739–749.

Gryaznov, V. K., Iosilevskii, I. L., and Fortov, V. E. (1973). Calculation of shock adiabats of argon and xenon. *J. Appl. Mech. Techn. Phys.*, **14**, No. 3, 70–76.

Gryaznov, V. K., Zhernokletov, M. V., Zubarev, V. N., Iosilevskii, I. L., and Fortov, V. E. (1980). Thermodynamic properties of a nonideal argon or xenon plasma. *JETP*, **51**, 288–295.

Hall, T. A., Djaoui, A., Eason, R. W., Jackson, C. L., Shiwai, B., Rose, S. L., Cole, A., and Apte, P. (1988). Experimental observation of ion correlation in a dense laser–produced plasma. *Phys. Rev. Lett.*, **60**, 2034–2037 .

Hess, H. (1989). (private communication).

Hornung, K. and Michel, K. W. (1972). Equation–of–state data of solids from shock vaporization. *J. Chem. Phys.*, **56**, 2072–2078.

Iakubov, I. T. (1977). Thermal instability of nonideal current–carrying plasmas of metal vapors. *Beitr. Plasma Phys.*, **17**, 221–227.

Ikezi, H., Schwarzenegger, K., Simons, A. L., Passner A. L., and McCall, S. L. (1978). Optical–properties of expanded fluid mercury. *Phys. Rev. B*, **18**, 2494–2499.

Ivanov, Y. V., Mintsev, V. B., Fortov, V. E., and Dremin, A. N. (1976). Electric conductivity of nonideal plasma. *JETP*, **44**, 112–116.

Keeler, R. N., van Thiel, M., and Alder, B. J. (1965). Corresponding states at small interatomic distances. *Physica*, **31**, 1437–1440.

Kerley, G. I. and Wise, J. L. (1987). Shock-induced vaporization of porous aluminum. In *Shock waves in condensed matter*. Shmidt, S. C. and Holmes, N. C. (eds), pp. 155–158. Elsever, New York.

Kirzhnits, D. A., Lozovik, Y. E., and Shpatakovskaya, G. V. (1975). Statistical model of matter. *Sov. Phys. Uspekhi*, **18**, 3–48.

Kondratenko, M. M., Lebedev, U. F., Ostashev, V. E., Safonov, V. I., Fortov, V. E., Ul'yanov, A. V. (1988). Experimental investigation of magneto–plasma acceleration of a dielectric plunger in a railotron. *High Temp.*, **26**, 159–164.

Kormer, S. B. (1968). Optical study of the characteristics of shock–compressed condensed dielectrics. *Sov. Phys. Uspekhi*, **11**, 229–254.

Kormer, S. B., Urlin, V. D. and Popova, L. T. (1961). Interpolation equation of state and its application to description experimental data on shock compression of metals. *Phys. Solid State*, **3**, 2131–2141.

Kormer, S. B., Funtikov, L. I., Urlin, V. D., and Kolesnikova, A. N. (1962). Dynamic compression of porous metals and the equation of state with variable specific heat at high temperatures. *JETP*, **42**, 686–702.

Kunavin, A. G., Kirillin, A. V., and Korshunov, Y. S. (1974). Investigation of cesium plasma by method of adiabatic compression. *High Temp.*, **12**, 1302–1305.

Leont'ev, A. A. and Fortov, V. E. (1974). About wave of melting and evaporation of metals in release wave. *J. Appl. Mech. Techn. Phys.*, **15**, No. 3, 162–167.

Leont'ev, A. A., Fortov, V. E., and Dremin, A. N. (1977). *Combustion and explosion.* Nauka, Moscow.

Lepouter, M. (1965). *Metal ammonia solutions.* Academic Press, New York.

Likal'ter, A. A. (2000). Critical points of condensation in Coulomb systems. *Phys. Uspekhi*, **43**, 777–797.

Lomakin, B. N. and Fortov, V. E. (1973). Equation of state of a nonideal cesium plasma. *JETP*, **36**, 48–53.

Lomakin, B. N. and Lopatin, A. D. (1983). Electrical conductivity of compressed vapors of cesium and potassium. *High Temp.*, **21**, 190–193.

Lysne, P. C. and Hardesty, D. R. (1973). Fundamental equation of state of liquid nitromethane to 100 kbar. *J. Chem. Phys.*, **59**, 6512–6523.

Manzon, B. M. (1981). Acceleration of macro–particles for controlled fusion. *Sov. Phys. Uspekhi*, **24**, 611–614.

Minaev, V.N. and Ivanov, A. G. (1976). Electromotive force produced by shock compression of a substance. *Sov. Phys. Uspekhi*, **19**, 400–419.

Mintsev, V. B. and Fortov, V. E. (1979). Electric conductivity of xenon under supercritical conditions. *JETP Lett.*, **30**, 375–378.

Mintsev, V. B., Fortov, V. E., and Gryaznov, V. K. (1980). Electric conductivity of a high–temperature nonideal plasma. *JETP*, **52**, 59–63.

Model', I. T. (1957). Measurement of high temperatures in strong shock waves in gases. *JETP*, **32**, 589–601.

Model', I. S., Narozhnyi, A. T., Kharchenko, A. T., Kholin, S. A., and Khrustalev, V. V. (1985). Equation of state for graphite, aluminum, titanium, and iron at pressures > 13 Mbar. *JETP Lett.*, **41**, 332–334.

More, R. M. (1981). Laser–driven shockwave at extreme high pressures. In *Laser interaction and related plasma phenomena. Vol. 5*, Schwarz, H. J., Hora, H., Lubin, M. J., and Yaakobi, B. (eds), pp. 253–277. Plenum, New York.

Nellis, W. J., Ross, M., Mitchell, A. C., van Thiel, M., Young, D. A., Ree, F. H., and Trainor, R. J. (1983). Equation of state of molecular hydrogen and deuterium from shock–wave experiments to 760 kbar. *Phys. Rev. A*, **27**, 608–611.

Nellis, W. J., van Thiel, M., and Mitchell, A. C. (1982). Shock compression of liquid xenon to 130 GPa (1.3 Mbar). *Phys. Rev. Lett.*, **48**, 816–818.

Ng, A., Parfeniuk, D., and DaSilva, L. (1985). Hugoniot measurements for laser–generated shock waves in aluminum. *Phys. Rev. Lett.*, **54**, 2604–2607.

Ng, A., Parfeniuk, D., and Celliers, P., DaSilva, L., More, R. M., and Lee, Y. T. (1986). Electrical conductivity of a dense plasma. *Phys. Rev. Lett.*, **57**, 1595–1598.

Perry, F. C. and Widner, M. M. (1976). Energy deposition of superpinched relativistic electron beams in aluminum targets. *J. Appl. Phys.*, **47**, 127–134.

Pfeifer, H. P., Freyland, W. F., and Hensel, F. (1973). Absolute thermoelectric power of fluid cesium in metal–nonmetal transition range. *Phys. Lett. A*, **43**, 111–112.

Prokhorov, A. M., Anisimov, S. I., and Pashinin, P. P. (1976). Laser thermonu-clear fusion. *Sov. Phys. Uspekhi*, **19**, 547–560.

Ragan, C. E., Silbert, M. G., and Diven, B. C. (1977). Shock compression of molybdenum to 2.0 TPa by means of a nuclear explosion. *J. Appl. Phys.*, **48**, 2860–2870.

Rashleigh, S. C. and Marshall, R. A. (1978). Electromagnetic acceleration of macroparticles to high velocities. *J. Appl. Phys.*, **49**, 2540–2542.

Roman, J. P., Cottete, F., Hallouin, M., Fabbro, R., and Perin, H. (1986). Laser shock experiments at pressures above 100 Mbar. *Physica B*, **139**, 595–598.

Rosen, M. D., Phillipson, D. W., Price, R. H., Campbell, E. M., Obenschain, S. P., Whitlock, R. R., McLean, E. A., and Ripin, B. H. (1984). Creation of ultra high pressure shocks by the collision of laser accelerated disks: Experiment and theory. In *Shock waves in condensed matter–83,* Asay, J. R., Graham, R. A., and Straub, G. K. (eds), pp. 323–326. Elsever, New York.

Ross, M., Nellis, W., and Mitchell, A. (1979). Shock–wave compression of liquid argon to 910 kbar. *Chem. Phys. Lett.*, **68**, 532–536.

Ryabinin, Y. N. (1959). *Gases at high densities and high temperatures.* Fizmat-giz, Moscow.

Scidmore, C. and Morris, E. (1962). (unpublished).

Sechenov, V. A., Son, E. E., and Shchekotov, O. E. (1977). Electrical conduc-tivity of a cesium plasma. *High Temp.*, **15**, 346–349.

Shaner, J. W. and Gathers, G. R. (1979). (unpublished).

Shatzman, E. (1977). (private communication).

Simonenko, V. A., Voloshin, N. P., Vladimirov, A. S., Nagibin, A. P., Nogin, V. N., Popov, V. A., Vasilenko, V. A., and Shoidin, Y. A. (1985). Absolute measurements of shock compressibility of aluminum at pressures $p \geq 1$ TPa. *JETP*, **61**, 869–873.

Thom, K. and Schneider, R. T. (eds). (1971). *Research of uranium plasmas and their technological applications.* NASA, Washington.

Tkachenko, B. K., Titarov, S. I., Karasev, A. B. and Alipov, S. V. (1976). Experimental determination of parameters behind strong reflected shock waves in air. *Comb., Expl., Shock Waves*, **12**, 763–768.

Trainor, R. J., Graboske, M. C., and Long, K. S. (1978). (unpublished).

Trainor, R. J., Holmes, N. C., and Anderson, R. A. (1982). Ultrahigh pressure laser–driven shock wave experiments. In *Shock waves in condensed matter–81.* Nellis, W. J., Seaman, L., and Grolham, R. A. (eds), pp. 145–149. North Holland, New York.

Trainor, R. J., Shaner, J. W., Auerbach, J. M., and Holmes, N. C. (1979). Ultrahigh–pressure laser–driven shock–wave experiments in aluminum. *Phys. Rev. Lett.*, **42**, 1154–1157.

Trunin, R. F., Podurets, M. A., Moiseev, B. N., Simakov, G. B., and Popov, L. V. (1969). Relative compressibility of copper, cadmium, and lead at high pressures. *JETP*, **29**, 630–631.

Trunin, R. F., Podurets, M. A., Simakov, G. B., Popov, L. V., and Moiseev, B. N. (1972). An experimental verification of the Thomas–Fermi model for metals under high pressure. *JETP*, **35**, 500–552.

Tsukulin, M. A. and Popov, E. G. (1977). *Radiative properties of shock waves in gases*. Nauka, Moscow.

Van Kessel, C. G. M. and Sigel, R. (1974). Observation of laser–driven shock waves in solid hydrogen. *Phys. Rev. Lett.*, **33**, 1020–1023.

Veeser, L. R. and Solem, J. C. (1978). Studies of laser–driven shock waves in aluminum. *Phys. Rev. Lett.*, **40**, 1391–1394.

Vladimirov, A. S., Voloshin, N. P., Nogin, V. N., Petrovtsev, A. V., and Simonenko, V. A. (1984). Shock compressibility of aluminum at $p \geq 1$ Gbar. *JETP Lett.*, **39**, 82–85.

Volkov, L. P., Voloshin, N. P., Mangasarov, R. A., Simonenko, V. A., Sin'ko, G. V., and Sorokin, V. L. (1980a). Shock compressibility of water at pressure of ~ 1 Mbar. *JETP Lett.*, **31**, 513–515.

Volkov, L. P., Voloshin, N. P., Vladimirov, A. S., Nogin, V. N., and Simonenko, V. A. (1980b). Shock compressibility of aluminum at pressure of 10 Mbar. *JETP Lett.*, **31**, 588–592.

Volkov, V. A., Titarov, S. I., and Tkachenko, B. K. (1978). Study of argon at high particle concentrations. *High Temp.*, **16**, 342–344.

Zababakhin, E. I. (1970). Unbounded cumulation effects. In *Mechanics in USSR for 50 years*. Nauka, Moscow.

Zaporozhets, Yu. B., Mintsev, V. B., and Fortov, V. E. (1987). Creation of the metal phase during compression of silicon by shock waves. *JTP Lett.*, **13**, 204–207.

Zel'dovich, Y. B. and Raizer, Y. P. (1966). *Physics of shock waves and high–temperature hydrodynamic phenomena*. Academic Press, Dover–New York.

Zhernokletov, M. V., Zubarev, V. N., Sutulov, Y. N. (1984). Adiabats of porous samples and isentropes of expansion of continuum matter. *J. Appl. Mech. Techn. Phys.*, **25**, No. 1, 119–123.

IONIZATION EQUILIBRIUM AND THERMODYNAMIC PROPERTIES OF WEAKLY IONIZED PLASMAS

4.1 Partly ionized plasma

4.1.1 *Classical low–temperature plasma*

Let us consider a degenerate low–temperature plasma, for which the following inequality is valid:

$$\beta \text{Ry} \gg 1, \quad \text{Ry} = me^4/2\hbar^2, \quad \beta = 1/kT, \tag{4.1}$$

where Ry is the ionization energy of the hydrogen atom. Such a plasma is classical, because inequality (4.1) implies that $e^2\beta \gg \lambda$. The quantity $e^2\beta$ is the mean scattering amplitude (Landau's length), which is equal to within an order of magnitude to the square root of the mean effective scattering cross–section. Therefore, not only does the system follow classical statistics, $n_e\lambda^3 \ll 1$, but the motion of individual particles in the system may be described within the framework of classical mechanics.

From the standpoint of problems treated in this book, partly ionized plasma is of most interest. It is also referred to as three–component plasma, bearing in mind the atoms and free electrons and ions that remain unbound. The emergence of atoms in the initial electron–ion system is due to the electron/ion interaction which, if $\beta \text{Ry} \gg 1$, becomes strong even at relatively low values of the density. This interaction results in the formation of bound electron–ion pairs, i.e., atoms.

It is important that the electron and ion bound in an atom are spaced from each other to a distance of the order of Bohr's radius a_0, which is the least characteristic dimension in the system, $a_0 \ll n^{-1/3}$, $a_0 \ll e^2\beta$, $a_0 \ll \lambda$. Therefore, the electron and ion, which make up an atom, shield one another such that, to a good approximation, the interaction of such pairs with surrounding particles may be fully ignored.

4.1.2 *Three–component electron–ion–atomic plasma*

The concentrations of components are related by the condition of ionization equilibrium. Let us write this condition – the Saha's equation and first corrections to it. This may be done within the second virial approximation developed in the initial electron–ion system.

The second virial correction to Gibbs' thermodynamic potential of an electron–ion system $\beta\Delta\Omega$ has the form (Vedenov and Larkin 1959; Kopyshev 1968)

$$-\beta \Delta \Omega = \frac{1}{2} \sum_{i,j} \xi_i \xi_j \left(\frac{2\pi \hbar^2 \beta}{m_{ij}} \right)^{3/2} [Z_{ij}(e^2) - Z_{ij}(0)],$$

$$Z_{ij} = \frac{1}{g_i g_j} \sum_{\nu} e^{-\beta E_{ij\nu}},$$

(4.2)

where ξ_i and ξ_j denote the fugacity of the plasma particles,

$$\xi_i = g_i e^{\beta \mu_i} \left(\frac{M}{2\pi \hbar^2 \beta} \right)^{3/2}, \quad \xi_j = g_j e^{\beta \mu_j} \left(\frac{m}{2\pi \hbar^2 \beta} \right)^{3/2}.$$

Here μ_j, μ_i, g_j, g_i, m, M denote their chemical potentials, statistical weights, and masses, respectively, m_{ij} is the reduced mass, ν is the quantum number of relative motion of the i– and j–particles, and $E_{ij\nu}$ stands for the corresponding energy of continuous and discrete spectra.

We retain in (4.2) the major term under condition $\beta \mathrm{Ry} \gg 1$, and write $\beta \Delta \Omega$ in the form

$$-\beta \Delta \Omega = \xi_i + \xi_e + \xi_i \xi_e (2\pi \hbar^2 \beta / m_{ei})^{3/2} (g_a / g_i g_e) \exp(\beta E_1).$$

(4.3)

Here $m_{ei} = mM(M + m)^{-1}$ and g_a is the statistical weight of the state with energy $E_1 = \mathrm{Ry}$.

Now we proceed to the three–component description of the plasma, by taking atoms into account. For this purpose, we define the activity of atoms by the following expression:

$$\xi_a = \exp(\beta \mu_a)[(M + m)/2\pi \hbar^2 \beta]^{3/2} g_a \exp(\beta E_1),$$

and then use the condition of chemical equality $\mu_a = \mu_e + \mu_i$, where μ_a is the chemical potential of the new particle. Then, we derive, instead of (4.3) the expression

$$-\beta \Delta \Omega = \sum_j \xi_j,$$

(4.4)

describing a three–component ideal plasma ($j = \mathrm{e, i, a}$). By using the formula $n_j = -\xi_j \partial(\beta \Delta \Omega)/\partial \xi_j$, this yields the particle density, $n_j = \xi_j$, and the equation of ionization equilibrium – Saha's formula,

$$\frac{n_e n_i}{n_a} = \frac{g_e g_i}{g_a} \left(\frac{m_{ie}}{2\pi \hbar^2 \beta} \right)^{3/2} e^{-\beta \mathrm{Ry}}.$$

Because an atom and an ion are multilevel systems, it is natural that g_i and g_a should be replaced by the corresponding internal partition functions, Σ^+ and

Σ. The ionization potential of a complex atom I is different from Ry, and the reduced mass is $m_{ei} \simeq m$. Saha's equation has the final form

$$\frac{n_e n_i}{n_a} = \frac{2\Sigma^+}{\Sigma} \left(\frac{m}{2\pi\hbar^2\beta}\right)^{3/2} e^{-\beta I}, \tag{4.5}$$

where $\sum = \sum_k g_{ka} \exp(-\beta E_{ka})$ and $\sum^+ = \sum_{ki} g_{ki} \exp(-\beta E_{ki})$. The sums are taken over all bound states ofthe atom (ion), where E_{ka} and E_{ki} are the energies and g_{ka} and g_{ki} are the statistical weights of the kth state of the atom and ion, respectively.

However, the three–component model in this form is inadequate. First, the contribution to the second virial coefficient due to the interaction between free electrons and ions – which was not taken into account – diverges. Second, the partition functions over bound states diverge as well. This is because of the long–range nature of the Coulomb potential.

4.1.3 Second virial coefficient and partition function of atom

Let us discuss the origin of the divergence of the second virial coefficient corresponding to interactions between free electrons and ions at large distances,

$$T\sum_{k,j} \xi_k \xi_j \int r^2 dr(e^{-e_k e_j \beta/r} - 1) \sim T\sum_{k,j} \int (e_k e_j \beta)^2 dr/2,$$

where k, j assume the values of e, i. This divergence is eliminated if one takes into account the many–particle aspect of the interaction. For this purpose, integration must be restricted to a distance less than the Debye length r_D. As a result, we have the Debye–Hückel correction $\Delta(\beta\Omega)_D$,

$$\beta\Delta\Omega_D = (12\pi)^{-1}(4\pi\beta e^2 \sum_j \xi_j)^{3/2}, \quad j = e, i. \tag{4.6}$$

If (4.6) is included in constructing the ionization equilibrium, the following correction (Likal'ter 1969) appears to the ionization potential in (4.5):

$$\Delta I = -2kT \ln[1 + \Gamma\varphi(\Gamma)/2],$$

where $\Gamma = e^2/r_D kT$ is the Debye parameter of the interaction and the function $\varphi(\Gamma)$ is derived from the expression $\Gamma = 2(1 - \varphi^2)/\varphi^3$. Thus, the ionization potential decreases.

The sum over bound states

$$\sum = \sum_{k=1}^{\infty} g_k \exp(-\beta E_k)$$

diverges at small energies, which are hydrogen–like, $E_k = -\text{Ry}k^{-2}$ and $g_k = k^2$. This divergence, occurring in a discrete spectrum, is compensated, in the case of correct inclusion of the electron/ion interaction, by similar divergence in a continuous spectrum (Vedenov and Larkin 1959; Vorob'ev 2000). Rather than calculating $\sum_{k=1}^{\infty} g_k \exp(-\beta E_k)$, one should consider the quantity

$$\xi_i \xi_e \left\{ \left(\frac{2\pi \hbar^2 \beta}{m} \right)^{3/2} \sum_k g_k e^{-\beta E_k} - \int d\mathbf{r} \left[1 + \frac{\beta e^2}{r} + \frac{1}{2} \left(\frac{\beta e^2}{r} \right)^2 \right] \right\}.$$

This quantity emerges if the terms that were already included in expression (4.6) are subtracted from the electron–ion component of the virial correction. The first of those terms cancel out in deriving (4.6) due to the plasma electroneutrality, and the second term has already made its contribution to (4.6). As a result of integration, a new (renormalized) partition function emerges (for which we use Σ as before). Thus

$$\sum = \sum_{k=1}^{\infty} g_k [\exp(-\beta E_k) - 1 + \beta E_k]. \tag{4.7}$$

One can see that $\sum$ converges when the energy $|E_k|$ is of the order of the thermal energy.

The problem of eliminating the divergence has been long discussed in the literature. A rigorous result was derived by Vedenov and Larkin (1959). A correct description of the solution of this problem can be found in (Vorob'ev 2000; Larkin 1960; Kraeft et al. 1985). Expression (4.7) is valid, provided the second virial approximation is sufficient. Under these conditions, the values of temperature exceed considerably the characteristic values of the interaction energy. Otherwise, no such correct solution is available at present.

The second virial coefficient of the plasma is calculated in a number of studies (see Vorob'ev 2000; Larkin 1960; Kraeft et al. 1985 and references therein). In addition to the main components discussed above, this coefficient includes high–order density terms such as $n^2 \ln n$ (this term vanishes in a symmetrical plasma), n^2, $n^{5/2} \ln n$, $n^{5/2}$.

Quantum corrections appear in a term that is quadratic with respect to density. They are proportional to λ/r_D and hence are minor, because $\lambda/r_D \sim \gamma^{3/2}(\beta \mathrm{Ry})^{-1/2}$.

4.2   Anomalous properties of a metal plasma

When the degree of ionization is low, the nonideality of the plasma may only lead to small corrections to the thermodynamic properties of the material. However, its effect on the free electron concentration may be very strong. The electrical conductivity, for instance, varies by orders of magnitude as a result of variation of the number of carriers rather than of their mobility. Below we discuss the basic theoretical principles, detailed description can be found in Alekseev and Iakubov (1983); and Alekseev et al. (1981).

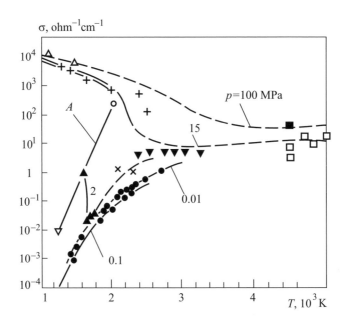

FIG. 4.1.   Electrical conductivity of cesium. Measurements on isobars: • — $p = 0.01$
and 0.1 MPa (Ermokhin *et al.* 1971); ▲ — $p = 2$ MPa (Alekseev *et al.* 1975); × —
$p = 2$ MPa (Renkert *et al.* 1971); ▼ — $p = 2$ MPa (Isakov and Lomakin 1979); + —
$p = 15$ MPa (Iermokhin *et al.* 1978); □ — $p = 15$ MPa (Sechenov *et al.* 1977); Δ —
$p = 100$ MPa (Borzhievsky *et al.* 1987); ■ — $p = 100$ MPa (Iermokhin *et al.* 1978);
∇ — $p = 0.5$ MPa (Borzhievsky *et al.* 1987). Solid lines are plotted over the measured
points, dashed lines are guides to the eye. A — conductivity on the phase coexistence
curve (Renkert *et al.* 1971); ○ — vicinity of the critical point (Freyland 1979).

4.2.1 *Physical properties of metal plasma*

Fig. 4.1 plots the electrical conductivity of cesium versus temperature in the
entire range of pressures and temperatures, and Fig. 4.2 shows the data on
thermal e.m.f. Let us now discuss the salient features.

A plasma may be considered ideal at low pressures, for instance, at $p =
0.01$ MPa $\ll p_c$. Therefore, at low temperatures the electrical conductivity grows
with temperature because the electron concentration rises in accordance with
Saha's equation, $n_e \sim \exp[-I/(2kT)]$. At high temperatures, the degree of ionization increases and collisions with ions begin to prevail over those with atoms.
In this case, the conductivity is defined by Spitzer's formula and has the power
dependence on temperature, $\sigma \sim T^{3/2}/\ln\Lambda$, where $\ln\Lambda$ is the "Coulomb" loga-

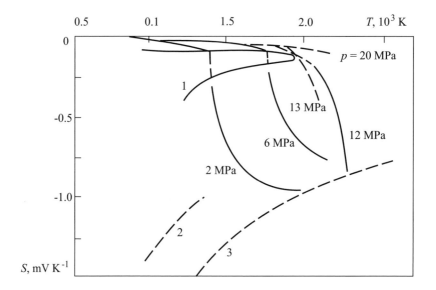

FIG. 4.2. Isobars of the thermal e.m.f. of cesium: $p = 2$ and 6 MPa (Iakubov and Likal'ter 1987); $p = 12$ MPa (Alekseev *et al.* 1975); $p = 13$ and 20 MPa (Pfeifer *et al.* 1973): 1 — thermal e.m.f. on the phase coexistence curve; 2 and 3 — limiting relations (6.37) and (6.38), respectively.

rithm.

We can say that the $p = 0.1$ MPa isobar separates the regions of ideal and nonideal cesium plasma. At lower pressures, the electrical conductivity grows with decreasing pressure, $(d\sigma/dp)_T < 0$. At $p > 0.1$ MPa, in contrast, σ increases with pressure, $(d\sigma/dp)_T > 0$. At pressures above the critical, the electrical conductivity varies monotonically with temperature. Upon transition from a liquid metal to a plasma, the electrical conductivity decreases upon heating. Especially sharp is decrease of σ in the $T \sim T_c$ region at pressures comparable with p_c, for example, $p = 15$ MPa. The metal–nonmetal transition occurs which results in the emergence of a nonideal plasma. Its electrical conductivity upon heating passes through a minimum. Further heating leads to an increase of the ionization fraction and the conductivity tends to the Spitzer dependence.

At high temperatures, all isobars sooner or later approach the Spitzer dependence $\sigma_{Sp}(T)$ (see Chapter 6), because the nonideality decreases under conditions of strong heating. The σ_{Sp} values depend on p very weakly (via the Coulomb logarithm).

The most interesting is the dependence $\sigma(T)$ at $p < p_c$, for instance, at $p = 2$ MPa. This isobar has common features both with subcritical ($p \ll p_c$) and supercritical isobars. On the saturated vapor curve, the conductivity reaches anomalously high values (Fig. 2.9) which are five to six orders of magnitude higher than the results of conventional ideal–gas estimates. However, a slight

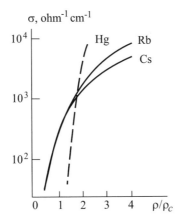

FIG. 4.3. Isotherms of the electrical conductivity of cesium, rubidium, and mercury, $T = 1.025\,T_c$ (Hensel 1976; Hensel and Warren 1999).

temperature increase leads to a sharp drop in conductivity. Upon heating, the electrical conductivity on the isobar passes through a minimum at some value of temperature, and then starts growing and tends to Spitzer values.

Let us now turn to the discussion of the thermal e.m.f. of cesium, S, as shown in Fig. 4.2. In ideal weakly ionized plasmas the thermal e.m.f. has very weak dependence on p and is close to $S = -I/(2eT)$, i.e., it increases monotonically upon heating. At high temperatures, all S isobars should approach this dependence because the plasma becomes ideal.

Starting from the values typical for liquid metals, isobars of thermal e.m.f. fall off sharply with heating (Fig. 4.2). In the nonideal plasma region, upon heating the thermal e.m.f. passes through a minimum and tends to a $[-I/(2eT)]$ dependence.

Therefore, the experimental results obtained for the electrical conductivity and thermal e.m.f. of nonideal cesium plasma do not fit the ideal–gas approximation limits (conventional for plasma physics). Equally inapplicable is the "almost free" electron approximation, which is successfully used in the theory of the liquid metal state.

4.2.2   Lowering of the ionization potential

The electrical conductivity on the supercritical isobars decreases by orders of magnitude with a small temperature increase. Under these conditions, one can speak about an exponential density dependence of the electrical conductivity (Fig. 4.3). This was interpreted by Vedenov (1968), and Alekseev and Vedenov (1970) as a result of the electron concentration growth due to a decrease of the ionization potential.

The lowering of the ionization potential, ΔI, is the sum of two quantities, $\Delta I = \Delta I_1 + \Delta I_2$. First of all, the free electron energy decreases, due to the strong polarization interaction with neutral atoms, by the quantity $\Delta I_1 = n_a \hbar^2 \sqrt{\pi q}/2m$. Here, q is the total electron–atom scattering cross–section and n_a is the atomic concentration (see Eq. 1.12). Second, the ion energy in a gas is also reduced by the quantity $\Delta I_2 = 2\pi n_a \alpha e^2/r_a$ due to the polarization of the neighboring atoms (see Eq.1.10). The decrease of ionization potential leads to an exponential growth of the electron concentration with density (in this section, we take the absolute value of ΔI). Saha's equation gives

$$n_e \sim \exp[-I_m/(2kT)], \quad I_m = I - \Delta I_1 - \Delta I_2. \tag{4.8}$$

Let us assume that the electron mobility has a weak dependence on density. Then the electrical conductivity, following variations of n_e, grows exponentially with density,

$$\sigma \sim \exp[-(I/2kT) + (n_a/n_a^*)]. \tag{4.9}$$

Here n_a^* is the density value starting from which this growth becomes substantial. These formulas yield the following expression for n_a^*:

$$(n_a^*)^{-1} = (2mkT\hbar^2/\sqrt{\pi q})^{-1} + (kT r_a/2\pi\alpha e^2)^{-1}. \tag{4.10}$$

The calculated n_a^* (4.10) is in good agreement with the experimental value, $n_a^* \simeq 1.7 \cdot 10^{20}$ cm^{-3} ($\alpha = 400\,a_0^3$, $r_a = 5a_0$, $q = 2 \cdot 10^{-13}$ cm^2). Smirnov (1970) and Popielawski and Gryko (1977) independently drew the same conclusion.

It should be noted that the mean field approximation used by these authors to calculate ΔI ignores the correlation of atoms in the neighborhood of charged particles. This is valid if the depth of the charge–neutral interaction potential, g, is not too large. However, the value of g/kT is in fact much larger than unity. For estimation, one can use D_2^+ (the bond energy of Cs_2^+ or Hg_2^+ ions) or the polarization well depth $\alpha e^2/(2r_a^4)$. It is easy to see that the correlation is substantial.

4.2.3   Charged clusters

Above, we have already discussed the necessity to take into account the correlation of atoms around charged particles. Indeed, in a dense medium the state of charged particles experiences radical changes. Because of attractive forces acting between the charged and neutral particles, charged clusters are formed. The clusters are stabilized due to the presence of charge. The structure of clusters depends strongly on the degree of nonideality. Let us follow the variation of the cluster structure in a vapor upon isobaric heating. On the saturation line ($T = T_s$) one observes charged droplets. For temperatures somewhat higher than T_s, these are less dense formations which can no longer be regarded as liquid but which are still in dynamic equilibrium with the vapor. At substantially higher temperatures, these are cluster ions containing several atoms and are bound by polarization forces. Finally, in a weakly nonideal plasma the cluster ions dissociate and there remain ordinary molecular ions. The existence of clusters might strongly affect the properties of matter. For instance, in saturated cesium vapors the thermal emission of electrons from droplets raises the degree of ionization and the electrical conductivity of the plasma by several orders of magnitude. This is responsible for the anomalously high electrical conductivity of high–temperature metallic vapors near the saturation line.

The clusters and similar particles were observed in different temperature ranges in numerous gases and liquids, see review by Khrapak and Yakubov (1979, 1981) and Khomkin et $al.$ (1995). Based on the review by Lagar'kov and Yakubov (1980), we shall discuss below the ionization equilibrium in a metal vapor plasma.

We shall start with the simplest case. Let us consider a plasma consisting of electrons, atoms A, ordinary ions A^+, and molecular ions A_2^+. The electron density is defined by Saha's equation

$$n^+ n_e / n_a \sim \lambda_e^{-3} \exp(-\beta I), \qquad (4.11)$$

and the quasineutrality condition,

$$n^+ + n_2^+ = n_e. \qquad (4.12)$$

The concentration of the molecular ion A_2^+ is given by the equation

$$n^+ n_a / n_2^+ \sim \lambda_a^{-3} \exp(-\beta D_2^+). \qquad (4.13)$$

In Eqs. (4.11)–(4.13), n^+, n_e, n_a, n_2^+ stand for the concentration of ordinary ions, electrons, atoms and molecular ions, respectively, λ_e and λ_a are the thermal wavelengths for the electron and atom, and D_2^+ is the bond energy of the atom in the ion A_2^+. Let us assume that the degree of ionization is low. Then, by excluding n_2^+ and n^+ from (4.13) with Eqs. (4.11) and (4.12), we obtain

$$n_e^2 = n_a k_1 \exp(-\beta I)[1 + n_a k_2 \exp(\beta D_2^+)], \qquad (4.14)$$

where k_1 is the constant of ionization equilibrium in $A \rightleftarrows A^+ + e$ and k_2 the constant of dissociation equilibrium in $A_2^+ \rightleftarrows A^+ + A$. For cesium, $k_2^{-1} = 1.45 \cdot 10^{21} T^{-1/2}$ cm^{-3} where T is expressed in eV and $D_2^+ \simeq 0.7$ eV. It follows from Eq. (4.14) that the inclusion of the molecular ions led to a shift of ionization equilibrium towards increased electron concentration. The higher the molecular ion dissociation energy, the larger is the shift. Under conditions when $T = 1400$ K and $n_a = 10^{20}$ cm^3 (near the saturation point for cesium vapors at $p = 2$ MPa), the inclusion of Cs_2^+ ions gives an approximately fivefold increase of n_e. The electrical conductivity increases accordingly. The important role played by molecular ions in alkali plasmas was confirmed by measurements of the electrical conductivity of sodium vapors near the saturation line by Morrow and Craggs (1973).

The inclusion of Cs_2^+ ions is not sufficient at high pressures and low temperatures, because under these conditions we have $n_a k_2 \exp(D_2^+ \beta) \gg 1$. This means that $n_2^+ \gg n^+$. One can naturally expect that a substantial number of Cs_3^+ ions is present in a plasma. Similarly to Eq. (4.13), for the $A_3^+ \rightleftarrows A_2^+ + A$ equilibrium we have the equation $n_3^+ = n_2^+ n_a k_3 \exp(\beta D_3^+)$. Taking these ions into account, we obtain for n_e^2 instead of Eq. (4.14),

$$n_e^2 = n_a k_1 \exp(-\beta I)[1 + n_a k_2 \exp(\beta D_2^+)$$
$$+ n_a^2 k_2 k_3 \exp(\beta D_2^+ + \beta D_3^+)]. \qquad (4.15)$$

Evidently, as the formation of more and more complex ions (referred to as cluster ions) is taken into account, subsequent terms will appear in the square brackets in Eq. (4.8). Assuming that $D_3^+ = D_2^+$ and $k_3 = k_2/2$, the right side of Eq. (4.15) can be regarded as the beginning of a series whose sum is expressed by the exponent. Such summation yields (Likal'ter 1978)

$$n_e^2 = n_a k_1 \exp[-\beta I + n_a k_2 \exp(\beta D_2^+)]. \qquad (4.16)$$

One of the first papers containing the discussion of heavy cluster ions and the summation similar to that performed above, was the paper by Magee and Funabashi (1959).

The procedure of summation in Eq. (4.15) is heuristic. Indeed, even the third term in the series, calculated with due regard for the real values of D_3^+ and k_3, does not correspond to the third term of the exponential series. Nevertheless, Eq. (4.16) represents the major features of the phenomenon. It describes the presence on the isobar of a sharp temperature minimum of free electron concentration and, thereby, of conductivity. Indeed, if $\sigma \propto n_e$, then

$$\sigma \simeq \sigma_0 \exp[-\beta I/2 + (n_a k_2/2) \exp(\beta D_2^+)], \qquad (4.17)$$

where σ_0 is some constant. The position of the minimum of electrical conductivity on the isobar approximately corresponds to the value of

$$T_{\min} \simeq D_2^+/k \big| \ln(n_a k_2 D_2^+/I) \big|.$$

If $p = 2$ MPa, then $T_{\min} = 1500$ K for cesium, which is in good agreement with experiment. The emergence of the minimum of electrical conductivity is due to the competition between the dissociation of cluster ions and the thermal ionization of atoms.

4.2.4   Thermodynamics of multiparticle clusters

As the nonideality becomes stronger, the number of particles in the cluster increases. Therefore, we shall approach the formation of ion clusters from the standpoint of macroscopic thermodynamics. Let an ion be placed in a gas with a mean atomic concentration n_a and temperature T. Then, if the density of particles attracted by the ion reaches a value corresponding to the loss of thermodynamic stability, the vapor–liquid phase transition occurs in the neighborhood of the ion, in the $r < R$ region. Let us now estimate the value of R. In a steady state, the gradient of the vapor pressure in the neighborhood of the ion must be balanced by the force of attraction of an atom by the ion located at the origin of coordinates,

$$(M/\rho)\,\text{grad}\,p = -dV/dr, \qquad (4.18)$$

where $V(r)$ is the potential of the ion–atom interaction and M the atomic mass. Given the equation of state for the vapor, one can integrate Eq. (4.18) to find the density distribution $n_a(r)$ in the neighborhood of the ion,

$$M\int\limits_{p}^{p(r)} \rho^{-1}(p,T)\,dp = -V(r). \qquad (4.19)$$

In an ideal gas, Eq. (4.19) gives the obvious result,

$$n_a(r) = n_a\exp[-\beta V(r)].$$

Near the saturation line the interatomic attraction is substantial, $p = n_a kT - an_a^2$, where a is the parameter characterizing the attraction in van der Waals' equation. In this case, Eq. (4.19) gives, after the integration,

$$n_a(r) = n_a\exp\{-\beta V(r) + 2\beta a n_a[n_a(r)/n_a - 1]\}. \qquad (4.20)$$

For $n_a = 10^{20}$ cm^{-3} and $T = 1400$ K at a distance from the ion equal to $R = 17a_0$, the density $n_a(R)$ exceeds the mean density by a factor of two. The number of particles in a drop of the calculated radius is close to 20. As the number of atoms in a complex increases, the valence electrons will be located on molecular orbitals around the entire complex and will form a band whose width increases as yet another atom is added.

Properties of an ion cluster containing a few dozens of particles are close to metallic. On the other hand, of course, it might still be far from having all the inner properties of a large metal sample (Montano *et al.* 1986). If equal numbers of positive and negative ion clusters were formed, no anomalous growth of free electron concentration would be observed. In fact, the electron–atom interaction is weaker than the ion–atom interaction (although their polarization asymptotes are identical). This is reflected even by the fact that the bond energy of a positive ion D_2^+ is substantially higher than that of a negative ion E. For cesium, $D_2^+ \simeq 0.7$ eV and $E = 0.47$ eV, and for mercury the bond energy of the ion is $D_2^+ \simeq 1$ eV while the mercury atom does not bind an electron at all. Therefore, it turns out that the number of positive ion clusters in a plasma is much higher than the number of negative clusters. Consequently, a large number of free electrons is present in the plasma. The electronic transport coefficients are proportional to the free electron concentration and, therefore, they are very sensitive to the concentration ratio between the positive and negative clusters.

Let us now consider how the emergence of ion clusters shifts the ionization equilibrium. We follow a simple model described by Khrapak and Yakubov (1981). This model takes into account the ion–atom interactions and ignores the interatomic and electron–atom interactions. Assume that F_0 is the free energy of a mixture of ideal gases of electrons, atoms, and ions. The free energy of the system, F (with due regard for interaction), is lower than F_0 by the quantity kTn^+g, where g is the number of ions in a cluster. This quantity is close to the excess number of atoms around the ion,

$$g = n_a \int d\mathbf{r}[\exp(-\beta V) - 1].$$

Thus,

$$F = F_0 - kTn^+ n_a \int d\mathbf{r}[\exp(-\beta V) - 1].$$

By implementing for F a conventional procedure of variation with respect to the concentrations of particles involved in the ionization reactions and taking into account the quasineutrality, we obtain the equation of ionization equilibrium,

$$n_e^2 = n_a k_1 \exp\{-\beta I + n_a \int d\mathbf{r}[\exp(-\beta V) - 1]\}. \tag{4.21}$$

Equation (4.21) includes the exponential density dependence of electron concentration, which describes the increase of n_e upon compression. Of considerable importance is the presence in the exponent of strong temperature dependence, $\Delta I = n_a kT \int d\mathbf{r}[\exp(-\beta V) - 1]$. The potential $V(r)$ can be characterized by two parameters, namely, by the (spherical) volume of the force action between the ion and atom, v_0, and the potential well depth, q. At low temperatures, the principal relations are reproduced by a simple formula analogous to Eq. (4.17),

$$\sigma = \sigma_0 \exp[-\beta I/2 + (n_a v_0/2) \exp(\beta q)]. \tag{4.22}$$

At low pressures, the densities n_a are low and therefore the conductivity grows monotonically with heating. At pressures of the order of $p \sim p_c$ the nonideality leads to the emergence of a minimum of conductivity on the isobar. This occurs at temperature

$$T_{\min} = q/k[\ln(I/qn_a v_0)]. \tag{4.23}$$

As in the case discussed above, let us assume $q = 0.7$ eV for cesium. The radius of the ion–atom forces is assumed to be equal to $10a_0$ (which is close both to the polarization radius $(\alpha e^2 \beta/2)^{1/4}$ and the internuclear distance of the Cs_2^+ ion). Then, Eq. (4.23) gives $T_{\min} = 1700$ K for the 2 MPa isobar, and $T_{\min} = 2500$ K for the 15 MPa isobar.

4.2.5 "Vapor–liquid" phase transition and metallization

It follows from diagrams showing regions of the existence of nonideal plasma of alkali metals and mercury (see Fig. 1.4) that these regions adjoin the high–temperature part of the two–phase region.

The question of how the metallization is related to the liquid–vapor phase transition was already discussed by Zel'dovich and Landau (1944). Two possibilities have been mentioned – the coincidence of the thermodynamic critical point with the metal–dielectric transition, and the loss of metallic properties in the liquid phase region. The latter was apparently revealed in mercury experiments by Kikoin and Senchenkov (1967), and Hensel and Franck (1968). The temperature coefficient of electrical conductivity lost the value characteristic of metal at a density of $\rho \simeq 9$ g cm^{-3} (Fig. 1.4) exceeding considerably the critical density $\rho_c \simeq 5.8$ g cm^{-3} (Kikoin and Senchenkov 1967). In the density range of 4–6 g cm^{-3}, i.e., near and below the critical point, gaseous mercury has typical properties of semiconductors (Kikoin and Senchenkov 1967). Conductivity decreases much below the minimum metal value (although it is still ten orders of magnitude higher than that calculated for an ideal gas) and grows exponentially with temperature at constant density. The value of the dielectric permittivity varies between a few and ten, i.e., is typical of semiconductors. In cesium, the metal–dielectric transition apparently occurs in the critical density region. This is indicated by measurements of electrical conductivity (Alekseev and Vedenov 1970; Renkert et al. 1969) and data on magnetic susceptibility (Franz et al. 1980).

Figure 4.3 demonstrates the electrical conductivity of cesium and mercury at temperatures slightly above the critical. There is a qualitative difference between the values of electrical conductivity for mercury and cesium. At $\rho \simeq \rho_c$, the electrical conductivity of mercury (10^{-1} ohm^{-1}cm^{-1}) is three orders of magnitude smaller than that of cesium (200 ohm^{-1}cm^{-1}). The latter value corresponds to the level of minimum metallic conduction in accordance with Mott's theory (Mott and Devis 1971) while the former value is too low. This (along with the data on the temperature coefficient of conductivity, thermal e.m.f., etc.) leads to the following conclusion: In mercury, in the neighborhood of the critical point, the phase transition is similar to the transition in nonmetallic media, i.e., liquid mercury at high density exists in the metallic state and at lower density, in the low–conducting state. For cesium (as well as for potassium and rubidium), the "liquid–gas" and "metal–nonmetal" transitions coincide.

The isotherm of electrical conductivity for mercury, shown in Fig. 4.4, gives a clear indication of the behavior of the dielectric–metal transition. The transition range in density is stretched out by an order of magnitude. This indicates that the first stage of metallization is due to the presence of long–range forces showing up even at moderate densities. Such an interaction is the polarization interaction of ions and electrons with atoms. Many years ago, Herzfeld (1927) proposed the criterion of metallization discussed by Frenkel (1955). The ionization potential (the bond energy of electron and atom) in a dense medium, $I_m = I/\varepsilon$, decreases according to the microscopic theory because of the increase of dielectric permeability ε. In accordance with the Clausius-Mossotti equation,

$$\varepsilon = 1 + 4\pi n_a \alpha (1 - 4\pi \alpha n_a /3)^{-1}. \qquad (4.24)$$

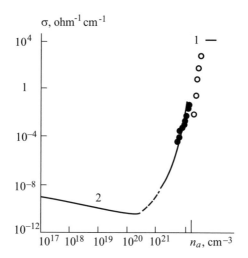

FIG. 4.4.   Isotherm of electrical conductivity for mercury, $T = 1973$ K (Khrapak and Yakubov 1981). Measurement results: ○ — (Hensel and Franck 1968); • — (Kikoin and Senchenkov 1967); 1 — level of electrical conductivity for metal; 2 — electrical conductivity for ideal plasma.

If the density increases to approach the value of $3/(4\pi\alpha)$ (where α is the polarizability in the gaseous state) the dielectric permeability increases indefinitely, which is typical for metals. Case $n_a > 3/(4\pi\alpha)$ corresponds to spontaneous ionization, i.e., the transition of electrons to the free or, rather, collectivized state. In other words, if $n_a \geqslant 3/(4\pi\alpha)$, the material is a metal. The value of $3/(4\pi\alpha)$ equals approximately $4 \cdot 10^{21}$ cm^{-3} for cesium and $4 \cdot 10^{22}$ cm^{-3} for mercury. These values coincide to within the order of magnitude with the observed transition values.

High–temperature vapors of some metals in the vicinity of the saturation line are nonideal because of the charge–neutral and charge-charge interactions. However, the results of measurements of p–ρ–T-dependencies for cesium and mercury, given in Figs. 2.20 and 2.21, failed to point explicitly to any peculiarity in the equation of state that would be due to the plasma effects. Nevertheless, it was suggested by Khrapak and Yakubov (1970), and Starostin (1971) that the vapor condensation in the high–temperature region is largely due to plasma effects. The critical parameters of cesium shown in Table 4.1 were estimated by using the methods of the nonideality description developed by Kraeft *et al.* (1985).

Apparently, these assumptions about the nature of phase transition in metals are valid. Then one can hardly expect that, at higher temperatures, plasma interactions will once again cause the phase separation. The interesting possibility of a new "plasma" phase transition was discussed by Norman and Starostin (1970) (see Chapter 5 below).

Table 4.1 *Estimates for critical parameters of cesium (Kraeft et al. 1985)*

	T_c, K	p_c, MPa	n_{ec}, cm^{-3}
Experiment	1925	9.25	$1.7 \cdot 10^{21}$
Calculations by Redmer and Röpke (1985)	2200	22.0	$1 \cdot 10^{21}$
Calculations by Richert *et al.* (1984)	2600	180.0	$3 \cdot 10^{21}$

4.3   Lowering of ionization potential and cluster ions in weakly nonideal plasmas

In a moderately dense plasma, the interaction of electrons and ions with atoms and molecules results in two effects, namely, lowering of the ionization potential and formation of molecular ions to be followed by heavier cluster ions. At moderate temperatures, these effects show up most clearly in a metal plasma. Below, data on molecular ions of alkali metals are given and the ionization equilibrium in a multicomponent mixture is considered.

4.3.1   *Interaction between charged particles and neutrals*

The first corrections to the free energy of the system, F, due to electron–atom and ion–atom interactions, can be calculated given $B_{ae}(T)$ and $B_{ai}(T)$, the virial coefficients of these interactions,

$$F = F_0 - kT n_e n_a B_{ae} - kT n^+ n_a B_{ai}.$$

Let us consider first the electron–atom interaction. The second virial coefficient $B_{ae}(T)$ should be calculated from the Bethe–Uhlenbeck formula (Landau and Lifshitz 1980),

$$B = (2\sqrt{\pi}\lambda_e)^3 \left[\sum_n e^{-\beta E_n} + \frac{1}{\pi} \int_0^\infty \sum_l (2l+1) \frac{d\delta_l}{dE} e^{-\beta E} dE \right] = B^d + B^c, \quad (4.25)$$

where $\lambda_e = \hbar(2mkT)^{-1/2}$ is the electron thermal wavelength, δ_l is the partial phase of electron–atom scattering, and E_n is the energy of electron–atom bound states. The sums are taken over all states of the negative ion and over all scattering moments. A set of scattering phases enables one to determine the effect of the interaction in the continuous spectrum while the bond energies provide the possibility of finding the second virial coefficient of the interaction in discrete spectra. Let us first discuss the interaction in the continuous spectrum.

Within the framework of scattering theory in an effective radius approximation, the electron–atom interaction is characterized by only two parameters, namely, the scattering length L and atomic polarizability α (Smirnov 1968). If k is the electron wavenumber, then

$$\delta_0(k) \simeq -Lk - \pi\alpha k^2/(3a_0),$$

$$\delta_l(k) \simeq \pi\alpha k^2[(2l+1)(2l+3)(2l-1)a_0]^{-1}, \quad l \geqslant 1.$$

By substituting $\delta_l(k)$ in Eq. (4.25) and performing the integration and summation over l, we obtain the virial coefficient of interaction in the continuous

Table 4.2 *Length of electron scattering by inert gas atoms (in a_0)*

He	Ne	Ar	Kr	Xe	Reference
1.16	0.45	−1.63	−3.8	−6.8	Gus'kov *et al.* (1978)
1.12	0.14	−1.40	−3.1	−5.7	Massey and Burhop (1969)

spectrum, $B^c = -4\pi L\lambda_e^2$, in a medium of molecules with a dipole moment, $B^c = 2q(T)\hbar(2mkT)^{-1/2}$, where q is the transport cross–section of electron–molecule scattering (Polishchuk 1985). This value should be substituted in the expression for the plasma free energy $F = F_0 - kTn_e n_a B^c$, where F_0 is the free energy of an ideal plasma. From this expression, Saha's equation is derived with the potential of atomic ionization in the medium,

$$I_m = I + \Delta I, \quad \Delta I = 2\pi\hbar^2 Ln_a/m.$$

The conditions of applicability of the effective radius theory are

$$|L| \ll \lambda_e, \quad \alpha/a_0 \ll \lambda_e^2.$$

This is the low–temperature region.

The lengths of electron scattering by inert gas atoms are listed in Table 4.2 and by atoms of alkali metals in Table 4.3.

Given the scattering length L of the order of $\pm a_0$ and $T = 2000$ K, we have $|\Delta I| \geqslant kT$ only at $n_a \geqslant 7 \cdot 10^{21}$ cm^{-3}. This can be a dense plasma of inert gas with an easily ionizable, such as alkali, seed. If $L = -10a_0$, then $|\beta\Delta I| \geqslant 1$ already at $n_a \geqslant 10^{20}$ cm^{-3} and $T = 2000$ K. Large absolute values of the scattering length are typical for atoms of alkali metals. However, the effective radius theory is not applicable in this case for a number of other reasons. Therefore, one needs for calculations a set of experimental values of δ_l. We shall use the values calculated by Karule (1972) and Norcross (1974).

The density–temperature diagram for cesium shown in Fig. 4.5 represents the $n_a B^d = 1$ and $n_a B^c = 1$ curves. These curves separate the regions of strong and weak nonideality due to the electron–atom interaction in discrete and continuous spectra. At low temperatures $B^d \gg B^c$ and, with an increase of temperature, the difference between them becomes smaller. It has been shown above that the contribution to the free energy by the term $kTn_e n_a B^c$ can be interpreted as a shift of the continuous spectrum boundary, that is, as a reduction of the ionization potential. In turn, the interaction in the discrete spectrum can be taken into account by introducing negative ions into the plasma composition. Reduction of the ionization potential is a minor correction and becomes important at high densities.

The ion–atom interaction in the continuous spectrum can be regarded as classical. Therefore, we employ a quasiclassical approximation for scattering phases,

$$\delta_l(E) = \sqrt{M/\hbar^2} \int d\mathbf{r}[\sqrt{E - V(r) - \hbar^2 l^2/Mr^2} - \sqrt{E - \hbar^2 l^2/Mr^2}], \quad (4.26)$$

Table 4.3 *Singlet, L_+, and triplet, L_-, length of electron scattering by atoms of alkali metals (in a_0) (Karule 1972)*

Scattering length	Li	Na	K	Cs
L_+	3.65	4.23	0.45	4.04
L_-	−5.66	−5.91	−15	−25.3

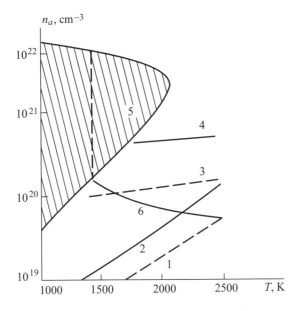

FIG. 4.5.   Density–temperature diagram for cesium (Khrapak and Yakubov 1981): Curves $n_a B^c = 1$ and $n_a B^d = 1$ are for electron–atom (1, 2) and ion–atom (3, 4) interaction, respectively; 5 — vapor–liquid coexistence curve; 6 — isobar at 2 MPa.

where $V(r)$ is the ion–atom interaction potential, E is the energy of colliding particles, and M is the ion mass equal to the atomic mass. By substituting Eq. (4.26) in Eq. (4.25), we replace summation by integration. The integration is performed in a classically accessible region where the expressions under the square roots in Eq. (4.26) are positive. In calculation, it is convenient to integrate first over l, and then over E. The result is (Hill 1956),

$$B^c = 4\pi \int_0^\infty r^2 \, dr \left[e^{-\beta V} \frac{\Gamma(3/2, -\beta V)}{\Gamma(3/2)} - 1 \right], \qquad (4.27)$$

$$\Gamma(3/2, -\beta V) = \int_{-\beta V}^\infty x^{1/2} e^{-x} dx, \quad \Gamma(3/2) = \sqrt{\pi}/2.$$

The obtained expression resembles the classical virial coefficient, but deviates from the latter by the factor with the gamma function. This difference is due to the contribution of bound states, that is, the states of diatomic molecular ions that contribute to B^d. The expression for B^d reduces to the constant of dissociation equilibrium in the reaction $A_2^+ \rightleftarrows A^+ + A$,

$$B^d = (2\sqrt{\pi}\lambda_a)^3 (\Sigma_2^+/\Sigma\Sigma^+) \exp(\beta D_2^+),$$

where Σ_2^+, Σ, and Σ^+ denote the internal partition functions of ion A_2^+, atom A, and ion A^+, respectively, and $\lambda_a = \hbar(MkT)^{-1/2}$. Since small distances in the integral (4.27) yield no substantial contribution, one can use for $V(r)$ the polarization potential. Then (Likal'ter 1969),

$$B^c = 4\pi C(\alpha e^2 \beta/2)^{3/4}, \quad C \simeq 1.61.$$

Plotted in Fig. 4.5 are the $n_a B^c = 1$ and $n_a B^d = 1$ curves for the ion–atom interaction. The nonideality of cesium vapors in the case of the ion–atom interaction occurs earlier than in the case of the electron–atom interaction.

Thus, in a moderately dense plasma, the nonideality due to the charge–neutral interaction is reflected by the emergence of negative and positive molecular ions and by the lowering of the ionization potential by

$$\Delta I = 2\pi\hbar^2 L n_a/m - 4\pi C n_a kT(\alpha e^2/2kT)^{3/4}. \tag{4.28}$$

4.3.2   Molecular and cluster ions

Now we consider the region where $n_a B^d > 1$. This means that the concentration of molecular ions A_2^+ exceeds that of atomic ions A^+. In other words, the interaction between ions and atoms becomes strong. The heavier ions A_3^+ should be taken into account. One can expect that, with an increase of density or decrease of temperature, the concentrations of heavier ions will grow.

Data on the bond energy and structure of molecular and cluster ions has been accumulating intensively in recent years (Lagar'kov and Yakubov 1980; Smirnov 1983, 2000a, 2000b). Listed in Table 4.4 are the parameters of alkali diatomic ions A_2^+ and triatomic ions A_3^+, dissociation energies D_2^+ and D_3^+ (for the $A_2^+ \rightarrow A^+ + A$ and $A_3^+ \rightarrow A_2^+ + A$ processes), equilibrium internuclear distances R_e and vibrational frequencies ω_e, as suggested by Gogoleva et al. (1984). Table 4.5 presents the same quantities for inert gas ions.

Table 4.6 gives D_2 (dissociation energy of molecules A_2), D_2^+, D_3^+, and q (heat of vaporization of metals per atom). The most important is the fact that $D_2^+ > D_2$. For instance, at $T = 1500$ K the ratio n_2^+/n^+ for sodium and cesium is an order of magnitude larger than n_2/n. It is known that the concentrations of trimers and dimers in alkali metal vapors are low. In spite of this, trimer and dimer ions can play an important role in ionization equilibrium.

Table 4.4 *Parameters of alkali ions A_2^+, A_3^+ (Gogoleva et al. 1984)*

Parameter	Li_2^+	Na_2^+	K_2^+	Rb_2^+	Cs_2^+
D_2^+, eV	1.28	1.02	0.84	0.77	0.66
R_e, a_0	5.91	6.69	7.9	8.5	9.1
ω_e, cm^{-1}	273	119	69	43	31

Parameter	Li_3^+	Na_3^+	K_3^+	Rb_3^+	Cs_3^+
D_3^+, eV	1.37	1.12	0.95	0.88	0.79
R_e, a_0	5.67	6.48	7.94	8.76	9.40
ω_e, cm^{-1} $\begin{cases} \\ \end{cases}$	342 / 239	149 / 104	86.7 / 60.6	53.7 / 37.5	39.4 / 27.5

Table 4.5 *Parameters of molecular ions of inert gases A_2^+, A_3^+*

Parameter	He_2^+	Ar_2^+	Kr_2^+	Xe_2^+
D_2^+, eV	2.23	1.33	1.15	1.03
R_e, 10^{-8} cm	1.08	2.50	2.75	3.22
ω_e, cm^{-1}	1627	300	160	123

Parameter	He_3^+	Ar_3^+	Kr_3^+	Xe_3^+
D_3^+, eV	0.17	0.20	0.27	0.27
R_e, 10^{-8} cm	—	2.67	2.97	3.47

The inequality $D_2^+ > D_2$ implies also that the ionization energy I (A_2) of a molecule is less than that of an atom. Indeed, from the energy conservation in the Born cycle, $A_2 \rightarrow A_2^+ + e \rightarrow A^+ + A + e \rightarrow A + A \rightarrow A_2$, it follows that

$$I(A_2) = I(A) - (D_2^+ - D_2).$$

Another important conclusion that follows from Table 4.6 is that the values of D_2^+, D_3^+, and q are close to each other. Hence, one can consider the separation energy of an atom and heavy ion as almost constant and, in the first approximation, suppose that the value of D_m^+ for heavier ions is known.

Very important information about heavy ions A_m^+ was provided by experiments where ions A_m^+ emerged upon passage of a molecular beam through an ionization chamber. Figure 4.6 shows the facility sketch by Forster *et al.* (1969).

Table 4.6 *Dissociation energy and heat of vaporization, eV (Khrapak and Yakubov 1981)*

Atom	D_2	D_2^+	D_3^+	q
Li	1.03	1.28	1.37	1.69
Na	0.73	1.02	1.12	1.13
K	0.514	0.84	0.95	0.87
Rb	0.49	0.77	0.88	0.78
Cs	0.45	0.66	0.79	0.74

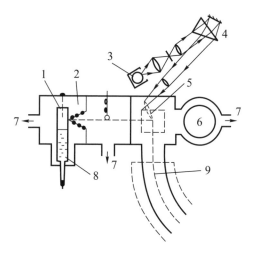

FIG. 4.6.   Experimental setup of Forster *et al.* (1969): 1 — boiler; 2 — heater; 3 — xenon
     lamp; 4 — monochromator; 5 — ionization chamber; 6 — trap; 7 — evacuation; 8 — liquid
     metal; 9 — magnetic field region.

Boiler 1 was loaded with liquid metal 8. A source of metal vapors was provided
by a heater. The molecular beam was generated as a result of molecular effusion
through a hole of small diameter. The beam was collimated with the aid of di-
aphragms and entered the ionization chamber 5. A high–pressure xenon lamp 3
served as a UV–radiation source. By varying the wavelength of radiation coming
from a monochromator 4, one could measure the photoionization thresholds. The
emerging ions were pulled out to a mass-spectrometer 9. Herrmann *et al.* (1978)
performed the experiments up to $m = 14$ for sodium, and Kappes *et al.* (1986)
up to $m = 66$ for sodium, and up to $m = 34$ for potassium. The results are listed
in Table 4.7. Similar data were obtained for clusters of lead, iron, nickel, and
aluminum (see the list of references in Rademann *et al.* 1987).

   Let us now discuss the dependence of the measured ionization potentials I_m
on the number of particles m in a complex at high m. In the case of $m \to \infty$, the
ionization potential must tend to the electron work function for a macroscopic
metal sample. A good description of the measurements is provided by a simple
macroscopic model. One can assume that a heavy ion is a metallic droplet with
a radius R, the value of which depends on the number of atoms in the droplet,
m. Then, the ionization potential I_m is the electron work function for a drop
with a finite radius $I(R)$. It can be related to the electron work function for a
plane surface, $I(\infty)$ (in the limit $R \to \infty$), by the following relation:

$$I(R) = I(\infty) + e^2/2R. \tag{4.29}$$

   Plotted in Fig. 4.7 are the experimental and calculated dependencies I_m for
sodium. The relation $R = R_m m^{1/3}$ between the drop radius and the number

Table 4.7 *Photoionization thresholds for sodium and potassium clusters, eV (Forster et al. 1969; Herrmann et al. 1978; Kappes et al. 1986)*

Cluster	I_m	Cluster	I_m	Cluster	I_m	Cluster	I_m
Na	5.15	Na_{11}	3.8 ± 0.1	K	4.34	K_{15}	3.21 ± 0.10
Na_2	4.934 ± 0.01	Na_{12}	3.6 ± 0.1	K_2	4.05 ± 0.05	K_{20}	3.23 ± 0.10
Na_3	3.97 ± 0.05	Na_{13}	3.6 ± 0.1	K_3	3.3 ± 0.1	K_{28}	3.05 ± 0.10
Na_4	4.97 ± 0.05	Na_{14}	3.5 ± 0.1	K_4	3.6 ± 0.1	K_{34}	3.01 ± 0.10
Na_5	4.5 ± 0.05	Na_{15}	3.7 ± 0.1	K_5	3.4 ± 0.1		
Na_6	4.12 ± 0.05	Na_{19}	3.6 ± 0.1	K_6	3.44 ± 0.10		
Na_7	4.04 ± 0.05	Na_{35}	3.4 ± 0.1	K_7	3.40 ± 0.10		
Na_8	4.10 ± 0.05	Na_{45}	3.3 ± 0.1	K_8	3.49 ± 0.10		
Na_9	4.0 ± 0.1	Na_{55}	3.3 ± 0.1	K_9	3.40 ± 0.10		
Na_{10}	3.9 ± 0.1	Na_{65}	3.2 ± 0.1	K_{10}	3.27 ± 0.10		

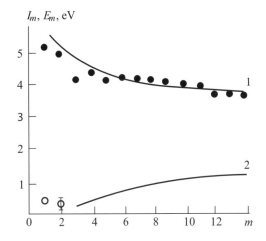

FIG. 4.7.   Ionization potential I_m (1) and electron affinity E_m (2) of sodium clusters Na_m: •— experiment Na_m; ∘ — Na^- and Na_2^-; curves are calculations with Eqs. (4.29) and (4.30).

of particles was employed for this, where $R_m = R_{inf} \exp[-2k\gamma/(3R_m m^{1/3})]$ and R_{inf} is the atomic radius in a macroscopic metal sample, and k and γ denote the macroscopic compressibility and surface tension, respectively. Inclusion of the surface tension somewhat reduces the value of R_m for small clusters.

The experimental results reveal good stability of positively charged heavy cluster ions and provide some information on their structure, which apparently approaches the structure of liquid metals.

At low m, considerable fluctuations are observed in the dependence I_m. These fluctuations are well described by the results of rather complicated quantum–mechanical calculations performed by a number of authors.

Table 4.8 *Parameters of negative alkali ions (Gogoleva et al. 1984)*

Parameter	Li	Na	K	Rb	Cs
E, eV	0.620	0.548	0.501	0.486	0.470
E_2, eV	0.55	0.49	0.45	0.43	0.42
$D_e(A_2^-)$, eV	0.96	0.66	0.47	0.43	0.40
$R_e(A_2^-)$, a_0	6.1	7.2	8.9	9.8	10.5
$\nu(A_2^-)$, cm^{-1}	213	92	63	38	28
$D_e(A_3^-)$, eV	0.90	0.59	0.49	0.40	0.38
$R_e(A_3^-)$, a_0	11.2	13.2	16.3	18.2	19.3

While the situation appears very simple for alkalis and a number of other metals, it is very different in the case of mercury. As demonstrated by the measurements of Rademann *et al.* (1987) (Fig. 4.8), clusters of mercury are metallized only at $m \gtrsim 70$. In this case, as in a number of other situations, this is due to the fact that mercury is an uncommon metal (see Section 4.5).

The ionization potential I_m and electron affinity E_m of Al_m clusters were measured by Seidl *et al.* (1991); see, also, the review by De Heer (1993).

Considerably less experimental information is available on negative alkali complexes A_m^- although in a number of experiments in gas discharge plasmas quite heavy complexes were reliably identified, for instance, complexes K_m^- up to $m = 7$ (Korchevoi and Makarchuk 1979). It is known that the binding energies of an electron with an atom and diatomic molecule are substantially smaller than the binding energy of positive ion with the same particles. Table 4.8 lists the parameters of ions A^-, A_2^-, A_3^-. Here E_1 and E_2 are the energies of electron affinity for an atom and molecule, $D(A_2^-)$ is the dissociation energy of $A_2^- \rightleftarrows A^- + A$, and $D(A_3^-)$ is the dissociation energy of $A_3^- \rightleftarrows A_2^- + A$. Also, R_e and ν are internuclear distances and vibrational frequencies.

For the energy of the electron affinity to a heavy complex, one can write an expression similar to Eq. (4.29),

$$E(R) = E(\infty) - e^2/2R. \qquad (4.30)$$

Obviously, $E(\infty) = I(\infty) = A_e$, where A_e is the electron work function for the metal. This dependence is plotted in Fig. 4.7.

4.3.3 Ionization equilibrium in alkali metal plasma

In order to determine the composition of a plasma formed by electrons (e), atoms (A), diatomic molecules (A_2), di- and triatomic positively charged ions (A_2^+ and A_3^+), and diatomic negatively charged atoms (A_2^-), let us consider the following processes:

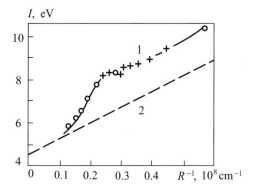

FIG. 4.8.   Ionization energy of mercury clusters I versus cluster radius R (Rademann et $al.$ 1987): 1 — plotted by experimental points, 2 — calculated with Eq. (4.29).

$$
\begin{aligned}
\mathrm{e} + \mathrm{A}^+ &\rightleftarrows \mathrm{A}, & n_e n^+ / n_a &= k_1; \\
\mathrm{e} + \mathrm{A} &\rightleftarrows \mathrm{A}^-, & n_e n_a / n^- &= k_2; \\
\mathrm{A} + \mathrm{A} &\rightleftarrows \mathrm{A}_2, & n_a^2 / n_2 &= k_3; \\
\mathrm{A} + \mathrm{A}^+ &\rightleftarrows \mathrm{A}_2^+, & n_a n^+ / n_2^+ &= k_4; \\
\mathrm{A}_2 + \mathrm{A}^+ &\rightleftarrows \mathrm{A}_3^+, & n_2 n^+ / n_3^+ &= k_5; \\
\mathrm{A}_2 + \mathrm{e} &\rightleftarrows \mathrm{A}_2^-, & n_2 n_e / n_2^- &= k_6
\end{aligned}
\tag{4.31}
$$

where k_i are chemical equilibrium constants. Equations (4.31), along with the conservation of charge and total number of particles, enable one to determine the concentration of any component. After simple transformations we get

$$
n_e^2 = k_1 n_a \frac{1 + n_a/k_4 + n_a^2/(k_3 k_5)}{1 + n_a/k_2 + n_a^2/(k_3 k_6)}.
\tag{4.32}
$$

It is readily seen that the primary fraction in Eq. (4.32) represents the effect of molecular ions on n_e. Individual components in the numerator (denominator) correspond to the contribution of a positive (negative) ion. Obviously, the inclusion of cluster ion A_4^+ would lead to the emergence of a subsequent series term in the numerator, whereas the inclusion of A_3^- ion would lead to that in the denominator, and so on.

Figure 4.9 shows the composition of charged components in a plasma of cesium vapors, as calculated on the $p = 2$ MPa isobar. The chemical equilibrium constants (4.31) were used for this (their parameters are listed in Table 4.9). The temperature dependencies of the equilibrium constants k_i are

$$
k_i = c_i T^{\eta_i} \exp(-E_i/kT).
$$

The peculiarities mentioned above show up in Fig. 4.9. As the temperature decreases, heavy ions play an increasingly important role: an A^+ ion, prevailing

Table 4.9 *Parameters of negative ions of alkali metals (Gogoleva et al. 1984)*

i	η_i	Na		Cs	
		c_i, 10^{21}cm^{-3} eV$^{-\eta_i}$	E_i, eV	c_i, 10^{21}cm^{-3} eV$^{-\eta_i}$	E_i, eV
1	3/2	3	5.14	3	3.89
2	3/2	12	0.54	12	0.47
3	−1/2	29	0.73	5	0.45
4	−1/2	4.22	1.14	1.45	0.85
5	1	2.5	1.63	0.63	1.2
6	3/2	4.9	0.25	0.45	0.30

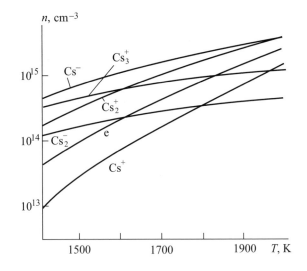

FIG. 4.9.   Charged particle concentration in a plasma of cesium vapors on the 2 MPa isobar by Khrapak (1979).

among positive ions at $T \gtrsim 2200$ K, is replaced by an A_2^+ ion which, in turn, gives way to an A_3^+ ion. One can expect the emergence of heavier positive ions on further cooling. The A^- ion prevails among negatively charged components at $T \lesssim 2000$ K. However, the A_2^- ion remains inconspicuous.

4.4   Droplet model of nonideal plasma of metal vapors. Anomalously high electrical conductivity

4.4.1   *Droplet model of nonideal plasma*

The model was proposed by Iakubov 1979. A vapor of alkali metals close to the saturation line represents a strongly nonideal plasma. Outside the close proximity of the critical point, this plasma can be treated as weakly ionized. In this case, the nonideality is mainly caused by the strong interaction between charged and neutral particles. This interaction promotes the formation of heavy charged complexes – clusters – in the plasma. The plasma of metal vapors may be treated as a mixture of the electron gas and a gas of clusters of different charges and sizes. The effect of clusters on the degree of plasma ionization was taken into account by Likal'ter (1978), Khrapak (1979), Iakubov (1979), Lagar'kov and Sarychev (1979), and Hernandez *et al.* (1984) in the framework of different models. Inclusion of this effect enables one to understand the reason why the electrical conductivity of cesium and mercury vapors in the vicinity of the saturation line is anomalously high. For instance, the Saha'sand Lorentz formulas yield the electrical conductivity $\sigma = 3 \cdot 10^{-5}$ ohm^{-1}cm^{-1} at $p = 2$ MPa and $T = 1400$ K. The value of $\sigma = 10$ ohm^{-1}cm^{-1} obtained in experiments is five orders of magnitude higher than these (typical for ideal plasmas) estimates.

Likal'ter (1978), Khrapak (1979), Iakubov (1979), Lagar'kov and Sarychev (1979), and Hernandez *et al.* (1984) share the common assumption of a sharp increase in the concentration of charged polyatomic clusters when approaching the saturation line. Moreover, the concentration of positively charged clusters exceeds considerably that of negatively charged ones. The plasma electroneutrality is ensured by the corresponding increase in the electron number density, which leads to an increase in the electrical conductivity of a plasma. Differences in theoretical models occur when one tries to specify the cluster properties.

It was proposed by Iakubov (1979) to treat clusters as small droplets. The advantages of this model include the possibility of using the characteristics of metals, such as surface tension and electron work function, to determine the cluster properties.

Changes in the aggregate state of matter, which are caused by electrostriction in the neighborhood of charged particles, are well–known in physics (Frenkel 1955; Jortner and Kestner 1974; Iakubov and Khrapak 1982). Charged droplets can form in vapors, and icicles can form in liquids. On the one hand, in a plasma of metal vapors the conditions for droplet stabilization by a charge are less favorable, because of high temperatures. On the other hand, the very high polarizability enhance electrostriction.

The elementary droplet model by Iakubov (1979) enables one to put forth the basic ideas and make rough estimates. Let us assume that a plasma consists of atoms, positively charged droplets and electrons. The concentration of these particles in a volume V is N_a, N^+, and N_e, and the total number of all particles is $\tilde{N} = N_a + N^+ + N_e$. The thermodynamic potential of the system has the following form:

$$\Phi = N_a \varphi_G + N^+ (g\varphi_L + 4\pi\gamma R^2 + W + e^2/(2R)) + kT\sum_i N_i \ln(N_i/\tilde{N}). \quad (4.33)$$

Here φ_L and φ_G are the thermodynamic potentials of liquid and vapor per atom, so that the quantity $gkT\ln(p_s/p) = g(\varphi_L - \varphi_G)$ is the work done to form a neutral droplet of radius R, with $g = (4\pi/3)R^3 n_L$ being the number of particles in the droplet, n_L is the concentration of particles in the liquid, and p_s is the saturation pressure. Also, $4\pi\gamma R^2$ is the surface energy (γ is the surface tension), and $W + e^2/(2R)$ is the electron work function for a droplet. In the last entropy

Table 4.10 *Parameters of ion droplets in saturated cesium vapor*

Parameter	T, K			
	1290	1430	1600	1690
p_s, 0.1 MPa	10.8	20.6	40.3	51.3
n_s, 10^{22} cm^{-3}	0.95	1.6	3.0	3.9
γ, g s^{-2}	23.9	17.5	11.7	8.9
R, a_0	16.4	17.6	19.2	20.5
g	15	18	21	23
n_e, 10^{16} cm^{-3}	0.13	1.5	11	36

term in (4.33), the summation is performed over all sorts of particles. Such a simple approach – when we are not interested in droplets with radii other than the most probable radius R, and do not describe the internal degrees of freedom of the droplets – reduces to Frenkel's theory of pretransition phenomena (Frenkel 1955).

We minimize, with respect to the radius, the work required to produce a single charged droplet, $gkT \ln(p_s/p) + 4\pi\gamma R^2 + W + e^2/(2R)$, in order to derive the Kelvin equation for determining the radius of the most probable droplet,

$$kTn_L \ln(p/p_s) = 2\gamma/R - e^2/(8\pi R^4).$$

In saturated vapor, R is equal to the "electrocapillary" radius

$$R = [e^2/(16\pi\gamma)]^{1/3}. \tag{4.34}$$

This allows us to estimate the parameters of ion droplets along the saturation line of cesium vapors (Table 4.10). Of course, the emerging droplets are too small for the macroscopic description to be sufficiently correct (curvature corrections to γ provide only slight improvement). However, according to the nucleation theory the droplets containing more than 10 particles can be treated as macroscopic.

Below we consider only saturated vapor $p = p_s$. We derive the equation of ionization equilibrium, $gA \rightleftarrows A_g^+ + e$, by varying (4.33) and taking into account the stoichiometry, $g\delta n_a = -\delta n^+ = -\delta n_e$. Since the plasma is weakly ionized, $n_e = n^+ \ll n_a$, we finally derive

$$n_e = n_a \exp[-(W + 4\pi\gamma R^2 + e^2/R)/2kT]$$
$$= n_a \exp[-(W + 3e^2/(4R))/(2kT)]. \tag{4.35}$$

For a number of reasons, the pre–exponential factor in (4.35) has been very poorly determined. However, the major effect is due to the exponent itself. It contains W, the work function of an electron from the metal, which is known to decrease as the metal temperature rises (as the density decreases). This dependence was investigated by Iakubov *et al.* (1986). Let us extrapolate the derived values taking into account that the work function must reach zero at the critical temperature T_c. It turns out that one can write

$$W(T) = W_0(T_c - T)/T_c, \qquad (4.36)$$

where $W_0 = 1.8$ eV is the tabular value of the work function for cesium (see Appendix C).

The results of the calculations for n_e are listed in Table 4.10 (see above). The interaction effects led to an increase in n_e by orders of magnitude as compared to the ideal–gas approximation. However, such a simple analysis is too crude. The drop model was improved considerably by Pogosov and Khrapak (1988), and Zhukhovitskii (1989). Below we present these results, following mainly Pogosov and Khrapak (1988).

4.4.2 Ionization equilibrium

Let us consider the ionization equilibrium in a multicomponent mixture of ideal gases of electrons and charged complexes consisting of g atoms and a charge Z. For such a mixture, the thermodynamic potential Φ is

$$\Phi = \sum_{g,Z} \Phi_g^Z + \Phi_e$$

$$= \sum_{g,Z} N_g^Z(kT \ln p_g^Z + \chi_g^Z) + N_e(kT \ln p_e + \chi_e) \qquad (4.37)$$

$$= \sum_{gZ} N_g^Z(kT \ln p + \chi_g^Z + kT \ln(N_g^Z/F)) + N_e(kT \ln p + \chi_e + kT \ln(N_e/F)),$$

where

$$F = \sum_{g,Z} N_g^Z + N_e = pV/kT,$$

$$\chi = f - kT \ln T + kT \ln V, \quad f = -kT \ln \Sigma. \qquad (4.38)$$

In Eqs. (4.37) and (4.38), N_g^Z and N_e denote the number of particles in volume V, p_g^Z, and p_e are the partial pressures, p is the total pressure, and Σ is the partition function over internal and translational degrees of freedom of the species.

Electrons possess only translational degrees of freedom,

$$\Sigma_e = \exp(-\beta f_e) = 2V\lambda_e^{-3}\exp(-\beta W) \equiv Vn_{es}, \qquad (4.39)$$

where W is the electron work function from a flat surface, λ_e is the thermal wavelength of the electron, and n_{es} is the equilibrium number density of electrons in the vicinity of the metal surface. We derive the relation between the concentrations of charged and neutral components by treating the equilibrium $A_g^Z \rightleftharpoons A_g^{Z+1} + e$, which yields

$$n_g^Z = n_g^0 (n_{es}/n_e)^Z \exp\{-\beta[f_g^Z - f_g^0]\}. \tag{4.40}$$

Let us now study another reaction – atom condensation leading to a growth of complexes, $A_{g-1}^Z + A_1^0 \rightleftharpoons A_g^Z$. This gives

$$N_g^0 = \exp[-\beta(f_g^0 - g\mu_G)]. \tag{4.41}$$

The chemical potential of a gas μ_G is defined by the expression

$$\mu_G \equiv \mu_1^0 = kT \ln(n_1^0 \lambda_1^3) = f_1^0 + kT \ln n_1^0, \tag{4.42}$$

where λ_1 is the thermal wavelength of atom. One usually tries to express f_g^Z or f_g^0 in terms of the macroscopic characteristics of the disperse phase. Let us assume that the presence of charge has no effect on the internal properties of the particle, and that the dependence of f_g^Z on Z is defined by the electrostatic energy of the excess charge. The latter is only true for large particles, when the fraction of surface atoms is much smaller than the volume fraction. Hence

$$f_g^Z = f_g^0 + Z^2/(2\varepsilon R) = f_g^0 + hZ^2/2g^{1/3}, \tag{4.43}$$

$$h = (4\pi n_l/3\varepsilon)^{1/3}, \tag{4.44}$$

where n_l is the number density of atoms in a liquid and ε is the dielectric permeability of the vapor. The substitution of this expression into (4.40) gives

$$n_g^Z = n_g^0 (n_{es}/n_e)^Z \exp(-hZ^2/2g^{1/3}kT). \tag{4.45}$$

The positively charged complexes with

$$Z^* = RkT(n_{es}/n_e), \tag{4.46}$$

have maximum concentration. In the vicinity of the saturation line for alkali metals, the charge Z^* is close to unity.

Let us identify in f_g^0 the contributions made by the translational, $f_{g,tr}$, degrees of freedom of the complex as a whole,

$$f_{g,tr} = kT \ln(\lambda_g^3/V), \quad f_g^0 = f_{g,tr} + E_g^0,$$

where $\lambda_g = g^{1/2}\lambda_1$ is the thermal wavelength of a g–atom complex. The quantity E_g^0 includes the kinetic energy and the energy of interaction of atoms in the complex. It is evident that, in the limit of high values of g, the energy E_g^0 is proportional to g. The dependence of E_g^0 on g can be sought for in the form of the expansion over inverse powers of the complex radius,

$$E_g^0 = \mu_l g + \alpha g^{2/3} + \zeta g^{1/3}, \tag{4.47}$$

where μ_l is the chemical potential of atoms in a liquid. The second term in Eq. (4.47) corresponds to the free surface energy of the complex, and the third

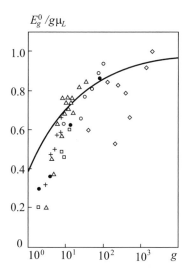

FIG. 4.10.   Energy E_g^0 as a function of the number of atoms in the complex (Pogosov and Khrapak 1988).

term can be treated as the inclusion of the curvature dependence of the surface tension.

In a large number of studies, E_g^0 is calculated from "first principles". Figure 4.10 gives the values of E_g^0 borrowed from papers cited by Pogosov and Khrapak 1988. These values correspond to complexes A_g formed by atoms of different elements (Li, Na, K, Fe and Cu), and are marked in Fig. 4.10 using different symbols. One can see that, with good accuracy, the quantity $E_g^0/(g\mu_L)$ is independent of atoms in the complex. This fact agrees with conclusions by Langmuir and Frenkel (see, Frenkel 1955) about the relation between the sublimation energy and specific surface energy. This justifies the choice of expression (4.47) for E_g^0. Indeed, according to Frenkel (1955), at fairly low temperature (when $\mu_L \simeq -q$, where q is the sublimation energy),

$$E_g^0/(g\mu_L) = 1 - \alpha/(qg^{1/3}),$$
$$\alpha/q \simeq (\eta - \eta')/\eta,$$

where $\alpha = 4\pi\gamma(3/4\pi n_L)^{2/3} = 4\pi\gamma r_s^2$ is the "surface energy" of a single atom, and η and η' are the numbers of the nearest neighbors for an atom deep inside the liquid and on the surface, respectively. It is clear that $\alpha/q \simeq 1/2 < 1$. It follows from Table 4.11 that the ratio α/q, taken at triple points of metals, is close to 0.64.

Now let us reconsider Eq. (4.41). We retain the first two terms of the expansion of E_g^0 over $g^{1/3}$. Taking into account that $\mu_L - \mu_G = -kT\ln(p/p_s)$, we derive the "classical" expression for the concentration of uncharged droplets,

Table 4.11 *"Surface energy" of atom α, heat of vaporization q, and ratio α/q at the triple point (Pogosov and Khrapak 1988)*

Parameter	Li	Na	K	Pb	Cs	Fe	Cu
α, eV	0.98	0.68	0.60	0.54	0.52	2.9	2.2
q, eV	1.6	1.1	0.93	0.85	0.82	3.5	4.4
α/q		0.60	0.63	0.64	0.65	0.63	0.69	0.64

$$n_g^0 = An_1^0 \exp\{-[-gkT\ln(p/p_s) + \alpha g^{2/3}]/kT\}. \tag{4.48}$$

The pre–exponential factor A, which is usually assumed to be equal to unity, will be determined below, with the aid of the experimental data for the equation of state of saturated vapor.

The condition of plasma neutrality and expression (4.45) for the concentration of charged droplets yield the equations for determining the electron number density,

$$n_e = \sum_{g,Z} Zn_g^Z = \sum_{Z=-\infty}^{\infty} Z(n_{es}/n_e)^Z F(Z^2)$$

$$= \sum_{Z=0}^{\infty} Z[(n_{es}/n_e)^Z - (n_e/n_{es})^Z]F(Z^2). \tag{4.49}$$

Because $F(Z^2) > 0$, it follows from (4.49) that the number density n_e is always less than the Richardson concentration. The latter corresponds to the equilibrium electron concentration near a plane emitting surface,

$$n_{es} = 2(h^2/2\pi mkT)^{-3/2}\exp(-W/kT). \tag{4.50}$$

4.4.3 Calculation of the plasma composition

Equation of state for matter and conservation of mass are $p = kT\sum_{g,Z} n_g^Z$ and $\rho = M\sum_{g,Z} gn_g^Z$, respectively. Using the known dependencies of pressure and density on the saturation curve (Vargaftic *et al.* 1985), now we can determine densities n_1^0 and n_e and constant A. For sufficiently small complexes, one must include the dependence of the surface tension on the droplet radius,

$$\gamma(R) = \gamma^0/(1 + \delta/R),$$

where γ^0 is the surface tension of a flat surface. The correction for the curvature δ for alkali metals was calculated in Iakubov *et al.* (1986). Both quantities, γ^0 and δ, depend on temperature. Figure 4.11 gives the calculation results for the quantity $g^*(T)$ in saturated cesium vapor (g^* is the value of g, at which n_g^+ attains maximum).

Figure 4.12 gives the results of the calculation for the electron number density in plasma of Na, K, Rb, and Cs vapors.

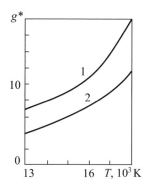

FIG. 4.11.   The temperature dependence of g^* in saturated cesium vapor (Pogosov and Khrapak 1988): 1 — δ is taken into account, 2 — for $\delta = 0$.

The validity of the droplet model is mainly determined by the values of $g^*(T)$ and by the half–width of the distribution $n_g^+(T)$. Figure 4.13 shows the normalized distribution of charged complexes for cesium versus the number of atoms, g, for $T = 1400$ K and $T = 1800$ K. As the temperature rises, the distribution of droplets over the number of particles becomes broader and the positions of the maxima shift towards larger values of g.

The magnitude of A turned out to be temperature dependent. For cesium, $A = 5.3 \cdot 10^2$ at $T = 1400$ K, and $A = 3.3 \cdot 10^3$ at $T = 1800$ K. Taking into account the very sharp variation of $n_e(T)$, such a dependence, as well as the magnitude of A itself, appears acceptable.

Figure 4.14 gives the results of the calculation and measurements for the specific electrical conductivity of saturated cesium vapors. The electrical conductivity was calculated with the Lorentz formula $\sigma = e^2 n_e/(\bar{q} m \bar{v} n_1^0)$, where $\bar{v} = \sqrt{8kT/(\pi m)}$ and $\bar{q}$ is the cross section of electron–atom scattering. In highly polarizable media, the density effects may cause a strong increase in mobility. This is due to the fact that for high densities, the long–range (polarization) components of electron–atom potentials overlap. The resulting potential field is smoothed out and the electron is scattered by the short–range components of potentials (see Chapter 6). Therefore, the cross–section $\bar{q}$ used by Pogosov and Khrapak (1988) is an order of magnitude smaller than the tabular value.

For the remaining alkali metals, the qualitative picture is similar. Pogosov and Khrapak (1988) do not describe the transition from anomalous to normal electrical conductivity. Based on the results of measurements by Borzhievsky et $al.$ (1988), this transition might occur at $T \simeq 1200$ K. By Zhukhovitskii (1989), yet another version of the droplet model was suggested, with the attempt of a unified description of the vapor equation of state and electrical conductivity, including its transition from the anomalous to normal.

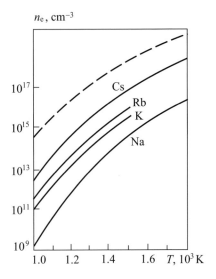

FIG. 4.12. Calculations of $n_e(T)$ for saturated vapors of alkali metals by Pogosov and Khrapak (1988). The broken line shows $n_{es}(T)$ for cesium.

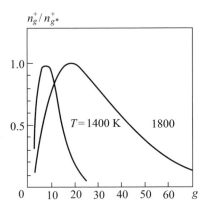

FIG. 4.13. Distribution of Cs_g^+ clusters over the number of bound atoms g (Pogosov and Khrapak 1988).

4.5 Metallization of plasma

4.5.1 Mott's transition

A continuous decrease in the density of metals during their heating causes, in the long run, a sharp decrease in electrical conductivity. Reference is usually made to the metal–dielectric transition or to Mott's transition (Mott and Devis 1971). Mott and Hubbard have demonstrated that the metal–dielectric transition is caused by the electron–electron interaction, which leads to band splitting. As a

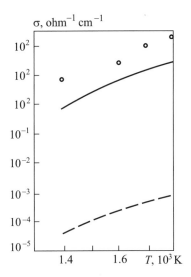

FIG. 4.14. Specific electrical conductivity of saturated cesium vapors (Pogosov and Khrapak 1988). The solid curve represents the droplet model, broken curve is for ideal plasma, points are the experimental data.

result, the empty and filled bands become separated by an energy gap.

The initial Mott estimation of the parameters of the dielectric–metal transition is very simple: The spectrum of bound states of the shielded potential $V(r) = -(e^2/r)\exp(-r/r_0)$ is investigated, and it is demonstrated that the bound states disappear completely when r_0 decreases to a value close to

$$r_0^M = 0.84a_0. \tag{4.51}$$

Depending on the absence or presence of electron degeneracy, r_0 is the Debye–Hückel or Thomas–Fermi screening radius.

Unlike a phase transition, Mott's transition in a metal occurs at high temperature. Therefore, the transition is smeared out and has no critical point. Upon the expansion of a metal, the electrical conductivity falls off sharply but continuously. Mott has introduced the concept of the minimum electrical conductivity of a metal, which is estimated from the condition that the mean free path of the electron is equal to its wavelength. Now, it is commonly assumed that this value is approximately 200 ohm^{-1}cm^{-1}.

Let us discuss the transition from a different point of view, namely, as the metallization of the plasma due to the compression. The increase of the electrical conductivity upon compression is due to a decrease in the ionization potential. The rate of this decrease rises during compression. It is often suggested that nonthermal ionization or ionization by pressure occurs. These assumptions, however, remain valid as long as one can speak about the individual atoms as well as bound and free electrons. At higher values of density, the shells of the ground atomic states overlap. The description of such a system requires a different approach developed by Likal'ter (1982, 2000); see, also, Vorob'ev and Likal'ter (1988).

4.5.2   Quasi–atomic model

Atoms of a metal come so close together that the classically accessible regions of motion of valence electrons start overlapping (the radius of such a region is e^2/I). Moreover, the classic transition of electrons from one atom to another becomes possible. The quasi–atomic state is formed when the density reaches a value corresponding to the percolation threshold of

$$\zeta_0 = (4\pi/3)(e^2/I)^3 n_{\mathrm{i}} \simeq 1/3.$$

At this point, electrons can cross the entire metal sample, passing from one ion to another. Each quasi–atom occupies a spherical cell with a volume of $\sim n_{\mathrm{i}}^{-1}$. A valence electron spreads out over its own cell and, to some extent, over the neighboring cells. Each ion is mainly shielded by its own electron and, partly, by other electrons as well. This constitutes the essential difference from the liquid–metal state, when the entire plasma volume is accessible to free electron motion. Roughly speaking, this is realized when the close packing of accessible regions is achieved, that is, at $\zeta_0 = 0.74$. Therefore, the discussed region is narrow with respect to density (approximately from one–half to two times the value of the critical metal density n_{c}), but important changes occurring there govern the critical state of matter.

Because $\beta I \gg 1$, it follows from the foregoing that the range of validity of the theory corresponds to the condition of strong nonideality, $\gamma \gg 1$. These are the conditions in the neighborhood of the critical points of a number of metals. In cesium, for example, the total ionization is achieved, $n_{\mathrm{e}} = n_{\mathrm{c}}$, so that $\gamma_{\mathrm{c}} = e^2 n_{\mathrm{c}}^{1/3} \beta \simeq 10$.

The plasma energy is the sum of the interaction energy, ΔE, and the kinetic energy of ions and electrons,

$$E = \Delta E + c_v nkT = -(1/2)nI - \alpha e^2 n^{4/3} + (3/2)nkT, \qquad (4.52)$$

where $n = n_{\mathrm{e}} + n_{\mathrm{i}}$. The term $-nI/2$ in ΔE is peculiar to the quasi–atomic model. This suggests that the renormalization of the Madelung constant α is required. The latter is usually calculated for a system of ions against the background of uniformly smeared electrons and is equal to 0.57. In our case, the value of α should be smaller. It takes into account the electrostatic interaction between an ion and the periphery of the cell containing the electric charge brought about by "extraneous" electrons. According to the estimate by Vorob'ev and Likal'ter (1988), the Madelung constant in (4.52) is $\alpha \simeq 1/4$.

Based on the model discussed above, one can write the expression for free energy and derive the equation of state,

$$p = nkT - \alpha e^2 n^{4/3}/3. \qquad (4.53)$$

This equation reveals thermodynamic instability. The finite compressibility of the liquid phase is included in (4.53) in Van der Waals' approximation,

$$p = nkT(1 - nb)^{-1} - \alpha e^2 n^{4/3}/3. \tag{4.54}$$

The parameter b corresponds to the minimum volume per particle and is defined by the quantity I, via $b \sim (e^2/I)^3$.

4.5.3 Phase transition in metals

The degrees of ionization of a number of metals in the neighborhood of the critical points are so high that plasma interactions are predominant. Therefore, a number of researchers suggested that the phase transition in its high–temperature region might be due to plasma interactions, i.e., be a "plasma" phase transition. More generally, the issue of plasma phase transition is discussed in Section 5.8.

Equation (4.54) provides a qualitative description of the phase transition in metals and helps to estimate the critical parameters,

$$n_c = (1/7)b, \quad T_c = (48/7^{7/3})(a/b^{1/3}), \quad p_c = 7^{-7/3}(a/b^{4/3}), \tag{4.55}$$

where $a = \alpha e^2/3$. Parameter a, which describes attraction, is universal for all monovalent metals. It is only parameter b that is specific. This is an important peculiarity of the equation of state (4.54), which distinguishes it from the Van der Waals' equation. The similarity relations for the critical parameters (Likal'ter 2000; Vorob'ev and Likal'ter 1988) follow from Eq. (4.55),

$$n_c \sim I^3, \quad T_c \sim I, \quad p_c \sim I^4.$$

The theory was generalized by Likal'ter (1985) to metals with Z valence electrons, which yielded more general similarity relations,

$$n_c \sim (Z+1)I^3, \quad T_c \sim \frac{Z^2}{Z+1}I, \quad p_c \sim Z^2 I^4. \tag{4.56}$$

For alkali metals, Eq. (4.55) yields fairly good estimates. For instance, for cesium with $\alpha = 0.21$ (Likal'ter 2000) the deviation of the calculated critical temperature from the experimental value (Jüngst et al. 1985) is about 15%, and the deviation of pressure is about 5%. One can see from the experimental data shown in Table 4.12, that the similar correspondence holds for other alkali metals (with the only exception being the sodium density, which has an unreliable experimental estimate). One can conclude that the Van der Waals' equation provides a rather good description of the major properties of "good" metals in the vicinity of the vapor–liquid phase transition.

The similarity relations (4.56) make it possible to express the critical parameters of a number of metals (which maintain the metallic state in the critical point) in terms of the parameters measured for one of them. It is natural to use the critical parameters of cesium for this purpose. By taking into account that cesium is monovalent, and using a bar to indicate the parameters related to the analogous parameters of cesium, we have

$$\bar{T}_c = \frac{Z^2}{Z+1}\bar{I}, \quad \bar{p}_c = Z^2\bar{I}^4. \tag{4.57}$$

The compressibility at the critical point is (Likal'ter 2000)

$$z_c = p_c/(n_c k T_c) = \frac{7}{48}(Z+1).$$

The value of this factor is approximately equal to 0.29, which is about 45% larger than the experimental value 0.2 for alkali metals (Jüngst et al. 1985). Likal'ter (1996) proposed to use a three–parameter modification of the Van der Waals' equation, which allowed him to approximate critical values of three thermodynamic variables and make direct comparison of theoretical estimates of the hard–sphere radius and the Madelung constant with the experimental values. By replacing the Van der Waals' denominator with the fourfold excluded volume, $b = 4\eta$, with the Carrnahan–Starling (1969) function,

$$F(\eta) = \frac{1 + \eta + \eta^2 - \eta^3}{(1-\eta)^3}, \tag{4.58}$$

and taking into account the next after the Madelung term in the expansion of the interaction energy, the following equation of state was derived:

$$\pi = 10\nu[\tau F(\eta) - A_0\eta^{1/3} - B_0\eta^{-1/3}], \tag{4.59}$$

where $A_0 = 2.854$, $B_0 = 0.03643$, and $\pi = p/p_c$, $\nu = n_i/n_{ic}$ and $\tau = T/T_c$ are the reduced pressure, number density, and temperature.

Critical parameters determined from the equation of state (4.59) can be expressed in terms of the functions I and Z used for the Van der Waals's equation, but with prefactors normalized by the parameters of cesium at the critical point (Likal'ter 2000),

$$T_c \simeq \frac{0.085}{k}\frac{Z^2 I}{Z+1}, \quad p_c \simeq 0.405 \cdot 10^5 Z^2 I^4,$$

$$n_{ic} \simeq 2.92 \cdot 10^{19} I^3, \quad Z_c = 0,1(Z+1), \tag{4.60}$$

where I and T are measured in eV, p_c is in Pa, and n_{ic} is in cm^{-3}. Similarity relations (4.60) allow us to estimate critical parameters without performing detailed calculations for a particular metal.

The range of validity of the theory is discussed by Vorob'ev and Likal'ter (1988). As mentioned above, this range is not wide. Nevertheless, it has been verified that a number of metals satisfy the condition of the metallic state at the critical point. These include all alkali metals, copper and silver of the noble

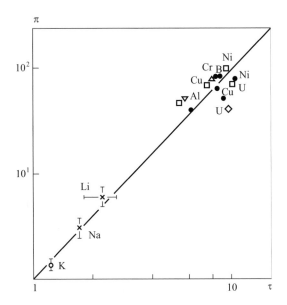

FIG. 4.15. Relative critical pressure of metals π versus relative critical temperature τ: points indicate the data from a number of works (see references in the work by Likal'ter 1985) normalized by the critical parameters of cesium.

metals, beryllium of the elements of Group II, and all elements of the main subgroup of Group III, including aluminum. The transition metals include all elements found in the iron group, some elements in the palladium and platinum groups, including molybdenum, and uranium. However, there is no sharp dividing line between the different types of states. In a number of metals, which are not on this list, the interaction at the critical point is not of a pure Coulomb type, but makes a transition to the Van der Waals interaction due to the mutual polarization of atoms. This transition is terminated in the case of uncommon metals, such as mercury, arsenic, bismuth, and tellurium.

The critical temperature and pressure of some simple and transition metals, as calculated with formulas (4.57) and (4.60) on the basis of the critical parameters of cesium, are given in Table 4.12. Figure 4.15 shows the relative critical pressure as a function of the relative critical temperature. These quantities correspond to formula (4.56) if the values of T_c and p_c are normalized by the corresponding values known for cesium. The estimates of the critical parameters, available from the literature and made on the basis of experimental data and semiempirical relations, correlate well with the dependencies given above. One can conclude that metallic states of this type with Coulomb interaction at the critical point are quite typical for metals.

Table 4.12 *Critical parameters of metallic liquids (Likal'ter 2000)*

Metal	T_c, K	p_c, bar	ρ_c, g cm^{-3}	Method	Reference
Cs (6s)	1600	87	0.3	Van der Waals' equation	Likal'ter and Schneidenbach 2000
	2000	460	0.55	Partially ionized gas	Redmer 1997
	2350	60	0.47	MC simulations	Chacon et al. 1995
	1924	92.5	0.38	Experiment	Jüngst et al. 1985
Rb (5s)	1720	115	0.24	Van der Waals' equation	Likal'ter and Schneidenbach 2000
	2200	650	0.45	Partially ionized gas	Redmer 1997
	2475	73	0.35	MC simulations	Chacon et al. 1995
	2060	123	0.3	Scaling with cesium normalization	Likal'ter 1996
	2017	124	0.29	Experiment	Jüngst et al. 1985
K (4s)	1790	134	0.12	Van der Waals' equation	Likal'ter and Schneidenbach 2000
	2350	690	0.21	Partially ionized gas	Redmer 1997
	2550	70	0.22	MC simulations	Chacon et al. 1995
	2140	144	0.16	Scaling with cesium normalization	Likal'ter 1996
	2178	150	0.17	Experiment	Hensel et al. 1991
Na (3s)	2115	263	0.12	Van der Waals' equation	Likal'ter and Schneidenbach 2000
	2400	1400	0.27	Partially ionized gas	Redmer 1997
	2970	128	0.22	MC simulations	Chacon et al. 1995
	2535	282	0.15	Scaling with cesium normalization	Likal'ter 1996
	2485	248	0.30	Experiment	Hensel et al. 1991

Table 4.12 (*continued*)

Metal	T_c, K	p_c, bar	ρ_c, g cm^{-3}	Method	Reference
Li (2s)	3700	530	0.042	Van der Waals' equation	Likal'ter 2000
	3500	770	0.025	Partially ionized gas	Redmer 1997
	2660	342	0.053	Scaling with cesium normalization	Likal'ter 1996
	3225	690	0.1	Extrapolation	Fortov *et al.* 1975
Cu ($4s^2$)	7620	5770	1.4	Scaling with cesium normalization	Likal'ter 1996
	8390	7460	2.4	Extrapolation	Fortov *et al.* 1975
Al ($3s^2\,3p$)	8860	4680	0.28	Scaling with cesium normalization	Likal'ter 1996
	8000	4470	0.64	Extrapolation	Fortov *et al.* 1975
Be ($2s^2$)	9200	12200	0.35	Scaling with cesium normalization	Likal'ter 1996
	8100	11700	0.55	Extrapolation	Fortov *et al.* 1975
U ($7s^2\,7p$)	9000	5000	2.6	Scaling with cesium normalization	Likal'ter 1997
	11600	6100	5.3	Extrapolation	Fortov *et al.* 1975
La ($6s^2\,6p$)	8250	3500	1.2	Scaling with cesium normalization	Likal'ter 1997
	11000	3500	1.8	Extrapolation	Fortov *et al.* 1975
Y ($5s^2\,5p$)	9500	6000	1.1	Scaling with cesium normalization	Likal'ter 1997
	10800	3700	1.3	Extrapolation	Fortov *et al.* 1975

References

Alekseev, V. A. and Iakubov I. T. (1983). Nonideal plasmas of metal vapors. *Phys. Rep.*, **96**, 1–69.

Alekseev, V. A. and Vedenov, A. A. (1970). Electric conductivity of dense cesium vapor. *Sov. Phys. Uspekhi*, **13**, 830–831.

Alekseev, V. A., Vedenov, A. A., Ovcharenko, V. G., Krasitskaya, L. S., Rizhkov, Y. F., and Starostin, A. N. (1975). The effect of saturation on the thermo–e.m.f. of caesium at high temperatures and pressures. *High Temp. – High Press.*, **7**, 677–679.

Alekseev, V. A., Fortov, V. E., and Iakubov I. T. (1981). Current status of physics of nonideal plasma. In *Proceedings of the 15th international conference on phenomenon in ionized gases*. Minsk. Invited papers, pp. 73–85.

Borzhievsky, A. A., Sechenov, V. A., and Horunzhenko, V. I. (1987). Experimental investigation of electrical conductivity of cesium vapour. In *Proceedings of the 18th International Conference on Ionization Phenomena in Gases*. Contributed Papers, Swansea. Vol. 1, pp. 250–251.

Borzhievsky, A. A., Sechenov, V. A., and Khoruzhenko, V. I. (1988). Electrical conductivity of cesium vapors. *High Temp.*, **26**, 722–726.

Carrnahan, N. F. and Starling, K. E. (1969). Equation of state for nonattracting rigid spheres. *J. Chem. Phys.*, **51**, 635–636.

Chacon, E., Hernandez, J. P., and Tarazona, P. (1995). Theoretical models for the liquid–vapor and metal–nonmetal transitions of alkali fluids. *Phys. Rev. B*, **52**, 9330–9341.

De Heer, W. A. (1993). The physics of simple metal clusters: Experimental aspects and simple models. *Rev. Mod. Phys.*, **65**, 611–676.

Ermokhin, N. V., Ryabyi, V. A., Kovalev, B. M., and Kulik, P. P. (1971). Experimental investigation of coulomb interaction in dense plasma. *High Temp.*, **9**, 611–620.

Forster, P. J., Leckenby, R. E., and Robbins, E. J. (1969). The ionization potentials of clustered alkali metal atoms. *J. Phys. B*, **2**, 478–483.

Fortov, V. E., Dremin, A.N., and Leont'ev, A. A. (1975). Evaluation of the parameters of the critical point. *High Temp.*, **13**, 984–992.

Franz, G., Freyland, W., and Hensel, F. (1980). Thermodynamic and electric transport properties of fluid cesium and rubidium in the M–NM transition region. *J. de Phys.*, **41**, No. C8, 70–74.

Frenkel, J. (1955). *Kinetic theory of liquids*. Dover, New York.

Freyland, W. (1979). Magnetic susceptibility of metallic and nonmetallic expanded fluid cesium. *Phys. Rev. B*, **20**, 5104–5112.

Gogoleva, V. V., Zitserman, V. Y., Polishchuk, A. Y., and Yakubov, I. T. (1984). Electrical conductivity of an alkali–metal–vapor plasma. *High Temp.*, **22**, 163–170.

Gus'kov, Y. K., Savvov, R. V., and Slobodnyuk, V. A. (1978). Total cross-section of slow electro elastic scattering on He, Ne, Ar, Kr, and Xe atoms. *JETP*, **48**, 277–284.

Hensel, F. (1976). Metal–nonmetal transitions in fluid systems. *Ber. Bungenges. Phys. Chem.*, **80**, 786–792.

Hensel, F. and Franck, E. U. (1968). Metal–nonmetal transition in dense mercury vapor. *Rev. Mod. Phys.*, **44**, 697–703.

Hensel, F. and Warren, W. W. (1999). *Fluid metals.* Princeton University Press, Princeton.

Hensel, F., Stolz, M., Hohl, G., Winter, R., and Gotzlaff, W. (1991). Critical phenomena and the metal–nonmetal transition in liquid metals. *J. Phys. IV (Paris) Coll. C5.*, **1**, 191–205.

Hernandez, J. P., Schonherr, G., Gotzlaff, W., and Hensel, F. (1984). Density dependence of the electrical conductivity of slightly ionised mercury vapour. *J. Phys. C*, **17**, 4421–4427.

Herrmann, A., Schumacher, E., and Wöste, L. (1978). Preparation and photoionization potentials of molecules of sodium, potassium, and mixed atoms. *J. Chem. Phys.*, **68**, 2327–2336.

Herzfeld, K. F. (1927). On atomic properties which make an element a metal. *Phys. Rev.*, **29**, 701–705.

Hill, T. L. (1956). *Statistical mechanics.* McGraw–Hill, New York.

Iakubov, I. T. (1979). Toward theory of heightened electrical conductivity of dense metal vapors near saturation. *DAN USSR*, **247**, 841–844.

Iakubov, I. T. and Khrapak, A. G. (1982). Self–trapped states of positrons and positronium in dense gases and liquids. *Rep. Progr. Phys.*, **45**, 697–752.

Iakubov, I. T. and Likal'ter, A. A. (1987). New results in field of physics of nonideal plasma: experiments and interpretation. *Contr. Plasma Phys.*, **27**, 479–490.

Iakubov, I. T., Khrapak, A. G., Pogosov, V. V., and Triger, S. A. (1986). Energy characteristics of liquid metal drops. *Solid State Commun.*, **60**, 377–380.

Iermokhin, N. V., Kovaliov, B. M., Kulik, P. P., and Ryabyi, V. A. (1978). Condactivité électrique des plasmas denses. *J. de Phys.*, **39**, No. C1, 200–206.

Isakov, I. M. and Lomakin, B. N. (1979). Measurement of electric conductivity during adiabatic compression of cesium vapors. *High Temp.*, **17**, 222–225.

Jüngst, S., Knuth, B., and Hensel, F. (1985). Observation of singular diameters in the coexistence curves of metals. *Phys. Rev. Lett.*, **55**, 2160–2163.

Jortner, J. and Kestner, N. R. (eds). (1974). *Electrons in fluids.* Springer, Berlin.

Kappes, M. M., Schär, M., Radi, P., and Schumacher, E. (1986). On the manifestation of electronic structure effect in metal clusters. *J. Chem. Phys.*, **84**, 1863–1875.

Karule, E. (1972). The spin polarization and differential cross–sections in the elastic scattering of electrons by alkali metal atoms. *J. Phys. B*, **5**, 2051–2060.

Khomkin, A. L., Iakubov, I. T., and Khrapak, A. G. (1995). Ionization equilibrium, equation of state, and electric conductivity of partially ionized plasma. In *Transport and optical properties of nonideal plasma,* Kobzev, G. A., Iakubov, I. T., and Popovich, M. M. (eds), pp. 78–130. Plenum Press, New York.

Khrapak, A. G. (1979). Conductivity of a weakly nonideal multicomponent alkali–metal vapor plasma. *High Temp.*, **17**, 946–951.

Khrapak, A. G. and Yakubov, I. T. (1970). On the phase transition and negative clusters in an imperfect plasma of metal vapor. *JETP*, **31**, 945–950.

Khrapak, A. G. and Yakubov, I. T. (1979). Electrons and positrons in dense gases. *Sov. Phys. Uspekhi*, **22**, 703–726.

Khrapak, A. G. and Yakubov, I. T. (1981). *Electrons in dense gases and plasma.* Nauka, Moscow.

Kikoin, I. K. and Senchenkov, A. P. (1967). Electrical conductivity and equation of state of mercury in temperature range 0–2000°C and pressure range of 200–5000 atm. *Phys. Metals and Metallography–USSR*, **24**, 74–87.

Kopyshev, V. R. (1968). Second virial coefficients of a plasma. *JETP*, **28**, 684–686.

Korchevoi, Yu. P. and Makarchuk, V. N. (1979). Generation of ionized clusters in discharge potassium plasma. *Ukr. Phys. J.*, **24**, 799–808.

Kraeft, W. D., Kremp, D., Ebeling, W., and Röpke, G. (1985). *Quantum statistics of charged particle systems.* Akademic Verlag, Berlin.

Lagar'kov, A. N. and Yakubov, I. T. (1980). Ion complexes in vapors of alkali metals. In *Chemistry of plasma,* Smirnov, B. M. (ed.), pp. 75–109. Atomizdat, Moscow.

Lagar'kov, A. N. and Sarychev, A. K. (1979). (private communication).

Landau, L. D. and Lifshitz, E. M. (1980). *Statistical physics, Part 1.* Pergamon Press, New York.

Larkin, A. I. (1960). Thermodynamic functions of a low–temperature plasma. *JETP*, **11**, 1363–1364.

Likal'ter, A. A. (1969). Interaction of atoms with electrons and ions in a plasma. *JETP*, **29**, 133–135.

Likal'ter, A. A. (1978). Electrical conductivity of dense vapors of alkali metals. *High Temp.*, **16**, 1039–1045.

Likal'ter, A. A. (1982). Electrical conductivity alkali metal vapors in vicinity of critical point. *High Temp.*, **20**, 1076–1080.

Likal'ter, A. A. (1985). About critical parameters of metals. *High Temp.*, **23**, 1076–1080.

Likalter, A. A. (1996). Equation of state of metallic fluids near the critical point of phase transition. *Phys. Rev. B*, **53**, 4386–4392.

Likalter, A. A. (1997). Electric conductivity of expanded transition metals. *Phys. Scripta*, **55**, 114–118.

Likalter, A. A. (2000). Critical points of condensation in Coulomb systems. *Phys.–Uspechi*, **43**, 777–797.

Likalter, A. A. and Schneidenbach, L. H. (2000). Critical points of metallic fluids. *Physica A*, **277**, 293.

Magee, J. L. and Funabashi, K. (1959). The clustering of ions in irradiated gases. *Radiat. Res.*, **10**, 622–631.

Massey, H. S. W. and Burhop, E. H. S. (1969). *Electronic and ionic impact phenomena*. Clarendon Press, Oxford.

Montano, P. A., Shenoy, G. K., Alp, E. E., Schulze, W., and Urban, J. (1986). Structure of copper microclusters isolated in solid argon. *Phys. Rev. Lett.*, **56**, 2076–2079.

Morrow, R. and Craggs, J. D. (1973). The electrical conductivity of sodium vapour. *J. Phys. D*, **6**, 1274–1282.

Mott, N. F. and Devis, E. A. (1971). *Electronic processes in noncrystalline materials*. Clarendon Press, Oxford.

Norcross, D. W. (1974). Application of scattering theory to the calculation of alkali negative–ion bound states. *Phys. Rev. Lett.*, **32**, 192–196.

Norman, G. E. and Starostin, A. N. (1970). Thermodynamics of strongly non-ideal plasma. *High Temp.*, **8**, 381–408.

Pfeifer, H. P., Freyland, W. F., and Hensel, F. (1973). Absolute thermoelectric power of fluid cesium in metal–nonmetal transition range. *Phys. Lett. A*, **43**, 111–112.

Pogosov, V. V. and Khrapak, A. G. (1988). Drop model of alkali metal valor plasma. *High Temp.*, **26**, 209–218.

Polishchuk, A. Y. (1985). Thermodynamics of the interaction of a charge and a dipole in the quantum and classical cases. *High Temp.*, **23**, 10–15.

Popielawski, J. and Gryko, J. (1977). The semiconductor model of electric conductivity applied to the supercritical mercury vapor. *J. Chem. Phys.*, **66**, 2257–2261.

Rademann, K., Kaiser, B., Even, U., and Hensel, F. (1987). Size dependence of gradual transition to metallic properties in isolated clusters. *Phys. Rev. Lett.*, **59**, 2319–2321.

Redmer, R. (1997). Physical properties of dense, low–temperature plasmas. *Phys. Rep.*, **282**, 35–157.

Redmer, R., and Röpke, G. (1985). Quantum statistical approach to the equation of state and the critical point of cesium plasma. *Physica A*, **130**, 523–552.

Renkert, H., Hensel, F., and Franck, E. U. (1969). Metal–nonmetal transition in dense cesium vapor. *Phys. Lett. A.*, **30**, 494–495.

Renkert, H., Hensel, P., and Franck, E. U. (1971). Electrical conductivity of liquid and gaseous cesium up to 2000 °C and 1000 bar. *Ber. Bunsenges. Phys. Chem.*, **75**, 507–512.

Richert, W., Insepov, S. A., and Ebeling, W. (1984). Thermodynamic functions of nonideal alkali plasmas. *Ann. Phys.*, **41**, 139–147.

Sechenov, V. A., Son, E. E., and Shchekotov, O. E. (1977). Electrical conductivity of a cesium plasma. *High Temp.*, **15**, 346–349.

Seidl, M., Meiwes–Broer, K.–H., and Brack, M. (1991). Finite–size effects in ionization potentials and electron affinities of metal clusters. *J. Chem. Phys.*, **95**, 1295.

Smirnov, B. M. (1968). *Atomic collisions and elementary processes*. Atomizdat, Moscow.

Smirnov, B. M. (1970). Transition of vapor in state with metallic conductivity. *DAN USSR*, **195**, 75–78.

Smirnov, B. M. (1983). *Cluster ions.* Nauka, Moscow.

Smirnov B. M. (2000a). *Clusters and small particles in gases and plasma.* Springer, New York.

Smirnov, B. M. (2000b). Cluster plasma. *Phys. Uspekhi*, **43**, 453–491.

Starostin, A. N. (1971). Some questions of the strongly coupled plasma theory., D. Phil. thesis. MPTI, Moscow.

Vargaftic, N. B., Voliak, L. D., and Stepanov, V. G. (1985). Thermodynamic parameters of cesium vapor. In *Handbook of thermodynamic and transport properties of alkali metals.* Ohse, R. W. (ed.), pp. 641–666. Blackwell, Oxford.

Vedenov, A. A. (1968). On one-dimensional weak plasma turbulence. In *Abstracts of the international conference on quiscent plasma.* Fraskatti, pp. 107–108.

Vedenov, A. A. and Larkin, A. I. (1959). Equation of state of a plasma. *JETP*, **9**, 806–811.

Vorob'ev, V. S. (2000). Asymptotic methods of description of low temperature plasma thermodynamics. In *Encyclopedia of low temperature plasma. Introductory volume 1,* Fortov, V. E. (ed.), pp. 293–299. Nauka, Moscow.

Vorob'ev, V. S. and Likal'ter, A. A. (1988). Physical properties of strongly coupled plasma. In *Plasma chemistry. Vol. 15*, Smirnov, B.M. (ed.), pp. 163–207. Energoatomizdat, Moscow.

Zel'dovich, Y. B. and Landau, L. D. (1944). On correlation between liquid and gaseous states of metals. *Acta Phys.–Chim. USSR*, **18**, 194–198.

Zhukhovitskii, D. I. (1989). Coalescence in nonisotermal turbulent plasma. *High Temp.*, **27**, 15–22.

5

THERMODYNAMICS OF PLASMAS WITH DEVELOPED IONIZATION

5.1 One–component plasma with the neutralizing charge background

We shall begin the description of the thermodynamics of strongly ionized plasma by discussing the most popular and well–studied model of the one–component plasma. A one–component plasma (OCP) represents a system of point ions placed in a homogeneous medium of charges of opposite sign (Brush *et al.* 1966; Hansen 1973; Pollock and Hansen 1973; Ng 1974; Lieb and Narnhofer 1975; DeWitt and Hubbard 1976; Galam and Hansen 1976; Zamalin *et al.* 1977; DeWitt 1977; Gann *et al.* 1979; Fisher *et al.* 1979; Baus and Hansen 1980; Slattery *et al.* 1980; Ichimaru 1982, 1992; Dubin and O'Neil 1999). Such a model serves as a good approximation for the plasma at ultrahigh pressure realized in the center of white dwarfs and heavy planets of Jupiter type (see Table 5.1). In these cases, matter is ionized under the effect of pressure and degenerate electrons have sufficient kinetic energy, $\varepsilon_{\mathrm{kin}} \approx \varepsilon_{\mathrm{F}} = (3\pi^2)^{2/3}\hbar n_{\mathrm{e}}^{2/3}/(2m)$, to produce an almost uniform background density distribution. Due to the small electron mass, the kinetic energy of electrons at high density ($r_{\mathrm{s}} \to 0$) is $\varepsilon_{\mathrm{F}} \gg kT$, and its pressure is much higher than that of an ion subsystem. In fact, two systems are realized in this case: the Coulomb system of point nuclei described by the Boltzmann statistics, and the quantum electron fluid. The weak interaction between these components results in an insignificant increase of the electron density in the neighborhood of nuclei (polarization), and most attention is concentrated on the analysis of the Coulomb internuclear interaction. The OCP model is the simplest, nontrivial model of a plasma, since there is no doubt concerning the form of the interaction potential, and the absence of quantum effects enables one to exclude from the treatment the formation of bound states (of molecules, atoms, and ions) (Rushbrooke 1968; Kovalenko and Fisher 1973) and the effect of degeneracy and interference (Rosenfeld and Ashcroft 1979; Springer *et al.* 1973; DeWitt 1976; Gorobchenko and Maksimov 1980; Ebeling *et al.* 1991). Therefore, the OCP model has been comprehensively studied, both analytically and numerically, in a broad range of nonideality parameters.

5.1.1 *Monte Carlo method*

A great deal of results have been obtained by Zamalin, Norman, and Filinov (1977) in the framework of the OCP model by using the Monte Carlo method. This enables one to perform "computer" experiments with a plasma of any den-

169

Table 5.1 *One–component plasma of astrophysical objects*

Parameter	Jupiter	White dwarf	Neutron star
Z	1 (H)	6 (C)	26 (Fe)
n_i, cm^{-3}	$6 \cdot 10^{24}$	$5 \cdot 10^{30}$	10^{32}
T, K	10^4	10^8	10^8
γ	50	10–200	870
r_s, a_0	0.65	$0.4 \cdot 10^{-2}$	$0.8 \cdot 10^{-3}$

sity. The method employs the first principles of statistical physics and is based on direct computer calculations of mean thermodynamic values,

$$\langle F \rangle = Q^{-1}(V, N, T) \int \ldots \int F(\mathbf{q}) \exp\{-\beta U_N(\mathbf{q})\} d^N \mathbf{q}, \qquad (5.1)$$

where $Q(V, N, T) = \int \ldots \int \exp\{-\beta U_N(\mathbf{q})\} d^N \mathbf{q}$ is the configuration integral defining the equilibrium properties of the thermodynamic system and $\mathbf{q} \equiv \mathbf{q}_1, \ldots, \mathbf{q}_N$ are the coordinates of the particles. The interparticle interaction potential is assumed to be pre–assigned and, in most calculations, binary,

$$U_N(\mathbf{q}) = \sum_{a,b,i<j} \Phi_{ab}(\mathbf{r}_{ij}), \quad \mathbf{r}_{ij} = \mathbf{q}_i - \mathbf{q}_j. \qquad (5.2)$$

By using a pseudo–random number generator, the Monte Carlo method in this case is reduced to the generation of a Markov chain which represents a set of space configurations, A_1, A_2, ..., A_N, with probability ω_{ij} of the $A_i \to A_j$ transition,

$$\omega_{ij} = \exp\{-\beta[U(A_j) - U(A_i)]\}. \qquad (5.3)$$

In doing so, the thermodynamic functions are averaged along the obtained Markov chain, which is equivalent to averaging over the canonical ensemble.

Direct calculations are usually performed for a single cubic elementary cell containing N particles. In order to take into account the interaction of these particles with those of the neighboring cells, the Ewald procedure and periodic boundary conditions were employed.

In the pioneering study by Brush *et al.* (1966), which involved the use of the Monte Carlo method in the OCP, the values of energy, heat capacity c_V, and binary correlation function $g(r)$ were calculated as the mean over $\alpha \sim 10^5$ configurations, with the number of particles in a cell being $N \simeq 108$ (for tests, N varied from 32 to 500).

Having the dense plasma of Jupiter in mind, DeWitt and Hubbard (1976) performed numerical calculations of a hydrogen plasma as well as a plasma of a mixture of light elements, taking into account the polarization of the background ($r_s \neq 0$) by the method of linear response theory.

The calculations of Brush *et al.* (1966) describe OCP in a liquid state in the range $0.05 \lesssim \gamma \lesssim 100$ (from 32 to 500 particles per cell). The calculations of

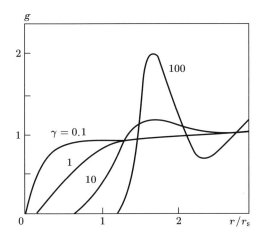

FIG. 5.1.   The binary correlation function of the OCP for different values of the nonideality
parameter γ (Ichimaru 1982).

Hansen (1973) cover the range $1 \lesssim \gamma \lesssim 160$ (from 16 to 250 particles per cell).
The results of these studies are in good agreement, except for the region of the
highest values of γ which is described more accurately by Hansen (1973). OCP
in a crystalline state was used by Pollock and Hansen (1973) for the range $140 \lesssim
\gamma \lesssim 300$ (128 and 250 particles per cell). Detailed calculations of OCP properties
were performed also by Ng (1974), Slattery $et\ al.$ (1980), and Stringfellow $et\ al.$
(1990).

5.1.2   Results of calculation

The binary correlation function $g(r)$ is presented in Fig. 5.1. Figure 5.2 demonstrates the behavior of the static structure factor,

$$S(\mathbf{q}) = 1 + n_i \int d\mathbf{r}[g(r) - 1]e^{-i\mathbf{q}\mathbf{r}}. \qquad (5.4)$$

The static dielectric permeability $\varepsilon(q,0)$ is given in Fig. 5.3,

$$\varepsilon(q,0) = [1 - q_D^2/q^2 S(q)]^{-1}, \quad q_D^2 = 4\pi n_i(Ze)^2/kT. \qquad (5.5)$$

For $\gamma \ll 1$, these functions vary monotonically and are described by the linearized
Debye approximation,

$$S_D(q) = q^2(q^2 + q_D^2)^{-1}, \quad \varepsilon(q,0) = (q^2 + q_D^2)/q^2. \qquad (5.6)$$

The pair correlation function is monotonic up to $\gamma \simeq 2.5$ and is in qualitative
agreement with the Debye approximation,

$$g(r) = 1 - \Gamma\frac{r_D}{r}\exp(-r/r_D),$$
$$\Gamma = (eZ)^2/(r_D kT), \quad r_D = (4\pi Z^2 e^2 n_i/kT)^{-1/2}. \qquad (5.7)$$

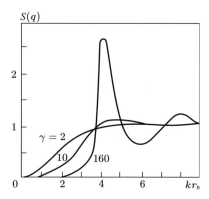

FIG. 5.2. The structure factor $S(q)$ of the OCP for different values of the nonideality
parameter γ (Ichimaru 1982).

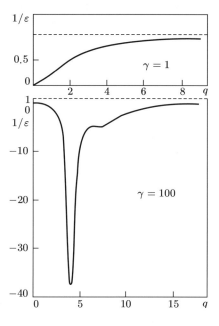

FIG. 5.3. The static dielectric function of the OCP for different values of the nonideality
parameter γ (Ichimaru 1982).

At $\gamma \gtrsim 2.5$, oscillations of $g(r)$ appear, suggesting the emergence of short–
range order: the system changes from the ideal gas to a liquid state. With an
increase of γ, the oscillations increase and an effective hard core is formed.

Given $g(r)$, one can calculate the internal energy,

$$u = 3n_{\rm i}kT/2 + u_{\rm ex},$$

$$u_{\rm ex}/(n_{\rm i}kT) = (n_{\rm i}/2kT) \int d\mathbf{r}(Z^2e^2/r)[g(r) - 1]. \tag{5.8}$$

In the presence of the neutralizing background, the potential energy is not equal to zero solely because of the correlations in the charge positions. Therefore, the mean potential (excess) energy of interaction, $u_{\rm ex}$, is often referred to as the "correlation energy". For $\Gamma \ll 1$, by substituting (5.7) in (5.8), one can easily obtain the classic a Debye–Hückel result,

$$u_{\rm ex}/(n_{\rm i}kT) = -\frac{\Gamma}{2} = -\frac{\sqrt{3}}{2}\gamma^{3/2}.$$

Although the potential of the interparticle interaction is repulsive, the correlation energy is negative because of the neutralizing background. A "correlation hole" is formed around each charge, where the presence of another charge is improbable because of the repulsion. Then the major contribution to the potential energy is due to the interaction of the point–like charge with the background within the hole, which obviously corresponds to attraction.

The results of Monte Carlo calculations for $1 \leqslant \gamma \leqslant 160$ were approximated by Slattery *et al.* (1980) with an accuracy of $3 \cdot 10^{-5}$ by the following expression:

$$u_{\rm ex}/(n_{\rm i}kT) = a\gamma + b\gamma^{1/4} + c\gamma^{-1/4} + d, \tag{5.9}$$

where $a = -0.897\,52$, $b = 0.945\,44$, $c = 0.179\,54$, $d = -0.800\,49$. In subsequent work by Stringfellow *et al.* (1990) and Dubin and O'Neil (1999) the accuracy increased substantially, which caused some changes in the approximations. By integrating Eq. (5.9) over γ, one can obtain the free energy of the OCP,

$$f = F/(n_{\rm i}kT) = a\gamma + 4(b\gamma^{1/4} - c\gamma^{-1/4}) + (d+3)\ln\gamma - (a + 4b - 4c + 1.135). \tag{5.10}$$

The referencing of f in Eq. (5.10) to the point $\gamma = 1$ was done by Slattery *et al.* (1980) by using expressions for $u_{\rm ex}$ valid at $\gamma \lesssim 1$.

Note that in accordance with the virial theorem, $pV = u/3$, a negative value of u at large γ leads to the negative pressure of the ion component. This, however, does not make the OCP unstable – the total pressure is, of course, positive because of the high value of the electron background pressure. The isothermal compressibility, χ_T, behaves similarly,

$$\frac{1}{n_{\rm i}kT\chi_T} = 1 + \frac{1}{3}\left[\frac{u_{\rm ex}(\gamma)}{n_{\rm i}kT}\right] + \frac{\gamma}{9}\frac{d}{d\gamma}\left[\frac{u_{\rm ex}(\gamma)}{n_{\rm i}kT}\right]. \tag{5.11}$$

The results of the calculation for solidified OCP (Fig. 5.4) were given for the bcc lattice, which has the lowest energy. Slattery *et al.* (1980) approximated the excess energy at $160 \lesssim \gamma \lesssim 300$ by the expression

$$u_{\rm ex}/n_{\rm i}kT = a_{\rm bcc}\gamma + \frac{3}{2}b\gamma^{-2} = 0.895\,929\gamma + 1.5 + 2980/\gamma^2. \tag{5.12}$$

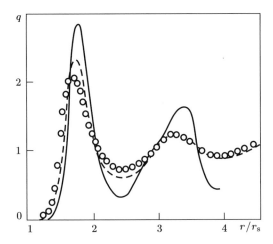

FIG. 5.4.   Binary correlation function for $\gamma = 160$. Solid line is for the solid phase (Pollock and Hansen 1973), broken line is for the liquid phase (Hansen 1973), dots are for the HMC approximation (Ng 1974).

For the physical interpretation of the individual terms in the approximate expressions, such as Eqs. (5.9) and (5.10), simplified models were used by Lieb and Narnhofer (1975), Ichimaru (1977), and Dubin and O'Neil (1999). We discuss them below.

5.1.3   Ion–sphere and hard–sphere models for OCP

The ion–sphere model (valid for $\gamma \gg 1$), or the cell model, breaks the system up into spheres with radius r_s. Each sphere (called the Wigner–Seitz cell) accommodates an ion surrounded by a homogeneous cloud of neutralizing electrons the density $3Ze/4\pi r_s^3$. Since the cells are not overlapped and the total charge is equal to zero, the interaction between them is absent and the potential energy is simply equal to the sum the energies of individual cells. Within this model, one can easily calculate the mean excess energy,

$$u_{\mathrm{ex}}/(kn_iT) = -\frac{9}{10}\gamma + \frac{3}{2} \geqslant -\frac{9}{10}\gamma. \tag{5.13}$$

Note that Eq. (5.13) satisfies one of the global inequalities imposed on OCP by the singularity of the Coulomb potential (Lieb and Narnhofer 1975). Equation (5.13) should be compared with Eq. (5.10), as well as with Eq. (5.12). In the latter, a_{bcc} is the Madelung constant for the bcc lattice. Therefore, the dominant term in Eq. (5.12) is static, and the other terms represent the "thermal" part associated with harmonic oscillations of ions. The contribution of the thermal part to u_{ex}, p, and χ proves to be less than 3% at $\gamma \gtrsim 100$.

As we can see, the leading term in expressions (5.9) and (5.10) is close to the Madelung term for the bcc lattice of Coulomb charges. OCP yields the bcc lattice in the limit of strong interaction, making use of the short–range effect of the interparticle interaction, which is characteristic for cell models.

In order to clarify the role of this short–range effect and to describe the second ($\propto \gamma^{1/4}$) term in Eq. (5.9), the hard–sphere model popular in the theory of liquids was employed by DeWitt and Rosenfeld (1979). By treating the hard spheres as "zero" approximation models (H_0, F_0) in describing OCP (H, F), using the Gibbs–Bogoliubov inequality,

$$F \leqslant F_0 + \langle H - H_0 \rangle \,,$$

and employing solutions of the Percus–Yevick equations, we derive after the variational procedure (see Section 5.4.3 for details):

$$\frac{u_{\mathrm{ex}}}{n_{\mathrm{i}} kT} = -\frac{9}{10}\gamma + \left[\frac{8\gamma}{9}\right]^{1/4} - \frac{1}{2} + \frac{7}{80}\left[\frac{18}{\gamma}\right]^{1/4} - \ldots$$

In fact, this relation reproduces the first term in the ion–sphere model (5.13) and explains the appearance of the second term in (5.9), which can be interpreted as a manifestation of the short–range effects in highly compressed plasmas.

5.1.4 Wigner crystallization

The crystallization was first considered by Wigner (1934) for a degenerate plasma, where it was shown that at fairly low concentrations the electron gas with the compensating background charge should become ordered. Indeed, as the plasma expands the Coulomb energy that stabilizes the lattice, $V_{\mathrm{C}} \sim e^2 n_{\mathrm{e}}^{1/3} \sim e^2/r_{\mathrm{s}}$, decreases slower than the kinetic energy that destroys the lattice, the scale of the latter being the Fermi energy $\varepsilon_{\mathrm{kin}} \sim \varepsilon_{\mathrm{F}} \sim \hbar^2 n_{\mathrm{e}}^{2/3}/2m$. Therefore, at sufficiently low density the kinetic energy $\varepsilon_{\mathrm{kin}} \sim n_{\mathrm{e}}^{2/3}$ becomes smaller than the potential energy $V_{\mathrm{C}} \sim n_{\mathrm{e}}^{1/3}$, and is incapable destroying of the ordered electron structure formed due to the repulsion.

The problem of determining the conditions of the Wigner crystallization is treated in numerous studies, which are analyzed by Tsidil'kovskii (1987). We will briefly describe the results of this analysis. A simple comparison of the potential and kinetic energy of electrons gives the limits of crystallization,

$$r_{\mathrm{s}}^c \sim (1.1 - 3.7)a_0, \quad n_{\mathrm{c}}^{1/3}a_0 \sim 0.56 - 0.17,$$

where the lower estimate for r_{s}^c was derived by neglecting the exchange and correlation effects, and the upper estimate by taking them into account in the Hartree–Fock approximation. Based on the Lindemann melting criterion, one derives the estimate

$$r_{\mathrm{s}}^c \sim 20a_0, \quad n_{\mathrm{c}}^{1/3}a_0 \sim 3,1 \cdot 10^{-2},$$

which is very sensitive to the details of this empirical criterion.

Assuming that the melting corresponds to the disappearance of bound states of the electron, one can derive the stability limit,

$$50 \leqslant r_{\mathrm{s}}^{\mathrm{c}}/a_0 \leqslant 100, \quad 6.2 \cdot 10^{-3} \leqslant n_{\mathrm{c}}^{1/3} a_0 \leqslant 1.2 \cdot 10^{-2}.$$

Detailed calculations of the ground state yield

$$r_{\mathrm{s}}^{\mathrm{c}} \sim (5\text{--}6)a_0, \quad n_{\mathrm{c}}^{1/3} a_0 \sim 0.1\text{--}0.12.$$

The Monte Carlo calculations for paramagnetic and ferromagnetic Fermi liquids result in the critical values

$$r_{\mathrm{s}}^{\mathrm{c}} \sim (75 \pm 5)a_0, \quad n_{\mathrm{c}}^{1/3} a_0 \sim (8.3 \pm 0.5) \cdot 10^{-3},$$

$$r_{\mathrm{s}}^{\mathrm{c}} \sim (100 \pm 20)a_0, \quad n_{\mathrm{c}}^{1/3} a_0 \sim (6.2 \pm 1.3) \cdot 10^{-3}.$$

It is believed that the latter values of r_{s} are the most reliable ones (Tsidil'kovskii, 1987).

For a nondegenerate plasma, the estimates of the melting parameters are obtained by comparing the free energy of the gas and the crystal phases. From Eq. (5.12), the expression for the free energy density (Slattery *et al.* 1980) is derived,

$$f(\gamma) = -0.895\,929 + 9\gamma/2 - 1.8856 - 1490/\gamma^2. \tag{5.14}$$

The free energy curves for gaseous (5.10) and solid (5.14) phases intersect at $\gamma_{\mathrm{m}} = 165$ (Slattery *et al.* 1980). The Wigner crystallization in OCP occurs for this value of nonideality parameter, which is very sensitive to the details of the calculation procedure. Different authors give, in their opinion, the most probable values of γ_{m}, which lie in the range between 155 and 171.

Apparently, Wigner crystallization was experimentally observed in a two–dimensional system of electrons localized at the surface of liquid helium by Grimes and Adams (1979). A classical electron crystal with triangular lattice occurred at $\gamma_{\mathrm{m}} \simeq 140$, with $\gamma = (Ze)^2(\pi n_{\mathrm{e}})^{1/2}/kT$. This value is close to the calculated value, $\gamma = 125 \pm 15$ (Gann *et al.* 1979).

The quasiperiodic structure of an ion cloud with $n_i \sim 10^8$ cm^{-3} and $T \sim 10^2$ K in a magnetic field was also interpreted by Gilbert *et al.* (1988) as a manifestation of the Wigner crystallization. Oscillations of the plasma binary function, which are typical for ordered structures, were measured by Hall *et al.* (1988) in laser-driven shock waves and were treated from the same perspective. It is possible that at $\gamma > \gamma_{\mathrm{m}}$, supercooled OCP is in an amorphous glassy state (Ichimaru 1977), and the phase transition is accompanied by pronounced hysteresis phenomena. This state is a disordered system of monocrystals with a size of several lattice spacings. It is interesting to note that this might be the state of matter in the interior of white dwarfs because, according to some estimates, the parameter γ there may even exceed γ_{m}. Properties of Wigner crystals are discussed in detail in Chapter 10.

5.1.5   *Integral equations*

In the theory of liquids, integral equations relate the binary correlation function to the two–particle interaction potential $V(r)$ (Rushbrooke 1968; Kovalenko and Fisher 1973). The Percus–Yevick equation, which had shown itself to be advantageous in describing the interaction of the system properties with the short–range potential (Rosenfeld and Ashcroft 1979), turned out to be hardly suitable for the OCP (Springer *et al.* 1973). For plasmas, the hypernetted chain (HMC) approximation is, apparently, the most appropriate one. One can conclude this from the qualitative analysis of the HMC equations as well as from the good agreement between the results of the Monte Carlo simulations and the results of numerical integration of hypernetted chain equations (Ng 1974; Springer *et al.* 1973). These equations have the following form:

$$g(r) = 1 + h(r) = \exp\{-V(r)/kT + h(r) - c(r)\}, \qquad (5.15)$$

$$h(r) = c(r) + N \int d\mathbf{r}' c(|\mathbf{r} - \mathbf{r}'|) h(r'), \qquad (5.16)$$

where $h(r)$ and $c(r)$ are the complete and direct correlation functions, respectively, and $V(r) = Z^2 e^2/r$. The solution procedure is as follows: Starting from some initial $h(r)$, one calculates its Fourier transform $h(k)$ and obtains $c(r)$,

$$c(r) = \int \frac{d\mathbf{k}}{(2\pi)^3} e^{i\mathbf{k}\mathbf{r}} \frac{h(k)}{1 + h(k)}. \qquad (5.17)$$

Upon substituting Eq. (5.17) in Eq. (5.16), one derives the next interpretation of $h(r)$, and so on.

Springer *et al.* (1973) found $g(r)$ and $u_{ex}(r)$ in the $0.05 \lesssim \gamma \lesssim 50$ range. Ng (1974) did that with high accuracy for $20 \lesssim \gamma \lesssim 7000$. The binary correlation functions in the HMC approximation describe well the values obtained with the Monte Carlo method (see, e.g., Fig. 5.4).

Numerical HMC results were approximated by DeWitt (1976) with the following expression:

$$u_{ex}/(n_i kT) = a\gamma + b\gamma^{1/2} + c \ln \gamma + d, \qquad (5.18)$$

where $a = -0.900\,470$, $b = 0.268\,8263$, $c = 0.071\,9925$, and $d = 0.053\,7919$. Because of the predominance of the static term, Eqs. (5.9) and (5.18) coincide with an accuracy better than 1%. As for the thermal terms in u_{ex}, they differ both qualitatively [scaling $\propto \gamma^{1/2}$ in (5.18) instead of $\propto \gamma^{1/4}$ in (5.9)] and numerically (45 % at $\gamma = 150$).

At the highest values of γ, the strict constraint (5.13) is not satisfied. In spite of the considerable advantages of the HMC approximation in general, it proves unsatisfactory in situations when the thermal part plays a major role. This is the case, for instance, in calculating the heat capacity,

$$c_V/n_i - 1 = -\gamma^2 d(u_{ex}/n_i kT\gamma)/d\gamma. \qquad (5.19)$$

Here, the error is as high as 20%. The shortcomings of the HMC approximation arise because the contribution of the ladder diagrams is neglected in calculating u_{ex}. The semiempirical inclusion of this contribution proved successful by De-Witt (1976). It was performed in such a manner as to satisfy the sum rule for compressibility violated by the HMC approximation. As a result, the thermodynamic quantities and correlation functions were found to be very close to those obtained using the Monte Carlo method.

5.1.6   *Polarization of compensating background*

The degenerate electronic gas, which provides the compensating background, is polarized due to the nonuniformity of the ion charge distribution. An electron cloud is formed around each ion, which modifies the ion–ion potential. In the Fourier representation,

$$V(q) = 4\pi(Ze)^2 q^{-2} \varepsilon_e^{-1}(q). \tag{5.20}$$

The static dielectric permeability $\varepsilon_e(q)$ is known in various approximations (Gorobchenko and Maksimov 1980) and has the form

$$\varepsilon_e(q) = 1 - \frac{4\pi(Ze)^2 \chi_0(q) q^{-2}}{1 + 4\pi(Ze)^2 G(q)\chi_0(q)q^{-2}}, \tag{5.21}$$

where $\chi_0(q)$ is the static dielectric permeability of the free electronic gas (Lindhard function), and $G(q)$ is the local field correction due to the exchange and correlation effects (DeWitt and Rosenfeld 1979). In the limit of high electron density ($r_s \to 0$), the exchange and correlation can be neglected and $\varepsilon_e(q)$ is given by a random phase approximation (RPA),

$$\begin{aligned}
\varepsilon_e(q) &= 1 - \frac{4\pi(Ze)^2 \chi_0(q)}{q^2} \\
&= 1 + \frac{1}{(q\lambda_{TF})^2}\left\{ \frac{1}{2} + \frac{q_F}{2q}\left[1 - \frac{q^2}{4q_F^2} \right] \ln\left| \frac{q + 2q_F}{q - 2q_F} \right| \right\}.
\end{aligned} \tag{5.22}$$

Here $\lambda_{TF} = r_s(\pi/12Z)^{1/3}$ is the Thomas–Fermi screening distance and q_F is the Fermi momentum, $q_F^{-1} = r_s(4/(9\pi Z))^{1/3}$. Actually, Eq. (5.22) works fairly well as long as $r_s \lesssim a_0$.

To the rough Thomas–Fermi approximation, the interion potential is shielded as follows:

$$V(r) = \frac{(Ze)^2}{r} \exp\left(-\frac{r}{\lambda_{TF}} \right). \tag{5.23}$$

If $\lambda_{TF} \gg r$, screening is not important, that is, the background polarization can be neglected. In the opposite case, the results of the calculation depend on two, rather than one, parameters, that is, γ and r_s (Galam and Hansen 1976).

For $r_s \leqslant 0.5a_0$, the results of the Monte Carlo calculations are approximated as follows (Ichimaru 1982):

$$u_{ex}/(n_i kT) = -(0.8946 + 0.0543 r_s/a_0)\gamma$$
$$+(0.8165 - 0.1853 r_s/a_0)\gamma^{1/4} - (0.5012 - 0.0659 r_s/a_0). \quad (5.24)$$

At $r_s > a_0$, the results of the calculation are very sensitive to the form of $G(q)$.

Usually, the background polarization effect on the thermodynamic properties of the plasma is relatively small. It is mainly the static part of the energy that varies. The inclusion of correlations changes the energy by several per cent at $r_s \sim 0.5a_0$. The plasma pressure varies even less, because the polarization correction, $\sim r_s\Gamma = \beta e^2$, does not depend on the system volume.

5.1.7 Charge density waves

In OCP, the charge density waves emerge because the interelectron interactions make the spatially oscillating density profile more favorable than the uniform distribution. This leads to the density modulation of the background positive charge, in order to provide the electroneutrality (Tsidil'kovskii 1987). This effect is realized (Ichimaru 1982) with the emergence of a "soft mode" instability associated with the formation of the Wigner crystal. Therefore, the state with the charge density waves may be treated as an intermediate phase between the Wigner crystal and the common metal with a uniform charge distribution.

The condition for the emergence of charge density waves is that the dielectric permeability, $\varepsilon(q, 0)$, which is related to the structure factor, $S(q)$, via the fluctuation–dissipation theorem (5.5), is equal to zero.

Figure 5.3 demonstrates that, according to the Monte Carlo calculations, such states may emerge in OCP under conditions of considerable plasma nonideality, $\gamma \gtrsim 160$.

The real possibility for the emergence of charge density waves in a three–dimensional electron gas depends largely on how close is the "deformability" of the ion lattice to that of the deformable "jelly" model. Apparently, alkali metals are the most suitable three–dimensional objects in the search for charge density waves (Tsidil'kovskii 1987). This is because the weakness of the ion–ion interaction promotes the deformability of the ion lattice (Overhauser 1968, 1978; Gilbert et al. 1988; Hall et al. 1988). At present, the question of charge density waves in three–dimensional metals remains open, while in the case of quasi–one–dimensional structures (such as chalcogenides of transition metals, TaS_2, $NbSe_3$, and so on) those waves are detected, for instance, in electron–microscopic observations.

5.1.8 Sum rules

The numerical results obtained using the Monte Carlo method serve as the "experimental" data that enable one to assess the range of application of different analytical approximations. Direct comparison shows, for instance, that the methods developed for weakly nonideal plasmas remain quite applicable upon extrapolation to $\gamma \simeq 1$ (Iosilevskii 1980). It is important that in the case of such an extrapolation, general constraints should be met, which are valid for OCP with any nonideality. These include the following relations (Deutsch *et al.* 1981; Ebeling *et al.* 1991):

(i) condition of positiveness of the binary correlation function:

$$g(r) \geq 0; \tag{5.25}$$

(ii) condition of exponentiality, which restricts the value of $g(r)$ at small r:

$$g(r) \sim \exp[-\beta V(r)], \quad r \to 0; \tag{5.26}$$

iii) condition of local electroneutrality in a Coulomb system:

$$n_{\mathrm{i}} \int d\mathbf{r}[g(r) - 1] = -1; \tag{5.27}$$

(iv) condition of screening obtained by Stillinger and Lovett (1968):

$$n_{\mathrm{i}} \int d\mathbf{r}\, r^2[g(r) - 1] = -6k_{\mathrm{D}}^{-2}, \quad k_{\mathrm{D}} = (4\pi Z^2 e^2 n_{\mathrm{i}}/kT)^{1/2}. \tag{5.28}$$

We now consider in more detail the condition of exponentiality, by rewriting it as

$$g(r) \sim \exp[-\beta V(r) + a + br^2 + \ldots], \quad r \to 0. \tag{5.29}$$

The constant a is positive (Jankovic 1977). This is especially important for the theory of nuclear reaction rates in the interiors of stars.

A great deal of effort has been spent to determine the constants entering exponent (5.29). The point is that the behavior of $g(r)$ at small r determines the lowering of the Coulomb barrier whose magnitude and shape define the fusion rates in the interior of stars. The reaction rate depends on the probability of two nuclei approaching each other at distances of 10^{-13}–10^{-12} cm. In strongly nonideal plasmas, the screening effects increase $g(r)$ at small r. Consequently, they increase the rates of the nuclear reactions (Salpeter 1954; Graboske *et al.* 1973; Jankovic 1977). The lowering of the barrier is given by the expression

$$\exp(a) = \lim\{g(r)\exp[\beta V(r)]\}. \tag{5.30}$$

This effect can also be important for some exotic schemes of laser fusion.

5.1.9 *Asymptotic expressions*

The equilibrium properties of OCP are defined by the free energy $F = F_{\mathrm{id}} - F_1$, where

$$\beta F_1 = \ln \int \ldots \int \frac{\prod\limits_{i=1}^{N} d\mathbf{r}_i}{V^N} \exp\left\{-\beta\left[\sum_{i<j}^{N} \frac{(Ze)^2}{|\mathbf{r}_i - \mathbf{r}_j|} - \frac{N}{V}\sum_{i=1}^{N} \int d\mathbf{r} \frac{Ze^2}{|\mathbf{r}_i - \mathbf{r}|}\right.\right.$$
$$\left.\left. + \frac{N^2}{2V^2} \iint d\mathbf{r}d\mathbf{r}' \frac{e^2}{|\mathbf{r} - \mathbf{r}'|}\right]\right\}.$$

Here and below, N is the number of charged particles and V is the volume of the system.

Analytical estimates for βF_1 can only be obtained in the limit of weak ($\gamma \ll 1$) and moderate ($0.3 \leqslant \gamma \leqslant 1$) nonideality. The technique of estimating βF_1 on the basis of Mayer's diagrams, which is well–developed for rapidly decreasing (faster than $\propto r^{-3}$) interaction potentials, is inapplicable for Coulomb systems, because of the divergence of the resulting integrals. In order to avoid these divergences, it is necessary to perform regrouping and selective summation of series in the perturbation theory. This leads to (Abe 1959)

$$\beta F_1 = N\left\{ -\frac{1}{3}(\sqrt{3}\gamma^{3/2}) + S_2(\gamma) + S_3(\gamma) + \ldots \right\},$$

where the first term corresponds to a summation of the "ring" diagrams (Debye approximation). The terms S_2 and S_3, which are of higher order with respect to γ, are nonpower density functions and effectively describe the interaction of groups of particles in terms of the shielded Coulomb potential,

$$\beta V(r) = \frac{\beta e^2}{r} e^{-r/\lambda_{\mathrm{TF}}}.$$

The expression for S_2,

$$S_2 = \frac{N}{2} \int d\mathbf{r}\left[e^{-\beta V} - 1 + \beta V - \frac{1}{2}(\beta V)^2 \ldots \right],$$

is $O(\gamma^3 \ln \gamma)$ in the limit $\gamma \to 0$. Instead of using cumbersome analytical expressions for S_2 and S_3, a numerical calculation of the respective integrals was performed by Rogers and DeWitt (1973), with a subsequent approximation of the results. The relation for βF_1 leads, in a standard manner, to equations for the internal energy and thermal conductivity,

$$\frac{\beta U}{N} = -\frac{1}{2}\left(\sqrt{3}\gamma^{3/2} \right) + \frac{3}{2}\gamma \frac{d}{d\gamma}[S_2(\gamma) + S_3(\gamma) + \ldots],$$

$$\frac{c_V}{NkT} = \frac{1}{4}\left(\sqrt{3}\gamma^{3/2} \right) - \frac{3}{2}\gamma^2 \frac{d^2}{d\gamma^2}[S_2(\gamma) + S_3(\gamma) + \ldots].$$

In the absence of electron shielding ($r_s = 0$) and in accordance with the virial theorem,

$$\frac{p - p_0}{p_0} = \frac{1}{3}\frac{\beta U}{N}, \quad p_0 = NkT/V.$$

In the limit of weak nonideality ($\gamma \ll 1$), the dominant term of these expressions is $O(\gamma^{3/2})$, which corresponds to the well–known Debye–Hückel approximation,

$$\frac{\beta F_1}{N} = -\frac{1}{3}\varepsilon; \quad \beta\frac{U}{N} = -\frac{1}{2}\varepsilon,$$

$$\frac{c_V}{NkT} = \frac{1}{4}\varepsilon; \quad \varepsilon = \sqrt{3}\gamma^{3/2}.$$

However, as the nonideality increases ($\gamma \to \infty$), the term $O(\gamma^{3/2})$ is compensated by the contributions of other integrals in the βF_1 expansion. The dominant terms in this limit scale as $\propto \gamma$, which readily follows from the results of Monte Carlo calculations. The transition from the $\propto \gamma^{3/2}$ dependence to a weaker, $\propto \gamma$, scaling occurs in the range of $0.3 \lesssim \gamma \lesssim 0.75$. Therefore, the value of $\gamma \sim 0.75$ is taken to be the lower limit for the range of existence of a strongly nonideal plasma. Note that the asymptotic energy estimates can be formulated for OCP,

$$\frac{\beta U}{N} = -\frac{\sqrt{3}}{2}\gamma^{3/2} \quad \text{for} \quad \gamma \ll 1,$$

$$\frac{\beta U}{N} = -\frac{9}{10}\gamma \quad \text{for} \quad \gamma \gg 1.$$

The first one corresponds to the linear Debye approximations, whereas the second one corresponds to the model of nonoverlapping ion spheres.

The range of existence of a strongly nonideal Coulomb plasma (Coulomb fluid) is limited by the maximum value of $\gamma \sim 155$–171 and by the crystallization limit (see Section 5.1.4). Therefore, at $\gamma \lesssim 0.75$, the plasma is described by asymptotic expressions (Rogers and DeWitt 1973) based on perturbation theory, and at $\gamma \gtrsim 155$–171 it forms a Coulomb crystal. In the intermediate range, as revealed by the analysis of numerical experiments, the plasma may be treated as a disordered ion lattice described by relations (5.8) and (5.9).

Analysis by Iosilevskii (1980) shows that even the most simple correction of approximations applicable at $\gamma < 1$ substantially increases their capabilities at $\gamma \simeq 1$.

A series of known approximations give the correlation function of the form

$$g(x) = 1 - \exp(-Ax)/(Bx), \quad x = r/r_\mathrm{D}. \tag{5.31}$$

Such a linearized form violates the condition (5.26). The transition to the nonlinear form,

$$g(x) = \exp[-\exp(-Ax)/(Bx)], \tag{5.32}$$

violates condition (5.27). A method was proposed by Gryaznov and Iosilevskii (1973, 1976), and Ebeling et al. (1991), which consists of the simultaneous inclusion of conditions (5.25)-(5.27), where only the positive part of $g(r)$ is employed while the parameter B becomes a normalizing constant selected from condition (5.27). For example, a linearized Debye–Hiickel approximation (LDH) with $A = 1$ and $B = \Gamma^{-1}$ assumes the form LDH*,

$$g(x) = 1 - \frac{R}{x}\exp(R - x), \quad x \geqslant R; \quad g(x) = 0; \quad x < R, \tag{5.33}$$

where $R = (1 + 3\Gamma)^{1/3} - 1$ and $\Gamma = (Ze)^2/(kTr_\mathrm{D})$.

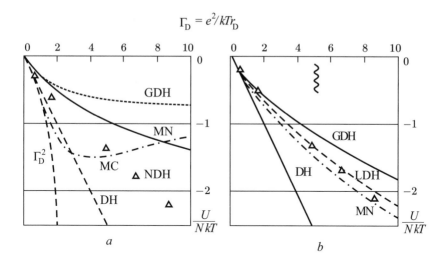

$$\Gamma_D = e^2/kT r_D$$

FIG. 5.5. Internal energy of the OCP, as given by different asymptotic approximations before (a) and after (b) the correction, in accordance with relations (5.25)–(5.27) (Gryaznov and Iosilevskii 1973).

In this approximation,

$$U/(NkT) = (-1/4)[(1 + 3\Gamma)^{2/3} - 1]. \qquad (5.34)$$

Figures 5.5 and 5.6 show the results of four approximations before and after the correction, which is analogous to that described above: Linearized and non-linearized Debye approximations (LDH and DH), the Debye approximation in a grand canonical ensemble (GDH), and approximations which asymptotically take into account the dependence of the screening distance on Γ (MN) (Mitchell and Ninham 1968; Guttman et al. 1968; Cohen and Murphy 1969). One can see that this correction improves considerably the agreement with computer experiment (MC), even in the cases when the oscillations of the correlation functions occur in the numerical solutions (Hirt 1967; Hansen 1973; Carley 1974) (behind the wavy line in Fig. 5.5).

Gryaznov and Iosilevskii (1976) have constructed a model describing the oscillations of $g(x)$ (LDH**):

$$g(x) = 1 - \frac{A}{x}\exp(-x)\cos Bx, \quad x \geqslant R, \quad g(x) = 0, \quad x \leqslant R, \qquad (5.35)$$

where A and B are the normalization constants, $R = R(A, B)$. The MSA theory (Lebowitz and Percus 1966), as well as the MN model, satisfies the criteria (5.25)–(5.28) and at the same time, as shown by Gillan (1974), describes well the results of the Monte Carlo calculations (see Fig. 5.6).

Therefore, the satisfaction of conditions (5.25)–(5.28) is an effective way to check the validity of the results of thermodynamic calculations obtained both

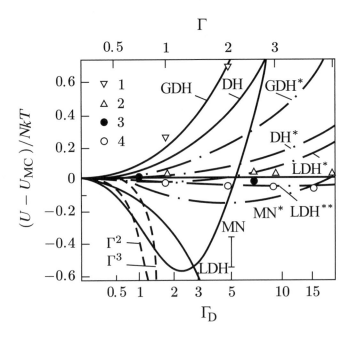

FIG. 5.6.   Coulomb interaction energy, U_{MC}, from the Monte Carlo method of expansion over Γ (Brush et al. 1966; Hansen 1973): LDH – first order (Debye–Hückel approximation); Γ^2 – second order (Abe 1959); Γ^3 – third order (Cohen and Murphy 1969). Simplest models: DH – nonlinearized form of Debye–Hückel approximation; GDH – Debye approximation in a grand canonical ensemble (Likal'ter 1969; Ebeling et al. 1991); MN – (Mitchell and Ninham 1968; Guttman et al. 1968); LDH*, LDH**, DH*, GDH*, MN* – the same approximations after correction with the local quasineutrality condition (5.25)–(5.27). More complex approximations: 1 – HMC (Carley 1974); 2 – HMC (Springer et al. 1973); 3 – solution of the BBGKY hierarchy equations (Hirt 1967); 4 – MSA (Gillan 1974). (After Gryaznov and Iosilevskii 1976).

with numerical and analytical techniques. For instance, the failure of Carley (1974) to satisfy condition (5.27) affected the results (see Fig. 5.6), and pointed to their considerable incorrectness.

5.1.10   OCP ion mixture

A mixture of charges Z_1 and Z_2 is interesting for astrophysics. In the ion–sphere model, the static energy of the binary mixture is given by the expression

$$u_{ex}/(nkT) = -0.9[x_1 Z_1^{5/3} + (1 - x_1)Z_2^{5/3}]\bar{Z}^{1/3}(e^2/r_s kT)$$
$$= -0.9\overline{Z^{5/3}}(\bar{Z})^{1/3}\gamma', \tag{5.36}$$

where $n = n_1 + n_2$, $x_1 = n_1/n$, $\bar{Z} = (Z_1 n_1 + Z_2 n_2)/n$, $r_s = (4\pi n/3)^{-1/3}$, and $\gamma' = e^2/r_s kT$.

If we use the concept of the mean ion charge, Eq. (5.36) suggests the following law of charge averaging (Salpeter 1954):

$$Z_{\text{eff}}^2 = \overline{Z^{5/3}}(\bar{Z})^{1/3}. \tag{5.37}$$

Note that this averaging technique differs substantially from the Debye–Hückel one, where $Z_{\text{eff}}^2 = \overline{Z^2}$. Since the Madelung term predominates in u_{ex}, it is Eq. (5.37) that corresponds to strong nonideality.

DeWitt and Hubbard (1976) and Hansen *et al.* (1977) have performed the Monte Carlo calculations of the thermodynamic properties of mixtures for $Z_2/Z_1 = 2$ and 3, and have obtained the solution to the hypernetted chain equations. It turned out that the excess free energy can be presented with very good accuracy as a linear superposition of free OCP energies, which are homogeneous with respect to charge (Slattery *et al.* 1980),

$$F_{\text{ex}}/(nkT) = f = x_1 f_0(Z_1^{5/3}\gamma') + (1 - x_1)f_0(Z_2^{5/3}\gamma'). \tag{5.38}$$

A binary ionic mixture can be separated into homogeneous phases (Stevenson 1975), such as the H^+ and He^{++} mixtures in the interior of giant planets. This occurs at temperatures below some pressure–dependent critical value T_{cr}. Hansen *et al.* 1977 found that in a hydrogen-helium OCP ($x_{\text{He}} = 0.28$) at pressure of 6 TPa the critical temperature is $T_c = 6300$ K. Under these conditions, $r_s = 0.85a_0$. Essential is that if the linear superposition (5.38) is quite accurate, the phase separation is impossible. Of considerable importance in describing the separation effect is the inclusion of the electron background correlation. As shown by Pollock and Alder 1977, the separation can occur because the local quasineutrality is more effective in homogeneous phases than in the mixture.

5.2   Multicomponent plasma. Results of the perturbation theory

The main disadvantage of the one–component plasma model resides in the extremely simplified inclusion of the opposite sign charge, which is treated as an unstructured compensating background. More sophisticated plasma models provide an explicit inclusion of the structure and interaction of charges of all signs, with the necessary description of quantum effects for the Coulomb interaction. The point is that if the quantum–mechanical features of the problem are neglected, this can lead to major difficulties with the classical description of the particle motion at small distances ($\sim \lambda_e$), because of the divergence of the coordinate part of the Gibbs probability for the oppositely charged particles (Norman and Starostin 1970). In quantum theory, such a divergence is naturally eliminated by the growth of the mean momentum and kinetic energy upon the localization of charges, which ensures the stability of the entire system. Yet the quantum effects lead to the formation of bound states (that is, molecules, atoms, and ions) which, in turn, affect the interaction of free charges. Numerous attempts have been made to maintain the classical formalism, by introducing a cutoff in the Coulomb potential at small distances with the elimination of configurations with closely spaced charges. In this case, the cutoff parameter is included in the final answer and the models become thermodynamically unstable at $\gamma \gtrsim 1$, when the cutoff radius becomes comparable with the interparticle distance.

A consistent quantum–mechanical treatment of the problem is based on the Hamiltonian description of the full interaction between all charges. This corresponds to the "physical" model of a multicomponent plasma, where the contribution of discrete spectrum is finite and occurs simultaneously with the free charge contribution (Ebeling et al. 1976). The "physical" model is the most general and consistent one to describe real plasmas. However, in practice the calculations with this model are very laborious and have not yet been implemented widely. The problem is that, being applied to partially ionized plasmas of multielectron elements, the physical model requires the quantum–mechanical calculations of the internal structure of bound states, which is analogous to the calculations of atoms and ions with the Hartree–Fock method.

In a rarefied plasma, where configurations with closely spaced particles are quite unlikely, considerable simplifications are possible. The principal simplification consists of the separate description of the discrete and continuous spectra states. The first spectrum corresponds to the internal structure of atoms and ions, and the second one to an electrically charged component. This is the essential approximation of the so–called "chemical" model (Ebeling et al. 1991), which currently is the most popular one in plasma physics. In this model, the number of particles of a different type, $\{N_i\}$, is governed by the conditions of chemical ionization equilibrium, $\sum \mu_i = 0$:

$$\left(\frac{\partial F(V, T, N_i)}{\partial N_k} \right)_{V,T} = 0. \tag{5.39}$$

Here, all hypotheses concerning the structure of particles and their interactions are contained in the expression for the free energy, $F(V, T, \{N_i\})$,

$$F(V, T, N) = F^k + F^c + F^b + F^l. \tag{5.40}$$

The contribution of the discrete spectrum, F^b, in this model is calculated independently of the continuous spectrum contribution, which is represented by the kinetic part, F^k, and various corrections for the interparticle interaction, F^c.

If radiation is in local thermodynamic equilibrium with matter, the contribution of the free energy of the photon gas becomes significant upon extremely high heating or strong rarefaction, $F^l = -(4\sigma/3c)VT^4$, where σ is the Stefan–Boltzmann constant and c is the speed of light (Landau and Lifshitz 1980). For instance, for tungsten of solid state density the contribution of thermal radiation becomes noticeable at $T \gtrsim 10^7$ K ($p \gtrsim 10^3$ TPa). The free energy of an ideal gas is

$$F^k = kT \sum_k N_k \left\{ \mu_k - \frac{F_{3/2}(\mu_k)}{F_{1/2}(\mu_k)} \right\}, \quad F_p(\mu_k) = \frac{1}{\Gamma(p+1)} \int\limits_0^\infty \frac{t^p dt}{e^{t-\mu_k} - 1}, \tag{5.41}$$

where the reduced chemical potential μ_k (in units of kT) is a measure of the degree of plasma degeneracy and is determined by the relationship

$$F_{1/2}(\mu_k) = \frac{N_k}{g_k V}\lambda_k^3 \sim n_k\lambda_k^3.$$

For heavy particles we have $n_k\lambda_k^3 \ll 1$ and $\mu_k \ll 1$, which provides the transition to the classical limit of Boltzmann statistics,

$$F^{\mathrm{k}} = kT\sum_k N_k\left[\ln\left(\frac{N_k}{g_k V}\lambda_k^3\right) - 1\right]. \qquad (5.42)$$

The inclusion of the electron component degeneracy reduces the degree of plasma ionization while the kinetic part of the electron pressure is increased.

The chemical model usually involves relations that are derived for a fully ionized plasma with the methods of modern perturbation theory (Ebeling *et al.* 1976). The perturbation theory is developed up to an arbitrary order with respect to the expansion parameter and is equipped with a diagram technique, which facilitates classification and identification of terms in the corresponding series. The major correction for plasmas is calculated by the summation of the so-called "ring" diagrams (Graboske *et al.* 1971; Ebeling *et al.* 1976). For degenerate plasmas, it corresponds to the Gell–Mann and Brueckner (1957) model, whereas in the Boltzmann limit it leads to the Debye–Hückel model,

$$\beta F_2^{\mathrm{c}} = -\sum_i N_i\frac{1}{3}\Gamma P(\gamma). \qquad (5.43)$$

Here, the parameter of nonideality of a multicomponent plasma is

$$\Gamma = 2\pi^{1/2}e^3\beta^{3/2}(N/V)^{1/2}\left(\sum_i Z_i^2\theta_i N_i/N\right)^{3/2}, \qquad (5.44)$$

where $N = \sum_i N_i$ and the parameter $\theta_i = F_{-1/2}(\mu_i)/F_{1/2}(\mu_i)$ counts the degeneracy of the i species. The additional quantum–mechanical correction in Eq. (5.43) takes into account the effects caused by the uncertainty principle in a high–temperature plasma (that is, when $\beta e^2\langle Z^2\rangle < \lambda_i$, see Fig. 5.7) and corresponds to the effective repulsion of charges at small distances (DeWitt 1962; Graboske *et al.* 1969, 1971):

$$P(\gamma) = 1 - (3/16)\pi^{1/2}[(Z_{\mathrm{e}}^4 N_{\mathrm{e}}^2\gamma_{\mathrm{ee}} + 2Z_{\mathrm{e}}^2 Z_{\mathrm{i}}^2 N_{\mathrm{e}}N_{\mathrm{i}}\gamma_{\mathrm{ei}}$$
$$+ Z_{\mathrm{i}}^2 Z_{\mathrm{j}}^2 N_{\mathrm{i}}N_{\mathrm{j}}\gamma_{\mathrm{ij}})/N^2\langle Z^2\rangle^2]$$
$$+ (1/4)\gamma_{\mathrm{e}}^2\{[Z_{\mathrm{e}}^4 N_{\mathrm{e}}^2 + Z_{\mathrm{e}}^2 Z_{\mathrm{i}}^2 N_{\mathrm{e}}N_{\mathrm{i}}(1 + m_{\mathrm{e}}/m_{\mathrm{i}})$$
$$+ Z_{\mathrm{i}}^4 N_{\mathrm{i}}^2(m_{\mathrm{e}}/m_{\mathrm{i}})]/N^2\langle Z^2\rangle^2\}, \qquad (5.45)$$

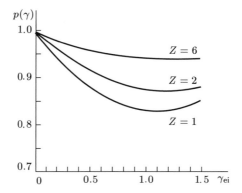

FIG. 5.7.   Effect of the uncertainty principle on the equations of state (5.43) and (5.45) (Graboske *et al.* 1971).

$$\gamma_i = \frac{\lambda_i}{r_D}; \quad \gamma_{ei} = \frac{\lambda_{ei}}{r_D} = \frac{\hbar/(2\mu_{ei}kT)^{1/2}}{r_D}; \quad \mu_{ei} = \frac{m_e m_i}{m_e + m_i},$$

where $r_D = [4\pi\beta e^2(N/V)\sum_j Z_j^2 N_j\theta_j]^{-1/2}$ is the Debye screening distance. The next (after Eq. 5.43) expansion term is referred to as the ladder term; it describes the binary interactions of charges in terms of the dynamically screened Coulomb potential. The general expression for the ladder correction is given by Ebeling *et al.* (1976) and includes multidimensional integrals of complex structure. In the high–temperature approximation, $\gamma_{ii} < \Gamma < \gamma_{ei}$ and $\gamma_{ee} < 1$ (for a hydrogen plasma, $50 \lesssim kT \lesssim 2000$ eV), the ladder correction takes the form (Ebeling *et al.* 1976)

$$\beta F_3^c = -\sum_i N_i \frac{1}{12} \frac{\Gamma^2}{N^2\langle Z^2\rangle^2} \Big\{ Z_e^6 N_e^2(\ln\gamma_{ee} + 0.887)$$

$$+2Z_e^3 N_e \sum_i Z_i^3 N_i(\ln\gamma_{ei} + 0.887)$$

$$+\sum_i\sum_j Z_i^3 Z_j^3 N_i N_j[\ln Z_i Z_j\beta e^2/r_D + 0.420]\Big\}. \quad (5.46)$$

The first term in this expression describes electron–electron interactions, the second one describes electron–ion interactions and the third one describes ion–ion interactions. In order to describe quantum and diffraction effects in this and subsequent terms of perturbation theory, a number of approximations of a much more complex structure have been proposed (Graboske *et al.* 1971).

    Along with the Coulomb interaction, perturbation theory methods allow one to calculate the corrections for exchange interaction between free charges of identical spins. The first–order correction for arbitrary statistics takes the form

$$\beta F^{\mathrm{c}}_{\mathrm{exch}_1} = -N_{\mathrm{e}} \frac{1}{2} \gamma_{\mathrm{e}}^2 \frac{N_{\mathrm{e}}}{\sum_i Z_i^2 N_i} \int_{-\infty}^{\mu_{\mathrm{e}}} \frac{d\mu'_{\mathrm{e}} F^2_{-1/2}(\mu'_{\mathrm{e}})}{F^2_{1/2}(\mu'_{\mathrm{e}})}. \tag{5.47}$$

The exchange integrals for the cases of a Boltzmann and degenerate plasma are calculated analytically, thus bringing about the following asymptotes:

$$\lim_{\mu_{\mathrm{e}} \ll -1} \beta F = N_{\mathrm{e}} \frac{1}{4} \gamma_{\mathrm{e}}^2 \frac{N_{\mathrm{e}}}{\sum_i Z_i^2 N_i}, \tag{5.48}$$

$$\lim_{\mu_{\mathrm{e}} \gg 1} \beta F = -N_{\mathrm{e}} \frac{1}{2} \gamma_{\mathrm{e}}^2 \frac{N_{\mathrm{e}}}{\sum_i Z_i^2 N_i} \frac{1}{2} \frac{\frac{4}{\pi}\mu_{\mathrm{e}}^2 - \frac{2}{3}\pi \ln \mu_{\mathrm{e}} + c_3}{\left[\left(\frac{4}{3}\sqrt{\pi} \right) \mu_{\mathrm{e}}^{3/2} \right]^2}. \tag{5.49}$$

One can see that the classical asymptote (5.48) is applicable for $-\infty \lesssim \mu_{\mathrm{e}} \lesssim 1.5$. Second–order exchange corrections (DeWitt 1962) can only be calculated for the Boltzmann case (for $\mu_{\mathrm{e}} \lesssim 4$),

$$\beta F^{\mathrm{c}}_{\mathrm{exch}_2} = -N_{\mathrm{e}} \frac{\pi^{3/2}}{2^{7/2}} \ln \left[2 \frac{\lambda_{\mathrm{F}} \gamma_{\mathrm{e}}}{2S_{\mathrm{e}} + 1} \left(\frac{Ze^2 N_{\mathrm{e}}}{\sum_i Z_i^2 N_i} \right)^{1/2} \right]$$
$$- \frac{N_{\mathrm{e}}}{12\pi} \left[\frac{1}{12\lambda_{\mathrm{F}}} \left(\frac{\beta e^2}{\lambda_{\mathrm{F}}} \right)^3 \right] \left(\frac{N_{\mathrm{e}}}{\sum_i N_i} \right)^2.$$

The general expression for the third–order exchange correction, as obtained by Ebeling *et al.* (1976) in the case of nondegenerate plasma, reduces to

$$\beta F^{\mathrm{c}}_{\mathrm{exch}_3} = -\frac{N_{\mathrm{e}}}{12\pi} \left[\frac{1}{12\lambda_{\mathrm{F}}} \left(\frac{\beta e^2}{\lambda_{\mathrm{F}}} \right)^3 \right] \left(\frac{N_{\mathrm{e}}}{\sum_i N_i} \right)^2.$$

The relative contribution of the corrections to the pressure of hydrogen plasma is illustrated in Fig. 5.8.

These models obtained with perturbation theory are asymptotic and hence applicable, strictly speaking, only for small values of the expansion parameters, $\Gamma \ll 1$ and $\gamma \ll 1$. According to Graboske *et al.* (1971), the range of validity of the ring model (5.43) is estimated as $\Gamma \leqslant 0.5$, whereas in Ebeling *et al.* (1976) the estimate is more moderate, $\Gamma \lesssim 0.1$. According to Graboske *et al.* (1971), the interference effects for hydrogen are adequately described by Eq. (5.45) at $\gamma < 1.1$. In the region of increased nonideality, $\Gamma \gtrsim 1$ and $\gamma \gtrsim 1$, the results of perturbation theory are inapplicable, and here one must either employ nonparametric methods or use extrapolations. For the latter case, the ring approximation in the grand canonical ensemble of statistical mechanics is convenient (Gryaznov and Iosilevskii 1976).

In the ring approximation, the expression for the thermodynamic potential of a plasma in the grand canonical ensemble takes the form (Gryaznov and Iosilevskii 1976)

$$-\beta\Omega = \beta pV = V\left[\sum_k \xi_0^k + \frac{1}{12\pi}\left(4\pi f\sum_k \xi_0^k Z_k^2\right)^{3/2}\right],\qquad (5.50)$$

where $\xi_0^k = \lambda_k^{-3}\exp(\mu_k\beta)$ is the activity and Z_k is the charge of the kth ion, and $f = e^2\beta$ is the Coulomb scattering amplitude. Using the relation

$$N_k = \left(\frac{\partial p}{\partial \mu_k}\right)_{T,V} = N_0^k\beta\left(\frac{\partial p}{\partial \xi_0^k}\right)_{T,V},\qquad (5.51)$$

one can readily obtain in the case of single ionization,

$$\beta p = N_0 + 2N_e\left\{\alpha^2\left(\frac{2}{\Gamma}\right) + \frac{\Gamma}{3}\alpha^3\left(\frac{2}{\Gamma}\right)\right\},\qquad (5.52)$$

where N_0 is the neutral number density and $\alpha(2/\Gamma)$ is the positive root of the equation

$$\alpha^3 + x\alpha^2 - x = 0;\quad x = 2/\Gamma.\qquad (5.53)$$

The correction to the enthalpy, $\beta\Delta H$, and the reduction of the plasma ionization potential, ΔI, in this approximation take the form

$$\beta\Delta H = -8N_e\left(1 - \alpha^2 - \frac{\Gamma}{3}\alpha^3\right),\qquad (5.54)$$

$$\beta\Delta I = \chi = \beta(\Delta\mu_a - \Delta\mu_e - \Delta\mu_i) = 2\ln[1 + (1/2)\Gamma\alpha].\qquad (5.55)$$

The ring approximation in the limit $\Gamma \to 0$ has the correct asymptotes, and for high Γ it possesses acceptable extrapolation properties which permit the description of shock–wave compression experiments up to $\Gamma \simeq 2.5$. This is the reason why model (5.50)–(5.55) was used in the engineering calculations of the thermophysical properties of working media in a gas–phase nuclear reactor (Gryaznov and Iosilevskii 1976).

5.3   Pseudopotential models. Monte Carlo calculations

Perturbation theory methods, as discussed in Section 5.2, are based on the expansion over small parameters and therefore are valid, strictly speaking, in the limit of a weakly nonideal plasma. In order to describe the thermodynamics of substantially nonideal plasmas, the Monte Carlo technique (Zamalin *et al.* 1977) is effective: it does not involve expansion over small parameters and, hence, is especially attractive in the case of dense gases and liquids as well as OCP (see

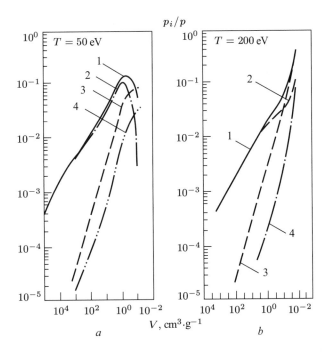

FIG. 5.8.   Coulomb and exchange corrections in hydrogen plasma for $kT = 50$ eV (a) and
$kT = 200$ eV (b): p_{3L} – ladder correction (5.46); p_{ring} – ring correction (5.43); p_{1ex} –
exchange correction (5.44); p_4 – total value of Coulomb corrections.

Section 5.1.2), where the form of the interparticle interaction is defined. Based
on the first principles of statistical physics, this method enables one to perform
direct computer calculations of mean thermodynamic values (see Section 5.1.1).
It is important that, within this method, the interparticle interaction potential
is assumed to be pre–assigned and, in most concrete calculations, two–particle.
Therefore, all physical hypotheses in this approach refer to a particular form of
interaction potential, and then the Monte Carlo technique enables one to carry
out all thermodynamic calculations through to the end.

However, the application of this general method to multicomponent plasmas
runs into specific difficulties of taking into account – within the classical Monte
Carlo formalism – quantum effects in the electron–ion interactions at short distances. Quantum effects play a decisive role in real plasmas, because they ensure
the stability and lead to the emergence of bound states (Zamalin et al. 1977;
Zelener et al. 1981; Filinov 2000; Filinov and Norman 2000).

The inclusion of the specific quantum–mechanical features within the classical Monte Carlo formalism is provided by a pseudopotential model, where the
electron–ion interaction is described via the effective potential $\Phi(r, T)$ which differs from the initial potential $\Phi(r)$ only at small distances $r \leqslant \lambda_e$. This difference

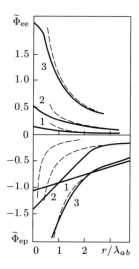

FIG. 5.9.   Pseudopotentials $\widetilde{\Phi}_{ep}$ and $\widetilde{\Phi}_{ee}$ (Zelener *et al.* 1981): $1 - T = 10^3$ K; $2 - T = 10^4$ K; $3 - T = 10^5$ K. The dashed line represents the Coulomb law.

is governed by the quantum spatial uncertainty of the electron within the thermal de Broglie wavelength and by exchange effects (for like particles), as well as by the possibility of bound state formation at low ($kT \lesssim$ Ry) temperatures. For the pseudopotential approach, the quantum partition function reduces to an expression that is classical in form, thus enabling one to employ the classical Monte Carlo calculation technique. In this case, it is possible to take into account rigorously the so–called pair quantum effects governing the deviation of the two–particle pseudopotential from the initial potential.

5.3.1   *Choice of pseudopotential*

Let us consider two particles with the interaction potential $\Phi(r)$. The probability density of finding these particles at a distance r in classical statistics is proportional to $\exp\{-\beta\Phi(r)\}$, while in quantum statistics it is expressed in terms of the Slater sum,

$$S_2(r, T) = 2\lambda_e^3 \sum_\alpha |\psi_\alpha(r)|^2 \exp(-\beta E_\alpha), \qquad (5.56)$$

where ψ_α and E_α are the wavefunctions and the respective eigenvalues of the energy of two particles, respectively, and the summation in Eq. (5.56) is performed over all states of discrete and continuous spectra.

By defining the pseudopotential $\widetilde{\Phi}(r, T)$ as a potential giving in the classical case the same particle distribution in space as the potential $\Phi(r)$ gives in the quantum case, we obtain

$$\widetilde{\Phi}(r, T) = -kT \ln S_2(r, T). \qquad (5.57)$$

In the limit $T \to \infty$ the pseudopotential $\tilde{\Phi}(r,T)$ coincides with $\Phi(r)$, while at finite temperatures they only converge at large distances. It is important that the difference $\tilde{\Phi}(r,T) - \Phi(r)$ is short–range, thus enabling one to construct a thermodynamic perturbation theory (Alekseev et al. 1972) for such deviations.

An important part in the development of the pseudopotential model of plasmas was played by the classic studies performed in Germany by the group of W. Ebeling. Ebeling et al. (1976) and Zelener et al. (1981) presented the results of calculations of pseudopotentials for electron–proton $\tilde{\Phi}_{ep}$ and electron–electron $\tilde{\Phi}_{ee}$ interactions. These were obtained by direct summation in (5.56) and using ψ_α and E_α of an isolated hydrogen atom (Fig. 5.9). In the high–temperature limit, also analytical results are available for $\tilde{\Phi}(r,T)$ (Ebeling and Sandig 1973; Ebeling et al. 1976; Zelener et al. 1981).

Unlike $\Phi(r)$, the pseudopotential at $r \to 0$ has a finite value. This value depends on the particular electronic structure of the atom, which, in turn, is governed by the self–consistent solution of the quantum–mechanical many–body problem and, therefore, cannot be described by the pair approximation (5.57). Neglect of this fact led to serious qualitative errors of the pseudopotential model (Barker 1971) and caused the emergence of nonphysical complexes, because of too large a value of the pseudopotential depth in Eq. (5.57). Indeed, the assumption of pair additivity of the interaction is, in principle, unfit to describe the interaction between dissociation and ionization products, because of the saturation. Therefore, in order to describe chemically reacting systems, one has to modify the pseudopotential model by introducing bound states explicitly.

For this purpose, the following relationship has been introduced (Ebeling and Sandig 1973; Zelener et al. 1981; Filinov and Norman 2000):

$$S_2(r) = S_2^b(r) + S_2^f(r), \tag{5.58}$$

where $S_2^f(r)$ corresponds to the states of the continuous spectrum, and $S_2^b(r)$ of discrete one. In the latter case, $S_2^b(r)$ is selected such as to obtain a converging expression for the partition function (Zelener et al. 1981),

$$\sum_b = \lambda_e^{-3} e^{\beta E_0} 4\pi \int S_2^b(r) r^2 dr = e^{\beta I} \sum_{n=0}^{\infty} \left(e^{-\beta E_n} - 1 + \beta E_n \right). \tag{5.59}$$

The pseudopotential in this case takes the form

$$\Phi^*(r,T) = -\beta^{-1} \ln S_2^f = \lambda_e^3 \left\{ \sum_{E_\alpha = E_0}^{0} |\psi_\alpha(r)|^2 (1 - \beta E_\alpha) \right.$$
$$\left. + \sum_{E_\alpha = 0}^{\infty} |\psi_\alpha(r)|^2 \exp[-\beta E_\alpha] \right\}, \tag{5.60}$$

where in the first term the summation is performed over the states of the discrete spectrum, and in the second term over the states of the continuous spectrum. For hydrogen, the following expression is valid (Zamalin *et al.* 1977) for the depth of the pseudopotential Φ_{ep}^* at $r = 0$:

$$\beta\Phi_{\mathrm{ep}}^*(0, T) = -\ln\{\pi^{1/2}\xi^3[\xi(3) + \beta\mathrm{Ry}\xi(5)] + 2\pi^{1/2}\xi\},$$

$$\xi = 2(\beta\mathrm{Ry})^{1/2}; \quad \xi(k) = \sum_{n=1}^{\infty} n^{-k}; \quad \xi(3) = 1.202; \quad \xi(5) = 1.0369.$$

Pseudopotentials Φ_{ei}^* constructed in this manner were calculated for a number of chemical elements and temperatures (Zamalin *et al.* 1977). Because of the weak dependence of Φ_{ei}^* on the temperature and the type of chemical element, Zamalin *et al.* (1977) proposed a simple approximation which forms the basis of the plasma pseudopotential model of "zero" approximation (the dependence of the pseudopotential on the density was not studied),

$$\beta\Phi_{\mathrm{ei}}^0(x, \beta) = \begin{cases} -\varepsilon, r \leqslant \sigma, \\ -x^{-1}, r \geqslant \sigma, \end{cases} \quad \sigma = \beta e^2\varepsilon^{-1}, \quad x = r/(\beta e^2), \qquad (5.61)$$

$$\beta\Phi_{\mathrm{ee}}^0(x) = \beta\Phi_{\mathrm{ii}}^0(x) = x^{-1},$$

where the numerical parameter of the model, $\varepsilon = 2$, is selected based on experimental data. Because of the simple form of Φ^0 in Eq. (5.61), this model satisfies the similarity relations, which allows one to present the results in compact form (Figs. 5.10 and 5.11) and greatly reduces the calculations. This fact simplifies implementation of the model for concrete thermodynamic calculations, where it is convenient to use approximation of Zamalin *et al.* (1977) with $\varepsilon = 4$,

$$\left.\begin{aligned} \frac{\beta p}{n_{\mathrm{e}}} &= 1 - 0.66\gamma^{3/2} + 0.59\gamma^3 - 0.2\gamma^{3/2}; \\ \frac{\beta F}{N_{\mathrm{e}}} &= -0.89\gamma - 0.45\gamma^2 + 0.54\gamma^3, \quad \gamma^3 = Z^2(Z+1)e^6 n_{\mathrm{e}}\beta^3. \end{aligned}\right\} \qquad (5.62)$$

Deviations between the real pseudopotential (5.60) and the zero approximation model (5.61) can then be taken into account with the thermodynamic perturbation theory developed by Zamalin *et al.* (1977) for arbitrary local potential.

As a whole, the range of validity of the pseudopotential Monte Carlo model proves limited because of the fact that the multiparticle interactions, degeneracy effects, and charge–neutral and neutral–neutral interactions are ignored. Moreover, one should know the discrete energy spectrum (5.51) and (5.57), which might be distorted in a dense plasma as a result of strong interaction and, generally speaking, is not known in advance (see Section 5.4).

Iosilevskii (1980) has proposed, for describing the thermodynamics of a partially ionized degenerate plasma, a model in which the interparticle interaction takes the form

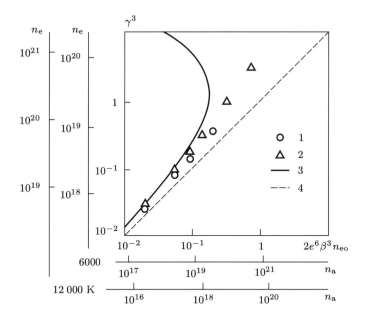

FIG. 5.10. Monte Carlo calculations of the equation of state for a plasma with the pseudopotential approximation (5.61) (Zamalin *et al.* 1977), $\gamma^3 = Z^2(Z+1)e^6 n_e \beta^3$: 1 – $\varepsilon = 2$ in (5.61); 2 – $\varepsilon = 4$; 3 – Debye dependence; 4 – ideal plasma

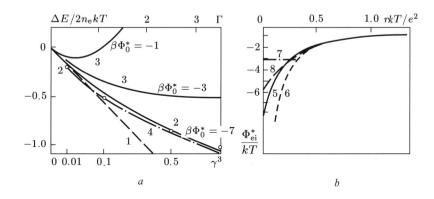

FIG. 5.11. Dimensionless plasma energy (Iosilevskii 1980) (a) and plasma pseudopotential (b). 1 – Debye approximation; 2 (points) – energy calculated using the Monte Carlo method for pseudopotential (5.61) with the depth $\beta\Phi_{ei}^*(0) = -3$ (Zamalin *et al.* 1977); 3 – linearized approximation (5.63)–(5.71) for different depths of pseudopotential (5.63); 4 – nonlinearized approximation with the depth $\beta\Phi_{ei}^*(0) = -6$ of pseudopotential (5.63); 5 – pseudopotential of hydrogen at $T = 10^5$ K (Zamalin *et al.* 1977); 6 – Coulomb potential; 7 – pseudopotential (5.61); 8 – pseudopotential (5.63).

$$\Phi_{\text{ei}}^*(r) = -\frac{e^2}{r}\left[1 - \exp\left(-\frac{r}{\sigma}\right)\right]; \quad \Phi_{\text{ee}}^* = \Phi_{\text{ii}}^* = \frac{e^2}{r}. \tag{5.63}$$

Following Gryaznov and Iosilevskii (1973), we shall write the binary correlation functions in the following form:

$$g_\pm(r) = g_{\alpha\beta}(r) = 1 \pm \psi_0 \exp(-\nu r)\left[\frac{\sinh(\omega r)}{\omega r}\right], \quad \alpha, \beta = \text{e}, \text{i}, \tag{5.64}$$

as derived within the self–consistent ring approximation for potential (5.63) in the limit $\Gamma \to 0$. The amplitude of the screening cloud, ψ_0, and the screening radii, ν^{-1} and $1/\omega$, are found from the screening condition and the approximate relation between ψ_0 and potential depth $\Phi_{\text{ei}}^*(0)$ (Iosilevskii 1985),

$$n_e \int [g_+(r) - g_-(r)]d\mathbf{r} = 1, \tag{5.65}$$

$$n_e \int [g_+(r) - g_-(r)]\left(\frac{r}{r_{\text{D}}}\right)^2 d\mathbf{r} = 3, \tag{5.66}$$

$$\psi_0 \simeq \beta(-\Phi_{\text{ei}}^*(0) + \Delta\mu_e + \Delta\mu_i). \tag{5.67}$$

Corrections to the potential energy, $\Delta\Pi^*$, and internal energy, ΔE^*, as well as to the pressure and chemical potential, have the following form (Iosilevskii 1980):

$$\Delta\Pi^* = N_e V^{-1} \int \frac{e^2}{r}[g_+(r) - g_-(r)]d\mathbf{r}, \tag{5.68}$$

$$\Delta E^* = N_e^2 V^{-1} \int [\Phi_{\text{ei}}^*(r)g_+(r) - \Phi_{\text{ii}}^*(r)g_-(r)]d\mathbf{r}, \tag{5.69}$$

$$\Delta p = (3V)^{-1}(2\Delta E^* - \Delta\Pi^*), \tag{5.70}$$

$$\Delta\mu_e = \Delta\mu_i \simeq (2N_e)^{-1}\Delta E^*, \quad N_e = N_i = n_e V. \tag{5.71}$$

Relation (5.70) follows from the virial theorem, whereas Eq. (5.71) highlights the condition that the correlation occurring upon inclusion of the additional charge is proportional to its magnitude. With this model, calculations are reduced to the solution of algebraic equations. In the weak nonideality limit, the obtained results tend to the Debye corrections. At higher densities the corrections are smaller than the Debye values, and become positive at $\sigma/r_{\text{D}} \geqslant 1$. Figure 5.11 shows the dimensionless energy of a free charge system (Iosilevskii 1980). One can see that for like potentials the results of simple calculations using model (5.63)–(5.71) are in adequate agreement with the Monte Carlo results. This is apparently because the general conditions of local electroneutrality, which proved very important (as in the case of the OCP model), are satisfied in model (5.63)–(5.71).

Selection of the single numerical parameter of the model (5.63)–(5.71) – the potential depth $\Phi_{\text{ei}}^*(0)$ – should be based on experimental data, as was done for the Monte Carlo model (5.61). It turned out that the optimal description is provided by choosing $\Phi_{ei}^*(0) \simeq \varepsilon = -kT$. This energy separates the electronic states into free and bound ones. Note that, with such a choice of $\Phi_{ei}^*(0)$, one manages to describe in a similar way experiments in cesium, argon, and xenon (only results for cesium were employed for this choice).

5.4   Bound state contribution. Confined atom model

An adequate inclusion of bound states represents one of the most complicated problems in describing nonideal low–temperature plasmas. The bound states appear separated from free states (5.40) and are described by partition functions,

$$F^b = -kT \sum_k N_k \ln \Sigma_k; \quad \Sigma_k = \sum_{n=1}^{\infty} g_n^k \exp\{-\beta E_n^k\}, \tag{5.72}$$

where g_n^k and E_n^k stand for the statistical weight and excitation energy, respectively, of the nth energy state of the kth particle type. In order to find these quantities, either the spectroscopic measurement data obtained for rarefied plasma, or the results of quantum–mechanical calculations of isolated atoms and ions are usually employed. The partition function diverges and requires cutoff which reflects the presence of the plasma environment. A large number of methods based on qualitative physical considerations have been proposed recently for such a cutoff: at the Debye screening length, at the mean interparticle distance, at the last realized quantum number (due to the effect of fluctuating microfields), etc. These models were reviewed by Armstrong et al. (1967) and Vorob'ev (2000). Usually, the energy of first excited states is comparable to the ionization potential, so that their contribution to Σ_k appears appreciable only at high temperatures, when the plasma is already ionized substantially and contains few neutrals. Because of this, for a rarefied plasma the concrete mechanism of constraint is of less importance than the very fact of the presence of such a constraint. This circumstance explains in part the extremely small number of available studies into the thermodynamics of the discrete spectrum of low–temperature plasma, as well as the particularly qualitative level of the models used. As the pressure rises, the degree of plasma ionization drops down, which increases the sensitivity of thermodynamic functions to specific methods of calculating Σ_k (Fortov et al. 1971) and requires more thorough analysis of the nonideality effects in the bound state contribution.

The inclusion of the quantum–mechanical interaction in perturbation theory leads (Larkin 1960) to the converging expression (hydrogen plasma),

$$\Sigma_k = \sum_n g_n \left(e^{-\beta E_n} - 1 + \beta E_n \right), \tag{5.73}$$

used in the pseudopotential model discussed in Section 5.3.

Furthermore, in a strongly compressed plasma the interparticle interaction causes considerable shift, deformation and splitting of energy levels, i.e., the

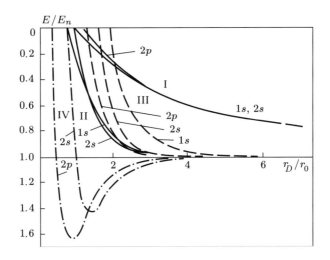

FIG. 5.12. Deformation of the hydrogen energy spectrum (Savukinas and Chizhunas 1974):
I – Debye–Hückel potential; II – cut-off Coulomb potential; III – boundary condition for
wavefunction $f(r_0) = 0$; IV – boundary condition $\partial f/\partial r|_{r=r_0} = 0$ (condition of periodic-
ity). r_D is the Debye screening distance and r_0 is the boundary atomic radius.

phenomena which cannot be described with perturbation theory and require
the solution of the complete quantum–mechanical problem, with all forms of
interaction taken into account.

The simplest problem is to calculate the bound states of a single electron in
potentials of different structures modeling the plasma environment. The results
of such calculations by Savukinas and Chizhunas (1974) are given in Fig. 5.12.
It shows a sharp variation of the energy structure in a relatively narrow range
of atomic compression.

The thermodynamic consequences of such deformation of the energy spectrum for model (5.74) have been analyzed by Graboske *et al.* (1969) and illustrated in Fig. 5.13, where the relative contributions of various thermodynamic
corrections are shown. One can see that in the range of increased plasma densities, even for hydrogen one should expect substantial changes in the plasma
compressibility due to deformation of bound states. Unfortunately, this simplest
model of a hydrogen plasma cannot be compared with experimental data which
are only available for multielectron atoms.

In order to describe the thermodynamic properties of strongly compressed
plasmas, one has to calculate the internal structure of atoms and ions as well as
to include the effects of compression on the position of energy spectra of bound
electrons. Such a problem can only be solved by employing numerical methods,
with the leading one being the Hartree–Fock method which is successfully used
for calculations of atomic structures.

This method deals with a system of N electrons in the field of the atomic

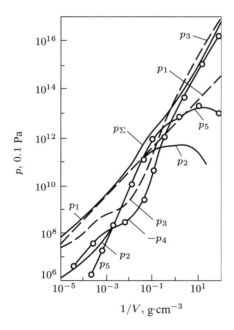

FIG. 5.13. Equation of state for the hydrogen plasma, $kT = 2$ eV (Graboske *et al.* 1969): p_1 is the ideal gas ion pressure; p_2 is the pressure where the deformation of the discrete spectrum is taken into account (Section 5.4); p_3 is the electron pressure (5.41); p_4 is the Coulomb correction; p_5 takes into account the atom size is taken into account; p_Σ is total pressure.

nucleus with a charge Z. In the atomic unit system ($m_e = 1$, $\hbar = 1$, $e = 1$), the Hamiltonian takes the form

$$\hat{H} = -\sum_{i=1}^{N} \left[\frac{1}{2} \Delta_i + U(r_i) \right] + \sum_{j<k}^{N} \frac{1}{r_{jk}}, \tag{5.74}$$

where $U(r_i)$ is the effective potential which is equal to Z/r_i for an isolated atom.

Based on the variational principle of quantum mechanics, the multielectron wavefunction is determined from the condition of minimum of the functional

$$E = \int \psi*(q^N) \hat{H} \psi(q^N) dq^N, \tag{5.75}$$

where q^N is a combination of $\{\mathbf{r}_1\xi_1, \mathbf{r}_2\xi_2, \ldots, \mathbf{r}_N\xi_N\}$ coordinates and spin variables of electrons, and the integration over dq^N includes, along with integration over coordinates, the summation over spin variables.

Selected as the approximate atomic wavefunction in the Hartree–Fock method is a determinant consisting of one–electron wavefunctions. Assuming that each electron is in the central field of the nucleus and of the remaining electrons, and also noting that the Hamiltonian (5.74) does not depend on spin variables, the one–electron wave function can be represented as

$$\Psi_i(q) = \frac{1}{r} f_{n_i l_i}(r) Y_{l_i}^{m_i}(\theta, \varphi) \chi(S_i). \tag{5.76}$$

Hence, in this approximation the motion of each electron is characterized by the following quantum numbers: the principal number, n, the orbital, l the magnetic, m, and the spin number, S. For given n and l, there exist $2(2l + 1)$ electron states corresponding to different m and S values. It is assumed that all electrons with preassigned n and l have identical radial parts $f_{nl}(r)$. In the absence of spin–orbit interaction, the atomic state is characterized by the complete orbital moment L and spin moment S which are equal to zero when, for any n and l, all $2(2l+1)$ states are occupied, i.e., all the nl–shells are filled. In the case when the atom has unfilled shells, in order to ensure that the atomic wavefunction is the eigenfunction of complete orbital and spin moment operators (i.e., the atom is characterized by given L and S), it is necessary to seek that function in the form of a linear combination of determinants. The coefficients of this combination are found from the condition that the complete orbital and spin moments are equal to the preassigned values.

The application of the variational principle yields the following set of integro-differential equations (Hartree–Fock equations):

$$\left[\frac{d^2}{dr^2} + U_{n_i l_i}(r) - \frac{l_i(l_i + 1)}{r^2} - \varepsilon_{n_i l_i} \right] f_{n_i l_i}(r)$$
$$+ \sum_{n_j \neq n_i} \varepsilon_{n_i l_i n_j l_j} f_{n_j l_j}(r) = G_{n_i l_i}(r), \tag{5.77}$$

where $U_{n_i l_i}(r)$ is the Coulomb potential of the electron interaction with the nucleus and with each other, $G_{n_i l_i}$ is the nonlocal part of the potential, or exchange term, and the quantum numbers n_i and l_i vary in accordance with the chosen atomic configuration. The eigenvalues $\varepsilon_{n_i l_i}$ and nondiagonal Lagrangian multipliers $\varepsilon_{n_i l_i n_j l_j}$, introduced to satisfy the condition of orthogonality of the radial functions $f_{n_i l_i}(r)$ and $f_{n_j l_j}(r)$ at $n_i \neq n_j$, are found from the condition of orthonormality,

$$\int f_{n_i l_i}(r) f_{n_j l_j}(r) d\mathbf{r} = \delta_{n_i n_j}, \tag{5.78}$$

and from the boundary conditions. One of these conditions follows from the finiteness of the wavefunction at the origin of coordinates,

$$f_{n_i l_i}(0) = 0,$$

and the other one is selected based on the particular statement of the problem. For instance, in the approximation of an isolated atom, the electron wavefunction should decrease exponentially at large distances. In the "confined" atom approximation (Gryaznov *et al.* 1980, 1989), which is one of the methods that include the influence of external effects on the bound states, the following condition is imposed on the radial wavefunction:

$$f_{n_i l_i}(r)|_{r=r_c} = 0, \qquad (5.79)$$

which corresponds to the interaction potential

$$U(r) = \begin{cases} -\dfrac{Ze^2}{r}, & r < r_c, \\ \infty, & r > r_c. \end{cases} \qquad (5.80)$$

Let us consider now a three–component plasma consisting of atoms, single-charged ions and electrons. We assume that the atoms are spheres of variable radius r_c, whereas the sizes of electrons and ions are ignored for simplicity. We shall treat the subsystem of finite–size atoms as a set of hard spheres which do not interact when the distance between them exceeds $2r_c$. The free energy of such a model can be written as

$$F(N_a, N_i, N_e, V, T) = F_{id} + F_{hs} + \Delta F_{coul}.$$

The first term is the free energy of the ideal plasma, with the only difference that now the atomic partition function depends, in accordance with the boundary condition (5.80), on the radius. The second term is the contribution of the hard sphere repulsion which also depends on r_c via the dimensionless parameter $\nu = n_a(4\pi r_c^3/3)$. In order to include this contribution, the thermodynamic results for the hard-sphere systems are used, as obtained from molecular dynamics calculations and described with the Pade approximation (Carnahan and Starling 1969),

$$\Delta F_{hs} = n_a kT \nu \frac{3\nu - 4}{(\nu - 1)^2}. \qquad (5.81)$$

For corrections to the pressure and chemical potential, this yields

$$\frac{\Delta p_{hs}}{n_a kT} = 2\nu \frac{\nu - 2}{(\nu - 1)^3}, \quad \frac{\Delta \mu_{hs}}{kT} = \nu \frac{3\nu^2 - 9\nu + 8}{(1 - \nu)^3}. \qquad (5.82)$$

Thus, the free energy in this model depends on r_c due to the compression of atoms, on the one hand, and because of their interaction as hard spheres, on the other. An equilibrium value of atomic radius can be determined from the condition of the minimum of the free energy,

$$\partial F/\partial r_c = 0. \qquad (5.83)$$

The dependence of the atomic partition function on r_c is found from the solution of the Hartree–Fock equations for the ground and excited atomic states,

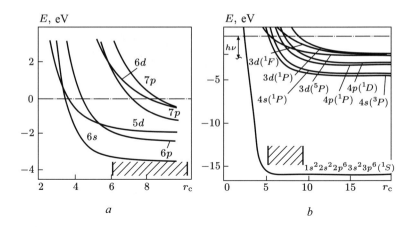

FIG. 5.14. Quantum–mechanical calculation of the energy spectrum of compressed cesium
(a) and argon (b) based on the bound atom model (5.77)–(5.83). The dashed regions
correspond to experiments (Bushman *et al.* 1975; Bespalov *et al.* 1975; Fortov 1982), r_c is
the atomic cell radius in units of the Bohr radius and $h\nu = 2.14$ eV is the energy of the
detected light radiation. The dot–dashed line shows the continuous spectrum boundary.

with boundary condition (5.79) at different r_c (see Fig. 5.14). The solution of
Eq. (5.83) gives an equilibrium value of $r_c(V, T)$, which makes the model thermodynamically closed.

In contrast to the solid–state cell models (Bushman and Fortov 1983), this
approximation is constructed in the framework of the quasichemical method
of description, with explicit inclusion of the translational degrees of freedom
for individual particles. Yet the electrons are separated into two types and are
located inside and outside the cell, with the volume $(3/4)\pi r_c^3$ being just a part
of the average volume per particle.

The thermodynamic calculations performed for model (5.77)–(5.83) demonstrate that the effective repulsion and deformation of the discrete atomic spectrum in the selected potential have a significant influence both on the thermal
and caloric plasma equations of state. Given the pressure and temperature, the
density calculated from this model has a smaller value than that obtained in the
ring Debye approximation; the same is true for the enthalpy at fixed pressure
and volume. The latter is important because it is one of the principal qualitative
results obtained so far experimentally in nonideal plasmas (see Section 5.6).

The effect of energy level deformation has significant influence on the optical
plasma properties, since atomic photoionization by visible radiation occurs from
highly excited energy levels that are distorted even at relatively low compressions.
In Chapter 7 we shall discuss this effect in detail.

5.5 Quasiclassical approximation

As the pressure increases, molecular and ion–molecular structures in the plasma become destroyed, the outer valence electrons leave atoms, the electron shells of atoms and ions reorganize themselves and, at extremely high pressures, disintegrate, which provides a quasi–uniform charge distribution in the atomic cell. In this case, it is possible to use a quasiclassical description expressed via the mean electron density distribution, $n_e(x)$, instead of a quantum–mechanical formalism in terms of the wavefunctions and discrete energy spectrum. The corresponding approximation is known as the Thomas–Fermi model and represents a quasiclassical limit ($\hbar \to 0$) with respect to the Hartree self–consistent field equations (5.77). A comprehensive analysis of this method has been given in the review by Kirzhnits *et al.* (1975).

In this approximation, the quasi–uniform degenerate electronic gas relations are used for the mean electron density (Kirzhnits *et al.* 1975; Shpatakovskaya 2000),

$$n_e(x) = \frac{\sqrt{2}}{\pi^2} T^{3/2} J_{1/2} \left(\frac{p_F^2(x)}{2T} \right)_{T \to 0} \to \frac{p_F^2(x)}{3\pi^2}, \tag{5.84}$$

where $J_{1/2}(y) = \displaystyle\int\limits_0^\infty \frac{\tau y^{1/2}}{\exp(\tau - y) + 1} d\tau$ is the Fermi–Dirac function (in this section we keep using the atomic units). The Fermi energy $p_F^2(x)/2$ is related to the chemical potential via

$$\mu = \frac{p_F^2}{2} + U(r). \tag{5.85}$$

In the thermodynamic Thomas–Fermi model the matter is broken down into electroneutral spherical Wigner–Seitz cells containing a nucleus and Z electrons surrounding the latter. The electrons and nucleus create a self–consistent potential which satisfies the Poisson equation,

$$-\nabla^2 U = 4\pi n_e, \tag{5.86}$$

with the boundary conditions,

$$U(r) \underset{r \to 0}{\to} \frac{1}{r}; \quad U(r) \underset{r \to R}{\to} (R - r)^2, \tag{5.87}$$

where the cell radius R is given by the condition of electroneutrality, $\int n_e(r) d\mathbf{r} = Z$. Numerical integration of (5.84)–(5.87) enables one to determine the intra-atomic density $n_e(x)$ from which all thermodynamic functions of the electronic gas of atomic cell are recovered in accordance with the expression for the free energy,

$$F = \frac{\sqrt{2}}{\pi^2} T^{5/2} \int d\mathbf{r} \left[\frac{p_F^2}{2T} J_{1/2} \left(\frac{p_F^2}{2T} \right) - \frac{2}{3} J_{3/2} \left(\frac{p_F^2}{2T} \right) \right] + \int U(r) n_e(r) d\mathbf{r}. \tag{5.88}$$

In order to obtain the total thermodynamic characteristics of the model, the contribution due to the motion of nuclei is added to the electron terms, as described either in the ideal–gas (5.42) or quasi–harmonic approximations (Kopyshev 1978), or by the OCP model (Kopyshev 1978) (see Section 5.1). At $kT \geqslant 1$ keV, one should take into account also the pressure and energy of equilibrium radiation.

The Thomas–Fermi model is characterized by self–similarity with respect to the nuclear charge Z, i.e., the results obtained for one particular element can be used for an arbitrary substance. The temperature, volume, chemical potential, pressure, and energy in the model have the following scalings on Z (Shpatakovskaya 2000):

$$T^{(Z)} = T^{(1)} Z^{4/3}, \quad V^{(Z)} = V^{(1)}/Z, \quad \mu^{(Z)} = \mu^{(1)} Z^{4/3},$$

$$p^{(Z)} = p^{(1)} Z^{10/3}, \quad E^{(Z)} = E^{(1)} Z^{7/3}.$$

Modifications of the Thomas–Fermi model are associated with the more detailed inclusion of correlation and quantum effects. The correlation corrections are caused by the difference between the Hartree self–consistent field and the true field within the atomic cell. The corrections that result from the antisymmetry of the electron wavefunctions are interpreted as exchange and correlation effects. In addition, force correlation effects occur because of inaccuracy of the independent particle model. The quantum–mechanical corrections are due to the quasiclassical formalism and are divided into a regular, with respect to $\hbar^2$, part (referred to as quantum) reflecting the presence of nonlocal correlation between $n_e(x)$ and the potential $U(x)$ due to the uncertainty principle, and an irregular correction, which reflects nonmonotonic physical quantities due to the discrete energy spectrum. It is important that the most modern modifications of the Thomas–Fermi model (Kirzhnits *et al.* 1975) are associated with the introduction of the oscillation correction, whereas the inclusion of exchange, correlation, and quantum corrections (Kalitkin 1960, 1989) are traditional for this model.

The relative magnitude of the correlation and exchange effects is controlled by dimensionless parameters (Kirzhnits *et al.* 1975) $\delta_{\text{corr}} \sim (n_e^{1/3}/p_F^2)^{\nu}$ and $\delta_{\text{exch}} \sim \delta_{\text{quant}} \sim n_e/p_F^4$. In the degenerate region (for $n_e^{2/3} \gg T$, $p_F \sim n_e^{1/3}$, $\delta_0 \sim n_e^{-1/3}$, $\nu = 2$) these are $\delta_{\text{corr}} \sim n_e^{-1/3}$ and $\delta_{\text{exch}} \sim n_e^{-2/3}$, whereas in the classical region ($n_e^{2/3} \ll T$, $p_F \sim T^{1/2}$, $\delta_0 \sim n_e^{-1/3}/T$, $\nu = 3/2$) we have $\delta_{\text{corr}} \sim n_e^{1/2}/T^{3/2}$ and $\delta_{\text{exch}} \sim n_e/T^2$. The dimensionless parameter δ_0 characterizes the nonideality and equals the ratio of the mean energy of the pair Coulomb interaction to the kinetic energy of the electrons. The range of validity of the quasiclassical method of describing the atomic cell electrons, estimated (Kirzhnits *et al.* 1975) on the basis of these criteria, is defined at $T = 0$ by the condition $n_e \gg 1$ and in hot matter by the condition $n_e \ll T^3$, which corresponds to the conditions of smallness of the binary interaction energy of electrons as compared with their kinetic energy.

The physical conditions of the validity of the quasiclassical model correspond to extremely high pressures $p \gg p_a$ (where $p_a = e^2/a_0^4 \sim 30$ TPa is the atomic unit of pressure) and temperatures $T \gg 10^5$ K, which are realized in the interior of superdense stars and in other astrophysical objects, but are still inaccessible in experiments under terrestrial conditions because they require extremely high local concentrations of energy (Fortov 1982). At present, record-breaking pressures and temperatures are attained by dynamic methods using the powerful shock wave technique. The data obtained in these experiments correspond to pressures from a few to a few tens of TPa, which is much below the range of formal validity of quasiclassical models. Therefore, the existing estimates of the lower range of validity of the Thomas–Fermi model (Altshuler *et al.* 1977) are based on the extrapolation of these experimental results. It is noted that the inclusion of quantum, exchange and correlation corrections (oscillation corrections were not included) substantially improves the extrapolation. According to Altshuler *et al.* (1977), in this case the extrapolation is possible for a cold ($T = 0$) plasma up to pressures of ~ 30 TPa and, at $T \gtrsim 10$ keV, up to about 5 TPa.

Recently, however, such estimates were questioned. The point is that the commonly used version of the quasiclassical model (Latter 1955, 1956; Kalitkin 1960, 1989; Kalitkin and Kuzmina 1975, 1976; Altshuler *et al.* 1977) describes "averaged" characteristics of matter and ignores the so–called shell effects caused by individual peculiarities of the population of atomic energy levels and bands. Elegant studies performed by Kirzhnits *et al.* (1975) and Kirzhnits and Shpatakovskaya (1995) have shown that the shell effects can be qualitatively described within the quasiclassical approximation by including an irregular (with respect to $\hbar^2$) correction corresponding to the oscillating part of the electron density which was previously erroneously discarded. The inclusion of shell effects changes noticeably the equation of state for superdense plasma and causes the emergence of sharp nonmonotonicities of thermodynamic functions in the high–pressure region (where this approximation is justified) (Kirzhnits *et al.* 1975). In addition, discontinuities emerge in the equations of state, associated with the first–order electron phase transitions caused by the "forcing out" of energy levels from the discrete spectrum to the continuous one. This, in turn, causes peculiarities in other physical quantities. Note that the substantial contribution by shell effects at ultra–high pressures follows from the results of quantum–mechanical calculations using the more accurate method of attached plane waves (Bushman and Fortov 1983). These effects are predicted in a wide range of parameters and must disappear only at $n_e \gg Z^4$ in the uniform region, when all atomic energy levels pass over to the continuous spectrum (Bushman and Fortov 1983).

Therefore, the question concerning the lower range of validity of the quantum–static model for superdense plasma is currently open, whereas the behavior of matter in the region of $p > 30$ TPa appears more diversified than previously assumed on the basis of simplified representations (Latter 1955, 1956; Kalitkin 1960, 1989; Kalitkin and Kuzmina 1975, 1976; Altshuler *et al.* 1977). At the same time, the simplicity and universality of the Thomas–Fermi model (5.88) make

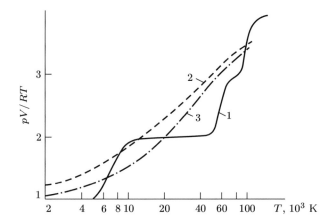

FIG. 5.15.   Thermal equation of state for lithium plasma (Iosilevskii and Gryaznov 1981) at $p = 0.1$ MPa: 1 – calculation from Saha's chemical model; 2 – Thomas–Fermi theory; 3 – Thomas–Fermi theory with corrections (Kalitkin 1960; Kalitkin and Kuzmina 1975, 1976).

it very attractive in the case of an "averaged" description of plasma properties at ultrahigh pressures, when the use of more complex approaches requires laborious quantum–mechanical calculations. Based on various modifications of the quasiclassical model, extensive tables of thermodynamic functions have now been published (Latter 1955, 1956; Kalitkin 1960, 1989; Kalitkin and Kuzmina 1975, 1976, 2000; Altshuler et al. 1977; Kirzhnits and Shpatakovskaya 1995) which cover the range of parameters far beyond the range of validity of this approximation. In particular, Kalitkin and Kuzmina (1975, 1976) propose to use their data for low–density "gas" plasmas, where the "chemical" model of a plasma (see Section 5.2) is traditionally used. Detailed analysis of the validity of the Thomas–Fermi model for this range of parameters is given by Iosilevskii and Gryaznov (1981).

Of special importance in using the quasiclassical method for the description of plasmas of reduced density (as compared with the density of solids) are specific errors introduced by the cell model itself. In this model, all correlations are automatically restricted by the atomic cell size and, therefore, cannot exceed the mean distance between the nuclei. This fact, in particular, sets limits on the applicability of the Thomas–Fermi model for the description of plasmas under conditions typical for this state of matter – when the Debye sphere contains a large number of charges, $\Gamma \lesssim 1$. Therefore, the limit of weakly nonideal (Debye) plasma cannot be retrieved from the Thomas–Fermi theory.

Figures 5.15 and 5.16 show the comparison between the thermal and caloric equations of state for the lithium plasma calculated from the "chemical" plasma model, with the ionization from Saha's equation (see Section 5.2) and in the Thomas–Fermi approximation with quantum and exchange corrections (Kalitkin

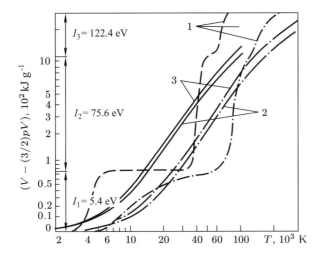

FIG. 5.16. Caloric equation of state for lithium plasma (Iosilevskii and Gryaznov 1981). The dashed curve is for a pressure of 100 Pa; solid curves are for 0.1 MPa; dashed–dotted curves are for 100 MPa. I_1, I_2, I_3 are the potentials of successive ionization. Other designations are the same as in Fig. 5.15.

1960; Kalitkin and Kuzmina 1975, 1976). The stepwise behavior of the thermodynamic functions in Figs. 5.15 and 5.16 is caused by ionization processes and the steepness of these steps increases as the pressure decreases. The Thomas–Fermi theory, which provides a "smoothed" description of thermodynamic functions, fails to convey peculiarities of the ionization. The characteristic maximum error of the quasiclassical theory in Fig. 5.15 increases and, at $\rho \to 0$, tends to unity. Even more dramatic are the deviations in the caloric equation of state (Fig. 5.16), where the maximum difference in energy, given by the Saha and Thomas–Fermi approximations in a rarefied plasma, is comparable with the ionization potential per atom and, thus, exceeds substantially the kinetic energy $(3/2)kT$ in the considered limit. Naturally, the deviations in differential characteristics (heat capacity, compressibility modulus, and so on) will be significant. Analysis by Iosilevskii and Gryaznov (1981) shows that in the limit of full ionization, the internal energy calculated by Kalitkin and Kuzmina (1975, 1976) is shifted relative to the exact value by some constant, which reaches 35–60 % for lithium. Due to the absence of long–range correlations in the cell model, the quasiclassical approximations yield in the caloric equation of state a negative term $\propto \rho^{4/3}$ corresponding to the nuclear charge interaction with electrons uniformly distributed over the cell volume, instead of giving the Debye term ($\propto \rho^{3/2}$).

5.6 Density functional method

The method of density functional (MDF) for the energy (thermodynamic potential) is presently the most effective way of describing strongly interacting dense systems. According to the Hohenberg-Kohn theorem and its generalizations, the energy (thermodynamic potential) and wavefunction of the ground state with a nonuniform electron system are single–valued functionals of the distribution of the electron number density $n_e(\mathbf{r})$. The consistent use of the MDF to describe the properties of dense electron–ion systems combines simplicity of a quasiclassical method with accurate inclusion of exchange-correlation effects and bound states. On the other hand, this method is free of considerable mathematical complications associated with the consistent use of the Hartree–Fock method. It is not surprising that the MDF and its modifications were successfully used in various fields of physics to describe the electronic structure of atoms and molecules, clusters, and other small particles, as well as surface phenomena and the properties of solid and liquid metals and nonideal plasma, etc. Also, the MDF is well suited to investigate equations of state of dense systems.

The description of MDF and some of its applications can be found in numerous reviews (see, for example, Lundqvist and March 1983; Rajagopal 1980; Klyuchnikov and Lyubimova 1987). Here, we dwell on the simplest formulation of MDF – a nonuniform electron fluid at $T = 0$ – and then discuss briefly the calculation of the equation of state for the cell plasma.

The energy of nonuniform electron system in external field $V(r)$ is a functional of its density,

$$E\{n_e(\mathbf{r})\} = \int V(\mathbf{r}) n_e(\mathbf{r}) d\mathbf{r} + \frac{1}{2} \int \frac{n_e(\mathbf{r}) n_e(\mathbf{r}')}{|\mathbf{r} - \mathbf{r}'|} d\mathbf{r} d\mathbf{r}' + G\{n_e(\mathbf{r})\}. \quad (5.89)$$

Here $G\{n_e(\mathbf{r})\}$ is the contribution of the kinetic energy and exchange-correlation effects. The functional G has a universal form, which is not known exactly, so that usually its relatively simple approximations are employed.

Let the electron number density vary slowly, that is, $d \ln n_e(r)/dr \ll k_F(r)$, where $k_F(r)$ is the local Fermi wavevector. Then, one can use the gradient expansion of G,

$$G = \int d\mathbf{r} \left[g_0(n_e) + g_2(n_e) |\nabla n_e|^2 + \ldots \right].$$

The kinetic and exchange-correlation energies are treated separately by writing G as the sum $G = E_{\text{kin}} + E_{\text{e-c}}$. For a degenerate electron gas,

$$E_{\text{kin}} = \frac{3}{10} (3\pi^2)^{2/3} \int n_e^{5/3}(\mathbf{r}) d\mathbf{r} + \frac{1}{72} \int |\nabla n_e(\mathbf{r})|^2 n_e^{-1}(\mathbf{r}) d\mathbf{r}, \quad (5.90)$$

where the second term is the Weizsaecker–Kirzhnitz gradient correction (we use atomic units). A large number of studies have been devoted to the investigation of $E_{\text{e-c}}$. In a local approximation, one of the simplest expressions for $E_{\text{e-c}}$ is the interpolation formula of Pines and Nosierez,

$$E_{\text{e-c}} = - \int n_e(\mathbf{r}) \left[\frac{3}{4} \left(\frac{3}{\pi} \right)^{1/3} n_e^{1/3}(\mathbf{r}) + 0.0474 \right.$$

$$\left. + 0.0155 \ln \left(3\pi^2 n_e^{1/3}(\mathbf{r}) \right) \right] d\mathbf{r}. \quad (5.91)$$

The functional (5.89)–(5.91) can be minimized by using a set of trial functions $n_e(\mathbf{r})$.

It is interesting that even the simplest approximations, such as (5.89)–(5.91), yield very satisfactory results. For example, the functional given above yields a qualitatively adequate description of the surface properties of alkali metals and even of small metal particles. For small particles, the role of the external potential $V(r)$ is played by the field of the ion core $n_e(r) = n_0\theta(R - r)$, where n_0 is the metal density, and R is the particle radius.

By writing the Euler–Lagrange equation for the functional $E[n_e(r)]$ and taking into account conservation $\int n_e(\mathbf{r})d\mathbf{r} = N_e$, one can derive the self–consistent system of Kohn–Sham equations,

$$\left[-\frac{1}{2}\nabla^2 + V_{\text{eff}}(\mathbf{r})\right]\psi_i(\mathbf{r}) = \varepsilon_i\psi_i(\mathbf{r}), \tag{5.92}$$

where

$$V_{\text{eff}}(\mathbf{r}) = \varphi(\mathbf{r}) + \frac{\delta E_{\text{e-c}}[n_e(\mathbf{r})]}{\delta n_e(\mathbf{r})}; \quad \varphi(\mathbf{r}) = V(\mathbf{r}) + \int\frac{n_e(\mathbf{r}')}{|\mathbf{r} - \mathbf{r}'|}d\mathbf{r}'.$$

The electron number density is given by n lower eigenstates,

$$n_e(\mathbf{r}) = \sum_{i=1}^{n}|\psi_i(\mathbf{r})|^2. \tag{5.93}$$

Therefore, the problem is reduced to the solution of Eq. (5.92), and the self–consistency is provided by (5.95). Unlike the Hartree–Fock equations, the Kohn–Sham equation is exact and the exchange-correlation potential $V_{\text{e-c}} = \delta E_{\text{e-c}}/\delta n_e$ is local and universal.

The theory is easily generalized to a two–component system at a finite temperature. The free energy F is the functional of the electron and ion distribution,

$$F = F[n_e(\mathbf{r}), n_i(\mathbf{r})] = \int f[n_e(\mathbf{r}), n_i(\mathbf{r})]d\mathbf{r}.$$

We restrict ourselves, as above, to the first terms of the gradient expansion and derive,

$$f[n_e(\mathbf{r}), n_i(\mathbf{r})] = g(n_e, n_i) + \frac{1}{2}\varphi(n_e - n_i) + \sum_{a,b}\Phi_{a,b}\nabla n_a\nabla n_b,$$

where $g(n_e, n_i)$ is the local free energy density (the quasihomogeneous part of the functional $f(n_e, n_i)$), and

$$\varphi(\mathbf{r}) = \int[n_e(\mathbf{r}') - n_i(\mathbf{r}')]|\mathbf{r} - \mathbf{r}'|^{-1}d\mathbf{r}'$$

is the electrostatic potential which satisfies the Poisson equation. In the gradient term, the subscripts a and b stand for e and i.

Liberman (1979), and Klyuchnikov and Lyubimova (1987) investigated the equation of state for metals in the framework of the cell model. A nucleus shielded by its Z electrons is located at the center of a spherical cavity of radius $r_s = (3/4\pi n_i)^{-1/3}$. The distribution of electrons inside the cell is defined by the set of solutions of the Kohn–Sham equation,

$$\frac{1}{2r^2}\frac{d}{dr}\left(r^2\frac{du_l}{dr}\right) + \left[E - V(r) - \frac{l(l+1)}{2r^2}\right]u_l = 0, \quad r \leqslant r_s. \tag{5.94}$$

For bound states, the boundary condition is $u_l(r_s) = 0$. For free states $E = k^2/2$, the solution of (5.94) should be matched at $r = r_s$ with the wavefunction of

$$u_{kl}(r) \sim \cos\delta_l(k)j_l(kr) - \sin\delta_l(k)n_l(kr), \quad r > r_s,$$

where j_l and n_l are the Bessel and Neumann functions, respectively. The potential $V(r)$ has the form

$$V(r) = -\frac{Z}{r} + \frac{4\pi}{r}\int_0^r n_e(r')r'^2dr' + 4\pi\int_r^R n_e(r')r'dr' + V_{e\text{-}c}^0(n_e(r)), \quad r < r_s.$$

The electron distribution is determined similarly to (5.93), but taking into account the weight factors described by the Fermi distribution at given temperature. The solution explicitly divides the electrons into free and bound ones. The pressure of the electron component is equal to that of the Fermi gas of density $n_e(r)$ at the spherical boundary of the cell. The ion contribution to the pressure must be calculated separately, which is an independent problem.

Liberman (1979) calculated the equation of state for cold copper, zinc and nickel. For copper and zinc, this agrees well with the results of dynamic experiments by Altshuler (1965) (see Fig. 5.17). Somewhat worse is the agreement with the measurement results for nickel.

Note the investigations of the thermodynamics of the cell plasma in the Hartree–Fock-Slater approximations (Nikiforov et al. 1979; Sin'ko 1983; Novikov 1985). These methods follow from the MDF. The mentioned papers give results of a broad–range calculations of the equation of state for a number of metals, which are compared with the available experimental data.

5.7   Quantum Monte Carlo method

In Section 5.3 we discussed the results of calculations of thermodynamic properties of nonideal plasmas performed with the classical Monte Carlo method. Pair quantum effects of the electron and ion interactions which caused, in particular, the formation of the bound states at low temperatures, were taken into account approximately in the framework of the pseudopotential model. Recently, the appearance of fast computers and the development of new efficient computational methods stimulated remarkable progress in modeling the thermodynamic

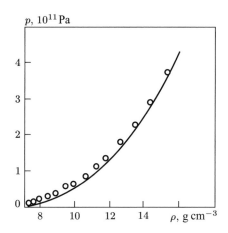

p, 10^{11} Pa

FIG. 5.17. Equation of state for zinc at zero temperature: Solid line shows experiment (Altshuler 1965), points are the calculations (Liberman 1979).

properties with the path–integral Monte Carlo (PIMC) method (Zamalin and Norman 1973; Zamalin, Norman and Filinov 1977; Ceperley 1995; Berne *et al.* 1998). However, such investigations of the Fermi systems were hampered by the so–called "sign problem", which was associated with the calculation of sum over all fermion permutations. This eventually resulted in a difference of two big numbers whose absolute values are practically equal to each other. Ceperley (1995) proposed to overcome this computational obstacle by taking into account only positive terms in the sum, with the simultaneous change of the integration region – the "fixed node approximation". However, it was not possible to control the computational accuracy with this approximation. One can rigorously prove that this method does not include properly the Fermi statistics, even for an ideal degenerate gas (Filinov 2001). Filinov *et al.* (2000a,b,c, 2001a,b) have proposed a method to solve the "sign problem". This method does not involve any approximations and allows us to broaden significantly the range of densities where the simulations can be performed.

The quantum Monte Carlo method is based on direct calculation of the partition function of interacting particles. In the simplest case of a hydrogen plasma, a binary mixture of classical (Boltzmann) N_i protons and N_e quantum electrons is considered. The partition function of such a system can be written as

$$Z(N_e, N_i, V, \beta) = Q(N_e, N_i, \beta)/N_e!N_i!,$$

$$Q(N_e, N_i, \beta) = \sum_\sigma \int_V dqdr \rho\,(q, r, \sigma; \beta). \tag{5.95}$$

Here $q = \{\mathbf{q}_1, \mathbf{q}_2, \ldots, \mathbf{q}_{N_i}\}$ are coordinates of the protons, and $\sigma = \{\sigma_1, \ldots, \sigma_{N_e}\}$ and $r = \{\mathbf{r}_1, \ldots, \mathbf{r}_{N_e}\}$ are electron spins and coordinates, respectively. The density matrix ρ in (5.95) is expressed in terms of the path integral,

$$\rho\left(q,r,\sigma;\beta\right)=\frac{1}{\lambda_i^{3N_i}\lambda_\Delta^{3N_e}}\sum_P(-1)^{\kappa_P}\int_V dr^{(1)}\ldots dr^{(n)}\times$$

$$\times\rho\left(q,r,r^{(1)};\Delta\beta\right)\ldots\rho\left(q,r^{(n)},\hat{P}r^{(n+1)};\Delta\beta\right)S\left(\sigma,\hat{P}\sigma'\right),\quad(5.96)$$

where $\Delta\beta\equiv\beta/(n+1)$, $\lambda_i^2=2\pi\hbar^2\beta/m_i$, and $\lambda_\Delta^2=2\pi\hbar^2\Delta\beta/m_e$. In Eq. (5.96), $r^{(n+1)}\equiv r$ and $\sigma'=\sigma$, i.e., electrons are represented by fermionic loops with coordinates $[r]\equiv\left[r,r^{(1)},\ldots,r^{(n)},r\right]$. The electron spin gives rise to the spin part of the density matrix S, and $\hat{P}$ is the permutation operator with parity κ_P. Dimensionless coordinates of quantum electrons are introduced for the calculations and then the density matrix is transformed, so that the sum over the permutations is reduced to the determinant of the exchange matrix $\psi_{ab}^{n,1}$,

$$\sum_\sigma\rho\left(q,r,\sigma;\beta\right)=\frac{1}{\lambda_i^{3N_i}\lambda_\Delta^{3N_e}}\sum_{s=0}^{N_e}\rho_s\left(q,[r],\beta\right),$$

$$\rho_s\left(q,[r],\beta\right)=\frac{C_{N_e}^s}{2^{N_e}}\exp\left\{-\beta U\left(q,[r],\beta\right)\right\}\prod_{l=1}^{n}\prod_{p=1}^{N_e}\phi_{pp}^l\det\left|\psi_{ab}^{n,1}\right|_s,\quad(5.97)$$

$$U\left(q,[r],\beta\right)=U^i\left(q\right)+\sum_{l=0}^{n}\frac{U_l^e\left([r],\beta\right)+U_l^{ei}\left(q,[r],\beta\right)}{n+1}.$$

Here U^i, U_l^e, and U_l^{ei} are energies of the pair interactions determined by the Kelbg (1964) pseudopotentials between protons, electrons in the lth vertex, as well as between electrons in the lth vertex and protons, respectively. In Eq. (5.97), the function $\phi_{pp}^l\equiv\exp\left[-\pi\left|\xi_p^{(l)}\right|^2\right]$ arises from the kinetic part of the density matrix presented as a product of high–temperature factors, $[r]\equiv[r;r+\lambda_\Delta\xi^{(1)};r+\lambda_\Delta(\xi^{(1)}+\xi^{(2)});\ldots]$, with $\xi^{(1)},\ldots,\xi^{(n)}$ being the dimensionless distances between the neighboring vertices on the loop. Components of the exchange matrix $\psi_{ab}^{n,1}$ are given by

$$\left\|\psi_{ab}^{n,1}\right\|_s\equiv\left\|\exp\left\{-\frac{\pi}{\lambda_\Delta^2}|(r_a-r_b)+y_a^n|^2\right\}\right\|_s,\quad y_a^n=\lambda_\Delta\sum_{k=1}^{n}\xi_a^{(k)}.$$

The index s denotes the number of electrons with the same spin projection.

Using the determined statistical function, on can derive the total energy and pressure,

$$\beta E=\frac{3(N_e+N_i)}{2}-\beta\frac{\partial\ln Q}{\partial\beta},\quad\beta p=\frac{\partial\ln Q}{\partial V}.$$

The standard Metropolis method is applied in order to calculate multiple integrals in such expressions. The accuracy of the results is determined by the number of factors n in the integrand in Eq. (5.96), the temperature T and the degeneracy parameter χ, via $\varepsilon \sim (\beta\mathrm{Ry})^2\chi/(n+1)$.

By employing the described method to overcome the "sign problem", one can describe with high precision thermodynamic properties of an ideal degenerate hydrogen plasma. Filinov *et al.* (2000a) calculated the pressure versus the degeneracy parameter $\chi = n_e\lambda_e^3$ (where λ_e is the thermal electron wavelength, $\lambda_e^2 = 2\pi\hbar^2\beta/m_e$) for different particle numbers. Even for $N_e = 32$ the agreement with the analytic dependence was remarkably good up to $\chi = 10$, and it improved as the particle number increased.

Similar calculations were performed by Filinov *et al.* (2000b,c) also for non-ideal plasmas in a fairly broad range of density, from $n_e = 10^{18}$ to 10^{26} cm^{-3}, at temperatures from $T = 10^4$ to 10^6 K. In this parameter range different phenomena occur, such as partial dissociation and ionization, the Mott transition, and ordering of ions at high densities. The analysis of correlation functions has been performed to investigate these phenomena. Figures 5.18 and 5.19 show the results for four most interesting points in the phase diagram. In Fig. 5.18a at $T = 20\,000$ K and $n_e = 10^{22}$ cm^{-3}, the proton–proton (g_{ii}) and electron–electron (g_{ee}) correlation functions have sharp maxima at the same coordinate, $r = 1.4a_0$, which suggests that the hydrogen molecules should exist for these conditions. Therefore, in comparison with the Bohr radius a_0, the position of the maximum of the electron–proton (g_{ie}) correlation function (multiplied by r^2) is shifted towards larger values. The molecules disappear as the density decreases, and then a maximum of g_{ie} emerges at $r = a_0$. Figure 5.18b shows that the maxima of g_{ii} and g_{ee} are smeared out as the temperature increases up to $T = 50\,000$ K, which indicates the dissociation of molecules.

The density increase at constant temperature $T = 50\,000$ K results in new physical phenomena. Figures 5.18b and 5.19 clearly demonstrate evolution of the proton–proton correlation function. The curve typical for a partially ionized plasma (see Fig. 5.18b) is modified to that typical for a liquid (Fig. 5.19a) and solid states (Fig. 5.19b).

Also very interesting is the evolution of the electron–electron correlation function as the density grows: The function g_{ee} in Fig. 5.18b is typical for partially ionized plasmas, whereas Fig. 5.19a reveals a steep increase at small coordinates. The subsequent density increase results in a more homogeneous electron distribution (see Fig. 5.19b). In order to analyze this phenomenon, the functions r^2g_{ee} and r^2g_{ii} are also plotted in Fig. 5.19a. They show that the most probable distance between electrons is two times smaller than that between protons. Analysis of electron trajectories for this case shows pairing of electrons with opposite spins. The "size" of electrons is approximately equal to the distance between protons, which results in partial overlap of the paired electrons.

The region of the hypothetical plasma phase transition is of special interest. To study this phenomenon, the calculations of the hydrogen plasma isotherms at $T = 10^4$, $2 \cdot 10^4$, $5 \cdot 10^4$, 10^5, $1.25 \cdot 10^5$ and 10^7 K in a broad range of densities from $n_e = 10^{18}$ to 10^{27} cm^{-3} were performed by Filinov *et al.* (2001b). The most

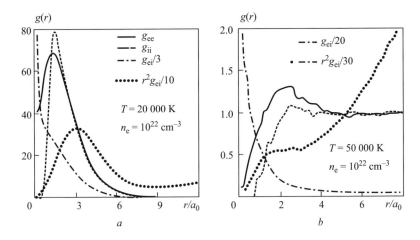

FIG. 5.18.   Electron–electron (g_{ee}, solid line), proton–proton (g_{ii}, dashed line) and electron–proton (g_{ie}, dashdotted line) pair correlation functions of dense hydrogen plasma. The values for the coupling, degeneracy, and Brueckner parameters are (a) $\gamma = 2.9$, $\chi = 1.46$, $r_s = 5.44$ and (b) $\gamma = 1.16$, $\chi = 0.37$, $r_s = 5.44$.

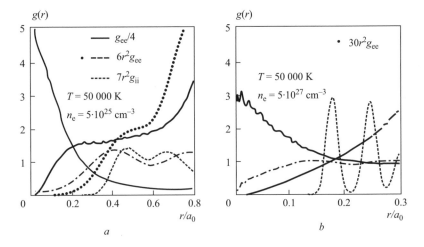

FIG. 5.19.   Correlation functions of dense hydrogen plasma. The notations are the same as in Fig. 5.18. The values for the coupling, degeneracy, and Brueckner parameters are (a) $\gamma = 19.8$, $\chi = 1848$, $r_s = 0.318$ and (b) $\gamma = 53.8$, $\chi = 37000$, $r_s = 0.117$

interesting results are shown in Fig. 5.20 and 5.21.

Figure 5.20 shows the pressure and energy of hydrogen plasma versus density at $T = 50\,000$ K. One can see that the pressure curve (Fig. 5.20a) is monotonic, and the results of this work in good agreement with calculations of Militzer

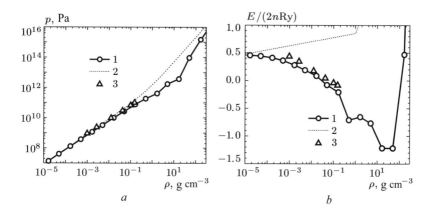

FIG. 5.20. Pressure (a) and energy (b) of a hydrogen plasma vs. density at $T = 50\,000$ K. 1
– direct PIMC simulation (Filinov *et al.* 2001b), 2 – analytical expression for ideal plasma,
3 – restricted PIMC calculations (Militzer and Ceperley 2001).

and Ceperley (2001) by the quantum Monte Carlo method with restrictions at moderate densities. The curve has a maximum, which suggests formation of bound states.

Another type of behavior is observed as the temperature decreases. Figure 5.21 shows pressure and energy of a hydrogen plasma versus density at $T = 10\,000$ K. Simulations revealed the region of poor convergence to the equilibrium at densities 0.1–1.5 g cm^{-3}, which might suggest the phase transition. Since such anomalies are absent at the $T = 50\,000$ K isotherm, the transition should have a critical point at $T \sim 30\,000$ K. There is a number of additional arguments which make the hypothesis of the phase transition favorable: Calculations with the density functional method performed in the stability region of crystalline hydrogen result in anomalies in the ion–ion correlation function as the density decreases down to $\rho = 0.8$ g cm^{-3} (Xu and Hansen 1998) (approximately, the upper density bound for the transition); the maximum density at which the restricted PIMC calculations (Militzer and Ceperley 2001) were possible was $\rho = 0.15$ g cm^{-3} (approximately, the lower density bound for the transition); PIMC simulations for a mixture of 33% helium and 67% hydrogen at the same temperature show no transition (curve 3 in Fig. 5.21a); an abrupt increase in the hydrogen electric conductivity (4–5 orders of magnitude) was observed in the shock–wave experiments in a very narrow range of densities, 0.3–0.5 g cm^{-3}, at temperatures 5000–15 000 K (points 6 and 7 in Fig. 5.21a). Finally, in the range of poor convergence at $T = 50\,000$ K and $\rho = 0.3346$ g cm^{-3}, the visualization of the Monte Carlo cell revealed the formation of proton clusters with the localized electrons, which is a clear indication of the transition to the highly conductive (quasimetallic) state.

In various chemical plasma models (Norman and Starostin 1970; Saumon

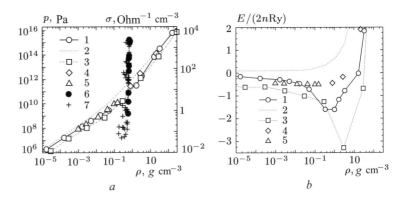

FIG. 5.21. Pressure and electrical conductivity (a) and energy (b) of a hydrogen plasma vs.
density at $T = 10\,000$ K. 1 – direct PIMC simulation (Filinov et al. 2001b), 2 – analytical
expression for ideal plasma, 3 – direct PIMC simulation of a mixture consisting of 33%
helium and 67% hydrogen (Filinov et al. 2001b), 4 – density functional theory (Xu and
Hansen 1998), 5 – restricted PIMC calculations (Militzer and Ceperley 2001), 6 and 7 –
experimentally measured electrical conductivity, (Weir et al. 1996) and (Ternovoi et al.
1999), respectively.

and Chabrier 1992), the plasma phase transition occurs approximately at the
same densities as in the calculations by the Monte Carlo method. According to
these models, this is a first–order transition between two phases with different
ionization fractions. However, such models become questionable in the region
of pressure–induced ionization and dissociation, where the consistent treatment
of all types of interaction is very important. Moreover, possible cluster formation is neglected. The approach developed by Filinov et al. (2000a,b,c, 2001a,b)
does not have all these disadvantages. Calculations in the region of the plasma
phase transition suggest alternative explanations of the anomalous increase of
the conductivity. Earlier, theoretical models predicted the increase either at lower
(hopping conductivity in molecular fluids), or at higher (free electron gas conductivity) densities (Redmer et al. 2001). However, the conductivity increase occurs
right in the region of the plasma phase transition, so that one has to take into
account one more conductivity mechanism – due to the electron hopping between
drops of metallic fluid. Obviously, this mechanism will be effective in between
the regions where the two other effects dominate. For a detailed discussion of
the hydrogen properties at high pressures and temperatures, see Chapter 9.

5.8 Comparison with experiments

By now, a fairly large body of experimental data has been accumulated on the thermodynamics of a nonideal plasma of various elements under conditions of developed ionization. The great majority of these data were obtained using dynamic methods (see Chapter 3), by shock and adiabatic compression of cesium, as well as compression of argon, xenon, and copper in powerful shock waves.

The data obtained for cesium plasma (Kunavin et al. 1974; Bushman et al. 1975; Iosilevskii and Gryaznov 1981; Alekseev et al. 1981; Iakubov 2000) relate to various regions of the phase diagram (see Fig. 3.2) that partly overlap at the boundaries and agree with each other within the claimed experimental accuracy. Experiments on adiabatic compression by Kunavin et al. (1974) enabled one to advance, as compared with static measurements (Kalitkin and Kuzmina 1975, 1976; Alekseev et al. 1981), to higher temperatures, $T \simeq 4000$ K. This, however, turned out to be insufficient for noticeable thermal ionization of the plasma. Under these conditions the charge–neutral interaction prevails, but its contribution to the equation of state remains within the measurement errors. The principal conclusion by Kunavin et al. (1974) drawn on the basis of these experiments consists in the absence of phase separation (see Section 5.9) caused by the metal–dielectric transition. Experiments on cesium compression by direct and reflected shock waves provided further extension of the temperature range, $T \simeq (2.6\text{–}20)\cdot10^3$ K (see Fig. 3.18), where the Coulomb interaction is strong, $\Gamma \simeq 0.2\text{–}2.2$, and defines the physical properties of the plasma with developed ionization.

Considerably higher plasma parameters were attained as a result of explosive compression of heavy rare gases (Bespalov et al. 1975; Fortov et al. 1976; Gryaznov et al. 1980; Fortov 1982), namely, pressures of up to 6 GPa and temperatures of up to $6\cdot10^4$ K. The obtained plasma densities of $\rho \sim 0.4$ g cm^{-3} and $n_e \sim 3\cdot10^{21}$ cm^{-3} approach the density of condensed xenon, and even exceed it ($\rho \sim 4.5$ g cm^{-3}) in experiment by Mintsev et al. (1980). The characteristic interparticle distances in plasma (Gryaznov et al. 1980) were about $(6\text{–}7)\cdot10^{-8}$ cm, which is comparable with the ion and atom size of $(3\text{–}4)\cdot10^{-8}$ cm, whereas the maximum nonideality parameters $\Gamma \sim 5$ were close to the maximum possible for nondegenerate plasmas value (see Fig. 1.1). The experimental data for argon and xenon (Bespalov et al. 1975; Fortov et al. 1976; Gryaznov et al. 1980; Fortov 1982) relate to the region of developed single ($x_{Ar} \leqslant 0.7$) and double ($x_{Xe} \leqslant 1.8$) ionization. These data allow us to advance into the region of condensed densities (see Fig. 1.5) and approach extremely high pressures ($p \leqslant 0.13$ TPa) and compressions ($\rho \leqslant 9.6$ g cm^{-3}) of a plasma obtained by shock compression of liquid xenon and heated to $T \leqslant 30 \cdot 10^3$ K (Lundqvist and March 1983).

All thermodynamic measurements (Figs. 5.22–5.26, Tables 5.2–5.5) exhibit a clear tendency: The measured enthalpy or internal energy (curves I in the figures) is lower than that obtained with the traditional plasma calculation represented by the ring Debye approximation in a grand canonical ensemble (5.50)–(5.52) (curves II). In Figures 5.22–5.26 and Tables 5.2–5.5, the experimental data are

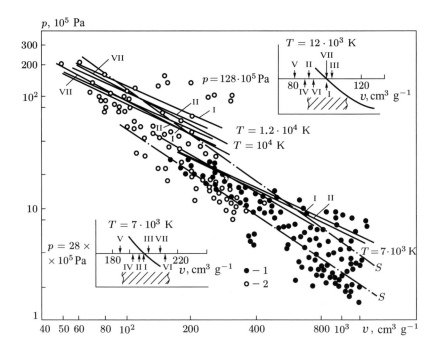

FIG. 5.22.   Results of the shock compression of cesium plasma. T – isotherms; S – isentropes; 1 and 2 – incident and reflected shock waves, respectively. Dashed regions show the measurements errors.

compared with a number of other plasma approximations: Ideal plasma model with $F_b = 0$ and $\Sigma_k = 2$ (curves III); $F_b = 0$ in Eq. (5.40) and Σ_k from Eq. (5.73) (curves IV); Debye theory in a small canonical ensemble, Eq. (5.43) (curves V); pseudopotential Monte Carlo model, Eqs.(5.58)–(5.61) (curves VI); and "bound" atom model, Eqs.(5.77)–(5.83) (curves VII). The relative contribution by other models to the equation of state was analyzed by Rajagopal (1980).

The performed analysis shows that, although the majority of the used theoretical models do not contradict the experimental isotherms of cesium plasma within the measurement accuracy, these models cannot provide a consistent description of the thermal and caloric data. Inert gas plasmas correspond to substantially higher parameters of nonideality and, in this case, the traditional plasma models, Eqs. (5.41)–(5.47), (5.50)–(5.52), (5.58)–(5.61), are in contradiction not only with the caloric, but also with the thermal measurements.

An analysis of thermodynamic data suggests the presence of interparticle repulsion in a strongly compressed plasma, which is not described by the plasma theories, Eqs. (5.41)–(5.47), (5.50)–(5.52), (5.58)–(5.61). Although one can attain good agreement with the thermal equation of state by introducing modifications to the correction F^c, such a procedure leads to even greater discrepancy with the caloric data. It was found by Bushman et al. (1975), Fortov et al. (1976),

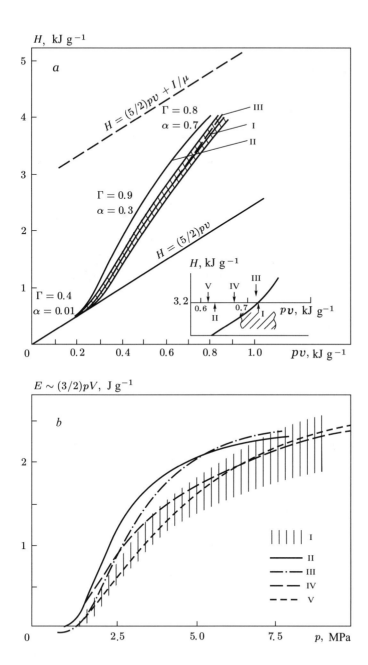

FIG. 5.23. Isochore of cesium plasma, $v = 10^3$ cm^3 g^{-1} (a) and $v = 200$ cm^3 g^{-1} (b), the error band is dashed. I – experiment; II – ring Debye approximation; III – bound atom model; IV – pseudopotential model (5.63); V – superposition of III and IV.

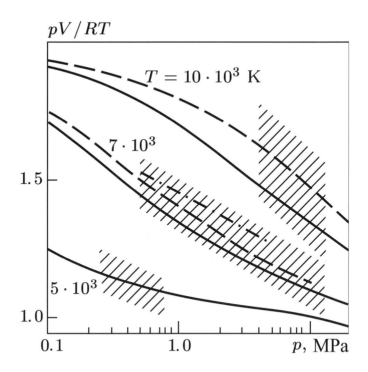

FIG. 5.24.   Thermal equation of state of cesium plasma (Iosilevskii 1980). The error band
is dashed. The solid line is the ring Debye approximation, the dashed line is the pseudopotential model (5.63), the dash–dotted line is the asymptotic model (Krasnikov 1977).

and Gryaznov *et al.* (1980) that the consistent description of thermal and caloric
data could be achieved by modifying the contribution of the bound states to the
thermodynamic functions of dense plasma – an effect ignored by most of the
traditional plasma approximations.

Irrespective of this conclusion, the experimental results indicate that the
Debye and similar theories, (5.43)–(5.47), overestimate considerably the corrections to thermodynamic functions for the interaction in a continuous spectrum.
The best properties of extrapolation to region $\Gamma \sim 1$ are exhibited by the theories whose nonideality corrections do not exceed the corrections of the ring
approximation in the grand canonical ensemble (5.50)–(5.52). This conclusion is
in qualitative agreement with the results obtained by the Monte Carlo method
for the one–component model (see Section 5.1), which indicates the validity of
asymptotic approximations up to $\Gamma \leqslant 1$.

In the considered region of high plasma densities, the mean interparticle distances are comparable with the characteristic sizes of atoms and ions. This fact,
along with a strong Coulomb interaction of free charges, can cause significant
deformation of energy levels.

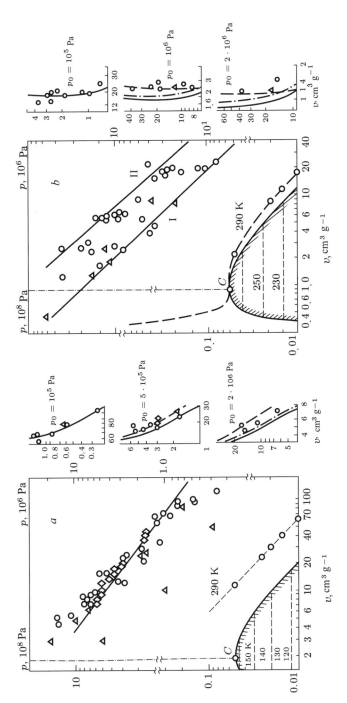

FIG. 5.25. Phase diagram of argon (a) and xenon (b). Dashes mark the two-phase region boundaries, C indicates the critical point. Experiments: o — (Gryaznov *et al.* 1980), △ — (Fortov *et al.* 1976), ◇ — (Bespalov *et al.* 1975), dashed lines show the isotherms of the initial states, solid lines show the boundaries of single (I) and double (II) ionization. Given on the right is a comparison between the experimental shock adiabats (dots) and theoretical data: Solid lines represent calculations with the ring Debye approximation in the grand canonical ensemble (5.50)–(5.52); dash–dotted lines represent the additional inclusion of the atomic interaction with the second virial coefficient; dashed lines in (a) represents the "bound" atom model (5.77)–(5.83), and, in (b), the pseudopotential model (5.63)–(5.71).

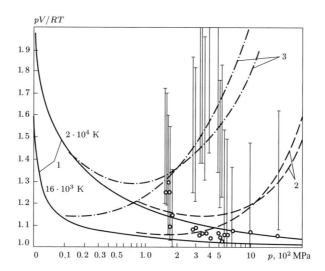

FIG. 5.26.   Thermal equation of state of argon plasma. Vertical lines represent experimental data (Bespalov *et al.* 1975; Fortov *et al.* 1976); o – calculation with the ring approximation in a grand canonical ensemble (5.50)–(5.52) for given $p_{\exp}$ and $T_{\exp}$; 1–3 stand
for calculated isotherms ($T = 16\,000$ K and $T = 20\,000$ K) in the approximations: 1 –
pseudopotential model (5.63)–(5.71), 2 – with additional inclusion of the second and third
virial coefficients, 3 – from the "bound" atom model (5.77)–(5.83).

The description of this effect required implementation of the quantum–mechanical model of a "bound" atom (5.78)–(5.83), which is unconventional for plasma
physics. The model takes into account the effect of the plasma environment on
the discrete spectrum of atoms and ions in a strongly compressed plasma. It is
seen from a comparison of this model with experiments that the model correctly
recovers the experimentally revealed tendency of the somewhat overestimated
repulsion effect at near–critical plasma densities. Under these conditions one
should apparently use the more adequate "soft" sphere model (Bushman and
Fortov 1983). Later, model (5.77)–(5.83) will be employed for the description of
the optical characteristics of nonideal plasmas.

The need to include in the thermodynamics of dense plasma the variation of
the discrete electron spectrum is clearly demonstrated in experiments with shock
compression of liquid argon and xenon at pressures up to 0.13 TPa (Lundqvist
and March 1983; Sin'ko 1983). The interpretation of these measurements has
shown (Fig. 5.27) that neglect of the discrete spectrum (excitation energy) leads
to a drastic overestimation of the calculated pressures (curve C). Curves A and
B take into account the thermodynamic effects of thermal excitation of electrons
from the valence band $5s$ to the conduction band $5d$, and the variation of the
forbidden band width from 7 to 4 eV with the density increase (Liberman 1979;
Lundqvist and March 1983).

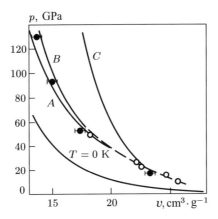

FIG. 5.27. Comparison of theoretical models with experimental data on the shock compressibility of xenon. • — (Lundqvist and March 1983); ∘ — (Klyuchnikov and Lyubimova
1987); C — calculations by Liberman 1979 without taking into account the electronic
excitation ; A and B – two examples where the discrete spectrum deformation is included.

A peculiarity of the quasi–chemical description of a plasma under conditions
of strong nonideality is a "conditional" division of particles into free and bound
ones. Therefore the effect regarded as a distortion of the excited states contribution may, in the case of a different separation into species, be interpreted as
a manifestation of the quantum nature of the electron–ion interaction at small
distances. The pseudopotential (5.63) was implemented by Iosilevskii (1980) and
Gryaznov et $al.$ (1989) in order to describe this interaction, and on this basis,
using the conditions of local electroneutrality (5.65) and (5.66), the semiempirical model of a nonideal plasma was constructed. By selecting an appropriate
value for the single parameter of the model – the pseudopotential depth $\Phi_{ie}^*(0)$,
which is equal to the energy separating the particles into free and bound ones –
a consistent description of currently available thermodynamic experimental data
has been performed for a "gaseous" plasma (Bushman et $al.$ 1975; Bespalov et
$al.$ 1975; Fortov et $al.$ 1976; Mintsev et $al.$ 1980; Iosilevskii and Gryaznov 1981).

In conclusion of this paragraph, we shall dwell on the comparison between
the quasiclassical model (Section 5.5) and plasma experiments, as shown in Fig.
5.28 for the cesium gaseous plasma from the shock–wave (Bushman et $al.$ 1975)
and electric explosion (Dikhter and Zeigarnik 1977) experiments.

In accordance with the conclusions of Section 5.5, the worst accuracy of the
Thomas–Fermi approximation is attained in the region of temperatures corresponding to the ionization from the filled electron shell (beginning of the secondary cesium ionization). Note that the plasma investigated by Bushman et $al.$
(1975) and Iosilevskii and Gryaznov (1981) has a noticeable degree of nonideality,
$\Gamma \leqslant 1$.

Table 5.2 Thermodynamic parameters of cesium behind an incident shock wave

| | | | Experiment I | | | | Theory | | | | | | |
| | | | | | | | II | | III | | | | |
p_0, 10^5 Pa	T_0, K	u, km s^{-1}	v_1, cm^3 g^{-1}	p_1, 10^5 Pa	H_1, 10^2 J g^{-1}	T_1, K	H, 10^2 J g^{-1}	T, K	H, 10^2 J g^{-1}	n_e, 10^{13} cm^{-3}	α	Γ	T, K
0.130	780	0.83	1200	1.40	4.20	2600	4.20	2600	4.10	0.006	0.002	0.02	2600
0.147	780	0.935	1000	2.00	5.10	3100	5.40	3100	5.10	3.03	0.007	0.34	3100
0.107	760	1.23	1030	2.80	8.3	4100	10.0	4200	8.6	0.250	0.056	0.62	4200
0.160	780	1.05	800	2.90	6.3	3500	6.6	3500	6.0	0.09	0.015	0.48	3500
0.074	750	1.53	1160	3.10	12.4	5100	15.7	5000	13.0	0.56	0.140	0.73	5000
0.240	830	0.95	630	3.30	5.30	3200	5.50	3200	5.20	0.046	0.006	0.41	3200
0.063	740	1.87	1060	4.20	18.0	5600	22.0	5700	18.0	1.05	0.250	0.83	5700
0.051	700	2.23	1130	5.1	25.0	6400	30.0	6600	25.6	1.60	0.41	0.85	6400
0.180	790	1.33	580	5.4	9.6	4600	9.5	4600	9.5	0.53	0.070	0.8	4600
0.380	860	1.03	390	6.0	6.2	3600	6.5	3600	6.0	0.13	0.01	0.59	3600
0.078	750	2.10	800	6.5	22.4	6100	26.4	6300	22.0	1.80	0.320	0.93	6200
0.080	740	2.05	770	6.6	21.6	6100	25.4	6200	21.0	1.70	0.290	0.94	6100
0.044	690	2.77	1270	6.9	39.0	7900	44.0	8500	41.0	2.50	0.70	0.76	8000
0.180	790	1.49	530	7.0	11.8	5200	15.6	5200	12.3	1.00	0.12	0.93	5200
0.056	710	2.50	950	7.1	32.0	7000	35.0	7300	31.0	2.40	0.50	0.89	7000
0.044	690	2.83	1260	7.2	40.0	8100	45.0	8700	43.0	2.60	0.73	0.75	8200
0.240	820	1.55	410	9.4	12.7	5500	17.0	5400	13.2	1.45	0.130	1.04	5400
0.290	840	1.49	350	10.4	12.0	5300	15.0	5200	11.7	1.30	0.10	1.05	5200
0.360	860	1.92	240	22.0	19.0	6700	25.0	6800	20.0	4.60	0.24	1.35	6700
0.440	870	1.87	200	25.0	18.0	6700	24.0	6600	18.3	4.80	0.21	1.40	6600

Table 5.3 *Thermodynamic parameters of cesium behind a reflected shock wave*

								Theory						
Experiment I								II		III				
p_0, 10^5 Pa	T_0, K	u, km s^{-1}	v_1, cm^3 g^{-1}	v_2, cm^3 g^{-1}	p_2, 10^5 Pa	H_2, 10^2 J g^{-1}	T_2, K	H, 10^2 J g^{-1}	T, K	H, 10^2 J g^{-1}	n_e, 10^{13} cm^{-3}	α	Γ	T, K
0.230	820	0.90	680	250	12.0	9.0	4400	10.4	4500	8.5	0.75	0.04	1.0	4500
0.155	790	1.10	820	260	15.0	13	5600	17	5600	13	2.1	0.12	1.2	5600
0.094	760	1.94	700	170	60	42	9600	45	10700	42	16	0.60	1.4	9900
0.047	690	2.52	1150	300	64	73	15600	73	17300	74	14	0.95	0.66	15700
0.166	790	1.70	450	110	70	32	8500	35	9000	30	16	0.38	1.7	8600
0.320	840	1.58	280	75	95	27	8700	33	8800	26	19	0.32	1.9	8500
0.180	800	1.96	420	100	100	42	10000	44	11000	41	25	0.56	1.6	10300
0.064	710	2.75	860	240	104	90	19700	85	21400	90	19	0.97	0.54	19500
0.130	790	2.30	490	144	130	64	15200	70	17400	71	28	0.90	0.95	15500
0.440	870	1.87	190	50	200	38	11000	41	11200	37	43	0.45	2.1	10500

Table 5.4 *Thermodynamic properties of shock-compressed argon*

	Experiment					Theory					Bounded atom (5.77)–(5.80)	
						Ring Debye approximation						
p_0, 10^5 Pa	D, km s^{-1}	u, km s^{-1}	p, 10^8 Pa	v, cm^3 g^{-1}	H, 10^{-3} J g^{-1}	v, cm^3 g^{-1}	T, 10^3 K	n_e, 10^{20} cm^{-3}	Γ	$n\sigma^3$	v, cm^3 g^{-1}	T, 10^3 K
0.78	7.78	6.94	0.692	86.0	30.1	88.2	22.0	0.642	0.843	0.00422	90.6	21.6
1.0	6.5	5.59	0.597	87	20.9	81.7	19.5	0.401	0.798	0.0057	85.5	19.1
3.0	7.93	6.97	2.73	25.0	31.1	24.0	24.3	2.23	1.35	0.016	25.6	23.0
5.0	6.51	5.52	2.96	18.8	20.9	17.45	21.3	1.63	1.4	0.0277	19.9	19.9
10.0	7.6	6.52	8.19	8.78	28.4	7.8	25.1	5.405	2.01	0.055	9.06	22.3
20.0	7.35	6.28	15.4	4.47	26.6	4.05	25.3	8.68	2.51	0.113	5.19	20.5
20.0	7.65	6.4	16.3	5.01	28.6	4.00	26.0	9.77	2.54	0.11	5.0	21.3

Table 5.5 *Thermodynamic properties of shock-compressed xenon*

	Experiment					Theory					Pseudopotential model (5.63)–(5.71)	
						Ring Debye approximation						
p_0, 10^5 Pa	D, km s^{-1}	u, km s^{-1}	p, 10^8 Pa	v, cm^3 g^{-1}	H, 10^{-3} J g^{-1}	v, cm^3 g^{-1}	T, 10^3 K	n_e, 10^{20} cm^{-3}	Γ	$n\sigma^3$	v, cm^3 g^{-1}	T, 10^3 K
1.0	4.58	4.11	1.02	19.3	10.4	18.8	21.0	1.33	1.31	0.0163	21.0	21.6
3.0	7.81	7.00	8.99	6.43	30.2	6.00	42.0	9.71	1.38	0.051	7.07	41.7
5.0	6.35	5.39	9.49	5.56	19.7	3.77	33.1	11.4	1.99	0.0812	4.62	33.7
10.0	6.23	5.4	19.2	2.38	19.1	1.83	34.0	22.0	2.64	0.168	2.34	35.1
10.0	8.93	7.69	39.2	2.48	39.1	1.94	55.8	34.9	1.78	0.158	2.40	55.4
10.0	4.11	3.52	8.27	2.56	8.32	2.04	21.6	7.2	2.91	0.150	2.4	23.4

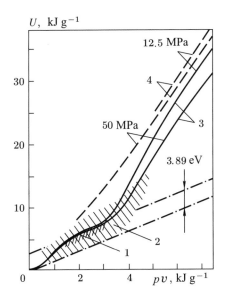

FIG. 5.28. Caloric equation of state for a cesium plasma (Iosilevskii and Gryaznov 1981), $p = 12.5 - 50$ MPa: 1 – experiment (Bushman *et al.* 1975); 2 – experiment (Dikhter and Zeigarnik 1977); 3 – calculation with the ring approximation in a grand canonical ensemble (5.50)–(5.52); 4 – Thomas–Fermi approximation with quantum and exchange corrections.

The ultrahigh–pressure region was investigated by Vladimirov *et al.* (1984) and Avrorin *et al.* (1987) using the technique of powerful shock waves. Comparison of these results and the results obtained with the plasma methods is given in Fig. 5.29 (Nikiforov *et al.* 1979). The plasma parameters are specified for some points. Curves 1, 2, 4, and 5 in Fig. 5.29 indicate the data by Nikiforov *et al.* (1979), 1 is the modified Hartree–Fock–Slater method (Nikiforov *et al.* 1979 and Novikov 1985), 3 is the self–consistent field method (Sin'ko 1983), and 4 is the Thomas–Fermi theory with quantum and exchange corrections (Kalitkin and Kuzmina 1975, 1976). Curve 2 indicates the plasma model of bound atom (see Section 5.4) which takes into account the electron degeneracy, Coulomb non-ideality, excitation of bound states, and short–range repulsion (Nikiforov *et al.* 1979), whereas 5 is the same, but with the contribution made by the equilibrium radiation.

One can see that the plasma model of the bound atom reproduces reasonably well the states of ultrahigh energy density. Unfortunately, the precision of measurements performed by Vladimirov *et al.* (1984) was insufficient for a thorough analysis of advantages and peculiarities of models describing the superdense compression and, in particular, the role of the shell effects in thermodynamics. However, these experiments enable one to follow the asymptotic properties of the theories. During the compression in experiments by Vladimirov *et al.* (1984), the parameter of plasma nonideality decreased from 4.3 to 0.05, although the ther-

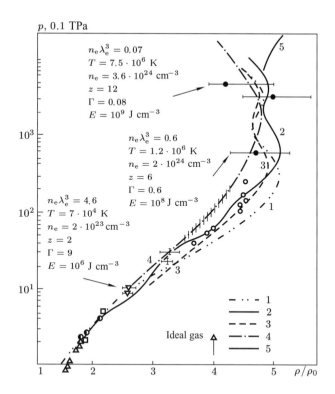

p, 0.1 TPa

$n_e \lambda_e^3 = 0.07$
$T = 7.5 \cdot 10^6$ K
$n_e = 3.6 \cdot 10^{24}$ cm^{-3}
$z = 12$
$\Gamma = 0.08$
$E = 10^9$ J cm^{-3}

$n_e \lambda_e^3 = 0.6$
$T = 1.2 \cdot 10^6$ K
$n_e = 2 \cdot 10^{24}$ cm^{-3}
$z = 6$
$\Gamma = 0.6$
$E = 10^8$ J cm^{-3}

$n_e \lambda_e^3 = 4.6$
$T = 7 \cdot 10^4$ K
$n_e = 2 \cdot 10^{23}$ cm^{-3}
$z = 2$
$\Gamma = 9$
$E = 10^6$ J cm^{-3}

Ideal gas

1 ··· - 1
—— 2
– – – 3
—·— 4
—— 5

ρ/ρ_0

FIG. 5.29.   Shock adiabats for aluminum at ultrahigh pressures (Vladimirov *et al.* 1984; Avrorin *et al.* 1987) (see text for notations): △ and □ – light–gas guns and explosives, respectively; ∘ — (Volkov *et al.* 1980; Akkerman *et al.* 1986); • — (Vladimirov *et al.* 1984; dashed region is from (Model' *et al.* 1985); ∘ — (Vladimirov *et al.* 1984); ∇ — (Simonenko *et al.* 1985).

modynamic parameters turned out to be exotically high, namely, $p \simeq 400$ TPa, $T \simeq 7 \cdot 10^6$ K, and $n_e \simeq 3.6 \cdot 10^{24}$ cm^{-3}, with the specific energy density close to 1 GJ cm^{-3}. One can see from Fig. 5.29 that the pressure range of thousands of TPa is the most interesting one for revealing the role of the shell effects, since the estimates of the bound states contribution based on different models have the worst accuracy in this range. Figure 5.30 gives the results of measurements by Avrorin *et al.* (1987) of the relative compressibility of aluminum and copper (taking iron as the standard), as well as the comparison of these results with the quasiclassical model of Kalitkin and Kuzmina (1975, 1976) (curve 2) and with the cell model of the self–consistent field of Sin'ko 1983 (curve 1). The experimental results clearly demonstrate the substantial contribution by the bound states that cause nonmonotonic behavior of the thermodynamic functions of dense plasma. Naturally, this cannot be described by the Thomas–Fermi model with the quantum and exchange corrections (Kalitkin and Kuzmina 1975, 1976).

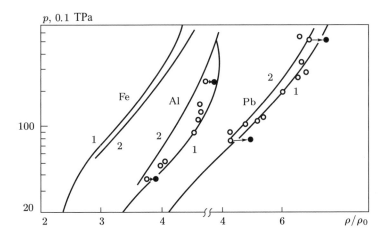

FIG. 5.30. Shock adiabats for aluminum and lead at ultrahigh pressures (Avrorin *et al.* 1987). Arrows indicate the shift of experimental data at the transition of the iron standard adiabat from curve 1 to curve 2.

Figure 5.31 illustrates the results of experiments by Trunin *et al.* (1969) with powerful shock–wave compression of porous copper enabling one to attain extremely high concentrations of thermal energy. At the maximum temperature of $\sim 2.5 \cdot 10^5$ K a superdense, quintuply ionized weakly degenerate ($n_e \lambda^3 \sim 0.7$) plasma is obtained with extremely high parameters of $p \sim 2.09$ TPa, $n_e \sim 2 \cdot 10^{23}$ cm^{-3}, and $\Gamma \sim 2$. Figure 5.31 demonstrates a reasonable description of the properties of such a medium with the "bound atom" model (5.77)–(5.83). We see that, as the Coulomb interaction decreases with the rise in temperature and pressure, the transition to a quasi–ideal multiply ionized plasma is realized. Yet the difference between the experimental results (as well as Saha's model (5.77)–(5.83)) and the Thomas–Fermi approximation increases gradually to reach, under extreme conditions, almost an order of magnitude in pressure, which supports the conclusion made by Iosilevskii and Gryaznov (1981) about the absence of the ideal–plasma asymptotes.

5.9 On phase transitions in nonideal plasmas

Because of serious difficulties in providing a consistent theoretical description of plasmas with strong interparticle interaction in the entire range of parameters, a number of researchers (see Wigner 1934; Norman and Starostin 1970; Ebeling *et al.* 1976; Zelener *et al.* 1981; Bushman and Fortov 1983; Redmer 1997; Likal'ter 2000; Iosilevskii and Starostin 2000, and references therein) make use of model consideration based on physical simplifications and extrapolations of views (developed for weakly nonideal plasmas) on the quantum and collective effects under conditions of Coulomb interaction (Dharma-wardana and Perrot 1982). While being purely approximate, such a "heuristic" approach is conve-

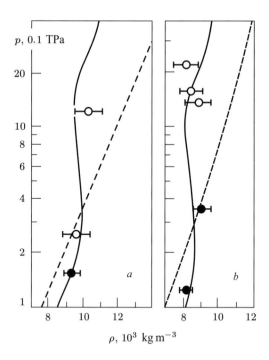

FIG. 5.31.   Thermodynamics of nonideal copper plasma at extremely high pressures (Gryaznov *et al.* 1982). Initial porosity is $m = \rho_0/\rho_{00} = 3$ (a) and $m = 4$ (b). Solid lines are shock adiabats calculated with the "bound" atom model (5.77)–(5.83); dashed lines are the Thomas–Fermi theory with corrections by Kalitkin and Kuzmina (1975, 1976); ○ and ● are experiments by Trunin *et al.* (1969) and Kormer *et al.* (1962), respectively.

nient and was widely used for qualitative analysis of the possible behavior of a plasma in situations characterized by the absence of measurements of its physical properties. Characteristically, a number of the proposed models in the region of increased nonideality lose thermodynamic stability, which is attributed to the possibility of a phase transition and the separation of the system into phases of different densities.

A model analysis (see Sections 5.1.4 and 10.2) of strongly compressed degenerate systems of electrons leads to the possibility of a phase transition with the formation of an ordered electronic structure (Wigner 1934) – a "Wigner crystal" (see Fig. 5.32). The properties of this crystal and its phase diagram have been studied in detail by numerous researchers (see, Carmi 1968; Norman and Starostin 1970; Norman 1971; Ebeling *et al.* 1976; Ceperley 1978; Maksimov and Dolgov 1982; Ichimaru 1992; Dubin and O'Neil 1999, and references therein) who considered the cases with zero and finite temperatures. In the latter case, the separation curve of an electron liquid and a Wigner crystal has a critical point while the stability limit of such a crystal at $T = 0$ is quite indeterminate

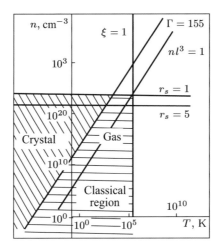

FIG. 5.32.   Region of existence of the electron Wigner crystal (Wigner 1934; Ebeling *et al.* 1976): $l = e^2/kT$

(see Section 5.1.4) (Carmi 1968; Norman 1971; Maksimov and Dolgov 1982). Ceperley 1978 related the phase transitions in OCP to the melting of metals. The dielectric permeability of strongly compressed Coulomb systems and the stability of a nonideal plasma were analyzed in reviews by Kirzhnits (1976) and Maksimov and Dolgov (1982).

More complex and diversified is the case of real multicomponent plasmas. Assuming that polarization is the principal effect in a dense plasma, Mott (1967) and Ebeling *et al.* (1988) considered a model according to which the interatomic electrons are in the screened potential $V(r) = -(e^2/r)\exp(-r/r_0)$. The numerical solution of Schrödinger's equation for this potential helps to find the critical value (Mott 1967) $r_c \sim 0.84a_0 = 0.84\hbar^2/me^2$, at which the ground state disappears and which corresponds to the so-called "Mott" metal–dielectric transition.

The possibility of anomalies in thermodynamic functions upon metallization of dense vapors was discussed by Zel'dovich and Landau 1944, who had assumed that metallic systems might have two interfaces, one for the liquid–gas transition and the other one for the metal–dielectric transition (see Fig. 5.33). The relation between these transitions and plasma anomalies was analyzed by Norman (1971).

There exists a substantial number of publications where the phase transformations in a nondegenerate, strongly nonideal plasma are investigated and numerous possibilities emerging in such plasmas are analyzed on the basis of simple models (Norman and Starostin 1970; Khrapak and Yakubov 1970; Ross and Greenwood 1971; Ebeling *et al.* 1976; Iosilevskii and Starostin 2000; Norman 2001). The principal effect responsible for the condensation of a multicomponent plasma is the polarization attraction of opposite charges, whereas the stabilization of the emerging phase is due to the quantum effects – interference and, at high densities, degeneracy and overlap of electronic shells of ions.

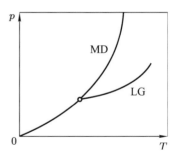

FIG. 5.33. Possible phase diagrams for metals (Zel'dovich and Landau 1944): LG – liquid–gas phase transition; MD – metal–dielectric transition

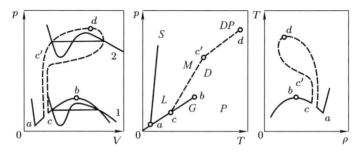

FIG. 5.34. Sketches illustrating the possible behavior of phase equilibria (Norman 1971): a, b, c, c′, d – critical and triple points; S – solid; L – liquid; G – gas; M – metal; D – dielectric; P – plasma; DP – dense plasma; 1,2 – isotherms.

Two phase transitions are predicted by the heuristic model of plasma metallization (Hansen *et al.* 1977). The determining factor at low densities and temperatures is the interatomic interaction leading to the gas–liquid transition (curve 1 in Fig. 5.34). With a density increase, the "ionization by pressure" occurs, along with a secondary violation of thermodynamic stability due to the Coulomb interaction. Therefore, the metal–dielectric transition in this model turns out to be the first–order phase transition which, as the temperature rises, changes over to a plasma phase transition (curve 2 in Fig. 5.34)

Among numerous calculations of the plasma phase boundaries, one should distinguish work by Ebeling *et al.* (1988) where the phase boundaries of a dense xenon plasma were derived. In these calculations, the Coulomb interaction was described by a broad–range Padé approximation (reproducing the asymptotes of Debye–Hückel and Hell–Mann–Brueckner), and the contribution of atoms and ions was described by the hard–sphere and Van der Waals models. It is demonstrated that this approach yields a first–order phase transition between weakly and strongly ionized states with the phase boundary of negative slope and the critical point at $T_c \simeq 12\,600$ K, $p_c \simeq 9.9$ GPa, and $\rho_c \simeq 3.7$ g cm^{-3}. So far, experiments with xenon (see Section 5.8 and 9.3.2) in the vicinity of these boundaries gave no evidence of the phase transition. Possibly, additional measurements are required here.

Apart from the phase anomalies caused by the strong Coulomb interaction, possibilities are discussed of a phase transition in a weakly ionized plasma due to strong charge–neutral interaction (Khrapak and Yakubov 1970). One of the methods to estimate conditions for the dielectric–metal transition is based on a very old idea by Herzfeld 1927 of the so–called "dielectric catastrophe" (Maksimov and Shilov 1999). By employing the Clausius–Mossotti formula for the dielectric permittivity of a dielectric,

$$\varepsilon = 1 + \frac{4\pi n\alpha}{1 - 4\pi n\alpha/3},\qquad(5.98)$$

where n is the atom or molecule number density and α is the polarizability, one can expect the transition when ε diverges. This occurs at $4\pi n\alpha/3 = 1$. According to the simplest estimates by Edwards and Ashcroft (1997) based on the magnitude of α for a free hydrogen molecule, this condition is satisfied for $p \sim 150$ GPa, which is in good agreement with other estimates performed with modern methods (Maksimov and Shilov 1999).

Being obviously approximate, the heuristic theories suffer from a great uncertainty of predictions and often lead to qualitatively different conclusions. On the other hand, they serve as a permanent stimulus for the experimental search of these exotic effects.

The hydrodynamic effects accompanying dynamic action on a plasma in the presence of phase transitions can be detected upon shock or isentropic variation of pressure (Fortov 1972). In the latter case, the phase transition leads to an abrupt change of isentropic compressibility depending on the properties of competing phases. If this change is sufficiently large, the rarefaction shocks are formed analogous to those recorded in iron and in ionic crystals (Altshuler 1965). In the general case, phase anomalies upon compression or expansion of metals would occur at the kinks of experimental and other thermodynamic curves discussed in Section 5.8. Especially characteristic in this respect are the experiments with the adiabatic expansion of shock–compressed metals (see Section 3.4), where in a single series of experiments one can follow the behavior of material over an extremely wide range of parameters and where one can expect to come across most of the phase anomalies predicted by theory (Figs. 5.35). The experimental data from isentropic compression and expansion of metals (see Figs. 3.17 and 3.18) as well as the data from rapid (Martynyuk 1977) and slow (Shaner and Gathers 1979) explosion of metals using an electric current have no clear hints of any anomalies which could be interpreted as phase transitions in nonideal plasma. Note that in the dynamic experiments one can detect both evaporation (Fig. 3.17 and 3.18) and melting (the latter produces little influence on the hydrodynamics of flow). In view of this, we would like to highlight studies by Ageev et al. (1988), where the optical measurements of the plasma temperature in an unloading wave also failed to reveal plasma anomalies.

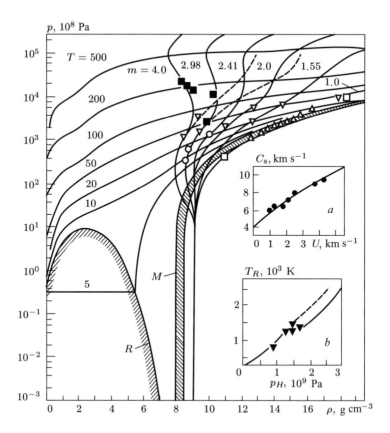

FIG. 5.35. Phase diagram of copper (semiempirical equation of state from Bushman and Fortov (1988): m – different values of porosity at the shock adiabats; T – isotherms (in 10^3 K); M and R – phase boundaries of melting and boiling, respectively. The two–phase regions are shaded, dashed lines indicate the adiabats from Kormer *et al.* (1962). Insets show velocity of sound on the shock adiabats (a) and temperature in the unloading wave (b).

Rusakov (1975) and Rusakov *et al.* (1977) presented the results of studies on the dynamics of expansion of shock–compressed materials in long (60–200 mm) cylindrical and cone–shaped channels. By measuring the velocities of luminescent plasma clusters and calculating the deviations from the results of self–similar solutions (obtained for an ideal gas with $c_p/c_v = 1.4$ on the assumption of stationarity of the incident wave), the authors draw the conclusion that phase anomalies are present in expanding plasma. Unfortunately, the plasma parameters in complex and obviously nonstationary flow were not measured directly, which forced the authors to make qualitative estimates. For instance, the density estimates were made assuming the plasma to be a noncompressible liquid, and the temperature estimates were based on the assumption of an ideal gas. According to Rusakov *et al.* (1977), the plasma temperature increases upon adiabatic expansion of the plasma.

Under the shock compression of a nonideal plasma, phase transitions would lead to a variation of shock compressibility whose increase, given some additional conditions (Fortov 1972), may cause the formation of continuous compression waves and multiwave structures, while a sharp decrease of compressibility may lead to the loss of shock wave stability. No such anomalies were revealed in experiments with the shock compression of cesium and heavy inert gases, and the recorded Hugoniot and Poisson adiabats (see Section 5.8) are smooth.

In addition, the experimentally found (see Sections 5.8 and 9.3) behavior of the variation of the thermodynamic functions of nonideal plasma corresponds to a one–phase case. Another indication of the absence of noticeable phase transformations in the range of parameters under investigation are the results of electrophysical (Chapter 7) and optical (Chapter 8) measurements, which can be well interpreted in the framework of one–phase plasma models.

Note that, based on existing knowledge (Fortov 1982; Bushman and Fortov 1983), phase transitions in dynamic experiments can occur over very short times (of the order of 10^{-10}–10^{-6} s) which are considerably less than in the static case. This makes unlikely (though does not rule it out completely) the phase "quenching" under nonstationary conditions.

Numerous studies are devoted to the search for the electronic Wigner crystal in a solid–state plasma, with a thorough review of these studies being given by Tsidil'kovskii (1987). In common metals, the high electron number density ($2 \leqslant r_s/a_0 \leqslant 7$) inhibits the formation of the Wigner crystal, so that the search was concentrated on highly compensated semiconductors and magnetic dielectrics.

By alloying n-type semiconductors with acceptors, one can substantially reduce the number density of free electrons and make the ion background close to uniform. In semiconductors of the type of ZnSb and $Hg_{0.8}Cd_{0.2}Te$, at $n = n_d - n_a \simeq 10^{14}$ cm^{-3} and $n_a/n_d \simeq 0.99$, approximately 300 impurity ions are found inside the electron orbit, which makes the OCP model applicable. However, no clear proof of the Wigner condensation is obtained in this case, apparently because of the earlier Mott localization of electrons on donors, or because of the destructive effect of fluctuating microfields.

In the case of magnetic isolators, the Wigner transition is apparently suppressed by the splitting of energy bands according to Hubbard's model.

The presence of a threshold magnetic field in ZnSb – below which no activation energy was revealed in measurements of the Hall effect and electrical conductivity – was treated as proof of the electron condensation. The results of these measurements are analyzed in detail by Tsidil'kovskii (1987), who produces arguments against the Wigner condensation, based on the fact that one deals with impurity electrons that are highly susceptible to the effect of a random potential. Also, the speculations of Ceperley (1978) and about the electron condensation in crystals of n-$Hg_{0.8}Cd_{0.2}Te$ are poorly grounded (Ceperley 1978; Ikezi *et al.* 1978), because apparently the electrons are localized on impurities due to the magnetic field.

Speaking about the plasma condensation, Norman and Starostin (1970), and Norman (1971) refer to a series of studies (Coock et al. 1955a, 1955b, 1981; Coock 1958; and Coock and McEwan 1958) which revealed "anomalous" effects upon the arrival of a detonation wave from condensed explosive to the free surface of an explosive charge. Coock et al. (1955a,b, 1981); Coock (1958); and Coock and McEwan (1958) give an interpretation of the observed effects, according to which a "Wigner crystal" is formed as a result of "chemical ionization" at the front of the detonation wave and, after this wave leaves the explosive charge, the crystal is in the metastable state. These plasma concepts formed the basis of the model of detonation of condensed explosives (Coock 1958), according to which the detonation-guiding process is the electron thermal conductivity rather than the shock wave.

A thorough analysis of the results of Coock et al. has shown (Davis and Campbell 1960; Fortov et al. 1974) that the proposed interpretation is not unambiguous and gives rise to objections, while the conclusions about the plasma phase transition lack substantiation, all the more that no direct measurements of density – which is the most affected by such a transition – were performed by Coock et al. (as well as by Rusakov 1975 and Rusakov et al. 1977).

The exposure of plasma to pulse a X–ray radiation with a wavelength of 0.1 nm and 9 nm revealed (Fortov et al. 1974) no anomalously high plasma density, while the measured values proved to be eight to ten times higher than the free air density, this being in good agreement with the calculation of states behind the air shock wave front performed by Kuznetsov (1945) for a quasi–ideal gas approximation. Detection of the optical radiation emerging from a plasma has shown (Alekseev and Arkhipov 1962; Arkhipov 1965; Fortov et al. 1974) that, in contrast to the conclusions drawn by Coock et al., this radiation is not anomalous and corresponds to a brightness temperature of $(8-10)\cdot10^3$ K, which is close to conventional calculations (Kuznetsov 1945). The optical absorption coefficient of the plasma also does not contain any qualitative anomalies and is described by the rarefied plasma model (see Chapter 7). Experiments involving detonation wave arrival into vacuum (Fortov et al. 1974) and helium (Davis and Campbell 1960) revealed no plasma glow, whereas, according to the concept of Coock et al., this glow must have intensified. In order to explain the observed velocities of the plasma clump, artificial reasoning about "reactive thrust during recombination" was used by Coock et al., while these velocities are easily calculated on the basis of the "discontinuity breakup" theory at the explosive–air interface.

Therefore, the performed experiments and analysis of available data demonstrate that the thermodynamic, optical, electrophysical, hydrodynamic and mechanical properties of the plasma clump are fully described by the ideal plasma model in the assumption that this clump is formed by compression and irreversible heating of air in the front of the wave emerging upon expansion of the detonation products. The wrong conclusion of Coock et al. about plasma condensation appears to be mainly due to the small velocity of the side expansion of the glass tube channels, because of their considerable mass, this being erroneously interpreted as the plasma "stability".

Hall *et al.* (1988) obtained a plasma with density several times higher than that of solids and temperature of about 1 eV, by collision of two shock waves excited by high–power $[(4\text{--}5)\cdot10^{12} \text{ W cm}^{-2}]$ laser radiation with a wavelength of 0.53 μm and pulse duration of about 100 ps. Peculiarities of the X–ray radiation, produced by an external source and transmitted through the plasma, were interpreted by Hall *et al.* (1988) as the presence of short–range order associated with the Wigner concentration in a highly compressed aluminum plasma. When interpreting the results of these interesting experiments, one should apparently perform numerical simulations of the complex process of plasma generation and assess the nonequilibrium effects during the melting of aluminum.

In a strongly compressed plasma, the occurrence of electronic phase transformations is possible due to the transition of inner electronic shells of an atom or ion from the discrete spectrum to the continuous one. Phase transitions associated with the electron redistribution in shells in the course of compression (Kirzhnits *et al.* 1975; Bushman and Fortov 1983) were theoretically analyzed using band theory methods (Alekseev and Arkhipov 1962; Arkhipov 1965; Royce 1967; Reichlin *et al.* 1989) and detected experimentally by Altshuler and Bakanova (1969), Trunin *et al.* (1969), Shatzman (1977), and Avrorin *et al.* (1987). Brush *et al.* (1963) and Kirzhnits *et al.* (1975) predicted electronic transformations at ultrahigh pressures, corresponding to the first–order phase transition, with an electronic shell being in a discrete spectrum in one phase and in a continuous spectrum in the other phase. Naturally, such electronic transitions correspond in fact to a series of phase transitions (Carmi 1968) caused by the "ionization by pressure" of a sequence of electronic shells. The evaluation of the parameters of these transformations using quasiclassical theory techniques (Kirzhnits *et al.* 1975) leads to ultrahigh pressures over 30 TPa which became experimentally accessible only recently (see Chapter 9).

References

Abe, R. (1959). Giant cluster expansion theory and its application to high temperature plasma. *Progr. Theor. Phys.*, **22**, 213–226 .

Ageev, V. G., Bushman, A. V., Kulish, M. I., Lebedev, M. E., Leont'ev, A. A., Ternovoi, V. Y., Filimonov, A. S., and Fortov, V. E. (1988). Thermodynamics of a dense lead plasma near the high–temperature boiling curve. *JETP Lett.*, **48**, 659–663.

Akkerman, A. F., Demidov, B. A., Fortov, V. E., *et al.* (1986). *Application of the heavy–current relativistic electron beams in dynamic physics of high temperatures and pressures.* Joint Institute of Chemical Physics, Chernogolovka.

Alekseev, E. S. and Arkhipov, R. G. (1962). Electron transitions in cesium and rubidium under pressure. *Phys. Solid State*, **4**, 1077–1081.

Alekseev, V. A., Starostin, A. N., Vedenov, A. A., and Ovcharenko, V. G. (1972). Nature of thermoelectric power of mercury in transcritical state. *JETP*

Lett., **16**, 49–53.

Alekseev, V. A., Fortov, V. E., and Iakubov I. T. (1981). Current status of physics of nonideal plasma. In *Proceedings of the 15th international conference on phenomena in ionized gases*, Minsk. Invited papers, pp. 73–85.

Altshuler, L. V. (1965). Use of shock waves in high–pressure physics. *Sov. Phys. Uspekhi*, **8**, 52–91.

Altshuler, L. V. and Bakanova, A. A. (1969). Electronic structure and compressibility of metals at high pressures. *Sov. Phys. Uspekhi*, **11**, 678–689.

Altshuler, L. V., Kalitkin, N. N., Kuzmina, L. V., and Chekin, B. S. (1977). Shock adiabats for ultrahigh pressures. *JETP*, **45**, 167–171.

Arkhipov, R. G. (1965). On possibility of existence of the critical point for electron transitions of the first type. *JETP*, **49**, 1601–1604.

Armstrong, B. H., Johnson, R. R., Kelly, R. S., DeWitt, H. E., and Brush, S. G. (1967). Opacity of high temperature air. In *Progress in High Temperature Physics and Chemistry. Vol. 1*, Rouse, C. A. (ed.), pp. 139–242. Pergamon Press, New York.

Avrorin, E. N., Vodolaga, B. K., Voloshin, N. P., Kovalenko, G. V., Kuropatenko, V. F., Simonenko, V. A., and Chernovolyuk, B. T. (1987). Experimental study of the influence of electron shell structure on shock adiabats of condensed materials. *JETP*, **66**, 347–354.

Barker, A. A. (1971). Effective potentials between the components of a hydrogeneous plasma. *J. Chem. Phys.*, **55**, 1751–1759.

Baus, M. and Hansen, J. P. (1980). Statistical mechanics of simple coulomb systems. *Phys. Rep.*, **59**, 228–237.

Berne, B. J., Ciccotti, G., and Coker, D. F. (eds). (1998). *Classical and quantum dynamics of condensed phase simulation.* World Scientific, Singapore.

Bespalov, V. E., Gryaznov, V. K., Dremin, A. N., and Fortov, V. E. (1975). Dynamic compression of nonideal argon plasma. *JETP*, **42**, 1046–1049.

Brush, S. G., DeWitt, H. E., and Trulio, H. E. (1963). Equation of state of classical systems of charged particles. *Nucl. Fusion.*, **3**, 5–22.

Brush, S. G., Sahlin, H. L., and Teller, E. (1966). Monte Carlo study of a one–component plasma. I. *J. Chem. Phys.*, **45**, 2102–2118.

Bushman, A. V. and Fortov, V. E. (1983). Model equations of state. *Phys. Uspekhi*, **26**, 465–496.

Bushman, A. and Fortov, V. (1988). Wide–range equation of state for matter under extreme conditions. *Sov. Tech. Rev. B Therm. Phys.*, **1**, 162–181.

Bushman, A. V., Lomakin, B. N., Sechenov, V. A., and Sharipdzhanov, I. I. (1975). Thermodynamics of nonideal cesium plasma. *JETP*, **42**, 828–831.

Carley, D. D. (1974). Parametric integral equation for radial distribution functions. *Phys. Rev. A*, **10**, 863–867.

Carmi, G. (1968). First–order phase transitions in quantum–Coulomb plasmas. *J. Math. Phys.*, **9**, 2120–2131.

Carnahan, N. F. and Starling, K. E. (1969). Equation of state for nonattracting rigid spheres. *J. Chem. Phys.*, **51**, 635–636.

Ceperley, D. (1978). Ground state of the fermion one–component plasma: A Monte Carlo study in two and three dimensions. *Phys. Rev. B*, **18**, 3126–3138.

Ceperley, D. (1995). Path integrals in the theory of condensed helium. *Rev. Mod. Phys.*, **67**, 279–355.

Cohen, E. G. D. and Murphy, T. J. (1969). New results in the theory of the classical electron gas. **12**, 1404–1411; (1970). Erratum. **13**, 216.

Coock, M. A. (1958). *The science of high explosives*. Reinhold, New York.

Coock, M. A. and McEwan, W. S. (1958). Cohesion in plasma. *J. Appl. Phys.*, **29**, 1612–1613.

Coock, M. A., Doran, R. L., and Morris, G. J. (1955a). Measurement of detonation velocity by doppler effect at three–centimeter wavelength. *J. Appl. Phys.*, **26**, 426–428.

Coock, M. A., Horsley, G. S., Partridge, W. S., and Ursenbach, W. O. (1955b). Velocity–diameter and wave shape measurements and the determination of reaction rates in TNT. *J. Chem. Phys.*, **24**, 60–67.

Coock, M. A., Keyes, R. T., and Udy, L. L. (1981). Propagation characteristics of detonation–generated plasmas. *J. Appl. Phys.*, **52**, 1881–1895.

Davis, W. C. and Campbell, A. W. (1960). Ultra–high–speed photographs refuting "cohesion in plasma". *J. Appl. Phys.*, **31**, 1225–1227.

Deutsch, C., Furutani, Y., and Gombert, M. M. (1981). Nodal expansions for strongly coupled classical plasmas. *Phys. Rep.*, **69**, 1358–1368.

DeWitt, H. E. (1962). Evaluation of the quantum–mechanical ring sum with Boltzmann statistics. *J. Math. Phys.*, **3**, 1216–1228.

DeWitt, H. E. (1976). Asymptotic form of the classical one–component plasma fluid equation of state. *Phys. Rev. A*, **14**, 1290–1293.

DeWitt, H. E. (1977). Equilibrium statistical mechanics of strongly coupled plasmas by numerical simulation. In *Strongly coupled plasmas*, Kalman, G. J. and Carini, P. (eds), pp. 81–115. Plenum Press, New York.

DeWitt, H. E. and Hubbard, W. B. (1976). Statistical mechanics of light elements at high pressure. IV – A model free energy for the metallic phase. *Astrophys. J.*, **205**, 295–301.

DeWitt, H. E. and Rosenfeld, Y. (1979). Derivation of the one component plasma fluid equation of state in strong coupling. *Phys. Lett.*, **75**, 79–80.

Dharma–wardana, M. W. C. and Perrot, F. (1982). Density–functional theory of hydrogen plasmas. *Phys. Rev. A*, **26**, 2096–2104.

Dikhter, I. Y. and Zeigarnik, V. A. (1977). Equation of state and conductivity of a highly ionized cesium plasma. *High Temp.*, **15**, 196–198.

Dubin, D. H. E. and O'Neil, T. M. (1999). Trapped nonneutral plasmas, liquids, and crystals (the thermal equilibrium states). *Rev. Mod. Phys.*, **83**, 87–172.

Ebeling, W. and Sandig, R. (1973). Theory of the ionization equilibrium in dense plasmas. *Ann. Phys.*, **28**, 289–305.

Ebeling, W., Kreft, W. D., and Kremp, D. (1976). *Theory of bound states and ionization equilibrium in plasmas and solids*. Akademie–Verlag, Berlin.

Ebeling, W., Forster, A., Richert, W., and Hess, H. (1988). Thermodynamic properties and plasma phase transition of xenon at high pressure and high temperature. *Physica A*, **150**, 159–171.

Ebeling, W., Förster, A., Fortov, V., Gryaznov, V., and Polishchuk, A. (1991). *Thermophysical properties of hot dense plasmas.* Teubner, Stuttgart–Leipzig.

Edwards, B. and Ashcroft, N. W. (1997). Spontaneous polarization in dense hydrogen. *Nature*, **388**, 652–655.

Filinov, V. S. (2000). Method Monte Carlo and method of molecular dynamics in the theory of nonideal quantum plasma. In *Encyclopedia of low temperature plasma. Introductory volume 3.*, Fortov, V. E. (ed.), pp. 243–252. Nauka, Moscow.

Filinov, V. S. (2001). Cluster expansion for ideal Fermi systems in the "fixed–node approximation". *J. Phys. A*, **34**, 1665–1677.

Filinov, V. S., Bonitz, M., Ebeling, W., and Fortov, V. E. (2001a). Thermodynamics of hot dense H–plasmas: Path integral Monte Carlo simulations and analytical approximations. *Plasma. Phys. Control. Fusion*, **43**, 743–759.

Filinov, V. S., Bonitz, M., and Fortov, V. E. (2000a). High–density phenomena in hydrogen plasma. *JETP Lett.*, **72**, 361–365.

Filinov, V. S., Fortov, V. E., Bonitz, M., and Kremp, D. (2000b). Pair distribution functions of dense partially ionized hydrogen. *Phys. Lett. A*, **274**, 228–235.

Filinov, V. S., Fortov, V. E., Bonitz, M., and Levashov, P. R. (2001b). Phase transition in strongly degenerate hydrogen plasma. *JETP Lett.*, **74**, 384–387.

Filinov, V. S., Levashov, P. R., Fortov, V. E., and Bonitz, M. (2000c). Thermodynamic properties of correlated strongly degenerate plasmas. In *Progress in nonequilibrium Green's functions*, Bonitz M. (ed.), pp. 513–520. World Scientific, Singapore.

Filinov, V. S. and Norman, G. E. (2000). Use of the Monte Carlo method for calculation of thermodynamical properties, chemical and ionization equilibrium. In *Encyclopedia of low temperature plasma. Introductory volume 3.*, Fortov, V. E. (ed.), pp. 236–243. Nauka, Moscow.

Fisher, D. S., Halperin, B. I., and Platzman, P. M. (1979). Phonon–ripplon coupling and the two–dimensional electron solid on a liquid–helium surface. *Phys. Rev. Lett.*, **42**, 798–801.

Fortov, V. E. (1972). Hydrodynamic effects in a nonideal plasma. *High Temp.*, **10**, 141–153.

Fortov, V. E. (1982). Dynamic methods in plasma physics. *Sov. Phys. Uspekhi*, **25**, 781–809.

Fortov, V. E., Leont'ev, A. A., Dremin, A. N., and Gryaznov, V. K. (1976). Shock–wave production of a nonideal plasma. *JETP*, **44**, 116–122.

Fortov, V. E., Lomakin, B. N., and Krasnikov, Y. G. (1971). Thermodynamic properties of a cesium plasma. *High Temp.*, **9**, 789–797.

Fortov, V. E., Musyankov, S. I., Yakushev, V. V. (1974). On "anomalous" effects on going out of detonation wave at free surface. *High Temp.*, **12**, 957–963.

Galam, S. and Hansen, J. P. (1976). Statistical mechanics of dense ionized matter. VI. Electron screening corrections to the thermodynamic properties of the one–component plasma. *Phys. Rev. A*, **14**, 816–832.

Gann, R. C., Chakravarty, S., and Chester, G. V. (1979). Monte Carlo simulation of the classical two–dimensional one–component plasma. *Phys. Rev. B*, **20**, 326–344.

Gell–Mann, M. and Brueckner, K. A. (1957). Correlation energy of an electron gas at high density. *Phys. Rev.*, **106**, 364–368.

Gilbert, S. L., Bollinger, J. J., and Wineland, D. J. (1988). Shell–structure phase of magnetically confined strongly coupled plasma. *Phys. Rev. Lett.*, **60**, 2022–2026.

Gillan, M. J. (1974). A simple model for the classical one–component plasma. *J. Phys. C*, **7**, L1–L4.

Gorobchenko, V. D. and Maksimov, E. G. (1980). The dielectric constant of an interacting electron gas. *Sov. Phys. Uspekhi*, **23**, 35–58.

Graboske, H. C., Harwood, D. J., and Rogers, F. J. (1969). Thermodynamic properties of nonideal gases. I. Free–energy minimization method. *Phys. Rev.*, **186**, 210–225.

Graboske, H. C., Harwood, Jr. D. J., and DeWitt, H. E. (1971). Thermodynamic properties of nonideal gases. II. The strongly ionized gas. *Phys. Rev. A*, **3**, 1419–1431.

Graboske, H. C., DeWitt, H. E., Grosmann, A. S., and Cooper, M. S. (1973). Screening factors for nuclear reactions. 2. Intermediate screening and astrophysical applications. *Astrophys. J.*, **181**, Part 1, 457–474.

Grimes, C. C. and Adams, G. (1979). Evidence for a liquid–to–crystal phase transition in a classical, two–dimensional sheet of electrons. *Phys. Rev. Lett.*, **42**, 798–801.

Gryaznov, V. K. and Iosilevskii, I. L. (1973). Problem of construction of interpolation equation of state of plasma. In *Numerical methods of continuous matter mechanics, No. 4*, pp. 166–171.Siberian Division AN USSR, Novosibirsk.

Gryaznov, V. K. and Iosilevskii, I. L. (1976). Some problems or thermodynamical calculation of multicomponent strongly coupled plasma. In *Thermophysical properties of low temperature plasma*, Ivlev, V.M. (ed.), pp. 25–29. Nauka, Moscow.

Gryaznov, V. K., Zhernokletov, M. V., Zubarev, V. N., Iosilevskii, I. L., and Fortov, V. E. (1980). Thermodynamic properties of a nonideal argon or xenon plasma. *JETP*, **51**, 288–295.

Gryaznov, V. K., Iosilevskii, I. L., and Fortov, V. E. (1982). Thermodynamics of strongly heated plasma of megabar pressure range. *Techn. Phys. Lett.*, **8**, 1378–1381.

Gryaznov, V. K., Ivanova, A. N., Gutsev, G. L., Levin, A. A., and Krestinin, A. V. (1989). The complex of programs ESCAPAK for calculations of the electronic structure adapted for the EC computers. *J. Structural Chem.*, **30**, 132–141.

Guttman, A. J., Ninham, B. W., and Tompson, C. J. (1968). Determination of critical behaviour in lattice statistics from series expansions. *Phys. Lett.*, **26**, 180–181.

Hall, T. A., Djaoui, A., Eason, R. W., Jackson, C. L., Shiwai, B., Rose, S. L., Cole, A., and Apte, P. (1988). Experimental observation of ion correlation in a dense laser–produced plasma. *Phys. Rev. Lett.*, **60**, 2034–2037.

Hansen, J. P. (1973). Statistical mechanics of dense ionized matter. I. Equilibrium properties of the classical one–component plasma. *Phys. Rev. A*, **8**, 3096–3109.

Hansen, J. P., Torrie, G. M., and Vieillefosse, P. (1977). Statistical mechanics of dense ionized matter. VII. Equation of state and phase separation of ionic mixtures in a uniform background. *Phys. Rev. A*, **16**, 2153–2168.

Herzfeld, K. F. (1927). On atomic properties which make an element a metal. *Phys. Rev.*, **29**, 701–705.

Hirt, C. W. (1967). Solution of the Born–Green–Yvon equation for a high density one–component plasma. *Phys. Fluids*, **10**, 565–570.

Iakubov, I. T. (2000). Electric conductivity of low temperature plasma in wide range of parameters. In *Encyclopedia of low temperature plasma. Introductory volume 1,* Fortov, V. E. (ed.), pp. 536–545. Nauka, Moscow.

Ichimaru, S. (1977). (private communication).

Ichimaru, S. (1982). Strongly coupled plasmas: High–density classical plasmas and degenerate electron liquids. *Rev. Mod. Phys.*, **54**, 1017–1059.

Ichimaru, S. (1992). *Statistical plasma physics. Vol. 1: Basic principles.* Addison–Wesley, Redwood City.

Ikezi, H., Schwarzenegger, K., Simons, A. L., Passner A. L., and McCall, S. L. (1978). Optical–properties of expanded fluid mercury. *Phys. Rev. A*, **18**, 2494–2499.

Iosilevskii, I. L. (1980). Equation of state of nonideal plasma. *High Temp.*, **18**, 355–359.

Iosilevskii, I. L. (1985). Phase transition in the simplest model of plasma. *High Temp.*, **23**, 1041–1049.

Iosilevskii, I. L. and Gryaznov, V. K. (1981). About comparative precision of thermodynamic description of gaseous plasma properties in Thomas–Fermi and Saha approximations. *High Temp.*, **19**, 1121–1124.

Iosilevskii, I. L. and Starostin, A. N. (2000). Problem of thermodynamical stability of low temperature plasma. In *Encyclopedia of low temperature plasma. Introductory volume 1,* Fortov, V. E. (ed.), pp. 327–339. Nauka, Moscow.

Jankovic, B. (1977). Pair correlation function in a dense plasma and pycnonuclear reactions in stars. *J. Stat. Phys.*, **17**, 357–362.

Kalitkin, N. N. (1960). The Thomas–Fermi model of the atom with quantum and exchange corrections. *JETP*, **11**, 1106–1110.

Kalitkin, N. N. (1989). Models of matter at extremal state. In *Mathematical modelling: Physicochemical properties of matter,* Samarskii, A. A. and Kalitkin, N. N. (eds), pp. 114–161. Nauka, Moscow.

Kalitkin, N. N. and Kuzmina, L. V. (1975). Tables of thermodynamic functions of matter at high energy concentration. *Preprint No. 35 of the Institute of Applied Mechanics*. Moscow.

Kalitkin, N. N. and Kuzmina, L. V. (1976). Quantum–statistical equation of state and shock adiabats. *Preprint No. 14 of the Institute of Applied Mechanics*. Moscow.

Kalitkin, N. N. and Kuzmina, L. V. (2000). Wide–range shock adiabats. In *Shock waves and extremal state of matter*, Fortov, V. E., Altshuler, L. V., Trunin, R. F., and Funtikov, A. I. (eds), pp. 107–120. Nauka, Moscow.

Kalitkin, N. N., Kuzmina, L. V., and Rogov, V. S. (1972). Tables of thermodynamic functions and transport coefficients of plasma. *Preprint No. 21 of the Institute of Applied Mechanics*. Moscow.

Kelbg, G. (1964). Klassische statistische Mechanik der Teilchen–Mischungen mit sortenabhängigen weitreichenden zwischenmolekularen Wechselwirkungen *Ann. Physik*, **14**, 394–401.

Khrapak, A. G. and Yakubov, I. T. (1970). Phase transition and negatively–charged complexes in a nonideal plasma of metal vapor. *JETP*, **32**, 514–516.

Kirzhnits, D. A. (1976). Are the Kramers–Kronig relations for the dielectric permittivity of a material always valid? *Sov. Phys. Uspekhi*, **19**, 530–537.

Kirzhnits, D. A. and Shpatakovskaya, G. V. (1995). Statistical model of matter, corrected in the neighborhood of nuclei. *JETP*, **81**, 679–686.

Kirzhnits, D. A., Lozovik, Y. E., and Shpatakovskaya, G. V. (1975). Statistical model of matter. *Sov. Phys. Uspekhi*, **18**, 3–48.

Klyuchnikov, N. I. and Lyubimova, I. A. (1987). Density functional method in thermodynamics of strongly compressed matter. In *Reviews on thermophysical properties of matter, No. 66*. IVTAN, Moscow.

Kopyshev, V. P. (1978). On thermodynamics of nuclei of monatomic matter. *Preprint No. 59 of the Institute of Applied Mechanics*. Moscow.

Kormer, S. B., Funtikov, L. I., Urlin, V. D., and Kolesnikova, A. N. (1962). Dynamic compression of porous metals and the equation of state with variable specific heat at high temperatures. *JETP*, **42**, 686–702.

Kovalenko, N. T. and Fisher, I. Z. (1973). Method of integral equations in statistical theory of liquids. *Sov. Phys. Uspekhi*, **15**, 592–607.

Krasnikov, Y. (1977). Thermodynamics of a nonideal low–temperature plasma. *JETP*, **73**, 516–525.

Kunavin, A. G., Kirillin, A. V., and Korshunov, Y. S. (1974). Investigation of cesium plasma by method of adiabatic compression. *High Temp.*, **12**, 1302–1305.

Kuznetsov, N. M. (1945). *Tables of thermodynamical functions and shock adiabats of air at high temperatures*. Mashinostroenie, Moscow.

Landau, L. D. and Lifshitz, E. M. (1980). *Statistical physics, Part 1*. Pergamon, New York.

Larkin, A. I. (1960). Thermodynamic functions of a low–temperature plasma. *JETP*, **11**, 1363–1364.

Latter, R. (1955). Temperature behavior of the Thomas–Fermi statistical model for atoms. *Phys. Rev.*, **99**, 1854–1870.

Latter, R. (1956). Thomas–Fermi model of compressed atoms. *J. Chem. Phys.*, **24**, 280–292.

Lebowitz, J. L. and Percus, J. K. (1966). Mean spherical model for lattice gases with extended hard cores and continuum fluids. *Phys. Rev.*, **144**, 251–258.

Liberman, D. A. (1979). Self–consistent field model for condensed matter. *Phys. Rev. E*, **20**, 981–4989.

Lieb, E. H. and Narnhofer, H. J. (1975). The thermodynamic limit for jellium. *J. Stat. Phys.*, **12**, 291–310.

Likal'ter, A. A. (1969). Interaction of atoms with electrons and ions in a plasma. *JETP*, **29**, 133–135.

Likal'ter, A. A. (2000). Critical points of condensation in Coulomb systems. *Phys. Uspekhi*, **43**, 777–797.

Lundqvist, S. and March, N. H. (eds). (1983). *Theory of the inhomogeneous electron gas.* Springer, New York.

Maksimov, E. G. and Dolgov, O. V. (1982). Transition temperature of strong–coupling superconductors. *Sov. Phys. Uspekhi*, **25**, 688–704.

Maksimov, E. G., and Shilov, Y. I. (1999). Hydrogen at high pressure. *Sov. Phys. Uspekhi*, **14**, 512–523.

Martynyuk, M. M. (1977). Phase explosion of metastable liquid. *Comb., Expl., Shock Waves*, **13**, 213–229.

Militzer, B. and Ceperley, D. M. (2001). Path integral Monte Carlo simulation of the low–density hydrogen plasma. *Phys. Rev. E*, **63**, 066404/1–10.

Mintsev, V. B., Fortov, V. E., and Gryaznov, V. K. (1980). Electric conductivity of a high–temperature nonideal plasma. *JETP*, **52**, 59–63.

Mitchell, D. J. and Ninham B. W. (1968). Asymptotic behavior of the pair distribution function of a classical electron gas. *Phys. Rev.*, **174**, 280–289.

Model', I. S. and Narozhnyi, A. T., Kharchenko, A. T., Kholin, S. A., and Khrustalev, V. V. (1985). Equation of state for graphite, aluminum, titanium, and iron at pressures > 13 Mbar. *JETP Lett.*, **41**, 332–334.

Mott, N. F. (1967). Electrons in disordered structures. *Adv. Phys.*, **16**, 49.

Ng, K. C. (1974). Hypernetted chain solutions for the classical one–component plasma up to Γ=7000. *J. Chem. Phys.*, **61**, 2680–2689.

Nikiforov, A. F., Novikov, V. G., and Uvarov, V. B. (1979). *Problems of atomic science and technics. Section: Methodics and programs*, **4**, No. 6, 16–26.

Norman, H. E. (1971). Thermodynamics (heuristics and Monte Carlo method). In *Sketches of physics and chemistry of low–temperature plasma*, Polak, L. S. (ed.), pp. 260–277. Nauka, Moscow.

Norman, H. E. (2001). Plasma phase transition. *Contrib. Plasma Phys.*, **41**, 127–130.

Norman, G. E. and Starostin, A. N. (1970). Thermodynamics of strongly non-ideal plasma. *High Temp.*, **8**, 381–408.

Novikov, V. G. (1985). Shock compression of lithium and iron according to model MCPS. *Preprint No. 133 of the Institute of Applied Mechanics.* Moscow.

Overhauser, A. W. (1968). Exchange and correlation instabilities of simple metals. *Phys. Rev.,* **167**, 691–695.

Overhauser, A. W. (1978). Charge–density waves and isotropic metals. *Adv. Phys.,* **27**, 343–364.

Pollock, E. L. and Alder, B. J. (1977). Phase separation for a dense fluid mixture of nuclei. *Phys. Rev. A,* **15**, 1263–1268.

Pollock, E. L. and Hansen, J. P. (1973). Statistical mechanics of dense ionized matter. II. Equilibrium properties and melting transition of the crystallized one–component plasma. *Phys. Rev. A,* **8**, 3110–3122.

Rajagopal, A. K. (1980). Theory of inhomogeneous electron systems: Spin–density–functional formalism. *Adv. Chem. Phys.,* **41**, 59–193.

Redmer, R. (1997). Physical properties of dense, low–temperature plasmas. *Phys. Rep.,* **282**, 35–157.

Redmer, R., Röpke, G., Kuhlbrodth, S., and Reinholz, H. (2001). Metal–non-metal transition in dense hydrogen. *Contrib. Plasma Phys.,* **41**, 163–166.

Reichlin, R., Brister, K. E., McMahan, A. K., Ross, M., Martin, S., Vohra, Y. K., and Ruoff, A. L. (1989). Evidence for the insulator–metal transition in xenon from optical, X–ray, and band–structure studies to 170 GPa. *Phys. Rev. Lett.,* **62**, 669–672.

Rogers, F. J. and DeWitt, H. E. (1973). Statistical mechanics of reacting Coulomb gases. *Phys. Rev. A,* **8**, 1061–1076.

Rosenfeld, Y. and Ashcroft, N. W. (1979). Theory of simple classical fluids: Universality in the short–range structure. *Phys. Rev. A,* **20**, 1208–1235.

Ross, R. G. and Greenwood, D. A. (1971). Liquid metals and vapours under pressure. *Progr. Mater. Sci.,* **14**, 173.

Royce, E. B. (1967). Stability of the electronic configuration and compressibility of electron orbitals in metals under shock–wave compression. *Phys. Rev.,* **164**, 929.

Rusakov, M. M. (1975). Evaporative expansion after passage of an intense shock wave. *High Temp.,* **13**, 17–19.

Rusakov, M. M., Ivanov, R. I., Shaidulin, B. K., and Shpak, S. G. (1977). Features of the expansion of condensed substances acted on by high–power shock waves. *High Temp.,* **15**, 381–385.

Rushbrooke, G. S. (1968). Equilibrium theories of liquid state. In *Physics of simple liquids,* Temperley, H. N. V., Rowlinson, J. S., and Rushbrooke, G. S. (eds). North–Holland, Amsterdam.

Salpeter, E. E. (1954). Electron screening and thermonuclear reactions. *Aust. J. Phys.,* **7**, 373–396.

Saumon, D. and Chabrier, G. (1992). Fluid hydrogen at high density: Pressure ionization. *Phys. Rev. A,* **46**, 2084–2100.

Savukinas, A. Y. and Chizhunas, A. R. (1974). Effect of screening and compression on atomic energy levels in plasma. *Lithuanian Phys. Collection,* **14**,

73–83.

Shaner, J. W. and Gathers, G. R. (1979). (unpublished).

Shatzman, E. (1977). (private communication).

Shpatakovskaya G. V. (2000). Cell approach to description of thermodynamical properties. In *Encyclopedia of Low Temperature Plasma. Introductory volume 1*, Fortov, V. E. (ed.), pp. 313–322. Nauka, Moscow.

Simonenko, V. A., Voloshin, N. P., Vladimirov, A.S., Nagibin, A.P., Nogin, V.N., Popov, V.A., Vasilenko, V.A., and Shoidin, Y. A. (1985). Absolute measurements of shock compressibility of aluminum at pressures $p \geq 1$ TPa. *JETP*, **61**, 869–873.

Sin'ko, G. V. (1983). Use of the self–consistent field method for calculation of thermodynamical electron functions of simple matters. *High Temp.*, **21**, 1041–1052.

Slattery, W. L., Doolen, G. D., and DeWitt, H. E. (1980). Improved equation of state for the classical one–component plasma. *Phys. Rev. A*, **21**, 2087–2095.

Springer, J. F., Pokrant, M. A., and Stevens, F. A. (1973). Integral equation solutions for the classical electron gas. *J. Chem. Phys.*, **58**, 4863–4867.

Stevenson, D. J. (1975). Thermodynamics and phase separation of dense fully ionized hydrogen–helium fluid mixtures. *Phys. Rev. B*, **12**, 3999–4007.

Stillinger, F. H. and Lovett, R. (1968). General restriction on the distribution of ions in electrolytes. *J. Chem. Phys.*, **49**, 1991–1994.

Stringfellow, G. S., DeWitt, H. E., and Slattery, W. L. (1990). Equation of state of the one–component plasma derived from precision Monte Carlo calculations. *Phys. Rev. A*, **41**, 1105–1111.

Ternovoi, V. Y., Filimonov, A. S., Fortov, V. E., Kvitov, S. V., Nikolaev, D. N., and Pyalling, A. A. (1999). Thermodynamic properties and electrical conductivity of hydrogen under multiple shock compression to 150 GPa. *Physica B*, **265**, 6–11.

Trunin, R. F., Podurets, M. A., Moiseev, B.N., Simakov, G. B., and Popov, L.V. (1969). Relative compressibility of copper, cadmium, and lead at high pressures. *JETP*, **29**, 630–631.

Tsidil'kovskii, I. M. (1987). Crystallization of a three–dimensional electron gas. *Phys. Uspekhi*, **30**, 676–698.

Vladimirov, A. S., Voloshin, N. P., Nogin, V.N., Petrovtsev, A. V., and Simonenko, V. A. (1984). Shock compressibility of aluminum at $p \geq 1$ Gbar. *JETP Lett.*, **39**, 82–5.

Volkov, L. P., Voloshin, N. P., Mangasarov, R. A., Simonenko, V. A., Sin'ko, G. V., and Sorokin, V. L. (1980). Shock compressibility of water at pressure of ~ 1 Mbar. *JETP Lett.*, **31**, 513–515.

Vorob'ev, V. S. (2000). Asymptotic methods of description of low temperature plasma thermodynamics. In *Encyclopedia of Low Temperature Plasma. Introductory volume 1*, Fortov, V. E. (ed.), pp. 293–299. Nauka, Moscow.

Weir, S. T., Mitchell, A. C., and Nellis, W. J. (1996). Metallization of fluid molecular hydrogen at 140 GPa (1.4 Mbar). *Phys. Rev. Lett.*, **76**, 1860–1863.

Wigner, E. P. (1934). On the interactions of electrons in metals. *Phys. Rev.*, **46**, 1002–1011.

Xu, H. and Hansen, J. P. (1998). Density–functional theory of pair correlations in metallic hydrogen. *Phys. Rev. E*, **57**, 211–223.

Zamalin, V. M. and Norman, G. E. (1973). The Monte Carlo method in Feynman's formulation of quantum statistics. *USSR Computational Mathematics and Mathematical Physics*, **13**, 169–183.

Zamalin, V. M., Norman, G. E., and Filinov, V. S. (1977). *Monte Carlo method in statistical physics*. Nauka, Moscow.

Zel'dovich, Y. B. and Landau, L. D. (1944). On correlation between liquid and gaseous states of metals. *Acta Phys.–Chim. USSR*, **18**, 194–198.

Zelener, V. B., Norman, G. E., and Filinov, V. S. (1981). *Perturbation theory and pseudopotential in statistical thermodynamics*. Nauka, Moscow.

6

ELECTRICAL CONDUCTIVITY OF PARTIALLY IONIZED PLASMA

6.1 Electrical conductivity of ideal partially ionized plasma

6.1.1 Electrical conductivity of weakly ionized plasma

The electrophysical properties of matter are defined primarily by the state of the electron component. During the transition of metal from the liquid to weakly conducting gas phase, the state of charged particles varies from the state of almost free electrons of liquid metal to that of electrons strongly bound in atoms. In so doing, the value of the coefficient of electrical conductivity varies by many orders of magnitude.

The electrical conductivity σ is defined by the concentration of electrons n_e and their mobility μ,

$$\sigma = e\mu n_e. \tag{6.1}$$

In an ideal plasma, these quantities are defined by expressions well known from kinetic theory. The concentrations of electrons, n_e, and ions, n_i, are related to each other by Saha's formula (4.5). Complex ions are absent in an ideal plasma, therefore, $n_i = n_e$. At low temperatures the degree of ionization is low, $n_e \ll n_a$, and

$$n_e = n_e^0 = \sqrt{K_1 n_a}. \tag{6.2}$$

In a weakly ionized plasma, one can ignore electron–ion and electron–electron interactions and consider only interactions of electrons with atoms (molecules) of a neutral gas. The latter can be regarded as a sequence of independent binary collisions. Such a system is well described by the Lorentz gas model (see, for example, Lifshitz and Pitaevskii 1981).

Under steady–state and spatially homogeneous conditions, Boltzmann's equation for the distribution function of electrons $f(\mathbf{v})$ in an electric field $\mathbf{F}$ takes the form

$$-(e\mathbf{F}/m)\partial f(\mathbf{v})/\partial \mathbf{v} = I_c(f).$$

The left–hand side of this equation describes the field effect and the right–hand side the variation of the number of electrons in an element of phase volume due to collisions, I_c being the collision integral. In the following we use the fact that the electron mass is much smaller than the atomic mass ($m \ll M$) and assume small deviations of f from equilibrium. Then, $f(\mathbf{v})$ should be close to spherically symmetric and linear on the electric field. This enables us to represent it as

$$f = f_0 + \delta f,$$

where $\delta f(v) = \cos\theta f_1(v)$ and θ is the angle between the directions of the velocity and electric field. The symmetric part of the distribution function, $f_0(v)$, under conditions of thermodynamic equilibrium is Maxwellian. The nonsymmetric part $f_1(v)$ is essential for calculating the electron mobility and hence plasma conductivity.

On substituting the expansion for $f(\mathbf{v})$ in the initial kinetic equation, we obtain the equation in $f_0(v)$:

$$-(e\mathbf{F}/m)\partial f_0/\partial\mathbf{v} = I_c(\delta f).$$

The collision integral $I_c(\delta f)$ in these conditions can be written as (Lifshitz and Pitaevskii 1981)

$$I_c(\delta f) = -\nu(v)\delta f(v), \qquad \nu(v) = n_a v q(v),$$

where $q(v)$ is the transport cross–section of electron–atom scattering and $\nu(v)$ the corresponding (velocity dependent) electron–atom collision frequency.

In the electric field $\mathbf{F}$ the electrons, which are mainly in chaotic thermal motion, on average drift in the direction opposite to the field. The drift velocity (mean electron velocity over the time exceeding greatly the time between individual collisions) is

$$\mathbf{u} = \int \mathbf{v}\delta f(v)\,d\mathbf{v} = \int \mathbf{v}\cos\theta f_1(v)\,d\mathbf{v},$$

since f_0 makes no contribution to $\mathbf{u}$. The mobility is determined from the equation $\mathbf{u} = -\mu\mathbf{F}$. Taking the projections on the direction of the electric field and using the fact that $\partial f_0/\partial\mathbf{v} = -f_0\mathbf{v}/v_{T_e}^2$, so that

$$\delta f(v) = -\frac{eFv\cos\theta f_0(v)}{mv_{T_e}^2\nu(v)}, \qquad f_1(v) = -\frac{eFvf_0(v)}{mv_{T_e}^2\nu(v)},$$

we obtain for the mobility (in the Lorentz plasma approximation)

$$\mu = \frac{e}{T}\int \frac{v^2 f_0(v)}{\nu(v)}\cos^2\theta d\mathbf{v}.$$

Integrating over the angles and substituting the Maxwellian distribution for $f_0(v)$ we finally get

$$\mu = \frac{1}{3}\sqrt{\frac{2}{\pi}}\frac{e}{mv_{T_e}^5}\int_0^\infty \frac{v^4}{\nu(v)}\exp(-v^2/2v_{T_e}^2)dv,$$

where $v_{T_e} = \sqrt{kT/m}$ is the thermal velocity of the electrons.

This expression is valid when certain conditions are met. These include the following: The neutral gas must be sufficiently rarefied for the binary collision approximation to be valid, $n_a q^{3/2} \ll 1$. The temperature must be sufficiently high and the thermal wavelength of the electron λ sufficiently small, so that one could ignore the interference of the electron on neighboring atoms ($n_a q \lambda \ll 1$). The potential energy of the Coulomb interaction between electrons must be much smaller than their kinetic energy, $e^2 n_e^{1/3}/kT \ll 1$. The plasma must be nondegenerate, $\hbar^2 n_e^{2/3}/mkT \ll 1$. The interatomic correlation can be neglected, $n_a b \ll 1$, $n_a a/kT \ll 1$, where a and b are the coefficients of the van der Waals equation of state for the neutral gas.

It is convenient to integrate over the electron energy, $E = mv^2/2$, instead of the velocity. Then, we have for the mobility

$$\mu = \frac{4}{3\sqrt{\pi}} \frac{e}{m(kT)^{5/2}} \int_0^\infty \frac{E^{3/2}}{\nu(E)} \exp(-E/T)\, dE, \tag{6.3}$$

where $\nu(E) = \sqrt{2E/m}\, n_a q(E)$. Introducing the mean (effective) collision frequency $\bar{\nu}$ and cross–section $\bar{q}$, one can write

$$\mu = e/m\bar{\nu}, \quad \bar{\nu} = \left(3\sqrt{2\pi}/4\right) n_a \bar{q}(T) v_{T_e}, \tag{6.4}$$

where

$$\frac{1}{\bar{q}(T)} = \frac{1}{(kT)^2} \int_0^\infty \frac{E}{q(E)} \exp(-E/kT)\, dE.$$

In the simplest case when the electron–neutral collision can be approximated as a scattering on a hard sphere of a diameter d, the transport cross–section is independent of energy, $q(E) = \pi d^2/4$. In this case averaging over energies yields

Table 6.1 *Averaged transport cross–sections of scattering of electrons from atoms of alkali metals, $\bar{q}(T)$ in units of $10^2 a_0^2$ (Gogoleva et al. 1984).*

$T, 10^3$ K	Li	Na	K	Cs	$T, 10^3$ K	Li	Na	K	Cs
1.0	16.5	15.0	15.3	14.1	2.6	6.88	7.31	7.00	8.98
1.2	14.4	14.0	13.6	12.8	2.8	6.41	6.73	6.52	7.63
1.4	12.6	12.9	12.1	11.7	3.0	5.99	6.23	6.10	7.32
1.6	11.1	11.7	10.9	10.8	3.2	5.63	5.79	5.73	7.04
1.8	9.91	10.6	9.84	10.1	3.4	5.32	5.41	5.41	6.79
2.0	8.93	8.73	8.96	9.42	3.6	5.05	5.07	5.12	6.57
2.2	8.11	8.73	8.20	8.89	3.8	4.80	4.77	4.87	6.37
2.4	7.46	7.97	7.56	8.41	4.0	4.59	4.51	4.64	6.20

the same value, $\bar{q} = \pi d^2/4$, and thus $\bar{\nu} = (3\pi^{3/2}/2^{7/2})n_a d^2 v_{T_e}$. In a more realistic situation the transport cross–section is a function of energy. If the dependence $q(E)$ is known, the mean cross–sections $\bar{q}(T)$ can also be readily calculated. A large amount of reference data on electron–atom and electron–molecule scattering cross–sections is available (e.g., Bederson and Kieffer 1971; Kieffer 1971; Gilardini 1972). Table 6.1 gives the values of averaged cross–sections for atoms of alkali metals.

One can readily perform numerical estimation of the coefficient of electrical conductivity using the following expression:

$$\sigma \cong 3.8 \cdot 10^6 \frac{n_e}{n_a} \frac{1}{\bar{q}\sqrt{T}} \quad \text{ohm}^{-1}\,\text{cm}^{-1}, \tag{6.5}$$

where $\bar{q}$ is the average cross–section in units of 10^{-16} cm^2, and T is the temperature in K.

6.1.2 Electrical conductivity of strongly ionized plasma

Equation (6.4) describes well the electron mobility until the electron–atom collision frequency ν becomes comparable with the electron–ion collision frequency ν_i. Let us now dwell on the peculiarities associated with the determination of ν_i (Spitzer 1962).

The cross–section of thermal electron scattering by an ion with charge number Z is defined by the square of the Coulomb (or Landau) length, $R_C = Ze^2/kT$. However, the Coulomb cross–section scale $4\pi R_C^2$ only accounts for the collisions which lead to large–angle scattering. The specific character of the long–range Coulomb interaction resides in the prevalence of *small-angle scattering*. As a result, to get the transport cross–section one should multiply $4\pi R_C^2$ by the so–called "Coulomb logarithm" $\ln \Lambda$. The quantity Λ in a classical plasma is often defined as the ratio between the Debye radius and the Coulomb length. The energy–dependent momentum transfer cross–section in the case of Coulomb scattering is

$$q(E) = \pi(Ze^2/E)^2 \ln \Lambda.$$

The expression for the electrical conductivity of a fully ionized plasma (ions with a charge Z and electrons) can be readily written assuming that the ions are immobile uncorrelated scatterers while the electrons do not interact with each other. Such a model corresponds with the Lorentz gas model used above to describe the conductivity of weakly ionized plasmas. Averaging over energies with the assumption that $\ln \Lambda$ does not depend on energy yields for the effective collision frequency

$$\bar{\nu}_i = \frac{\pi^{3/2}}{4\sqrt{2}} \left(\frac{Ze^2}{kT}\right)^2 n_i v_{T_e} \ln \Lambda. \tag{6.6}$$

Using Eqs. (6.1) and (6.4) and the neutrality condition $n_e = Zn_i$ one can readily obtain

$$\sigma = (4\sqrt{2}/\pi^{3/2})(Ze^2/kT)^{-1}v_{T_e}(\ln \Lambda)^{-1}. \tag{6.7}$$

The value of the Coulomb logarithm is uncertain to some extent, because variation of Λ with energy was neglected. However, the effect of such variations is no larger than that of other terms neglected in the evaluation of the transport cross–section. Thus, one may simply write $\Lambda \simeq r_D/R_C$, where $r_D \approx [4\pi e^2 n_e(1 + Z)/kT]^{-1/2}$ is the Debye screening radius and $R_C \simeq Ze^2/kT$. Using the plasma parameter $\Gamma = Ze^2/r_D kT$ we may write $\ln \Lambda \approx \ln(1/\Gamma)$. The approach described here is only applicable when $\Lambda \gg 1$, i.e., for ideal plasmas with $\Gamma \ll 1$. In this case the numerical factor, which is sometimes present under the logarithm, e.g., Spitzer form $\Lambda = 3/\Gamma$ (Cohen *et al.* 1950; Spitzer 1962), is of minor importance.

The specific nature of the Coulomb long–range interaction shows up in the strong effect produced by electron–electron correlations on the electrical conductivity even in the range of very small values of Γ. In order to account for these correlations, the integral of electron–electron collisions must be added to the integral of electron–ion collisions in the right–hand side of the kinetic equation. Such an equation was solved numerically by Spitzer and Härm (1953) and corrections were calculated to the Lorentz formula (6.7). These corrections are given by the factor γ_E dependent on the ion charge Z:

Ion charge, Z	1	2	4	16	∞
Factor γ_E	0.582	0.683	0.785	0.923	1.000

The electron–electron interactions cause a decrease of the coefficient of electrical conductivity. When the electric field is applied, the velocity distribution function of electrons, which was initially spherically symmetric, stretches along the field. In opposing this, the electron–electron interactions somewhat symmetrize the electron distribution in the velocity space, thus leading to a decrease of the transport coefficients. In the range of high values of Z, the factor γ_E tends to unity, because the importance of electron–electron collisions decreases.

The resulting expression for the electrical conductivity of fully ionized plasma, usually referred to as Spitzer's formula, is

$$\sigma_{Sp} = (4\sqrt{2}/\pi^{3/2})(Ze^2/kT)^{-1}v_{T_e}\gamma_E(Z)(\ln \Lambda)^{-1}. \tag{6.8}$$

For singly charged ions ($Z = 1$) we have $\gamma_E(1) = 0.582$. Then, numerically,

$$\sigma_{Sp} = 1.53 \cdot 10^{-4} \frac{T^{3/2}}{\ln \Lambda} \quad \text{ohm}^{-1}\,\text{cm}^{-1}, \tag{6.9}$$

where the temperature is expressed in K. In an ideal classical plasma, Eq. (6.9) is asymptotically exact in the limit of very large Coulomb logarithm, $\ln \Lambda \gg 1$.

At high temperatures, the Coulomb length, $R_C = Ze^2/kT$, becomes comparable with the proper radius of a complex ion. In such a case the scattering process becomes non–Coulomb. If an ion is fairly heavy (for example, an ion of xenon), its radius is large and, for $T \approx$ Ry, non–Coulomb corrections are significant (Podlubbnyi and Rostovskii 1980).

A plasma may be regarded as strongly ionized if the electrons collide more frequently with the ions rather than with the atoms. Equating the frequencies of these collisions, we have for singly charged ions

$$n_a \overline{q}(T) \sim (e^4/k^2 T^2)n_e \ln \Lambda. \tag{6.10}$$

For example, if $T \sim 1$ eV, $\overline{q}(T) \sim 10^{-16}$ cm^2, and $\ln \Lambda \sim 5$, then electron–ion collisions dominate at an ionization fraction greater than $\sim 10^{-3}$, i.e., already at relatively small ionization fraction. This is because the Coulomb cross–section for electron–ion collisions is appreciably larger than that for electron–neutral collisions.

In Fig. 6.1 the $\rho - T$ diagram for cesium is divided into a number of regions, with every one of those regions having a different charge transfer mechanism. Equation (6.10) describes the curve separating the regions VII and VI – the regions of strongly and weakly ionized plasma, respectively.

The electrical conductivity of the plasma in the region of intermediate degrees of ionization can be calculated using the Chapman–Enskog method of successive approximations. In the first approximation, the frequency $\nu(E)$ is averaged in the expression for mobility, instead of $\nu^{-1}(E)$. In this way the expression for the characteristic frequency of momentum transfer of the electron gas in collisions with ions

$$\tilde{\nu}_i = \frac{4\sqrt{2\pi}}{3} \left(\frac{Ze^2}{kT} \right)^2 n_i v_{T_e} \ln \Lambda \tag{6.11}$$

appears instead of the Lorentz frequency $\overline{\nu}_i$ given by Eq. (6.6).

The Chapman–Enskog method is characterized by poor convergence if the $q(E)$ cross–section depends nonmonotonically on E. In view of this, a number of interpolation formulas were proposed of the type of the Frost formula:

$$\sigma = \frac{4}{3\sqrt{\pi}} \frac{e^2 n_e}{m(kT)^{5/2}} \int_0^\infty \frac{E^{3/2}}{\nu(E) + \nu_i(E)/\gamma_E} \exp\left(-\frac{E}{kT}\right) dE. \tag{6.12}$$

Let us briefly explain the structure of the interpolation formula (6.12). It is based on the Lorentz approximation rather than on the Chapman–Enskog series. If one could ignore the electron–electron correlations, then, in order to determine the effective frequency of electron collisions, it would be sufficient to sum the frequencies of collisions with atoms and ions, $\nu(E) + \nu_i(E)$. The inclusion of the factor γ_E in expression (6.12) renders the latter exact in the weakly ionized and fully ionized limits and gives adequate interpolation to the intermediate region. Note that it follows from Eq. (6.12) that in ideal plasmas Spitzer's values of electrical conductivity at a fixed temperature limit the conductivity from above,

$$\sigma(T) \leq \sigma_{\mathrm{Sp}}(T).$$

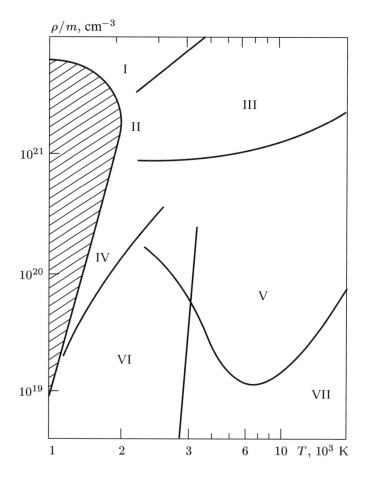

FIG. 6.1. $\rho - T$ diagram for cesium (Alekseev and Iakubov 1986): Region I represents liquid
metal; II – metal–dielectric transition; III – nonideal fully ionized plasma; IV – nonideal
plasma of metal vapors (weakly ionized plasma); V – weakly nonideal strongly ionized
plasma; VI – weakly nonideal plasma of metal vapors; VII – ideal strongly ionized plasma.

Figure 6.2 plots the electrical conductivity of the plasma as a function of
temperature. At high temperatures, plasmas of various compositions lose their
individuality, and all $\sigma(T)$ curves reach Spitzer values. They remain close to each
other until the temperatures are reached for which the second ionization becomes
important. At low temperatures, in a weakly ionized plasma the parameter I/T,
which determines the degree of ionization, is of principal importance. This is the
reason for the high values of σ for cesium plasma and a sharp increase in the
value of σ for hydrogen plasma when the latter is seeded with cesium.

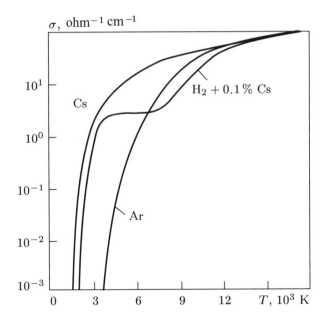

FIG. 6.2.   Electrical conductivity of plasma of three different compositions at a pressure of
    1 atm.

6.2   Electrical conductivity of weakly nonideal plasma

In a weakly ionized plasma, the nonideality is caused by charged particle–atom
interactions. The nonideality may affect both the concentration of free electrons
and their mobility. The effect of nonideality on n_e was discussed earlier in Section
4.3. This effect is due to the reduction of the ionization potential and the emergence of cluster ions. In moderately dense vapors the former effect is still small
($\Delta I \ll T$), however, the inclusion of molecular ions brings about an appreciable
variation of n_e. This region of the $\rho - T$ diagram for cesium in Fig. 6.1 is marked
as VI.

On substituting in Eq. (6.1) the expression (4.32) or (4.49) for n_e, derived
with due regard for cluster ions, one can observe how the emergence of cluster
ions affects the density dependence of the electrical conductivity. At high temperatures and low concentrations, the scaling $n_e \propto \sqrt{n_a}$ (see Eq. 6.2) is valid, and
the electrical conductivity decreases with an increase of density, $\sigma \propto n_a^{-1/2}$. With
a decrease of temperature and increase of concentration, cluster ions emerge. If
A_3^+ prevails among positive ions and A^- among negative ions (the binding energy of A_2^- is low), then it follows from Eqs.(4.32) and (6.5) that the electrical
conductivity is independent of n_a. If the A_4^+ ion prevailed, the value of σ would
increase with the growth of n_a as $\sigma \propto n_a^{1/2}$.

Since the "ion–atom" interaction in plasmas of metal vapors is stronger than
the "electron–atom" interaction, the nonideality parameter can be introduced as

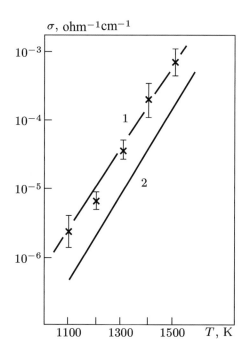

FIG. 6.3.   Coefficient of electrical conductivity σ of sodium plasma at a pressure $p \approx 0.5 p_s$:
the points correspond to experiment (Morrow and Craggs 1973); the curves correspond to
calculation (Khrapak 1979): 1 – eight–component composition of plasma; 2 – three–component composition.

the relation

$$\gamma_{ai} = n_3^+ / n_2^+ = n_a K_4 / K_3 K_5.$$

The condition $\gamma_{ai} = 1$ defines the boundary separating the regions V and VI
in Fig. 6.1. If $\gamma_{ai} > 1$, then $n_3^+ > n_2^+$. This means that one should expect the
emergence of A_4^+ and maybe then A_m^+, where $m \gg 1$.

The first experimental study, the results of which had pointed to the effect of
ionic complexes on electrical conductivity, was carried out by Morrow and Craggs
(1973). The measurements in sodium vapors were performed at $p \approx 0.5 p_s$ (p_s is
the saturated vapor pressure), $T = 1200$–1500 K. The results of calculation, performed by Khrapak (1979) using the procedure described in Section 4.3, are in
good agreement with experiment (see Fig. 6.3). In Fig. 2.11 the results of analogous calculations (Gogoleva et al. 1984) are compared with the measured values
of the electrical conductivity of cesium vapors. A comparison of the calculation
and measurement results points to their good agreement.

Figure 6.4 gives the isobars of the coefficient of electrical conductivity of
sodium and lithium plasma, with the coefficient of electrical conductivity for
cesium plasma being tabulated in Table 6.2. The pressure and temperature range

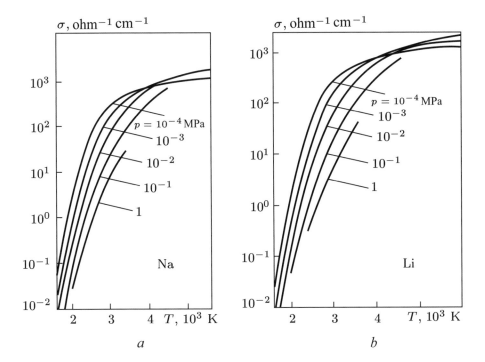

FIG. 6.4.   Isobars of electrical conductivity of sodium (a) and lithium (b) plasma (Khrapak 1979).

covered by calculations is limited by the condition $\gamma_{ai} \lesssim 1$ in the case of low temperatures and high pressures and by the condition $\ln \Lambda \gtrsim 1$ in the case of high temperatures. Therefore, the $p > 1$ MPa isobars are absent. Indeed, at $p = 2$ MPa, the description of the experimental isobar of the electrical conductivity of cesium vapor with the aid of analogous calculations is markedly poor. We speak of the portion of the isobar of electrical conductivity, shown in Fig. 2.9, which corresponds to high temperatures. An adequate description is attained (curve 2 in Fig. 2.9) only upon inclusion of density corrections to the electron mobility, which are discussed below.

As previously noted, the Lorentz formula for the electron mobility (Eq. 6.3) is derived assuming binary collisions, and hence is applicable under conditions when the characteristic radius of the interaction forces is much smaller than the electron free path. In other words a small number of particles must be present in the sphere of interaction on average, that is, the following inequality should be satisfied:

$$(4\pi/3)n_a q^{3/2} \ll 1. \tag{6.13}$$

At densities $n_a = 10^{20}$ cm^{-3}, the quantity $(4\pi/3)n_a q^{3/2}$ in a cesium plasma reaches a value close to unity ($q = 300 - 400\pi a_0^2$, see Table 6.1). Under conditions when the inequality (6.13) is violated, one must allow for the collisions between three, four, and more particles.

This, however, does not mean that in a plasma of metal vapors the electron mobility μ becomes, with a rise in density, less than the mobility μ_0 calculated for the same density from the Lorentz formula. It is not always the case that one can describe the electron–atom interaction as a scattering by a hard sphere with a radius of the order of $\sqrt{q}$. If the atomic polarizability is sufficiently high, the "electron–atom" interaction is mainly a polarization one. The negative sign of the scattering length may serve as an indication of this. Dense vapors of metals fall under the category of precisely such strongly polarizable media. Therefore, their compaction may cause an increase of mobility instead of its decrease. This effect is well known in the physics of dense fluids, the latter being referred to as "high mobility" fluids. A discussion of these effects can be found in the paper by Atrazhev and Yakubov (1980).

As a result of interferences of potentials from neighboring and more remote particles, the total field of scatterers is smoothed. In describing this effect, the

Table 6.2 *Electrical conductivity of cesium plasma, $ohm^{-1}\,m^{-1}$ (Gogoleva et al. 1984).*

T, K	$p = 10^{-4}$ MPa	10^{-3} MPa	10^{-2} MPa	10^{-1} MPa	1 MPa
1000	$4.97 \cdot 10^{-4}$	$1.57 \cdot 10^{-4}$	$4.95 \cdot 10^{-5}$	—	—
1200	$2.71 \cdot 10^{-2}$	$8.56 \cdot 10^{-3}$	$2.70 \cdot 10^{-3}$	$8.42 \cdot 10^{-4}$	—
1400	$4.84 \cdot 10^{-1}$	$1.54 \cdot 10^{-1}$	$4.87 \cdot 10^{-2}$	$1.54 \cdot 10^{-2}$	—
1600	4.19	1.37	$4.37 \cdot 10^{-1}$	$1.40 \cdot 10^{-1}$	$5.11 \cdot 10^{-2}$
1800	$2.09 \cdot 10^{1}$	7.38	2.43	$7.89 \cdot 10^{-1}$	$2.94 \cdot 10^{-1}$
2000	$6.56 \cdot 10^{1}$	$2.71 \cdot 10^{1}$	9.54	3.18	$1.21 \cdot 10^{-1}$
2200	$1.42 \cdot 10^{2}$	$7.23 \cdot 10^{1}$	$2.84 \cdot 10^{1}$	9.89	3.83
2400	$2.36 \cdot 10^{2}$	$1.49 \cdot 10^{2}$	$6.78 \cdot 10^{1}$	$2.53 \cdot 10^{1}$	1.00
2600	$3.32 \cdot 10^{2}$	$2.51 \cdot 10^{2}$	$1.35 \cdot 10^{2}$	$5.53 \cdot 10^{1}$	—
2800	$4.23 \cdot 10^{2}$	$3.67 \cdot 10^{2}$	$2.31 \cdot 10^{2}$	$1.06 \cdot 10^{2}$	—
3000	$5.09 \cdot 10^{2}$	$4.86 \cdot 10^{2}$	$3.53 \cdot 10^{2}$	—	—
3200	$5.87 \cdot 10^{2}$	$6.04 \cdot 10^{2}$	$4.92 \cdot 10^{2}$	—	—
3400	$6.58 \cdot 10^{2}$	$7.10 \cdot 10^{2}$	$6.42 \cdot 10^{2}$	—	—
3600	$7.20 \cdot 10^{2}$	$8.27 \cdot 10^{2}$	$7.98 \cdot 10^{2}$	—	—
3800	$7.75 \cdot 10^{2}$	$9.28 \cdot 10^{2}$	$9.56 \cdot 10^{2}$	—	—
4000	$8.27 \cdot 10^{2}$	$1.02 \cdot 10^{3}$	$1.11 \cdot 10^{3}$	—	—
4200	$8.74 \cdot 10^{2}$	$1.10 \cdot 10^{3}$	$1.26 \cdot 10^{3}$	—	—
4400	$9.21 \cdot 10^{2}$	$1.17 \cdot 10^{3}$	—	—	—
4600	$9.68 \cdot 10^{2}$	$1.24 \cdot 10^{3}$	—	—	—
4800	$1.02 \cdot 10^{3}$	$1.30 \cdot 10^{3}$	—	—	—
5000	$1.06 \cdot 10^{3}$	$1.36 \cdot 10^{3}$	—	—	—

following considerations are important. The potential of the "electron–isolated atom" interaction can be expressed as

$$V(r) = \frac{2\pi\hbar^2}{m} L_c \delta(r) - \frac{\alpha e^2}{2(r^2 + r_a^2)^2}. \tag{6.14}$$

Here, the short–range component is written in the form of the Fermi pseudopotential (1.15). The quantity L_c is the length of electron scattering by the potential $V(r)$. The polarization component in (6.14) features correct asymptotic behavior, α is the atomic polarizability and the cutoff parameter r_a corresponds to the size of the outer electron shell, i.e., the atomic radius. The amplitude of scattering, $f(\theta)$, and the length of scattering L by the potential (6.14) take the form

$$f(\theta) = -L - \frac{\pi\alpha}{2a_0\lambda}\sin\frac{\theta}{2}, \quad L = L_c - \frac{\pi\alpha}{4a_0 r_a}, \tag{6.15}$$

where $\lambda = \hbar(2mE)^{-1/2}$ is the wavelength of electron with energy E.

Equation (6.15) is valid for slow electrons while the following inequalities are satisfied:

$$\lambda \gg L_c, \quad \lambda \gg r_a, \quad \lambda^2 \gg \alpha/a_0.$$

An electron moving in a dense medium polarizes the latter. Internal fields occur, which reduce the electron–atom interaction. The polarization effect can be readily included by introducing the dielectric permeability $\varepsilon = 1 + (8\pi/3)\alpha n_a$ and replacing the potential (6.14) by the potential of electron–atom interaction in the medium,

$$V(r) = \frac{2\pi\hbar^2}{m} L_c \delta(r) - \frac{\alpha e^2}{2\varepsilon(r^2 + r_a^2)^2}.$$

If $(8\pi/3)\alpha n_a \ll 1$, the length of scattering in the medium is (Atrazhev and Yakubov 1980)

$$L_m = L + \frac{4\pi^2\alpha^2 n_a}{3a_0 r_a}. \tag{6.16}$$

This suggests that, if $L < 0$ (i.e., attraction prevails in the electron–atom interaction), $|L_m| < |L|$. In other words, the interaction weakens with an increase of density.

Atoms of alkali metals feature very high values of polarizability ($\alpha \sim 100a_0^3$). In order to produce a qualitatively correct result one can write the approximate expression for the electron scattering cross–section in a medium as

$$q_m(E) = q(E)(1 - \xi)^2, \quad \xi = \frac{4\pi^{5/2}\alpha^2 n_a}{3a_0 r_a \sqrt{q(E)}}, \tag{6.17}$$

where the parameter ξ allows for the dielectric screening of the polarization interaction. Under experimental conditions (Alekseev et al. 1975), the value of the parameter ξ is of the order of 0.2. As a result, the correction to electron mobility, $\mu/\mu_0 = (1 - \xi)^{-2}$, is appreciable.

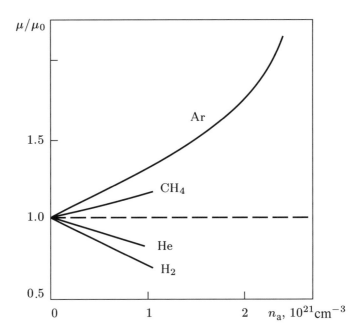

FIG. 6.5.   The measured values of μ/μ_0 at $T = 300$ K as a function of atom density n_a (Khrapak and Yakubov 1981).

In addition to the polarization effect, a whole number of factors affect the electron mobility. Let us discuss some of them. The interference effect (Iakubov and Polishchuk 1982) occurs when the de Broglie electron wavelength λ becomes comparable with the electron free path $\sim (qn_a)^{-1}$. Then, two successive scattering events overlap. The emerging correction can be interpreted as a correction increasing the collision frequency $\nu_{eff} = \nu(1 + 2n_a q\lambda)$. As a result, the mobility decreases as compared with the ideal–gas case:

$$\mu = \mu_0(1 - \sqrt{\pi}n_a q\lambda).$$

However, as the temperature increases, the interference effect disappears because the thermal wavelength $\lambda = \hbar(2mT)^{-1/2}$ decreases.

Therefore, as the density increases, the relative mobility may both decrease and increase. Figure 6.5 gives the results of measurements performed in different gases; see the list of references in Khrapak and Iakubov (1981). In polarizable gases, $\mu/\mu_0 > 1$, and in poorly polarizable gases, $\mu/\mu_0 < 1$. As follows from Fig. 6.5, the density effects in mobility for atomic and molecular gases, for which $q \sim 10^{-15}$ cm^2, occur at $n_a \sim 10^{21}$ cm^{-3}. In the plasma of alkali metals, considerable deviations of μ from μ_0 should occur at $n_a \gtrsim 10^{19}$ cm^{-3}.

As mentioned above, the values of the coefficient of electrical conductivity, which exceed the Spitzer values σ_{Sp}, cannot be attained under standard conditions. This fact is due to the additivity of the frequencies ν and ν_i in the formula for electron mobility $\mu = (e/m)(\nu + \nu_i)^{-1}$. It should be borne in mind, however, that the additivity of ν and ν_i in a very dense plasma may be disturbed. For this, in passing over the Landau length e^2/T, an electron must manage to repeatedly interact with many atoms, that is, the inequality $e^2 q n_a/T \gg 1$ must be satisfied.

Under these conditions, in a first approximation, the electron–ion collisions have no effect whatsoever on the mobility and may be ignored. Therefore, the values of σ in a dense medium are not restricted to Spitzer values.

Let us now treat the mobility in a medium of correlated scatterers. In a weakly ionized gas, the mobility may be strongly affected by interatomic correlations. The square of the amplitude of electron scattering in the medium must be averaged with due regard for the correlation between atoms rather than over their chaotic distribution. As a result, the expression for the averaged collision frequency varies. This frequency contains the structure factor of the medium; see, for example, Kirzhnits et al. (1975). This situation is typical of liquid metals and semiconductors.

Consider the electron scattering amplitude $f(k)$ in the field $U(r) = \sum_j V(\mathbf{r} - \mathbf{R}_j)$ produced by atoms located at points $\mathbf{R}_j$. In the Born approximation,

$$f(k) = -(m/2\pi\hbar)^2 \int \exp(-i\mathbf{k}\mathbf{r})[\mathbf{U}(\mathbf{r}) - \overline{U}]\,d\mathbf{r},$$

where $\mathbf{k}$ is the momentum transferred upon collision, and $\overline{U}$ is the mean field.

The scattering probability is proportional to the square of amplitude f^2, averaged over the distribution of scatterers,

$$\overline{f(k)^2} = \left(\frac{2\pi m}{\hbar^2}\right)^2 \left\{ V_k^2 \sum_i \sum_j \int \frac{d\mathbf{R}_1 \ldots d\mathbf{R}_N}{\Sigma\Omega^N} \times \right.$$

$$\left. \times \exp\left[-i\mathbf{k}(\mathbf{R}_j + \mathbf{R}_i) + U_a/T\right] - \delta^2(k)(n_a\overline{V})^2 \right\}. \tag{6.18}$$

Here, V_k is the Fourier component of the $V(r)$ potential;

$$\overline{V} = \int V(\mathbf{r})d\mathbf{r}, \qquad U_a = \frac{1}{2}\sum_{m,n} V_a\left(\mathbf{R}_m - \mathbf{R}_n\right),$$

where V_a is the potential of interatomic interaction; N is the total number of gas atoms, $N = n_a\Omega$; Ω is the system volume; and

$$\Sigma = \Omega^{-N} \int d\mathbf{R}_1 \ldots d\mathbf{R}_N \exp(-U_a/T)$$

is the partition function of the gas.

The term in curly brackets in (6.18), corresponding to $i = j$, is equal to NV_k^2. This is a usual result of the kinetic theory of gases. With due regard for correlation, the terms corresponding to $i \neq j$ are reduced to the emergence of a term proportional to the next degree of density. On introducing the pair correlation function of atoms $g(r)$ and making transformations in (6.18), the mean square ofthe scattering amplitude per atom may be represented as

$$(n_a\Omega)^{-1}\overline{f^2} = \left(\frac{m}{2\pi\hbar^2}\right)^2 V_k^2 \left\{1 + n_a \int \exp(-i\mathbf{k}\mathbf{r})\left[g(\mathbf{r}) - 1\right]d\mathbf{r}\right\}$$

$$= \left(\frac{m}{2\pi\hbar^2}\right)^2 V_k^2 S(k),$$

where $S(k)$ is the structure factor of the medium.

The expression for the collision frequency has the form

$$\nu(v) = 2\pi n_a v \int q(v,\theta)S\left(2v\sin\frac{\theta}{2}\right)(1 - \cos\theta)\sin\theta\, d\theta, \qquad (6.19)$$

where $q(v,\theta)$ is the differential cross–section of electron–atom scattering. Figure 2.22 shows the structure factor of rubidium. One can see that it is a nonmonotonic function of k and thermodynamic parameters of the medium. Therefore, the effect of correlations on mobility may vary considerably. Depending on the conditions, the mobility may decrease or increase, as compared with the mobility calculated in an ideal–gas approximation.

Podlubnyi *et al.* (1988) treated the coefficient of electrical conductivity of a multicomponent plasma. The electron scattering probability was averaged over the multicomponent system of scattering centers. The resultant collision frequency has the form

$$\nu(v) = \sum_{a,b} \sqrt{n_a n_b} Q_{ab} v, \qquad Q_{ab} = \int d\Omega(1 - \cos\theta) f_a f_b^* S_{ab}, \qquad (6.20)$$

where $S_{ab}(k) = \delta_{ab} + \sqrt{n_a n_b}\int d\mathbf{r}\exp(-i\mathbf{k}\cdot\mathbf{r})(g_{ab} - 1)$, the subscripts a and b number the sorts of particles, n_a and n_b denote the concentration, and f_a and f_b are the scattering amplitudes.

In the Debye–Hückel approximation, the ion–ion structure factor is given by expression (5.6). Formula (6.20) may be used to calculate the coefficient of electrical conductivity in a plasma, in which ions with different charge numbers are present.

6.3   Electrical conductivity of nonideal weakly ionized plasma

6.3.1   *The density of electron states*

In Chapter 4 we have treated the conditions in dense vapors of metals, when ions occur in clusters. This affects considerably the ionization equilibrium. As to the electrons, they remain free, that is, they are conduction electrons. However, a different situation is possible. If the density is high, and the energy of electron attachment to atom is sufficiently high, the fluctuations of the medium density

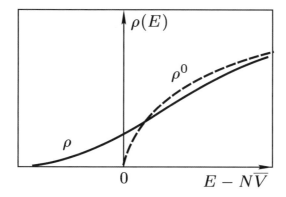

FIG. 6.6.   Qualitative dependencies of $\rho(E)$ and $\rho^0(E)$.

may cause deep potential wells. An electron is captured in such wells and making a transition to the negative-energy region stabilizes this fluctuation of density. An electron cluster emerges and, because it is characterized by low mobility, the electron is said to be localized.

The effect of self–trapping of electrons is characteristic of dense disordered systems; see, for example, Lifshitz (1965); Ziman (1979); Shklovskii and Efros (1984); Hernandez (1991). It is also possible that the electrons are localized in dense vapors of mercury (Khrapak and Iakubov 1981). The distribution of electrons in positive and negative energies is defined by the density of electron states in a dense medium. The main question, which arises, is as follows: the electrons of what energy have sufficient mobility to be regarded as conduction electrons? The theory may hardly provide an answer to this question. It permits the development of only rather crude models. The density of electron states $\rho(E)$ defines the electron concentration in the given energy range

$$n_e(E) = \rho(E)\exp\left[(\mu_e - E)/kT\right],\tag{6.21}$$

where μ_e is the chemical potential of the electronic gas. In an ideal gas,

$$\rho^0(E) = \begin{cases} 4\pi(2m)^{3/2}(2\pi\hbar)^{-3}\sqrt{E}, & E \geq 0, \\ 0, & E < 0. \end{cases}\tag{6.22}$$

It is well known that in a dense medium due to atom density fluctuations $\rho(E)$ may qualitatively differ from $\rho^0(E)$. First, the continuous spectrum boundary shifts through the value of mean field. In addition, there emerges a "tail" of the density of states $\rho(E)$, extending to the region of forbidden states with negative energy, Fig. 6.6. The states of the remote tail $\rho(E)$ correspond to electrons trapped in heavy clusters. At the same time the electrons with relatively low absolute values of negative energy may have a considerable mobility.

At present, one cannot regard all of the emerging problems as solved even on the qualitative level. The quantum effects are especially difficult to describe. Much simpler is the classical electron behavior in a dense medium.

The density of states of a classical electron placed in a potential field $U(\mathbf{r})$ generated by scatterers whose number is equal to N is defined by the expression

$$\rho(E) = \int \frac{d\mathbf{r}}{\Omega} \int \frac{d\mathbf{p}}{(2\pi\hbar)^3} \langle \delta\left[E - E(\mathbf{r}, \mathbf{p})\right] \rangle, \tag{6.23}$$

where

$$E(\mathbf{r}, \mathbf{p}) = \mathbf{p}^2/2m + U(\mathbf{r}); \qquad U(\mathbf{r}) = \sum_{i=1}^{N} V(\mathbf{r} - \mathbf{R}_i).$$

Here, $\mathbf{r}$ and $\mathbf{p}$ denote the electron coordinate and momentum, Ω is the system volume, and the averaging is performed over all possible scatterer configurations. For an arbitrary function $Y(\mathbf{R}_1, \mathbf{R}_2, \ldots, \mathbf{R}_N)$ its mean value is given by

$$\langle Y \rangle = \int Y \exp\left[T^{-1} \sum_{i,k} V_a(\mathbf{R}_i - \mathbf{R}_k)\right] \Omega^{-N} d\mathbf{R}_1 d\mathbf{R}_2 \ldots d\mathbf{R}_N, \tag{6.24}$$

where $V_a(\mathbf{R}_i - \mathbf{R}_k)$ is the energy of interaction between the ith and kth atoms.

In a rarefied gas, the potential energy of an electron in the atomic field $U(r)$ can be ignored. Then, it follows from Eq. (6.23) that $\rho(E) = \rho^0(E)$. In a dense gas, the calculation of $\rho(E)$ is related to the concrete form of the potential $V(r)$. The simple result can be obtained by including only the most probable fluctuations of the field U described by the Gaussian approximation. Performing an integration over momenta in Eq. (6.23) we get

$$\rho(E) = \frac{(2m)^{3/2}}{2\pi^2\hbar^3} \int \frac{d\mathbf{r}}{\Omega} \langle \sqrt{E - U(r)} \rangle. \tag{6.25}$$

The averaging over all atomic configurations in Eq. (6.24) can be replaced by averaging over a random field distribution $P(U)$. Then,

$$\rho(E) = \frac{(2m)^{3/2}}{2\pi^2\hbar^3} \int_{-\infty}^{\infty} dU P(U) \sqrt{E - U}. \tag{6.26}$$

Let $n\overline{V}$ be the magnitude of the mean field $\langle U \rangle$ and nW_0 its dispersion $n \int V^2(r) dr$, where $n = N/\Omega$ is the density of scatterers. For high negative energies $(E - n\overline{V})^2 \gg nW_0$ we then have

$$\rho(E) = 4\pi \frac{(2m)^{3/2}}{(2\pi\hbar)^3} \frac{8nW_0}{(E - n\overline{V})^2} \sqrt{|E - n\overline{V}|} \exp\left[-\frac{(E - n\overline{V})^2}{8nW_0}\right]. \tag{6.27}$$

Consequently, deep in the "tail" $\rho(E)$ decreases exponentially. In the region of renormalized zero, $\rho(E)$ is finite,

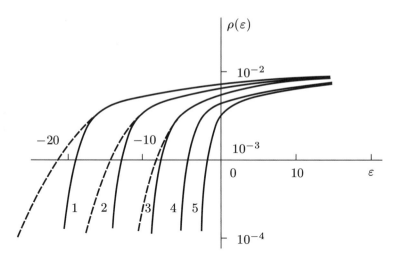

FIG. 6.7.   Electron state density $\rho(\varepsilon)$ as a function of relative electron energy ε for correlated (solid curves) and uncorrelated (dashed curves) scatterers (Lagar'kov and Sarychev 1975). Numbers correspond to different atomic densities: $1.33 \cdot 10^{22}$ cm^{-3} (1); $8.3 \cdot 10^{21}$ cm^{-3} (2); $4.8 \cdot 10^{21}$ cm^{-3} (3); $2.4 \cdot 10^{21}$ cm^{-3} (4); $1.05 \cdot 10^{21}$ cm^{-3} (5).

$$\rho(n\overline{V}) = 4\pi\Gamma(3/4)\frac{(2m)^{3/2}}{(2\pi\hbar)^3}\frac{(2nW_0)^{1/4}}{(2\pi)^{1/2}}. \qquad (6.28)$$

Given high positive energies, $\rho(E)$ assumes the regular root dependence but with an energy shift, $\rho(E) = \rho^0(E - n\overline{V})$. Such dependence of $\rho(E)$ is proper for strongly nonideal disordered systems.

It should be mentioned that Eq. (6.27) cannot correctly describe the $\rho(E)$ tail if the depth of the electron–atom interaction potential is comparable with or exceeds the temperature. In such cases, the $\rho(E)$ tail is formed by rare but large fluctuations of density which may be strongly affected by interatomic interactions. Interatomic repulsion restricts the $\rho(E)$ tail from the side of high negative energies while attraction extends this tail.

The density of states of a classical electron in a field of scatterers was studied by Lagar'kov and Sarychev (1975, 1978) using molecular dynamics simulations. A potential with attractive polarization asymptote, a depth g and rigid core of radius δ was selected as the interaction potential $V(r)$. In order to simulate a dense vapor of mercury, the parameter g was selected equal to 0.18 eV (the negative Hg ion does not exist and, therefore, the potential $V(r)$ cannot be very deep). The quantity $\delta = 3.3a_0$ is the radius of a mercury atom arises from the kinetic theory of gases.

Figure 6.7 illustrates the results of calculations by Lagar'kov and Sarychev (1975) of $\rho(E)$ as a function of the relative electron energy $\varepsilon = E/g$ for different atomic densities. The $\rho(E)$ dependence was constructed both allowing for the interatomic correlation and ignoring the latter. The scatterers were assumed to interact as hard spheres of radius δ. Interatomic repulsion limits the $\rho(E)$ tail.

6.3.2   Electron mobility and electrical conductivity

The linear response theory permits writing exact initial expressions for the kinetic coefficients. In particular, as long as the linear relationship between current and field is valid (Ohm's law), the electrical conductivity is expressed in terms of a correlation function of currents and, hence, velocities. Neglecting quantum effects we have

$$\sigma_{\mu\nu}(\omega) = (n_e e^2 \beta) \int \langle v_\nu(0)v_\mu(t)\rangle \exp(-i\omega t - \delta t)dt, \quad \delta = +0. \tag{6.29}$$

Here, n_e is the concentration of electrons, ω the external field frequency, and $\langle v_\nu(0)v_\mu(t)\rangle$ is the time autocorrelator of electron velocities. The averaging is performed over the electron velocity distribution and over the position of scatterers,

$$\langle\ldots\rangle = A \int d\mathbf{v} \exp\left(-\frac{\beta m v^2}{2}\right) \int d\mathbf{R}_1 \ldots d\mathbf{R}_N \Omega^{-N} \exp\left(-\beta V_a\right) \times (\ldots),$$

where A is the normalization constant of averaging. The expression for static electrical conductivity $\sigma \equiv \sigma_{xx}(0)$ can be rewritten as

$$\sigma = (e^2/m) \int_{-\infty}^{\infty} n_e(E)\tau(E)dE. \tag{6.30}$$

Here, $\tau(E)$ is the autocorrelation time of an electron with energy E (in a medium of massive scatterers the electron energy can be regarded as constant). The value of $\tau(E)$ is given by the following expression:

$$\tau(E) = \int_{-\infty}^{\infty} \varphi_E(t)dt, \qquad \varphi_E(t) = \langle v_x(0)v_x(t)\rangle / \langle v_x(0)^2\rangle, \tag{6.31}$$

where $\varphi_E(t)$ is the autocorrelation function of the electron velocity, and v_x is the velocity component in the direction of electric field.

    In a rarefied gas, $\tau(E)$ is inversely proportional to the free path between two successive collisions of the electron with atoms. The integration range is restricted by positive energies. Therefore, Eq. (6.30) reduces to Lorentz's formula. It is only for very fast electrons that an ideal–gas approximation of the free time of flight is valid. In a dense system, electrons with different energies make substantially different contributions to the electrical conductivity. Let us discuss these contributions and follow Hensel and Franck (1968).

    If $\sqrt{n_a W_0}/kT \gtrsim 1$ most of the conduction electrons are not free even if their energy is positive. They are constantly in the field of scatterers. This field is an alternation of "wells" and "hills" with the most probable depths (heights) close to $\sqrt{n_a W_0}$ and space dimensions of the order of l. The value of l corresponds to the correlation length of the field of scatterers; it is close to the effective range of the potential $V(r)$. Therefore, slow electrons with $E < \sqrt{n_a W_0}$ travel with velocities $v \sim (n_a W_0)^{1/4}/\sqrt{m}$ which exceed thermal velocities. The electrons are scattered from density fluctuations. Apparently, τ is close to the time of flight of the field correlation length, i.e., close to l/v.

The electrons of the $\rho(E)$ tail make no contribution to static conductivity because they correspond to low–mobility clusters. However, electrons with negative energies of lower absolute value, $|E| < \sqrt{n_a W_0}$, may be conduction electrons. The point is that field fluctuations lead to the emergence of "percolation channels". The possibility of formation of percolation channels permeating the entire macroscopic volume of the medium is a sharp, practically step, function of energy (Shklovskii and Efros 1984). The minimum energy, starting from which the electrons contribute to conductivity, is referred to as the percolation energy E_P. Let us now determine E_P for a classical particle in an arbitrary pattern of potential energy $U(r)$. It is equal to such a minimum value of energy at which it is still possible to find a region in space in which $U(r) < E_P$ and which permeates the entire plasma volume. Numerical modelling shows that $E_P \sim -0.3\sqrt{n_a W_0}$. Consequently, along with electrons of positive energies, electrons with negative energies $(E > E_P)$ are conduction electrons. If $\sqrt{n_a W_0}/kT \gg 1$, it is the electrons in the percolation energy range which make the major contribution to σ.

The most detailed information on the electron dynamics is contained in the correlation function $\varphi_E(t)$. It is only for fast electrons that $\varphi_E(t) = \exp(-t/\tau)$ where $\tau = \nu^{-1}$ and ν is the frequency of successive electron–atom collisions. The correlator $\varphi_E(t)$ in the case of low positive energies decreases nonexponentially, but under the Gaussian law $\varphi_E(t) = \exp(-t^2/\tau^2)$, where $\tau = l/v$. Finally, the correlator of trapped electrons oscillates, $\varphi_E(t) = \cos \omega_0 t$, where ω_0 is the natural frequency of electron oscillation in the potential well of the cluster. This gives the autocorrelation time τ close to zero.

The peculiarities of the behavior of the velocity autocorrelation function for different values of relative energy ε (calculated from the mean field level and normalized to the potential depth g) are well defined in Fig. 6.8, taken from Lagar'kov and Sarychev (1975). These results were obtained from molecular dynamics simulation of the electron dynamics. At positive energies, the electron is free and $\varphi_\varepsilon(t')$ decays monotonically (t' is dimensionless time). With an energy decrease, the function $\varphi_\varepsilon(t')$ acquires a minimum: The electron is as though trapped until it finds a passage (percolation channel) in a field of complex shape. In these conditions the autocorrelation time $\tau(\varepsilon)$ is still substantially different from zero. With further decrease of energy, the electron becomes localized.

Figure 6.9 illustrates the diffusion coefficient of electrons in the same medium, $D = \int_0^\infty \langle v_x(0) v_x(t) \rangle_\varepsilon \, dt$. This quantity goes to zero at the energy corresponding to the percolation threshold.

Based on $\rho(E)$ and $D(E)$, Lagar'kov and Sarychev (1979) calculated from

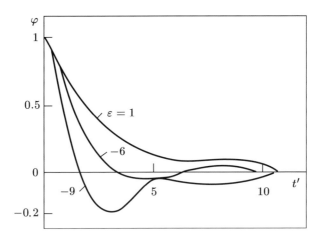

FIG. 6.8.   Autocorrelation function of electron velocity φ vs. dimensionless times t' for several values of relative energy ε (Lagar'kov and Sarychev 1975) and $n_a = 4.8 \cdot 10^{21}$ cm^{-3}.

Eq. (6.30) the conductivity of mercury plasma in a wide range of temperatures and densities (Fig. 6.10). Good agreement is observed both with experiment and with qualitative assumptions on the initial stage of the "metal–dielectric" transition.

6.4   The thermoelectric coefficient

In the presence of a temperature gradient, Ohm's law assumes the form

$$\mathbf{F} + (1/e)\nabla\mu = (1/\sigma)\mathbf{j} + S\nabla T. \tag{6.32}$$

The left–hand part includes $\nabla\mu$, the gradient of the chemical potential of electrons. The thermoelectric coefficient S appears in the right–hand part.

In a Lorentz gas, the thermoelectric coefficient is calculated analogously to the coefficient of electrical conductivity (see, for example, Section 6.1 and Lifshitz and Pitaevskii 1981). The electron distribution function is $f = f_0 + \delta f$, $f_0 \propto \exp\left[(\mu - E)/kT\right]$, where $E = mv^2/2$ is the electron energy and δf is a small correction which is linear on the field and gradients of the chemical potential and temperature. Proceeding from the kinetic equation

$$-\frac{e\mathbf{F}}{m}\frac{\partial f_0}{\partial \mathbf{v}} + \mathbf{v}\frac{\partial f_0}{\partial \mathbf{r}} = -\nu(v)\delta f, \tag{6.33}$$

we get for δf

$$\delta f = -\frac{f_0}{kT\nu(v)}(e\mathbf{F} + \nabla\mu)\mathbf{v} + f_0\frac{\mu - E}{kT^2\nu(v)}\mathbf{v}\nabla T.$$

The thermoelectric coefficient is calculated from the coefficient in the equation $\mathbf{j} = -s\sigma\nabla T$ at $e\mathbf{F} + \nabla\mu = 0$. The current is

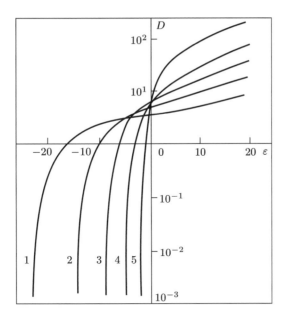

FIG. 6.9.   Energy–dependent electron diffusion coefficient $D(\varepsilon)$ (Lagar'kov and Sarychev 1975). Designations are the same as in Fig. 6.8; $n_a = 4.8 \cdot 10^{21}$ cm^{-3}.

$$\mathbf{j} = -en_e \int \mathbf{v}\delta f d\mathbf{v}.$$

Finally, we get for the thermoelectric coefficient

$$S = \frac{1}{eT} \left\{ \mu - \frac{\int_0^\infty \left[E^{5/2}/\nu(E)\right] \exp(-\beta E)dE}{\int_0^\infty \left[E^{3/2}/\nu(E)\right] \exp(-\beta E)dE} \right\}. \qquad (6.34)$$

In a weakly ionized plasma under conditions when $I \gg kT$, the thermoelectric coefficient is mainly defined by the temperature dependence of the degree of ionization. Because $\mu \approx kT \ln(n_e\lambda^3)$, then, in view of Saha's equation (4.5), we have

$$S \approx (1/eT) \left[-I/2 + (kT/2) \ln(n_a\lambda^3) - \text{const.}T\right] \approx -I/2eT. \qquad (6.35)$$

In a strongly ionized plasma, $\nu \propto E^{-3/2}$, and equation (6.34) yields

$$S = (1/eT)(\mu - 4kT). \qquad (6.36)$$

This expression is valid only if $Z \gg 1$. For small values of Z, electron–electron collisions affect the thermoelectric electrical conductivity coefficients. A discussion of this problem may be found in Kraeft *et al.* (1985).

Figure 6.11 gives the results of calculation by Redmer et al. (1990) of the thermoelectric coefficient of cesium on isobars. The calculation procedure is close to

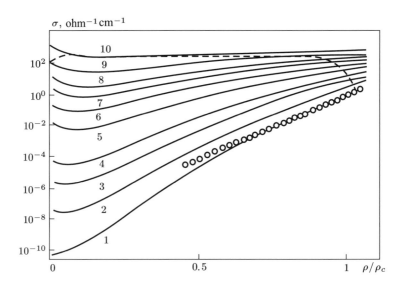

FIG. 6.10.   Isotherms of electrical conductivity of mercury plasma (Lagar'kov and Sarychev 1975). Dots indicate the measurements (Hensel and Franck 1968), and curves the calculation at $T = 1823$ K (1), 2200 K (2), 2600 K (3), 3000 K (4), 4000 K (5), 5000 K (6), 6000 K (7), 7000 K (8), 10 000 K (9), 15 000 K (10). The dashed line defines the region with the degree of ionization is smaller than 0.1.

that developed by researchers from the Rostock group to calculate the coefficient of electrical conductivity (see Section 7.3). Also shown in Fig. 6.11 is the behavior of the asymptotes for fully ionized (6.36) and weakly ionized (6.35) plasmas. These asymptotes correlate well with the numerical calculation results. Note that the calculation covers the $T \leq 3000$ K range, in which the effect of nonideality is still fairly small. At lower temperature (at about 2000 K), a sharp increase of the absolute magnitude of S occurs, due to the dielectric–metal transition (Fig. 4.2).

We will now turn to the discussion of the thermoelectric coefficient of a nonideal plasma. In a weakly nonideal plasma, one should use the equation of ionization equilibrium, which contains the quantity $(I - \Delta I)$ instead of the ionization potential I and allows for the presence of complex ions in the plasma composition. Then, we assume, for example, that the ionization equilibrium on the subcritical isobars of metal vapors is mainly defined by the A_3^+ and A^- ions, and derive

$$S = [-I - (E_3 + E_5 - E_2)]/(2eT), \tag{6.37}$$

where the designations of Table 4.9 are used. In Figure 4.2 results following from relations (6.36) and (6.37) are shown.

As was discussed in Section 4.2 on supercritical isobars the effect of nonideality shows up as a decrease of the ionization potential rather than as a variation

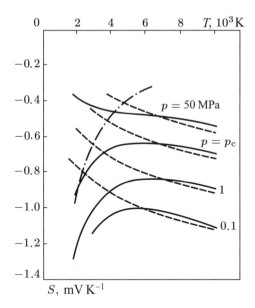

FIG. 6.11. Thermoelectric coefficient of cesium on isobars (Redmer *et al.* 1990). Solid curves indicate the isobars of a partly ionized plasma; dashed curves, asymptotes (6.36); dot–dashed curves, asymptote (6.35).

of the composition of charged components of plasma. Then, we use expression (4.9) to derive

$$S = -(I - \Delta I)/2eT = [-I + kT(n_a/n_a^*)]/(2eT). \tag{6.38}$$

Hence it follows that the quantity $S(T)$ must have a minimum on the isobars. Indeed, at low temperature $|S|$ increases with heating, because ΔI decreases. At high temperature, however, $|S|$ decreases inversely proportional to T. The existence of a minimum of S was predicted by Alekseev *et al.* (1970),

$$|S|_{\min} \approx 1 \text{ mV/K}.$$

Expression (6.35) may be transcribed to

$$S = -\Delta E/(2eT), \tag{6.39}$$

where $\Delta E = 2kT^2(d\ln\sigma/dT)$ is the temperature coefficient of conductivity. The quantity ΔE is sometimes referred to as the "transport energy gap". One can use formula (6.39) to obtain S if σ is known. In Figure 6.12, this is done for cesium. At low temperature, the thermoelectric coefficient of liquid cesium is proportional to temperature, $S \sim T/eE_F$, because $d\ln\sigma/d\ln E_F$ is almost independent of temperature (Lifshitz and Pitaevskii 1981). At high temperature, in accordance with (6.35), the thermoelectric coefficient of a weakly ionized plasma is

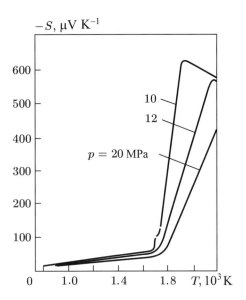

FIG. 6.12.   Isobars of the thermoelectric coefficient for cesium, calculated by Alekseev *et al.* (1970).

inversely proportional to T. The intermediate region, in which the thermoelectric coefficient varies sharply, corresponds to the metal–dielectric transition.

The pattern described above corresponds to the measurements results shown in Fig. 4.2. The subcritical isobars (2 and 6 MPa) are shown in this drawing in accordance with the theory of the metal–dielectric transition discussed in Section 4.5 (Iakubov and Likalter 1987).

Lagar'kov and Sarychev (1979) studied the thermoelectric coefficient using molecular dynamics simulation. The general expression

$$S = (1/eT) \left\{ \mu - \frac{\int_{-\infty}^{\infty} \rho(E)\tau(E)E \exp(-\beta E)dE}{\int_{-\infty}^{\infty} \rho(E)\tau(E) \exp(-\beta E)dE} \right\}$$

allows one to calculate S provided $\rho(E)$ and $\tau(E)$ are known. Figure 6.13 shows supercritical isobars of thermoelectric coefficient for mercury calculated in this way. Good agreement with experimental data (Alekseev *et al.* 1976; Schmutzler and Hensel 1973) is observed.

It is known that the thermoelectric coefficient is very sensitive to slight variations of the state of substances. Therefore, the value of the thermoelectric coefficient may suffer drastic variations in the neighborhood of the critical point (at $\rho \sim \rho_c$). Neale and Cusack (1979) registered such variations on the 170 MPa isobar for mercury.

Note that many of the interesting problems associated with the behavior of the thermoelectric coefficient in the region of the metal–dielectric transition

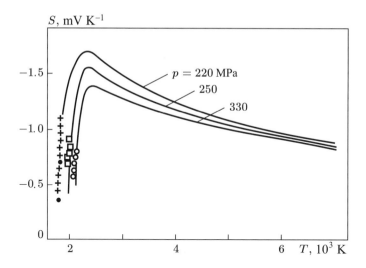

S, mV K^{-1}

$p = 220$ MPa

250

330

FIG. 6.13. Isobars of the thermoelectric coefficient for mercury (Lagar'kov and Sarychev 1979). The curves correspond to calculations. The symbols ● correspond to experiment by Schmutzler and Hensel (1973) and +, □, and ○ to the experiment by Alekseev *et al.* (1976).

remain in fact unclarified.

References

Alekseev, V. A. and Iakubov, I. T. Electrical conductivity and thermoelectromotive force of alkali metals. (1986). In *Handbook of thermodynamic and transport properties of alkali metals*, Ohse, R. W.(ed.), pp. 703–734. Blackwell, Oxford.

Alekseev, V. A., Vedenov, A. A., Krasitskaya, L. S., and Starostin, A. N. (1970). Thermoelectric power of cesium near critical temperatures and pressures. *JETP Lett.*, **12**, 351–354.

Alekseev, V. A., Vedenov, A. A., Ovcharenko, V. G., Krasitskaya, L. S., Rizhkov, Y. F., and Starostin, A. N. (1975). The effect of saturation on the thermo–e.m.f. of caesium at high temperatures and pressures. *High Temp.–High Press.*, **7**, 677–679.

Alekseev, V. A., Ovcharenko, V. G., and Ryzhkov, Y. F. (1976). Metal–dielectric transition in liquid–metals and semiconductors at high–temperatures and pressures in critical–point region. *Sov. Phys. Usp.*, **120**, 699–702.

Atrazhev, V. M. and Yakubov, I. T. (1980). Heating waves in high–current discharges in dense alkali–metal vapors. *High Temp.*, **18**, 14–24.

Bederson, B. and Kieffer, L. J. (1971). Total electron–atom collision cross sections at low energies – a critical review. *Rev. Mod. Phys.*, **43**, 601–640.

Cohen, R. S., Spitzer, L., Jr., and Routly P. McR. (1950). The electrical conductivity of an ionized gas. *Phys. Rev.*, **80**, 230–238.

Gilardini, A. (1972). *Low energy electron collisions in gases*. Wiley, New York.

Gogoleva, V. V., Zitserman, V. Y., Polishchuk, A. Y., and Yakubov, I. T. (1984). Electrical–conductivity of an alkali–metal–vapor plasma. *High Temp.*, **22**, 163–170.

Hensel, F., and Franck, E. U. (1968). Metal–nonmetal transition in dense mercury vapor. *Rev. Mod. Phys.*, **40**, 697–703.

Hernandez, J. (1991). Electron self–trapping in liquids and dense gases. *Rev. Mod. Phys.* **63**, 675–697.

Iakubov, I. T. and Polischuk, A. Y. (1982). Quantum density corrections to mobility and dispersion law for electrons in a medium of disordered scatterers. *J. Phys. B*, **15**, 4029–4041.

Iakubov, I. T., and Likalter A. A. (1987). New results in the field of a nonideal plasma–experiments and interpretation. *Contr. Plasma Phys.*, **27**, 479–490.

Khrapak, A. G. (1979). Conductivity of a weakly nonideal multicomponent alkali–metal vapor plasma. *High Temp.*, **17**, 946–951.

Khrapak, A. G. and Iakubov, I. T. (1981). *Electrons in dense gases and plasma*. Nauka, Moscow.

Kieffer, L. J. (1971). Low–energy electron–collision cross–section data. Part III: Total scattering: Differential elastic scattering. *Atomic Data*, **2**, 293–330.

Kirzhnits, D. A., Lozovik, Y. E., and Shpatakovskaya, G. V. (1975). Statistical model of substance. *Sov. Phys. Usp.*, **117**, 1–47.

Kraeft, W. D., Kremp, D., Ebeling, W., and Röpke, G. (1985). *Quantum statistics of charged particle system*. Akad. Verlag, Berlin.

Lagar'kov, A. N., and Sarychev, A. K. (1975). Dynamics of a classical electron in a dense medium of disordered scatterers. *JETP*, **41**, 317–320.

Lagar'kov, A. N., and Sarychev, A. K. (1978). Dynamics of an electron in a random field and the conductivity of a dense mercury plasma. *High Temp.*, **16**, 773–782.

Lagar'kov, A. N., and Sarychev, A. K. (1979). Effect of ion clusters on ionization equilibrium and conductivity of a dense weakly ionized cesium plasma at subcritical temperatures and pressures. *High Temp.*, **17**, 393–398.

Lifshitz, I. M. (1965). Energy spectrum structure and quantum states of disordered condensed systems. *Sov. Phys. Usp.*, **7**, 549.

Lifshitz, E. M. and Pitaevskii, L. P. (1981). *Physical kinetics*. Pergamon Press, Oxford.

Morrow, R. and Craggs, J. D. (1973). Electrical conductivity of sodium vapor. *J. Phys. D*, **6**, 1274–1282.

Neale, F. E., and Cusack, N. E. (1979). Thermoelectric–power near the critical–point of expanded fluid mercury. *J. Phys. F – Metal Physics*, **9**, 85–94.

Podlubnyi, L. I. and Rostovskii, V. S. (1980). Theory of the equilibrium and kinetic properties of plasmas incorporating the electron shells of ions. *Sov. J. Plasma Phys.*, **6**, 734–740.

Podlubnyi, L. I., Rostovskii, V. S., and Filinov, V. S. (1988). Electrical–conductivity of nonideal argon and xenon plasma. *High Temp.*, **26**, 153–159.

Redmer, R., Reinholz, H., and Röpke, G. (1990). (private communication).

Schmutzler, R. W. and Hensel, F. (1973). Thermoelectric power of fluid mercury in density range of metal–nonmetal transition. *Berichte der Bunsen–Gesellschaft fur Physikalische Chemie*, **76**, 531–535.

Shklovskii, B. I. and Efros, A. L. (1984). *Electronic properties of doped semiconductors*. Springer, Berlin.

Spitzer, L., Jr., and Härm, R. (1953). Transport phenomena in completely ionized gas. *Phys. Rev.*, **89**, 977–981.

Spitzer, L., Jr. (1962). *Physics of fully ionized gases*. Interscience Publishers, New York.

Ziman, J. M. (1979). *Models of disorder. The theoretical physics of homogeneously disordered systems*. Cambridge University Press, Cambridge.

ELECTRICAL CONDUCTIVITY OF FULLY IONIZED PLASMA

7.1 Kinetic equations and the results of asymptotic theories

Following the review by Gryaznov *et al.* (1980), we shall briefly dwell on the basic approaches to the calculation of kinetic coefficients for weakly nonideal plasma. The starting point of the kinetic theory is Liouville's equation for the N–particle distribution function describing the time evolution of the ensemble of N particles obeying Hamilton's equations. Integration over coordinates and particle momenta leads to the Bogoliubov, Born, Green, Kirkwood, Yvon (BBGKY) hierarchy of equations.

The BBGKY hierarchy of equations is a set of integro–differential equations for s–particle distribution functions, each containing in the right–hand side a function of $(s + 1)$th order. For uncoupling this hierarchy, use is made of an approximation which ignores triple correlations and, in addition, includes the plasma polarization indicative of the fact of charge interaction with a large number of neighbors. This leads to the kinetic equation for the distribution function (Klimontovich 1982, 1999):

$$\frac{\partial f_1(\mathbf{p}_1)}{\partial t} = -\frac{\partial I(\mathbf{p}_1)}{\partial \mathbf{p}_1},$$

$$I(\mathbf{p}_1) = \pi \int_{-\infty}^{\infty} \frac{d\omega}{2\pi} \int \frac{V^2(k)k^2\delta(\mathbf{k}\mathbf{v}_1 - \omega)\delta(\mathbf{k}\mathbf{v}_2 - \omega)}{|\varepsilon(k,\omega)|^2} \times$$
$$\times \left[f(\mathbf{v}_2)\frac{\partial f(\mathbf{v}_1)}{\partial \mathbf{v}_1} - f(\mathbf{v}_1)\frac{\partial f(\mathbf{v}_2)}{\partial \mathbf{v}_2} \right] \frac{d\mathbf{k}}{(2\pi)^3}. \tag{7.1}$$

The dielectric permeability of the plasma has the form (Lifshitz and Pitaevskii 1981)

$$\varepsilon(k,\omega) = 1 + \sum_{\alpha} \frac{4\pi e_{\alpha}^2}{k^2} \int \frac{d\mathbf{p}_{\alpha}}{\omega - \mathbf{k}\mathbf{v}_{\alpha} + i\delta} \mathbf{k}\frac{\partial f}{\partial \mathbf{p}_{\alpha}}, \tag{7.2}$$

were $\delta = +0$ and $V(k) = 4e^2/k^2$ is the Coulomb potential in the Fourier representation. For fast processes ($\omega \gg \omega_{pe} = \sqrt{4\pi e^2 n_e/m}$) the dielectric permeability is $\varepsilon(k,\omega) = 1$. In this case the screening does not have enough time to evolve and the collision integral has the form proposed by Landau. This expression diverges at large and small distances. For slow processes ($\omega \ll \omega_{pe}$), static screening is realized, $\varepsilon(k,\omega) = 1 + (k_D/k)^2$, where $k_D = r_D^{-1}$. This eliminates divergence at large distances. In order to eliminate divergence at small distances, the collision integral was represented (Hubbard 1961a, 1961b) as a combination of Boltzmann, I_B (infrequent "strong" collisions), Landau, I_L, and Lenard–Balescu, I_{LB} integrals:

$$I = I_B - I_L + I_{LB}. \tag{7.3}$$

This corresponds, in fact, to the inclusion of dynamic polarization of the plasma in the Boltzmann collision integral.

Along with the approaches based on the BBGKY hierarchy of equations, the method of time Green functions is employed to write the kinetic equations for the plasma. The emerging kinetic equations are written for advanced and retarded Green functions defining the particle density and the probability of allowed states with pre–assigned momentum and energy. It is by using this method equipped with the diagram technique for classification and regrouping of terms in the series of perturbation theory (ring and ladder fragments are included) that Gould and DeWitt (1967) obtained a convergent equation including the interparticle interaction via the screened Coulomb potential. The emerging approximation is, in a sense, analogous the combination (7.3).

The inclusion of collective effects in the Coulomb interaction was performed on to the basis of the Fokker–Planck equation, with the collision integral incorporating all moments of the distribution function. In such a procedure, the contribution made by close collisions is described by Boltzmann's integral and that made by distant collisions is described by the Lenard–Balescu integral with cutoff of the contributions from small and large distances.

In order to discuss concrete results, we shall write the expression for electrical conductivity in the form

$$\sigma = e^2 n_e / m \nu_i,$$

$$\nu_i = \frac{\pi^{3/2}}{4\sqrt{2}} \left(\frac{Ze^2}{kT} \right)^2 n_i v_{T_e} \gamma_E^{-1} \ln \Lambda, \tag{7.4}$$

where, in the roughest approximation, $\ln \Lambda = \ln(b_{max}/b_{min})$. The maximum impact parameter b_{max} is provided by the screening length. The minimum impact parameter is the maximum of the two lengths, namely, the classical distance of closest approach, e^2/kT, and the thermal electron wavelength, $\lambda = \hbar/\sqrt{2mkT}$.

Spitzer's formula (6.8), derived for a classical nondegenerate plasma, is only logarithmically correct. Spitzer's Coulomb logarithm,

$$\ln \Lambda_{Sp} = \ln(3/\Gamma), \quad \Gamma = e^2/r_D kT, \quad r_D = \sqrt{kT/8\pi e^2 n_e},$$

may serve as the starting point for further refinement.

The result of Williams and DeWitt (1969), who used the collision integral (7.3), corresponds to the Coulomb logarithm with numerical correction,

$$\ln \Lambda = \ln(1/\Gamma) - 2C + \ln 2, \quad C = 0.577. \tag{7.5}$$

Similar result can be obtained from the binary collision approximation assuming a Debye–Hückel interaction potential. In the limit $\Gamma \ll 1$, Liboff (1959) obtained

$$\ln \Lambda = \ln \left(1/\Gamma\right) - \frac{1}{2} - C + \ln 2.$$

The subsequent correction is analogous to the relaxation correction to the ion mobility in solutions of strong electrolytes (Lifshitz and Pitaevskii 1981). The external electric field affects the correlation functions leading to a spatial deformation of the screening cloud. This reduces somewhat the coefficient of electrical conductivity, because the electric field acting on a particle is compensated to some extent (Kadomtsev 1957, Ebeling and Röpke 1979). As a result, Röpke (1988) derived

$$\ln \Lambda = \ln \gamma^{-3/2} + 1.102 + \left(\sqrt{6} + \sqrt{3}\right)^{-1} \gamma^{3/2} \ln \gamma^{-3/2} + \dots,$$

where $\gamma = e^2 (4\pi n_{\mathrm{e}}/3)^{1/3} (kT)^{-1}$ $(\Gamma^2 = 3\gamma^3)$. The range of validity is limited by the inequalities $\gamma \ll 1$ and $\gamma^2 \theta^{-1} \ll \ln \Lambda$, where $\theta = kT/E_{\mathrm{F}}$.

The calculation of the coefficient of electrical conductivity for a quantum nondegenerate plasma, using the Lennard–Balescu collision integral, leads to the expression

$$\ln \Lambda = \int_0^\infty \frac{dq}{q} \frac{S_{\mathrm{ii}}(q)}{|\varepsilon(q)|^2} \exp\left(-\frac{\hbar^2 q^2}{8mkT}\right). \tag{7.6}$$

The use of Debye–Hückel static expressions for $S_{\mathrm{ii}}(q)$ and $\varepsilon(q)$ (5.6) leads to the series whose first terms are (Kivelson and Du Bois 1964)

$$\ln \Lambda = \int_0^\infty \frac{q^3 dq}{(q^2 + 2)(q^2 + 1)} \exp\left(-\eta^2 q^2/4\right) = \ln \eta^{-1} - (C - \ln 2)/2, \tag{7.7}$$

where $\eta = \lambda/r_{\mathrm{D}}$. The subsequent terms of this series are given in Ichimaru and Tanaka (1985).

It is not only that the resultant expressions (7.5) and (7.7) are logarithmically exact, but their extralogarithmic terms are exact as well. For a classical ideal plasma $(kT \ll \mathrm{Ry})$, Eq. (7.5) is valid, and for a quantum plasma, Eq. (7.7) is valid. The solution derived by Williams and DeWitt (1969) joins expressions (7.5) and (7.7),

$$\ln \Lambda = \ln \eta^{-1} - 2C + \ln 2 - \frac{1}{2} \exp(z) \mathrm{Ei}(z),$$

where $\mathrm{Ei}(z) = \int_z^\infty dt\, t^{-1} \exp(-t)$ and $z = \exp(2C)(e^2/2\lambda kT)$. The high–temperature limit corresponds to the limit when $z \ll 1$, and the low–temperature limit corresponds to that with $z \gg 1$. Later on, the result of Gould and DeWitt (1967) and Williams and DeWitt (1969) was improved by Rogers et al. (1981).

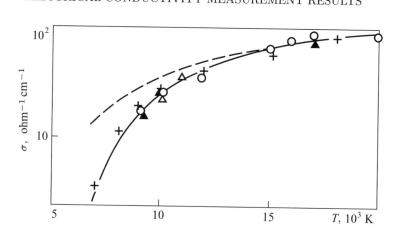

FIG. 7.1.   Electrical conductivity of argon plasma at $p = 1$ atm (Khomkin 1974). Symbols
correspond to the measurement results by Kopainsky (1971) – $\triangle$; Bues et al. (1967) – $\blacktriangle$;
Hackmann and Ulenbush (1971) – o. Solid line corresponds to the calculation by Devoto
(1973); dashed line shows the result from the Spitzer formula; symbols $+$ are calculated
using the expression (6.12).

7.2   Electrical conductivity measurement results

The electrical conductivity is the most revealing and readily observed characteristic of a plasma which governs the dissipative heating of the latter and its
interaction with the electromagnetic field. Experimenters are further attracted
by the relative simplicity of the well developed registration techniques, as well as
by the possibility of performing electrophysical measurements under most diverse
experimental conditions.

In the early experiments reviewed by Asinovsky et al. (1971), the attained
nonideality parameters did not exceed $\Gamma \approx 0.2$–0.3. Nevertheless, even at these
relatively small values of Γ, slight but systematic divergences of the measured
and calculated values were observed. The results of a series of investigations of
the electrical conductivity of a plasma of inert gases, mainly, in a plasma of arc
discharges with plasma density in the range of 10^{16}–10^{18} cm^{-3} and temperature
on the order of 10^4 K, demonstrated appreciable deviations of electrical conductivity from the Spitzer values. It has turned out that for the bulk of experimental
data the agreement with calculation results is attained if one allows in the calculations for the fact that the plasma is not fully ionized and that collisions with
atoms may appreciably reduce the electrical conductivity. Presented in Figs. 7.1
and 7.2 are the results of measurements of the electrical conductivity for argon
plasma and the calculation results by Devoto (1973).

However, at somewhat larger Γ, the inclusion of electron collisions did not
eliminate the difference between measured and calculated values of electrical
conductivity. The typical experimental conditions (Popovic et al. 1974; Bakeev
and Rovinskii 1970; Popovic et al. 1990; Günther et al. 1983) corresponded to

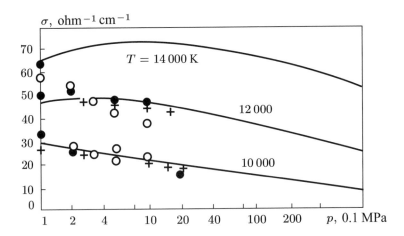

FIG. 7.2.   Electrical conductivity of argon plasma as a function of pressure at different
temperatures. Measurement results marked by dots: ∘ – Kopainsky (1971); • – Bauder *et
al.* (1973); + – Goldbach *et al.* (1978). Calculations by Devoto (1973) are shown by solid
lines.

pressures in the discharge, $p = 1$–2 MPa, and temperature $T = (1$–$1.8)\cdot10^4$ K
(Fig. 7.3). The inclusion of collisions with atoms (curve 2) leads to agreement
between theory and experiment in the low–temperature range. One can see, however, that an increase of temperature causes an appreciable deviation. Curve 3
is plotted with due regard for collision complexes and agrees well with the experimental data. The model of Vorob'ev and Khomkin (1977) introduced new
quasiparticle states into the classification of electron–ion interactions, namely,
quasibound states and collision complexes. They show up under conditions when
Γ is close to unity, and enable one to describe the observed decrease of the coefficient of electrical conductivity. Fortov and Iakubov (1989) outlined this theory,
and Khomkin (1978) discussed the role played by electron–atom collisions. An
analogous situation is observed in xenon plasma, see Fig. 7.4.

The dynamic methods discussed in Chapters 3 and 9 helped us to perform
measurements (Ivanov *et al.* 1976a; Ebeling *et al.* 1991; Sechenov *et al.* 1977;
Tkachenko *et al.* 1976; Fortov *et al.* 2003) of static electrical conductivity of a
plasma in a wide range of nonideality parameters from $\Gamma \approx 0.3$ to the region of
extremely high values of Γ, where most of theoretical approximations diverge and
the experimental results provide a basis for the construction of physical models
of electron transfer in a dense disordered medium. This range has been studied
using various media such as cesium, air, neon, argon and xenon.

The high values of plasma density (p of up to 11 GPa, n_e up to 10^{21} cm^{-3}, T
of the order of $(1$–$2)\cdot10^4$ K) were obtained by Ivanov *et al.* (1976a) and Ebeling *et
al.* (1991) for xenon (see Fig. 3.9). In these experiments it was possible to follow
the behavior of the electrical conductivity from the states of low density described

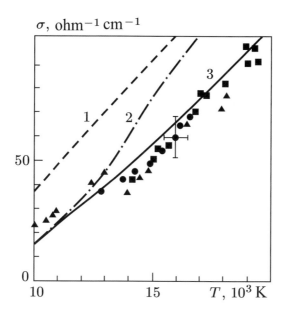

FIG. 7.3.   Electrical conductivity of argon plasma (Vorob'ev and Khomkin 1977). Experiment: • correspond to Bakeev and Rovinskii (1970); ▲ are from Popovic *et al.* (1990); ■
are from Günther *et al.* (1983). Theory: curve 1 corresponds to Spitzer conductivity; curve
2 is calculated taking into account collisions with atoms; curve 3 accounts for collision
complexes.

by the plasma models up to the states obtained by shock compression of liquid
xenon in the region of solid–state densities, where the shock compressibility is
described by the band theory of solids (Keeler *et al.* 1965). Figure 7.5 shows the
values of "reduced" dimensionless electrical conductivity

$$\overline{\sigma} = e^2 \sqrt{m} \sigma / (kT)^{3/2}. \tag{7.8}$$

The resulting combination of experimental data (see Fig. 7.5) definitely points
to an underestimation of the measured values of electrical conductivity as compared with Spitzer's theory. At the same time more stringent theories predict an
increase of conductivity with a rise of Γ, as compared with Eq. (7.4).

The existing quantitative divergence between different groups of experiments
in Fig. 7.5 is due to the physical peculiarities of the high–temperature plasma
behavior, as well as to the actual discrepancy of the primary data and to the
difficulties in determining the Coulomb component of electrical conductivity, i.e.,
derivation of the frequency ν_i from the total effective frequency ν. The latter
circumstance is most characteristic of experiments with alkali metals (Sechenov
et al. 1977; Barol'skii *et al.* 1972; Barol'skii *et al.* 1976) where the contribution
by neutrals to the collision frequency is especially large (Pavlov and Kucherenko
1977) in view of a large cross–sections of electron–atom scattering. Thus, at

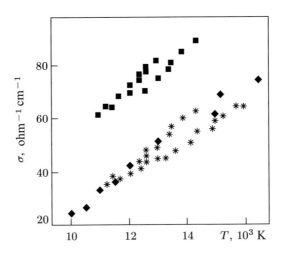

FIG. 7.4.   The coefficient of electrical conductivity of xenon plasma (Vitel *et al.* 1990). Symbols ◆ are from Popovic *et al.* (1990); ∗ are from Vitel *et al.* (1990); ■ correspond to Spitzer values.

maximum pressures in the experiment by Ermokhin *et al.* (1971), the degree of plasma ionization does not exceed 0.1%, thus making it impossible to single out the Coulomb component of σ.

The use of powerful shock waves (Chapter 3) enabled to produce a plasma with high degree of ionization (Ivanov *et al.* 1976a; Mintsev *et al.* 1980) for which there was no problem in singling out the Coulomb component. The results obtained using the dynamic methods can be conventionally divided into low–temperature with $T \leq 2 \cdot 10^4$ K (Ivanov *et al.* 1976a; Ebeling *et al.* 1991) and high–temperature with $T > 2 \cdot 10^4$ K (Ivanov *et al.* 1976b; Mintsev *et al.* 1980). The low–temperature points correspond to extremely high densities (up to 4 g cm^{-3}) which are close to the degeneracy limit of the electron component where the strong Coulomb interaction is realized ($\Gamma = 6{-}10$). In the high–temperature region, a plasma emerges with developed single and double ionization (see Section 7.5). As seen in Fig. 7.5 the results obtained for different gases (Ar, Xe, Xe, air) agree with each other and enable one to follow the effect of the Coulomb interaction on the electrical conductivity in a wide and continuously varying range of nonideality parameters, $\Gamma = 0.1{-}10$.

The highest degrees of nonideality are attained as a result of irradiation of the surface of condensed matter by powerful pulses of laser radiation. Speaking of relevant studies, that by Milchberg and Freeman (1990) deserves a special note.

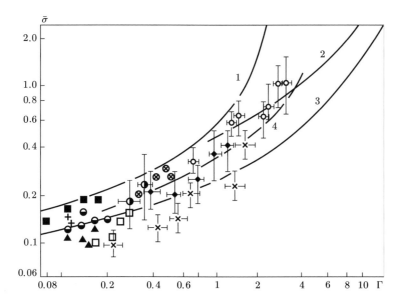

FIG. 7.5.   The dimensionless coefficient of electrical conductivity of a plasma $\bar{\sigma}$ as a function of the nonideality parameter Γ (Kraeft *et al.* 1985). The numbered curves correspond to theoretical results: $1 - \sigma_{Sp}$, $2 - t$-matrix, $3 -$ Born approximation, $4 -$ expression (7.10). Symbols correspond to experiments: ■ – argon, $11\,750$ K $\leq T \leq 159\,920$ K (Günther *et al.* 1976); ◆ – argon, ○ – xenon, and ◑ – krypton at $T \approx 25\,000$ K (Günther *et al.* 1983); ▲ – argon, $12\,800$ K $\leq T \leq 17\,400$ K (Bakeev and Rovinskii 1970); ◓ – xenon, 9000 K $\leq T \leq 13\,700$ K (Bakeev and Rovinskii 1970); × – cesium, 4000 K $\leq T \leq 25\,000$ K (Sechenov *et al.* 1977; + – hydrogen, $15\,400$ K $\leq T \leq 21\,500$ K (Günther *et al.* 1976); □ – air, $13\,500$ K $\leq T \leq 18\,300$ K (Andreev and Gavrilova 1975; Andreev 1975); ⊗ – polyethylene, $37\,000$ K $\leq T \leq 39\,000$ K (Günther *et al.* 1976).

7.3   The results of calculations of the coefficient of electrical conductivity

The asymptotic expressions for $\ln \Lambda$, given in Section 7.1, are justified only on condition that $\ln \Lambda \gg 1$ and, when they go into infinity (for example, $\sigma_{Sp} \to \infty$ at $\Gamma \to 3$), they lose their meaning. The simplest convergent value for σ may be derived if, in order to avoid formal divergences, one uses

$$\ln \Lambda = \frac{1}{2} \ln \left(1 + b_{max}^2 / b_{min}^2 \right), \tag{7.9}$$

where b_{max} and b_{min} are the maximum and minimum impact parameters, respectively. Unfortunately, this expression fails to solve the whole problem of constructing an expression for σ, which would turn into correct limiting formulas and describe well the available experimental data.

For this purpose, Gryaznov *et al.* (1976) proposed a model based on Ziman's approximation. The derivation is based on the expression for conductivity obtained in a Lorentz approximation as discussed in Section 6.1 and takes into account ion–ion correlations (see Section 6.3). The resulting expression is

$$\sigma = \frac{8}{3\sqrt{\pi}} \frac{e^2}{m(kT)^{3/2}\lambda^3} \int_0^\infty E^2 \nu_i(E)^{-1} \left(-\frac{\partial f_0}{\partial E} \right) dE. \qquad (7.10)$$

Here $f_0 = \{1 + \exp[(E - \mu)/kT]\}^{-1}$ is the electron distribution function and μ is the chemical potential. The following expression was used for the collision frequency:

$$\nu_i(E) = \sqrt{\frac{2}{\pi}} \frac{\pi e^4 n_e}{E^{3/2}} \int_0^{q_m} \frac{q^3}{(q^2 + q_D^2)^2} S(q)\, dq. \qquad (7.11)$$

Here, $n_e = (4/\sqrt{\pi})(2\pi\hbar^2/m)^{-3/2} \int \sqrt{E} f_0(E)\, dE$ and the quantity $q_m = 2E/e^2$ ensures for Eq. (7.10), with an accuracy to electron–electron interaction, Spitzer's asymptote for electrical conductivity. Limiting ourselves to a Debye approximation for the ion–ion correlation function

$$g(r) = 1 - (e^2/rkT) \exp(-q_D r), \qquad (7.12)$$

and substituting it in the expression for the structure factor, we get

$$\ln \Lambda = \frac{1}{2} \left[\ln(1 + \alpha) - \frac{\alpha}{\alpha + 1} - \frac{\alpha^2}{2(\alpha + 1)^2} \right], \qquad \alpha = (q_m r_D)^2. \qquad (7.13)$$

In the limit of weakly nonideal plasma ($\alpha \gg 1$), this expression reduces to the regular Coulomb logarithm. For strongly nonideal plasma ($\alpha \ll 1$), the expression for electrical conductivity (7.13) proves finite and contains no nonphysical divergences while reasonably describing experiments (cf. Fig. 7.5).

Apparently, the most systematic approach to constructing interpolation expressions is based on approximation (7.3). It consists in using the unscreened t-matrix approximation, subtraction of the unscreened Born approximation and addition of the Born approximation with dynamic screening. In addition, the electron–atom collisions and electron–electron interactions must be included. The result of such treatment would adequately describe exact limiting expressions. Kraeft et al. (1985) presented the results of extended studies and gave the results of calculations in a wide range of conditions.

Höhne et al. (1984) calculated the coefficient of electrical conductivity in a second Born approximation with dynamic screening (see Fig. 7.6). At low density, in order to make an effective transition to a t–matrix approximation, the thermal wavelength has been replaced (at the place where it acts as the cutoff length) by the Landau length. At low and high density, σ has the Lorentz and Ziman asymptotics, respectively. The equation of ionization equilibrium, whose solution gave the electron number density, describes ionization by pressure. The broken curve corresponds to Mott's transition which, in the approximation used, occurred in a jump rather than continuously. A minimum of the coefficient of

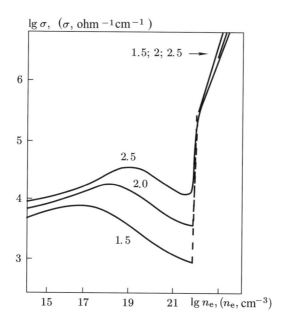

FIG. 7.6.   The coefficient of electrical conductivity σ of hydrogen plasma as a function of electon concentration n_e for different values of temperature T (in units of 10^4 K) (Höhne *et al.* 1984). The vertical dashed lines indicate Mott's transition.

electrical conductivity (whose value is affected by the electron–atom scattering) occurs in the region of Mott's transition.

Ziman's formula may be derived from Eqs. (7.10) and (7.11), with due regard for the fact that, under conditions of strong degeneracy, $df_0/dE = \delta(E - E_{\mathrm{F}})$, and the maximum transferred momentum is equal to $2\hbar k_{\mathrm{F}}$. Then,

$$\sigma_Z = \frac{\varepsilon_{\mathrm{F}}^{3/2}}{\pi\sqrt{2}Ze^2\sqrt{m}\ln\Lambda}, \qquad \ln\Lambda = \int_0^{2k_{\mathrm{F}}} \left[V(q)/4\pi Ze^2\right]^2 S(q)q^3\,dq,$$

where $V(q)$ is the Fourier component of the potential, and $E_{\mathrm{F}} = \hbar^2 k_{\mathrm{F}}^2/(2m)$.

The importance of electron–electron interactions is a maximum under conditions of low number density of charged particles, when it is defined by the Spitzer factor γ_{E}. It decreases as the nonideality increases, and is suppressed in the case of strong degeneracy due to Pauli's principle (Schlanges *et al.* 1984) (see Fig. 7.7).

In Fig. 7.5, the experimental data are compared with the results of calculations in a statically screened t–matrix approximation (curve 2) and statically screened Born approximation (curve 3). At small values of Γ, curve 2 tends to the Spitzer result (curve 1).

The calculations performed by Kraeft *et al.* (1985) may be classed as wide-range calculations of the coefficient of electrical conductivity. Such methods were

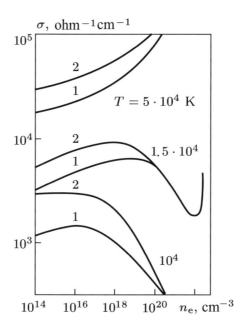

FIG. 7.7.   The coefficient of electrical conductivity σ of hydrogen plasma, calculated in view
(1) of and disregarding (2) the electron–electron interactions (Schlanges *et al.* 1984).

called into being by application requirements. Under conditions of a large pulsed
energy contribution, the matter may pass through the entire range of nonideality parameters. Ichimaru and Tanaka (1985) tabulated the Coulomb logarithm,
given by expression (7.6), in a very wide range of parameters. In a plasma with
arbitrary degree of degeneracy, the exchange–correlation correction with Lind-
hardt's dielectric permeability was included, and the value of the structure factor
was borrowed from the solution of hypernetted–chain equations (see Section 5.1).
The nonideality parameter is $\gamma = (Ze^2/kT)(4\pi n_i/3)^{1/3} \leq 2$.

Lee and More (1984) performed calculations with reference to the conditions
realized during laser–induced thermonuclear fusion. Simple calculations, in which
the Coulomb logarithm was provided by expression (7.9), covered a very wide
range of conditions (see Fig. 7.8). The screening distance in Lee and More (1984)
was provided by the Debye–Hückel or Thomas–Fermi radii in the region of weak
nonideality, or the mean interionic distance in the range where $\gamma \geq 1$ (see Section
7.4), while in the range where $\ln \Lambda$ turned to be less than two, it was assumed that
$\ln \Lambda = 2$. In the region of Mott's transition, the coefficient was assumed equal
to the minimum Mott value (see Section 4.5), while in the strongly correlated
degenerate system an expression of the type of Ziman expression was used for
the coefficient of electrical conductivity. Thereby, all of the simplest constructive
models were employed.

One can assume that, under conditions of isochoric heating from the melting

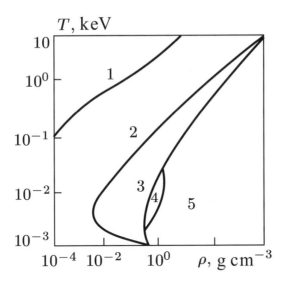

FIG. 7.8. The temperature and density regions treated by Lee and More (1984): Region 1 corresponds to Debye–Hückel and Thomas–Fermi screening; in region 2 screening occurs on the average interionic distance; in region 3 it is assumed $\ln \Lambda = 2$; In region 4 $\sigma = \sigma_{\min}$; and region 5 corresponds to Ziman's region.

point to 100 eV, the matter will pass through the entire set of nonideal–plasma states from liquid metal to ideal plasma. Milchberg and Freeman (1990) constructed the isochore of specific resistance ρ (see Fig. 7.9) as a result of treatment of the results of measurements of the coefficient of reflection of radiation from the aluminum surface heated by a powerful laser pulse with intensity I up to 10^{15} W cm^{-2}. During the period of 400 fs, the matter does not have enough time to expand, and retains the density of solid aluminum. The measured values were compared by Iakubov (1991) with those derived by the methods of a wide–range description of ρ. The effect of non–Coulomb scattering (see Section 7.5) from the core of complex Al^{+3} ions proves to be of extreme importance. The resistance of a degenerate plasma is described by the curve 2 accounting for the non–Coulomb scattering from a core of radius $R_c = 1.1a_0$. In the degeneracy region, the ρ_Z curve is constructed by Ziman's formula taking into account scattering by Ashcroft's pseudopotential with the core radius $R_c = 1.1a_0$. In the region of $T \sim 19$ eV, the free path length proved to be less than the interparticle distance. Here, the Ioffe–Regel formula is valid (curve IR),

$$\sigma_{\mathrm{IR}} = (n_e e^2/m) r_s/v. \qquad (7.14)$$

In the aggregate, the curves 2, ρ_{IR}, and ρ_Z give a qualitative description of experiments.

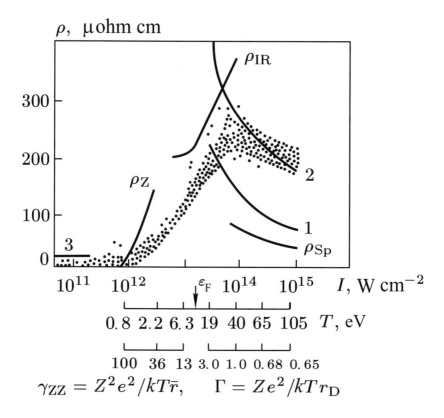

FIG. 7.9.   The measurements by Milchberg and Freeman (1990) (points) and calculation
by Iakubov (1991) (curves) of values of specific resistance ρ for a dense aluminum plasma.
Curve 1 corresponds to conversion from the data of Fig. 7.5; curve 2 accounts for non—
Coulomb scattering; and curve 3 corresponds to the specific resistance of liquid aluminum.

7.4   The coefficient of electrical conductivity of a strongly nonideal "cold plasma"

As the nonideality increases, considerable difficulties are encountered in substantiation of the input kinetic equations and methods of solving these equations.
In view of the strong collective interaction in a dense plasma, one cannot unam-
biguously separate the characteristic times of elementary processes, and the time
evolution of the system under the effect of an external field, generally speaking, is
no longer described by the Markovian process. The inclusion of bound states in a
partly ionized plasma presents a special problem (Klimontovich 1982) because of
the absence of appropriate kinetic equations. The results of qualitative analysis
point to the possibility of the emergence, under conditions of nonideality, of new
types of quasiparticle states. Their description is given below, largely in accordance with Likal'ter (1987, 1992) and Vorob'ev (1987). For $\gamma > 1$, the regions of

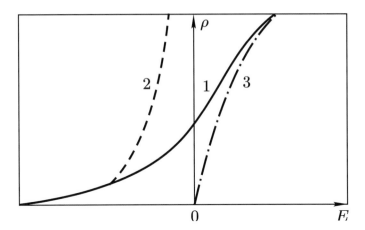

FIG. 7.10.   The density of electron states: 1, density of states in a nonideal plasma; 2, density of states of an isolated atom; 3, density of states of free electrons.

electron interaction with different ions overlap, because $e^2/kT > r_s$. An electron is permanently in the state of collision, making a transition from the field of one ion to that of another ion. The trajectory of an electron with $E > 0$ (and $e^2/E > r_s$) consists of segments of conjugate hyperbolas, and the trajectory of an electron with $E < 0$ consists of conjugate segments of ellipses $(e^2/|E| > r_s)$. Of course, no significant differences are observed between the trajectories of these two types.

In view of the foregoing, it is obvious that the electron density of states $\rho(E)$ must vary relatively little for low energy values, when $|E| \leq e^2/r_s$, and correspond qualitatively to curve 1 in Fig. 7.10. For high energy values, when $E \gg \gamma T$, the density of states is close to (6.22). For negative energy of high absolute values, when $|E| \gg \gamma T$, $\rho(E)$ reaches the asymptotic of quasiclassical density of states for a hydrogen–like atom, $\rho(E) = n_i \text{Ry}^{3/2} Z^3 / |E|^{5/2}$. The estimation of $\rho(E)$ in the vicinity of zero energy may be performed with the aid of formula (6.25), if one uses the distribution of potential energy of the electron in the field of the nearest ion,

$$P(U) = 4\pi n_i \int_0^\infty R^2 dR \exp\left(-\frac{4\pi}{3} n_i R^3\right) \delta\left(U + \frac{Ze^2}{R}\right). \qquad (7.15)$$

Indeed, the probability of finding the nearest ion at the distance R from the electron, $F(R) = 4\pi n_i \exp[-4\pi n_i R^3/3]$, decreases exponentially at $R > r_s$. We use Eq. (7.15) to derive from Eq. (6.25)

$$\rho(E) = 2\pi \frac{2(2m)^{3/2}}{h^3} \Gamma(5/6) \frac{\sqrt{Ze^2/r_s}}{2} \left[1 + \frac{Er_s}{2Ze^2} \frac{\Gamma(7/6)}{\Gamma(5/6)}\right]. \qquad (7.16)$$

Therefore, for low values of energy, the density of states is weakly dependent on energy.

The situation proves to be close to that discussed in Section 6.4 for the case of weakly ionized dense plasma. As in the latter case, the importance of the energy E_P at which the classically accessible regions of motion for negative-energy electrons overlap for the first time increases. The criterion of overlapping of randomly arranged spheres of radius R at concentration n_i has the form $1 - \exp(-4\pi n_i R^3/3) = 0.29$ (Shklovskii and Efros 1984). This gives the level of percolation energy

$$E_P = -e^2/(0.7r_s). \qquad (7.17)$$

Electrons with $E > E_P$, as well as free electrons, are conduction electrons. However, even if the energy, though exceeding E_P, is close to it, these electrons do not yet possess all characteristics of free electrons. Their mobility proves to be smaller.

Let us write the known relation

$$\sigma = e^2 n_e \left\langle D(E) \right\rangle /kT. \qquad (7.18)$$

The mean coefficient of electron diffusion is

$$\langle D(E)\rangle = \left[\int_{E_P}^{\infty} dE \rho(E) \exp(-E/kT) H(E)\tilde{D}(E) \Big/ \int_{E_P}^{\infty} dE \rho(E) \exp(-E/kT) \right].$$

Here, $\tilde{D}(E)$ is the diffusion coefficient for free electrons, and $H(E)$ takes into account the difference of the electronic diffusion coefficient in the vicinity of the percolation threshold $D(E)$ from $\tilde{D}(E)$. The function $H(E)$ may be determined accurately with the aid of the effective medium theory (Sarychev 1985). This function, however, has evident properties as well. It is zero at the point $E = E_P$, and it increases with E to unity when $E \geq 0$ is reached.

Under conditions of strong nonideality, in view of the properties of $H(E)$, we derive

$$\langle D(E)\rangle = \tilde{D}\left(E_P\right) \frac{dH}{dE}\bigg|_{E=E_P} kT \approx \tilde{D}\left(E_P\right)\frac{kT}{E_P}. \qquad (7.19)$$

In order to calculate the diffusion coefficient $\tilde{D}(E_P) \approx v(E_P)r_s$, one must know the mean velocity of an electron with energy close to E_P. It is believed that, during transition from one potential well to another, most of the time is spent by the electron in passing through the saddle point. In the saddle point region, the kinetic energy of the electron is not high, of the order of T (there is no other energy scale). Then, $\tilde{D}(E_P) = v_{T_e}r_s$. Substituting the written-out quantities in (7.18) we finally derive the following expression for the coefficient of electrical conductivity (accurate to a numerical factor):

$$\sigma \sim \omega_p/\sqrt{\gamma}, \qquad (7.20)$$

where $\omega_p = \sqrt{4\pi n_e e^2/m}$ is the plasma frequency, and n_e is the number density of conduction electrons, i.e., electrons with energy $E > E_P$.

The question of the coefficient of electrical conductivity for such a system may hardly be regarded as closed, although it is evident that σ is proportional to ω_P, because there is no other relevant parameter. If one assumes that an electron, on the contrary, passes through the saddle point very quickly, then $v(E_P) \approx \sqrt{E_P/m}$. In this case $\sigma \sim \omega_p$. This expression was proposed by Kurilenkov and Valuev (1984) on entirely different model grounds.

7.5 Electrical conductivity of high–temperature nonideal plasma. The ion core effect

Nonideal plasma experiments are characterized by relatively low $(T < 3 \cdot 10^4$ K) temperatures because these experiments are directed toward attaining developed Coulomb nonideality decreasing with a temperature rise. The use of cumulative shock tubes and the effects of shock wave reflection from obstacles (Mintsev *et al.* 1980) enables one to advance into the region of extremely high temperatures and obtain a highly heated, multiply ionized plasma with developed Coulomb nonideality of the order of $\Gamma \sim 1$–5 (Fig. 7.11). The electrophysical properties of such a plasma turned out to be unexpected to some extent (Ivanov *et al.* 1976b; Mintsev *et al.* 1980) because they point to the nonexistence of similarity: the dimensionless electrical conductivity of high–temperature plasma turns out to be smaller than that of low–temperature plasma at the same values of the Coulomb nonideality parameter Γ. Figure 7.12 shows the quantity

$$\overline{\sigma} = \left(1 + \overline{Z}\right)^{-1/2} \gamma_E^{-1}\left(\overline{Z}\right) \frac{\sigma}{\omega_p} \Gamma \qquad (7.21)$$

as a function of the nonideality parameter $\Gamma = \overline{Z}e^2/kT r_D$, where $\overline{Z}$ is the averaged ion charge, and $r_D^{-2} = 4\pi e^2(1 + \overline{Z})n_e/kT$. The experimental data at low and high temperatures clearly separate.

Indeed, in view of the long–range character of the Coulomb potential, the dominant contribution to the transport coefficient at moderate temperatures is made by electron scattering with large impact parameters, thus justifying the use of semiqualitative inclusion of close collisions by introducing various forced cutoffs. As the temperature rises, the amplitude of Coulomb scattering, $\overline{Z}e^2/kT$ turns out to be comparable with the proper size of ions, r_i, so that the high–energy conduction electrons on their scattering may approach rather closely the nucleus where the interaction potential is no longer purely Coulomb and appears distorted by inner electron shells. The estimates made for rarefied plasma (Ebeling *et al.* 1991; Maev 1970) indicate that this effect starts to show against the background of Coulomb scattering only at extremely high temperatures of the order of $2 \cdot 10^6$ K. An increase of the plasma density causes a strengthening of screening of the Coulomb interaction and, consequently, an increase of the importance of close collisions in the plasma which, in its turn, makes possible the manifestation of the effect of non–Coulomb scattering of electrons with small impact parameters in the region of lower temperatures accessible for experiment, $T \geq 5 \cdot 10^4$ K.

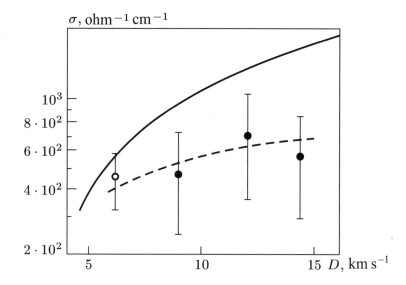

σ, ohm^{-1} cm^{-1}

FIG. 7.11.   Electrical conductivity σ as a function of shock wave velocity D in xenon at
initial pressure $p_0 = 0.1$ MPa. Symbols correspond to experiment: $\bullet$ – Mintsev *et al.* (1980);
$\circ$ – Ivanov *et al.* (1976a). Solid curve corresponds to calculation using Eq. (6.8), dashed
line accounts for non–Coulomb corrections.

In experiments by Mintsev *et al.* (1980), characterized by developed ionization
$(\overline{Z} = 3)$ at temperature up to 10^5 K, use was made of explosion cumulative
plasma generators, as well as of linear explosion systems for preparing a highly
heated plasma in reflected shock waves. The experiments were performed with
xenon gas because the high molecular weight makes the heating of this gas by
shock waves effective, while a large number of bound electrons causes a marked
distortion of the Coulomb potential at close distances. The cumulative systems
permitted the preparation of a plasma with $p \approx (0.5$–$3.5)$ GPa, $T \approx (5-10) \cdot 10^4$
K, $\rho \approx 0.05$–0.5 g cm^{-3}, $n_e \approx (0.4$–$3) \cdot 10^{21}$ cm^{-3}, $\Gamma \approx 1.1$–2.6, and $\overline{Z} \approx 2$–3. The
plasma parameters behind the reflected shock wave amounted to $p \approx (4$–$8)$ GPa,
$T \approx (3$–$8) \cdot 10^4$ K, $\rho \approx 0.3$–2 g cm^{-3}, $n_e \approx (2$–$6) \cdot 10^{21}$ cm^{-3}, $\Gamma \approx 2$–5, and $\overline{Z} \approx 2$.

In the region of low temperatures the obtained data agree well with the
previously obtained results (Fig. 7.5). However, the electrical conductivity of the
plasma increases with the shock wave velocity at a much slower rate than one
could expect on the basis of conventional plasma models. This points to the
absence of similarity of the Coulomb component of the electrical conductivity.
The potential of the electron–ion interaction at small distances is stronger than
$\overline{Z}e^2/r$, which leads to an increase of the scattering cross–section as compared
with the Coulomb one and, consequently, to a relative decrease of electrical
conductivity. The electron–ion interaction can be described by the effective two–
body potential (Maev 1970) allowing for the presence of the ion core,

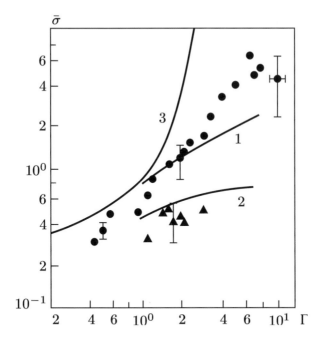

FIG. 7.12. The reduced coefficient of electrical conductivity $\bar{\sigma}$ as a function of the nonideality parameter Γ: Symbols ● correspond to experiment in argon, krypton, and xenon at
$T \approx 2.5 \cdot 10^4$ K (Ivanov et $al.$ 1976b); symbols ▲ correspond to experiment in xenon at
$T \approx 7 \cdot 10^4$ K (Mintsev et $al.$ 1980). Numbered curves correspond to theoretical calculations: curves 1 and 2 are obtained using pseudopotential (7.22) at the same temperatures;
curve 3 shows the Spitzer result σ_{Sp}.

$$V_{\mathrm{ei}}(r) = -\left[\frac{Z_{\mathrm{j}}}{r} - \frac{(Z - Z_{\mathrm{j}})}{r}\exp(-Br)\right]\exp\left(-\frac{r}{r_{\mathrm{D}}}\right), \qquad B = \frac{1.8 Z^{4/3}}{(Z - Z_{\mathrm{j}})a_0},$$
$$(7.22)$$

where Z, Z_{j} are the nuclear and ion charges, and r_{D} is the Debye screening
distance. At small values of r, the potential (7.22) coincides with the Thomas–
Fermi potential and, at $r \to \infty$, it reduces to the Debye–Hückel potential. This
potential was used in the numerical solution of Schrodinger's equation for the
radial part of the wavefunctions. The phases of scattering δ_l were determined.
The transport cross–sections for electron–ion scattering were then calculated
using these scattering phases. The results of these calculations show that for the
pseudopotential (7.22) in the region of electrons with the energy of the order
of 0.7 a.u. making the main contribution to the electrical conductivity of the
plasma at $T \approx 7 \cdot 10^4$ K, the presence of an interaction stronger than $(Z_{\mathrm{j}}e^2/r)$
approximately doubles the scattering cross–sections.

The results of electrical conductivity calculations in accordance with the employed model are shown in Figs. 7.11 and 7.12. It is seen that the ion core model

permits a reasonable description of the experimentally revealed effect of "separation" of the isotherms of electrical conductivity of a highly heated nonideal plasma.

References

Andreev, S. I. (1975). Radiation flow and spectral intensity of radiation of a pulsed discharge in a quartz tube with xenon. *Optika i Spektroskopiya*, **38**, 432–439.

Andreev, S. I. and Gavrilova, T. V. (1975). Measurement of electrical–conductivity of air plasma at pressures above 100 atm. *High Temperature*, **13**, 151–153.

Asinovsky, E. I., Kirillin, A. V., Pakhomov, E. P., and Shabashov, V. I. (1971). Experimental investigation of transport properties of low–temperature plasma by means of electric arc. *Proceedings of the IEEE*, **59**, 592–601.

Bakeev, A. A. and Rovinskii, R. E. (1970). Electrical properties of high pressure pulsed–discharge plasmas in inert gases. *High Temperature*, **8**, 1055–1061.

Barol'skii, S. G., Kulik, P. P., Ermokhin, N. V., Melnikov, V. M. (1972). Measurement of electric–conductivity of a dense strongly nonideal cesium plasma. *JETP 35*, 94–100.

Barol'skii, S. G., Ermokhin, N. V., Kulik, P. P., and Ryabyi, V. A. (1976). Energy–balance of a high–pressure cesium pulsed discharge in a transparent capillary. *High Temperature*, **14**, 626–633.

Bauder, U., Devoto, R. S., and Mukherjee, D. (1973). Measurement of electrical–conductivity of argon at high–pressure. *Phys. Fluids*, **16**, 2143–2148.

Bues, I., Patt, H. J., and Richter, J. (1967). Uber die elektrische Leitfahigkeit und die Warmeleitfahigkeit des Argons bei hohen Temperaturen. *Zeitschrift fur angewandte Physik*, **22**, 345–350.

Devoto, R. S. (1973). Transport coefficients of ionized argon. *Phys. Fluids*, **16**, 616–623.

Ebeling, W. and Röpke, G. (1979). Conductance theory of nonideal plasmas. *Annalen der Physik*, **36**, 429–437.

Ebeling, W., Forster, A., Fortov, V., Gryaznov, V., and Polishchuk, A. (1991). *Thermophysical properties of hot dense plasmas*. AVG, Leipzig.

Ermokhin, N. V., Ryabyi, V. A., Kovalev, B. M., Kulik, P. P. (1971). Experimental investigation of coulomb interaction in a dense plasma. *High Temperature*, **9**, 611–621.

Fortov, V. E. and Iakubov, I. T. (1989). *Physics of nonideal plasma*. Hemisphere, New York.

Fortov, V. E., Ternovoi, V. Y., Zhernokletov, M. V., Mochalov, M. A., Mikhailov, A. L., Filimonov, A. S., Pyalling, A. A., Mintsev, V. B., Gryaznov, V. K., and Iosilevskii, I. L. (2003). Pressure–produced ionization of nonideal plasma in a megabar range of dynamic pressures. *JETP*, **97**, 259–278.

Goldbach, C., Nollez, G., Popovic, S., and Popovic, M. (1978). Electrical–conductivity of high–pressure ionized argon. *Zeitschrift fur Naturforschung*, **33**, 11–17.

Gould, H. A., and DeWitt, H. E. (1967). Convergent kinetic equation for a classical plasma. *Phys. Rev.*, **155**, 68–74.

Gryaznov, V. K., Ivanov, Y. V., Starostin, A. N., and Fortov, V. E. (1976). Thermophysical properties of shock–compressed argon and xenon. *High Temperature*, **14**, 569–572.

Günther, K., Popovic, M. M., Popovic, S. S., and Radtke, R. (1976). Electrical conductivity of highly ionized dense hydrogen plasma .2. comparison of experiment and theory. *J. Phys. D – Applied Physics*, **9**, 1139–1147.

Günther, K., Lang, S., and Radtke, R. (1983). Electrical–conductivity and charge carrier screening in weakly nonideal argon plasmas. *J. Phys. D*, **16**, 1235–1243.

Hackhmann, J. and Ulenbusch, J. (1971). Determination of the electric conductivity of rare gases under normal pressure from arc measurements. In *Abstracts of 10th international conference on phenomena in ionized gases*, Francklin, R. N. (ed.), p. 260. Donald Parsons, Oxford.

Höhne, F. E., Redmer, R., Röpke, G., and Wegener, H. (1984). Linear response theory for thermoelectric transport–coefficients of a partially ionized plasma. *Physica A*, **128**, 643–675.

Hubbard, J. (1961a). Friction and diffusion coefficients of Fokker–Planck equation in a plasma. *Proceedings of the Royal Society of London: Series A– Mathematical and Physical Sciences*, **260**, 114.

Hubbard, J. (1961b). Friction and diffusion coefficients of Fokker–Planck equation in a plasma. 2. *Proceedings of the Royal Society of London: Series A– Mathematical and Physical Sciences*, **261**, 371–387.

Iakubov, I. T. (1991). On electric conductivity of superdense plasma. In *Proceedings of the 20th international conferences on phenomena in ionized gases*, Piza. Contributed papers, pp. 881–882.

Ichimaru, S. and Tanaka, S. (1985). Theory of interparticle correlations in dense, high–temperature plasmas. 5. Electric and thermal–conductivities. *Phys. Rev. A*, **32**, 1790–1798.

Ivanov, Y. V., Mintsev, V. B., Fortov, V. E., and Dremin, A. N. (1976a). Electric conductivity of nonideal plasma. *JETP*, **44**, 112–116.

Ivanov, Y. V., Mintsev, V. B., Fortov, V. E., and Dremin, A. N. (1976b). Electric conductivity of strongly nonideal plasma. *JTP Lett.*, **2**, 97–101.

Kadomtsev, B. B. (1957). On the effective field in plasma. *JETP*, **6**, 117–122.

Keeler, R. N., van Thiel, M., and Alder, B. J. (1965). Corresponding states at small interatomic distances. *Physica*, **31**, 1437–1440.

Khomkin, A. L. (1974). Electrical–conductivity of argon and xenon plasmas. *High Temperature*, **12**, 766–769.

Khomkin, A. L. (1978). Thermodynamic functions, composition, and electrical–conductivity of inert–gas plasmas. *High Temperature*, **16**, 29–34.

Kivelson, M. G. and Du Bois, D. F. (1964). Plasma conductivity at low frequencies and wavenumbers. *Phys. Fluids*, 7, 1578–1589.

Klimontovich, Y. L. (1982). *Kinetic theory of nonideal gases and nonideal plasmas*. Pergamon Press, Oxford.

Klimontovich, Y. L. (1999). *Statistical theory of open systems*. Vol. 2. Yanus, Moscow.

Kopainsky, J. (1971). Radiation transport mechanism and transport properties in ar high pressure arc. *Zeitschrift fur Physik*, **248**, 417–432.

Kraeft, W. D., Kremp, D., Ebeling, W., and Röpke G. (1985). *Quantum statistics of charged particle systems*. Akademic Verlag, Berlin.

Kurilenkov, Y. K. and Valuev, A. A. (1984). The electrical–conductivity of plasma in wide–range of charge–densities. *Beitrage aus der Plasmaphysik–Contributions to Plasma Physics*, **24**, 161–171.

Lee, Y. T. and More, R. M. (1984). An electron conductivity model for dense–plasmas. *Phys. Fluids*, **27**, 1273–1286.

Liboff, R. L. (1959). Transport coefficients determined using the shielded Coulomb potential. *Phys. Fluids*, **2**, 40–46.

Lifshitz, E. M. and Pitaevskii, L. P. (1981). *Physical kinetics*. Pergamon Press, Oxford.

Likal'ter, A. A. (1992). Gaseous metals. *Phys. Usp.*, **162**, 119–147.

Likal'ter, A. A. (1987). Electric–conductivity of a degenerate quasiatomic gas. *High Temperature*, **25**, 305–311.

Maev, S. A. (1970). Effect of finite ion size on electron collision frequency in a plasma. *Soviet Phys. – Technical Phys.*, **15**, 438–445.

Milchberg, H. M. and Freeman, R. R. (1990). Studies of hot dense–plasmas produced by an intense subpicosecond laser. *Phys. Fluids*, **2**, 1395–1399.

Mintsev, V. B., Fortov, V. E., and Gryaznov, V. K. (1980). Electric–conductivity of a high–temperature imperfect plasma. *JETP*, **79**, 116–124.

Pavlov, G. A., Kucherenko, V. I. (1977). Effect of excited atoms on conductivity of a dense–plasma of alkali–metals. *High Temperature*, **15**, 343–345.

Popovic, M. M., Popovic, S. S., and Vukovic, S. M. (1974). Distribution of electrical conductivity and density along section of plasma coulomb of high pressure arc. *Fizika*, **6**, 29–35.

Popovic, M. M., Vitel, Y., and Mihajlov, A. A. (1990). (private communication).

Rogers, F. J., DeWitt, H. E., and Boercker, D. B. (1981). Electrical–conductivity of dense–plasmas. *Phys. Lett. A*, **82** 331–334.

Röpke, G. (1988). Quantum–statistical approach to the electrical–conductivity of dense, high–temperature plasmas. *Phys. Rev. A*, **38**, 3001–3016.

Sarychev, A. K. (1985). Effective–medium method for calculating the conductivity of a dense weakly ionized plasma. *High Temperature*, **23**, 179–185.

Schlanges, M., Kremp, D., Keuer, H. (1984). Kinetic approach to the electrical–conductivity in a partially ionized hydrogen plasma. *Annalen der Physik*, **41**, 54–66.

Sechenov, V. A., Son, E. E., and Shchekotov, O. E. (1977). Electrical conductivity of a cesium plasma. *High Temp.*, **15**, 346–349.

Shklovskii, B. I. and Efros, A. L. (1984). *Electronic properties of doped semi-conductors*. Springer, Berlin.

Tkachenko, B. K., Titarov, S. I., Karasev, A. B. and Alipov, S. V. (1976). Experimental determination of parameters behind strong reflected shock waves in air. *Comb., Expl., Shock Waves*, **12**, 763–768.

Vitel, Y., Mokhtari, A., and Skowronek, M. J. (1990). Electrical conductivity and pertinent collision frequencies in nonideal plasma with only a few particles in the Debye sphere. *J. Phys. B: At. Mol. Phys.*, **23**, 651–660.

Vorob'ev, V. S. (1987). Electrical–conductivity of completely ionized nonideal plasma. *High Temperature*, **25**, 311–315.

Vorob'ev, V. S., Khomkin, A. L. (1977). Correlated electron–ion pairs in a plasma and their effect on electrical–conductivity. *High Temperature*, **15**, 157–160.

Williams, R. H. and DeWitt, H. E. (1969). Quantum–mechanical plasma transport theory. *Phys. Fluids*, **12**, 2326–2342.

8

THE OPTICAL PROPERTIES OF DENSE PLASMA

8.1 Optical properties

The optical properties are of considerable interest from the standpoint of plasma physics. This is because they enable one to follow the effect of nonideality on the dynamics and energy spectrum of electrons in a dense, disordered medium. The plasma radiation bears information on the temperature and concentration of particles, elastic and inelastic collisions, and ionization and recombination processes. Thus far, extensive data have been obtained on the optical properties of rarefied plasma where the elementary processes are readily separated. The theoretical models developed also provide an exhaustive description of numerous experiments (Section 7.1). As the density increases, the optical consequences of nonideality are registered before the corresponding variations of the thermodynamic and transport properties take place. They also show up as a spectral line shift and broadening, as well as shift of photoionization continua (Section 7.2).

With a further increase of nonideality, the effects do not change in their behavior but only increase quantitatively ("spectroscopic stability"). It is only under very high density that a marked rearrangement of the electron energy structure occurs in the plasma (primarily in the threshold region of the spectrum). This is described in various, and often still controversial, models. Numerous and clearly defined manifestations of the electron spectrum in a highly compressed plasma have been registered. This plasma emerges as a result of the contribution of great pulsed energy under laser irradiation of condensed matter in pinched electric discharges, as well as in a number of dynamic experiments. It was also revealed in static experiments that upon reaching near–critical values of density, a qualitative variation of the optical spectra of metal vapors takes place.

We will introduce the basic concepts of the radiation theory. The spectral absorptivity k_ν is determined in terms of the attenuation dI_ν, which is experienced by the radiation intensity I_ν whilst passing through a layer of matter of thickness dl:

$$dI_\nu = -k_\nu I_\nu \, dl. \tag{8.1}$$

Under conditions of thermodynamic equilibrium, k_ν is related to the radiation intensity I_ν by Kirchhoff's law:

$$I_\nu = k_\nu B_\nu (T), \quad B_\nu (T) = 2h\nu^3 c^{-2} \left[\exp\left(\frac{h\nu}{kT}\right) - 1\right]^{-1}, \tag{8.2}$$

where $B_\nu(T)$ is Planck's function. The quantity $I_\nu d\nu$ defines the energy emitted by volume dv per unit time in unit solid angle in the frequency interval $d\nu$.

These quantities are used to express the divergence of radiant energy flux q_R, which characterizes the energy losses (or release) due to radiation. In two limiting cases, that is, with high and low optical density, ∇q_R is written in a simple manner. If the plasma dimension l exceeds the radiation path length k_ν^{-1}, $k_\nu l \gg 1$ in the entire frequency interval covering the spectral region in which the bulk of radiant energy is transferred, the radiative energy transfer may be described in the approximation of radiant thermal conductivity,

$$q_R = -\frac{16}{3}\sigma_{SB} l_R(\mathbf{r}) T^3(\mathbf{r}) \nabla T(\mathbf{r}),$$

$$l_R(\mathbf{r}) = \int k_\nu^{-1}(\mathbf{r}) B_\nu(\mathbf{r}) d\nu \Big/ \int B_\nu(\mathbf{r}) d\nu, \tag{8.3}$$

where σ_{SB} is the Stefan–Boltzmann constant; and l_R is the Roseland length, which is the mean photon path length.

Conversely, in the limiting case of an optically thin medium, the divergence of energy flux is easily written:

$$\nabla q_R(\mathbf{r}) = 4\pi\sigma_{SB} k_{Pl}(\mathbf{r}) T^4(\mathbf{r}),$$

$$k_{Pl}(\mathbf{r}) = \int k_\nu(\mathbf{r}) B_\nu(\mathbf{r}) d\nu \Big/ \int B_\nu(\mathbf{r}) d\nu. \tag{8.4}$$

where the quantity k_{Pl} is referred to as the mean Planck absorptivity.

Although both these limiting cases are seldom realized in their pure form, the representation of the results of calculations of radiative properties in the form of the density and temperature dependences of mass absorption cross–sections yields important information. Figure 8.1 gives the Roseland and Planck mass absorption cross–sections as functions of the atomic nuclear charge.

8.2 Basic radiation processes in rarefied atomic plasma

The radiation processes in an ideal plasma have been studied fairly well (cf. Biberman and Norman 1967; Avilova *et al.* 1970; Raizer 1987; D'yachkov 2000). We will only present here the basic information along with some reference formulas.

Depending on the type of corresponding optical spectra, the electron radiation processes are divided into two groups:

(1) Bound–bound transitions in atoms. They provide a series of spectral lines that converge at the photoionization thresholds. Near the threshold, the convergence of lines occurs. As a result, the position of the threshold shifts to the long-wave side (Section 8.3).

(2) Bound–free and free–free transitions. The photoionization and bremsstrahlung processes defining the continuous spectrum manifest little dependence on

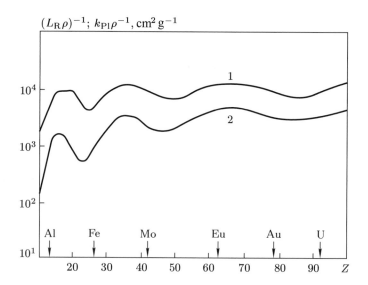

FIG. 8.1. The (1) Planck and (2) Rosseland mass absorption cross–sections as functions of the atomic nuclear charge (Klein and Meiners 1977); $T = 100$ eV, $\rho = 0.1 \, \mathrm{g\,cm^{-3}}$.

frequency and experience sharp discontinuities at frequencies which are equal to the threshold frequencies. Such a division is not absolute. For example, strongly broadened spectral lines overlap and are superimposed on the continuous spectrum. These effects are especially strong in a strongly nonideal plasma, that is, a high–pressure plasma.

The integrated intensity of the spectral line depends on the oscillator strength $f_{nn'}$:

$$\int k_\nu \, d\nu = (\pi e^2/mc) f_{nn'} n_n,$$

where k_ν is the coefficient of absorption in a spectral line due to the $n \rightarrow n'$ transition, and n_n is the concentration of absorbing atoms. Accordingly, the integrated intensity of the spectral line is

$$I = \int I_\nu \, d\nu = \frac{2\pi h e^2}{m\lambda^3} \frac{g_n}{g_{n'}} n_{n'} f_{nn'},$$

where $n_{n'}$ is the concentration of radiating atoms, and g_n and $g_{n'}$ are the statistical weights of the low–lying and high–lying states, respectively.

The absorption line oscillator strength depends on the Einstein probability of spontaneous transition,

$$f_{nn'} = \frac{g_{n'}}{g_n} \frac{mc^3}{8\pi^2 e^2 \nu^2} A_{n'n}.$$

The factor $A_{n'n}$ is equal to the inverse lifetime of an atom in the state n relative to the $n \to n'$ transition. The absorbing and radiating states are actually characterized by several quantum numbers such that the subscript n characterizes their entire required set. The fullest tables for $f_{nn'}$ and $A_{nn'}$ are provided by Wiese et al. (1966). The quasiclassical Kramers formula for hydrogen–like atom appears useful. The transition is considered between states with the main quantum numbers k and m averaged over the remaining quantum numbers. For such a transition,

$$f_{n'n} = \frac{32}{3\pi\sqrt{3}} \frac{1}{(n')^5 n^3} \left[(n')^{-2} - n^{-2} \right]^{-3} = \frac{1,96}{(n')^5 n^3} \left(\frac{E_{n'} - E_n}{Z^2 \text{Ry}} \right)^{-3},$$

where E_k and E_m are the binding energies of the m–th and kth levels. The error of f_{km}, when calculated using this formula for a hydrogen atom, does not exceed 30% for all values of k and m.

The line spectral intensity fully depends on the frequency dependence of the absorption coefficient. This is because the Planck function varies slightly within the line. The ν independence of k is defined by the behavior of the line broadening. In a rarefied plasma, it is defined by radiation damping and the Doppler effect. The broadening in a nonideal plasma, which is due mainly to interparticle interactions, will be discussed below. Continuous absorption and emission spectra are formed as a result of electron transitions from the bound to free state and vice versa (bound–free and free–bound transitions), and as a result of free–free transitions. Bound electrons are found in the atom while free electron may be in transit through the field of the atom or ion. Accordingly, for the emission of radiation, distinction is made between free–bound transitions in the field of a ion (recombination radiation),

$$\text{A}^+ + \text{e} \to \text{A} + h\nu,$$

and two types of free–free transitions: bremsstrahlung in the field of the ion,

$$\text{A}^+ + \text{e}(mv_1^2/2) \to \text{A}^+ + \text{e}(mv_2^2/2) + h\nu, \quad h\nu = mv_1^2/2 - mv_2^2/2,$$

and in the field of the atom,

$$\text{A} + \text{e}(mv_1^2/2) \to \text{A} + \text{e}(mv_2^2/2) + h\nu.$$

In the case of absorption, a distinction is made between bound–free transitions in the field of ions (phototonization of atoms),

$$\text{A} + h\nu \to \text{A}^+ + \text{e},$$

and free–free transitions in the field of the atom and ion. If an atom has a bound state with an electron, one should add the photodetachment of the electron (in the case of absoption) to the list of processes outlined above:

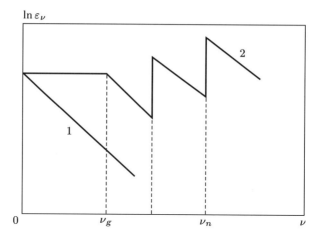

FIG. 8.2. The spectral intensity of continuous (1) bremsstrahlung and (2) recombination spectra.

$$A^- + h\nu \rightarrow A + e,$$

as well as the photoattachment (in the case of radiation)

$$A + e \rightarrow A^- + h\nu.$$

As a rule, the resultant continuous spectra represent the superposition of several continua due to individual processes. The number of such re–superposed continua may be considerable. Hence the determination and, especially, analysis of the resultant spectrum may prove to be a nontrivial problem.

In a plasma with developed ionization, the greatest contribution to the continuous spectrum intensity is made by free–free transitions of electrons in the fields of ions. The absorption coefficient, including the correction for stimulated radiation, is given by Kramers' formula:

$$k_\nu = \frac{2\sqrt{2}e^6 Z^2}{3\sqrt{3}\pi c\hbar m^{3/2}\sqrt{kT}} \frac{g}{\nu^3} n_e n_i \left(1 - e^{-h\nu/kT}\right), \tag{8.5}$$

where Z is the ion charge, and g is the Gaunt factor varying, as a rule, in the 1.1– 2 range. The bremsstrahlung contribution to the total intensity of the continuum prevails at $h\nu \leq 0.7\,kT$ (cf. Fig. 8.2). At higher frequencies, photorecombination radiation is significant.

In photoionization, the absorption coefficient is calculated by summation of the products of the cross–sections of photoabsorption $q_n(\nu)$, by various states n multiplied by the populations of these states, n_n,

$$k_\nu(T) = \sum_n q_n(\nu)n_n(T). \tag{8.6}$$

Photoionization is a threshold process: for the nth state, $q_\nu = 0$ if $\nu < \nu_n$, and $q_n = q_n(\nu)$ if $\nu \geq \nu_n$, where ν_n is the threshold frequency defined by the bonding energy of the nth state (cf. Fig. 8.2). The cross–section of absorption by a hydrogen–like atom in the n–th state is given by Kramers' formula:

$$q_n(\nu) = \frac{64\pi^4}{3\sqrt{3}} \frac{me^{10}Z^4}{ch^6 n^5} \frac{1}{\nu^3}. \tag{8.7}$$

The sum (8.6) can be calculated because the atomic level populations are known in the equilibrium plasma. The total absorption coefficient depends non-monotonically on the frequency. This is because given an increase of frequency, new levels begin to contribute to the sum (8.6) starting from their threshold frequencies. At low frequencies, however, the jumps of k_ν are small and the summation may be replaced with an integration. This leads to the Kramers–Unsöld formula. In doing so, it is convenient to add up the expression for k_ν, which is due to photoionization and the expression (8.5). In this manner, we obtain the summary coefficient of absorption which allows for both bound–free and free–free transitions:

$$k_\nu = \frac{2\sqrt{2}e^6 Z^2 g}{3\sqrt{3}\pi chm^{3/2}(kT)^{1/2}} n_e n_i \frac{1}{\nu^3} e^{h\nu/kT}\left(1 - e^{-h\nu/kT}\right), \quad \nu \leq \nu_g. \tag{8.8}$$

The value of the boundary frequency depends on the bonding energy of the lower excited state of the atom. The photoionization of the latter is included in Eq. (8.8).

In order to describe the plasma of complex atoms and ions, Biberman and Norman (1967) developed an approximate calculation technique. It enables one to derive unified and fairly simple formulas for the calculation of absorption and radiation coefficients in both free–free and bound–free transitions. The function $\xi(\nu, T)$, which is weakly dependent on temperature, is introduced and calculated. It takes into account that complex atoms are not hydrogen–like. The final expressions have the form

$$k_\nu = \frac{2\sqrt{2}e^6 Z^2 g}{3\sqrt{3}\pi chm^{3/2}(kT)^{1/2}} n_e n_i \frac{1}{\nu^3} e^{h\nu/kT}\xi(\nu)\exp[(h\Delta\nu - \Delta I)kT], \quad \nu \leq \nu_g, \tag{8.9}$$

$$k_\nu = k_\nu(\nu_g)\frac{\xi(\nu)}{\xi(\nu_g)}\left(\frac{\nu_g}{\nu}\right)^3 + \sum_{n < n_g} q_n(\nu)n_n(T), \quad \nu \geq \nu_g. \tag{8.10}$$

The factor $\exp[(h\Delta\nu - \Delta I)/kT]$ is due to two effects: the merging of spectral lines near the continuous spectrum boundary and the decrease of the atomic ionization potential ΔI. The latter has been discussed in detail in Chapters 4 and 5 above. The value of the shift in the continuous spectrum boundary, $\Delta\nu$, is discussed in Section 8.3. The sum in (8.10) is calculated in view of all the states that are not included in the integral formula, $h\nu_n > h\nu_g$. This expression defines the frequency ν_g, which is referred to as the boundary. The method of calculation of the function ξ and the absorption coefficient are discussed in detail by D'yachkov (1996) and D'yachkov, Kurilenkov, and Vitel (1998).

Using the equation of ionization equilibria, the absorption coefficient k_ν can be related to the atomic concentration n_a rather than to $n_e n_i$. Instead of Eq. (8.9), we have

$$k_\nu = \frac{32\pi^2 e^6 k}{3\sqrt{3}ch^4} n_a \frac{\Sigma^+}{\Sigma} \frac{Z^2}{\nu^3} T \exp^{(-I/kT)} \exp\left(\frac{h\nu + h\Delta\nu}{kT}\right) \xi(\nu, T), \quad \nu \le \nu_g. \tag{8.11}$$

where Σ^+ and Σ are the statistical sums of ion and atom.

At low degrees of ionization, the continuous spectrum can be defined by free–free transitions of electrons in the field of atoms. The probability of free–free radiative transitions in a certain approximation is related to the cross–section of elastic electron–atom scattering. Zel'dovich and Raizer (1966) have shown that the differential effective radiative cross–section in the $d\nu$ frequency range, upon the collision of an electron having a velocity v with an atom, takes the form

$$dq_\nu = (8e^2 v^2/3c^3 h\nu)qd\nu, \tag{8.12}$$

where q is the transport cross–section of electron–atom scattering.

If q does not depend on the energy of the electron, the absorption coefficient may be estimated by the formula

$$k_\nu = e^2(\pi mc\nu^2)^{-1} n_e n_a \langle q(mv^2/2)v \rangle, \tag{8.13}$$

where $\langle...\rangle$ implies an averaging over the Maxwellian distribution. Bremsstrahlung on atoms predominates over bremsstrahlung on ions if the following condition is satisfied:

$$\frac{1.5 \cdot 10^{13} T^{-5/4}}{q\sqrt{n_a}} \exp\left(-\frac{I}{2T}\right) \ll 1, \tag{8.14}$$

where T is in eV, the cross–section is in 10^{-16} cm^2, and the number density of atoms is in cm^{-3}. Therefore, the intensity of bremsstrahlung on an atom prevails in a weakly ionized plasma.

In the plasma of a number of chemical elements, an important part is played by photoattachment processes and by photodetachment from negative ions in the formation of continuous optical and ultraviolet spectra. We now compare the coefficients of absorption, which are due to photodetachment, k_ν^-, with the electron transitions in the fields of ions. We shall write the first coefficient as $k_\nu^- = n^- q(\nu)$, where n^- is the negative ion concentration and $q(\nu)$ is the photodetachment cross–section. The second coefficient k_ν^+ is given by Eq. (8.11). Thus (D'yachkov and Kobzev 1981),

$$k_\nu^-/k_\nu^+ = \left[3\sqrt{3}ch^4\left(128\pi^3\sqrt{2\pi}e^6m^{3/2}\right)^{-1}\right]n_e(kT)^{-5/2}(h\nu)^3\times$$
$$\times q(\nu)\xi^{-1}\left(g^-/\Sigma^+\right)\exp\left\{[I+E-h(\nu+\Delta\nu)]/kT\right\}, \tag{8.15}$$

where g^- is the statistical weight of the negative ion and E the bonding energy of the electron in the latter. The relation (8.15) is at its maximum near $\nu \simeq 3kT/h$. Upon substituting the characteristic values in Eq. (8.15), we derive

$$k_\nu^-/k_\nu^+ \cong 10^{-26}n_e\sqrt{T}\exp[(I+E)/kT].$$

Hence it follows that in the plasma of elements with the ionization potential $I \geq \mathrm{Ry}$, the continuum due to negative ions contributes greatly to the continuous spectrum.

8.3 The effect of weak nonideality: Spectral line broadening and shift; Phototonization threshold shift

The influence of weak interparticle interaction on the radiative properties of the plasma reduces the latter to the well–known effects of spectral line broadening and shift. The photoionization threshold also shifts to lower energies. In view of the ample literature available on these subjects (Baranger 1962; Griem 1964, 1974; Lisitsa 1977, 2000; D'yachkov 1996, 2000), we shall only discuss them briefly.

In a dense plasma, both broadening and shift are caused by the interaction between a radiating atom or ion and surrounding particles. In doing so, the interaction between radiator and plasma electrons and ions is of great importance.

The interaction with neutral atoms which are of the same type as the radiating system may lead to resonance broadening if one of the states in the line has an optically allowed transition to the ground state. The interaction with atoms of another type leads to van der Waals broadening. Both types of broadening in a plasma with developed ionization (at degrees of ionization of ≥ 0.01) are less important than the Stark broadening.

For the actual calculation of pressure broadening, use is made of impact (collision) and quasistatic approximations. The quasistatic approximation is applicable when the perturbing particles are moving relatively slowly. Hence the perturbation over the period of time which is of interest to us is practically constant. In the other limiting case, the radiating system does not experience any perturbations most of the time except for short impacts which are well separated in time. This approximation is well applicable for describing the perturbation of an atom by electrons. It is valid when the time lapse between the collisions is much longer than that of collision proper, that is,

$$n_a^{-1/3}v^{-1} \ll (n_e q v)^{-1},$$

where v is the electron velocity and q is the cross–section of electron–atom scattering. For long–range Coulomb interaction, which is responsible for line broadening in the plasma, the collisions are not separated in time. This is especially true for the conditions of nonideality. Therefore, the impact approximation is only valid when the interaction is weak on the whole.

The problem of calculating line broadening through the interaction with ions and electrons in a weakly nonideal plasma reduces to the following (Baranger 1962). The electric field generated by ions is assumed to be constant and the Stark broadening is determined for an atom in this field. The broadening of each Stark component is then calculated within the impact approximation. After that, the resultant distribution of intensity is averaged over all the possible values of intensity of the ion microfield.

Collisional broadening by electrons produces the dispersion line contour

$$k_\nu = k_0[1 + ((\nu - \nu_0) - \Delta)^2/(\gamma/2)^2]^{-1}.$$

Here, γ is the line width, and Δ is its shift. The absorption coefficient at the line center is $k_0 = (8\pi)^{-1}\lambda^2(g_2/g_1)A_{21}n_1(2/\pi\gamma)$, where λ is the radiation wavelength and n_1 is the number density of absorbing atoms.

The shift and width are expressed in terms of the amplitude of elastic forward scattering, $f(0)$:

$$\Delta = -\frac{h}{m}n_e \mathrm{Re}\,[f(0)]_{\mathrm{av}}\,, \quad \gamma = \frac{h}{m}n_e \mathrm{Im}\,[f(0)]_{\mathrm{av}} = \frac{1}{2}n_e\,(vq_{\mathrm{tot}})_{\mathrm{av}}\,.$$

The latter expression includes the averaged scattering frequency; q_{tot} is the total scattering cross–section. Averaging is performed over the electron energies and impact parameters. The maximum impact parameter is limited by the screening distance. This is important from the standpoint of the shift, but has little effect on the width as long as the transition frequency is much higher than the frequency of the plasma.

The spectral lines play an important part in the spectral and integral characteristics of dense plasma. Figure 8.3 illustrates the spectral absorptivity of air plasma (Kobzev 1984). The broken line indicates the contribution made by continua. As the density increases, the lines broaden considerably and merge to form quasi–continua, thus making a sizable contribution to the characteristics of the integral.

The methods of calculation of the absorption coefficient in lines involve the use of the Holtsmark distribution of ion microfields and the impact approximation for electrons. That is, the effects of strong correlation are ignored while other simplifying elements are employed which correspond to the conditions of weak nonideality. No systematic studies have yet been performed on the effect of strong nonideality on the spectral lines. Therefore, for estimations of boundaries of the domain of applicability of modern methods of calculation it is necessary to take into account results of experiments (Griem 2000).

Figure 8.4. gives the dependence of the shift in the spectral line of xenon $\lambda = 467.1$ nm on the electron number density. Günter, Hess, and Radtke (1985) have carried out measurements in a plasma of pulsed electric discharge and in an adiabatically compressed plasma up to $n_e = 3 \cdot 10^{18}$ cm^{-3}. It is only for the

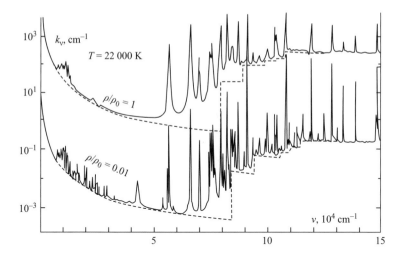

FIG. 8.3. The absorption coefficient of air plasma for $T = 2.2 \cdot 10^4$ K and two values of relative density ρ/ρ_0; ρ_0 is the normal density of air (Kobzev 1984).

maximum values of n_e that one can observe some deviations of the dependence Δn_e from the linear dependence $\Delta \sim n_e$, which is characteristic of an ideal plasma with lines broadened by the quadratic Stark effect. Sechenov and Novikov (1987) have performed measurements for the same line in shock–compressed xenon up to $n_e = 10^{19}$ cm^{-3}, $T = 1.4 \cdot 10^4$ K. These are the maximum parameters reached in xenon, which are discussed below (see Section 8.4).

The interparticle interaction leads to a shift of the boundary of continuous spectrum. This is interpreted as a decrease of the ionization potential. In addition, the highest states of discrete spectrum broaden strongly. In other words, an actual decrease of the ionization potential occurs and, in addition, the broadening of higher levels leads to an apparent decrease of the ionization potential. In spectrograms, the relative importance of these two effects is hard to identify. Therefore, they are combined and one speaks, bearing in mind both effects, of the transformation of lines into a continuous spectrum and of the respective shift in the ionization threshold.

In order to estimate the threshold shift, one usually proceeds from qualitative considerations. The photoionization spectrum shifts toward the red up to the position of the line where the total Stark width is equal to the distance to the nearest line in the series. The Inglis–Teller criterion is based on this principle which gives

$$n^{Z-1} = (1/2)Z^{3/5}(n_i a_0^3)^{-2/15},$$

for the main quantum number of the last line, and for the boundary shift,

$$\Delta E^{Z-1} = Z^2 \text{Ry}/(n^{Z-1})^2 = 4Z^{4/5}(n_i a_0^3)^{4/15}\text{Ry}. \tag{8.16}$$

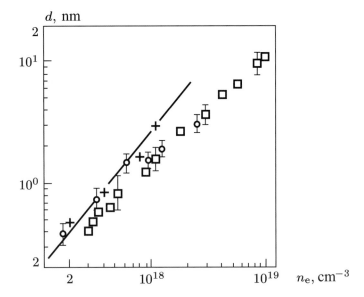

FIG. 8.4. The shift in the xenon line of 467.1 nm. Measurement results: curve, extrapolation of the data of Tsarikis and Eidman 1987, Truong-Bach *et al.* 1981; +, Kettlitz *et al.* 1985; o, Günter *et al.* 1985; □, Sechenov and Novikov 1987.

The models which treat the shift of photoionization threshold as a result of spectral line overlap are sufficiently valid only for the plasma of hydrogen, whose lines are closely arranged and strongly broadened due to the linear Stark effect. Even for ionized helium having a broad but wider (by a factor of four) spaced lines, such a treatment is not valid. This is even more true of the spectral series of complex atoms where the line broadening is much weaker. Therefore, a number of other methods have been proposed to estimate the shift of the photoionization threshold. Armstrong (1964) equated the interlevel energy distance $e^2(a_0 n^{*3})^{-1}$ with the level width given by the asymptotic Baranger formula which describes broadening by electron impacts. This yields

$$\omega_{n*} = 8\pi^2 \left(e^2 \alpha a_0^2 n_e\right) \left(me^2/2\pi kT\right)^{1/2} (n^*)^4,$$

where $\alpha = e^2/hc$ and n^* is the effective main quantum number of the level with the bond energy $E_{n^*} = \mathrm{Ry}/(n^{*2})$. The shift is

$$\Delta E = 4.71 \cdot 10^{-6} n_e^{2/7} T^{-1/7}, \tag{8.17}$$

where T and ΔE are expressed in eV, and n_e, in cm^{-3}.

In the spectra of complex atoms, the lines of different spectral series overlap. The spectra of inert gases of argon, krypton, xenon, and neon have a dense sequence of terms starting from the bond energy of 1 eV. The distance between the levels in this sequence is 0.01–0.05 eV. By introducing some averaged distance

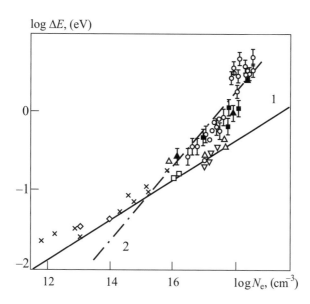

FIG. 8.5. The shift in the photoionization boundary as a function of the electron number density (Vorob'ev 1978). Theory: 1, Inglis–Teller formula; 2, (8.18), $\delta = 0.02$ eV. Experimental data are borrowed from Batenin and Minaev (1975). ΔE is expressed in eV and n_e in cm^{-3}.

among the levels of the dense sequence, δE, one can compare this distance with the width of the given higher level. The levels overlap if their bond energy is less than ΔE, which is defined by the expression (Vorob'ev 1978)

$$\Delta E = 3.5 \cdot 10^{-10} \left(n_e / \sqrt{T_e} \delta E \right)^{1/2}. \tag{8.18}$$

Figure 8.5 gives the values of δE derived in experiments for argon and xenon as a function of n_e with $T_e = 0.8$–1.5 eV.

The behavior of the near–threshold Balmer spectral region is shown in Fig. 8.6. The higher terms in the Balmer series are already absent: they have been transformed into a continuum (the undisturbed threshold is indicated in Fig. 8.6 by the arrow). The shift ΔE increases with n_e.

In conclusion, we emphasize the conditionality of the concept of "optical shift". At any frequency less the threshold frequency, an atom may radiate or absorb a quantum in both bound–bound and bound–free transitions. One can see from Fig. 8.6 that the optical shift, that is, the distance from the shifted photoionization threshold to the first line discernible in the spectrogram, is a conditional quantity dependent on subjective factors.

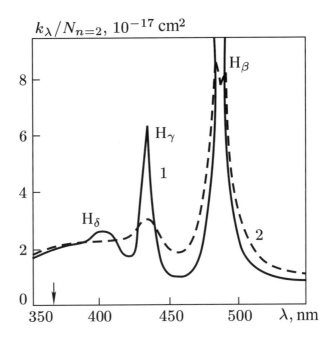

FIG. 8.6.   The absorption coefficient of the Balmer series per single absorbing atom as a function of wavelength λ (Günter *et al.* 1985). Measurement results: 1, $n_e = 1.7 \cdot 10^{17}$ cm^{-3}, $T = 1.6 \cdot 10^4$ K, $\gamma = 0.07$; 2, $n_e = 8.4 \cdot 10^{17}$ cm^{-3}, $T = 2.22 \cdot 10^4$ K, $\gamma = 0.09$.

8.4   The microfield distribution function in nonideal plasma

As a result of spontaneous violation of the electroneutrality of the plasma on scales of the order of the screening distance, microfields which fluctuate in time occur in the plasma (Ecker 1972; Sevastyanenko 1985; D'yachkov 1995). The microfield distribution function is used to calculate the optical properties of line broadening and shift, as well as the shift in photoionization thresholds. Microfields are divided into two groups: low–frequency microfields due to ions and high–frequency microfields due to electrons.

The distribution function for ion microfields is usually calculated within a one–component system of N ions against a uniform neutralizing electron background:

$$P(\mathbf{F}) = \int d\mathbf{R}_1 \ldots d\mathbf{R}_N \delta \left[\mathbf{F} - \sum_{j=1}^{N} \mathbf{F}(\mathbf{R}_j) \right] W(\mathbf{R}_1 \ldots \mathbf{R}_N) / \Omega^N. \qquad (8.19)$$

Here, $\mathbf{F}(\mathbf{R}_j)$ is the field intensity at the sampling point, developed by the jth ion; $\mathbf{R}_j$ are the coordinates of ions; Ω is the system volume, and W is the Gibbs probability defined by the potential energy of the charge interaction.

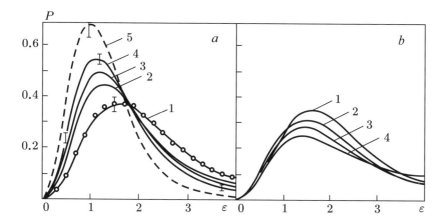

FIG. 8.7.   The ion microfield distribution function $P(\varepsilon)$ at (a) a positively charged point and (b) a neutral point (Kurilenkov and Filinov 1976): 1, Holtsmark distribution; distributions in a nonideal plasma: 2, $\gamma^3 = 0.01$; 3, $\gamma^3 = 0.1$; 4, $\gamma^3 = 1.0$; 5, $\gamma^3 = 5.0$.

The known Holtsmark distribution of ion microfield was derived for systems of uncorrelated charges, in which $W = 1$. In a nonideal plasma, the correlation affects the field distribution. A strong interion repulsion prevents ions from coming close together, thereby reducing the probability of emergence of strong fields. This, in turn, must affect the spectra, by reducing the shift and cutting off the wings of the spectral lines. A number of theoretical approaches were developed (Kurilenkov and Filinov 1976; Tighe and Hooper 1977; Iglesias, Lebowitx, and MacGowan 1983; Dharma-wardana and Perrot 1986) to include the correlation.

Kurilenkov and Filinov (1976, 1980) have calculated the microfield correlation functions at positively charged, neutral and negatively charged points of a plasma with $\gamma = 0.01$–5.0, where $\gamma = e^2(n_{\mathrm{i}} + n_{\mathrm{e}})^{1/3}/kT$. The classical expression (8.19) was generalized to the quantum case using the interaction pseudopotentials which enable one to include the quantum effects when the particles are drawing together (see Section 5.3). The Monte Carlo method was used for further calculations. The microfield distribution functions are given in Fig. 8.7. The correlation indeed leads to a reduction in the probability of emergence of strong fields.

In the experiments of Günter, Hess, and Radtke (1985), Bespalov, Kulish, and Fortov (1986), and Sechenov and Novikov (1987), deviations were observed from the linear dependence of the shift in xenon and argon spectral lines on the electron number density. In xenon, weak deviations were recorded at $n_{\mathrm{e}} \le 3 \cdot 10^{18}$ cm^{-3} (see Fig. 8.4). Sechenov and Novikov (1987) studied an argon spectrum corresponding to $4s - 4p$ transitions at $n_{\mathrm{e}} \le 10^{19}$ cm^{-3}, $T \le 1.7 \cdot 10^4$ K, $p \le 17$ MPa. The generation of the plasma and the recording of its thermodynamic parameters were performed during compression and heating in the shock front. For all lines investigated in the 730–780 nm range, the extrapolation of ideal–plasma calculations to the region of attained parameters showed that the observed shifts

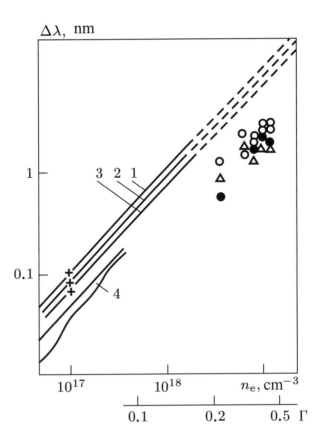

FIG. 8.8.   The shift of some $4s - 4p$ lines of argon as a function of n_e (Sechenov and Novikov
1987): theory: 1–3, for 750.4, 772.4, 763.5 nm (Griem 1974); experiment: +, 750.4, 751.5,
772.4 nm (Jones, Musiol, and Wiese 1983); ○, 750.4, 751.5 nm; □, 738.4 nm; •, 763.5 nm;
△, 772.4 nm (Sechenov and Novikov 1987); 4, experimental points in the 763.5 nm range
(Bober and Tankin 1970).

are four to five times less than the calculated values (Fig. 8.8). It is not improbable that the observed deviations towards reduced shifts reflect the tendency of
microfield reduction with an increase of nonideality.

The destruction of bound states by plasma microfields was investigated beginning with the study by Unzoeld (1948). The decay of the bound state occurs
with the values of field intensity $\mathbf{F}$ exceeding some critical value $\mathbf{F}^*$. The proba-
bility of the existence of this bound level is $\omega = \int_0^{\varepsilon^*} P(\varepsilon)\, d\varepsilon$, where $\varepsilon = \mathbf{F}/\mathbf{F}_0$. The
probability of decay of a given state is $(1 - \omega)$. Sevast'yanenko (1985) calculated
these integrals and used them to determine the populations of bound states,

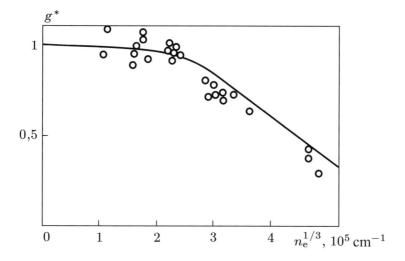

FIG. 8.9. Variation of the effective partition function of the upper level of $5d - 4p$ transition in argon ($\lambda = 560$ nm): points, experiment of Gündel (1971); curve, theory of Sevast'yanenko (1985).

$$n_n = n_a \left(g_n^* / \Sigma^* \right) \exp \left(-E_n / kT \right),$$

$$g_n^* = \omega_n g_n, \quad \Sigma^* = \sum_n \omega_n g_n \exp \left(-E_n / kT \right).$$

A simple analytic approximation of these integrals was proposed by D'yachkov (1997, 1998). Figure 8.9 illustrates the comparison of the calculation results of Sevast'yanenko (1985) with those of experiments in argon of Gündel (1971). From the drawing, one can see how important the microfields are. The clearly defined spectral lines lose their intensity. It will be shown below that the intensity of the continuous spectrum adjoining the spectral lines increases accordingly.

8.5 The principle of spectroscopic stability

The principle of spectroscopic stability consists of the conservation of the density of oscillator strengths df/dE of discrete and continuous spectra during the superposition of external perturbations (such as interparticle interactions). Physically, this implies that the interaction perturbs the radiators, but does not annihilate them and change their total number. According to this principle, it is assumed that the effect of interaction on the threshold spectra adds up to transformation of the higher terms of the spectral series to a continuous spectrum. This is in accordance with the unperturbed density of their oscillator strengths. Therefore, the photoionization cross–section is extrapolated to frequencies less than the ideal threshold frequency according to the behavior of df/dE. For higher frequencies, the photoionization cross–section is assumed to be unchanged. The

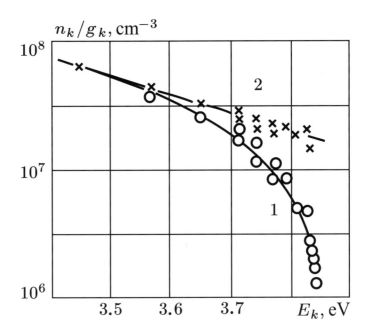

FIG. 8.10.   The measured population of atomic levels in a cesium plasma, $n_e = 9.7 \cdot 10^{13}$ cm^{-3}, $T = 0.274$ eV (Antonov and Popovich 1983): 1, without correction; 2, with correction for line broadening and slit width.

validity of this procedure is supported by a large body of available experimental data on weakly nonideal plasmas.

The effect of "line attenuation" due to the static destruction of radiating levels by plasma microfields (Fig. 8.9) does not contradict the principle of spectroscopic stability. A continuous background emerges instead of the levels being destroyed. The amplification of the background compensates for the attenuation of lines. Figure 8.10 gives two distributions of excited atoms over levels measured in a pulsed discharge in cesium vapors. Curve 1 demonstrates the manifestation of line attenuation which is analogous to that demonstrated in Fig. 8.9. Curve 2 corresponds to spectroscopic stability. It is constructed with correction to the results of data processing. This is necessary because of the strong broadening of lines and finiteness of the spectral width of monochromator slits as compared with the half–width of the line being recorded.

Kobzev et al. (1977) proposed that, with an increase in density, the spectroscopic stability would be lost and the density of oscillator strengths would cease to persist. The latter must show up in observations. At the same time, it was proved by Höhe and Zimmermann (1982); D'yachkov (1986); D'yachkov, Kobzev, and Pankratov, (1987) that this principle could not be violated in a one–particle potential approximation. Deviations from this principle must be associated with the effects of many–particle interaction. Any explicit replacement of the many–particle description by some effective interaction in a one–particle potential approximation leads to spectroscopic stability.

We will follow D'yachkov (1986) and cite the proof of spectroscopic stability within the model with central effective potential. The effective potential is a Coulomb potential with the tail deformed in some manner, which allows for the effect of surrounding particles. We will now treat the averaged density of oscillator strengths of the spectral series $df/dE = f_n dn/dE$.

If the effect of external perturbations in the region of localization of the wavefunction of the ground state is negligible, this approximation appears reasonable for the plasma density $n_e \leq 10^{19}$ cm^{-3} as regards the Lyman and Balmer series. Therefore, in the indicated region (the region of substantial integration in a dipole matrix element), the wavefunction of higher state with the energy E will be written in the form of an ordinary regular Coulomb function, but with some additional normalizing factor $B(E)$, which depends on the behavior of the perturbation of the potential tail. Consequently, the oscillator strengths differ from the respective Coulomb values by the square of this factor: $f_n = f_{nC} B^2(E)$.

As a result of perturbation, the density of states dn/dE will also vary. Therefore,

$$df/dE = (df_C/dE)\, B^2\, (dn/dE)\, (dn_C/dE)^{-1}.$$

It may be readily demonstrated that in a quasiclassical approximation, this yields

$$df/dE = df_C dE. \tag{8.20}$$

Indeed, the normalized quasiclassical wavefunction of the bound state has the form

$$\psi = a p^{-1/2} \cos\left((h/2\pi) \int_{r_1}^{r_2} p\,dr' - \beta \right), \quad a = (2m\Omega/\pi)^{1/2},$$

where p is the quasiclassical momentum, Ω is the frequency of classical periodic motion between the turning points r_1 and r_2, and β is some phase. Consequently, $B^2 = (a/a_C)^2 = \Omega/\Omega_C$. The density of states is the inverse distance between levels ΔE^{-1}. However, the Bohr–Sommerfeld quantum rule may readily yield $\Delta E = \hbar\Omega$. Therefore, $dn/dE = (\hbar\Omega)^{-1}$ and, consequently, $B^2(dn/dE)(dn_C/dE)^{-1} = 1$. As a result, we derive relation (8.20) that is indicative of the spectroscopic stability of spectral series.

The simplest way of extending the derived result to cover a continuous spectrum is by placing an atom into a spherical box of large radius. Relation (8.20) will then automatically extend to cover the region above the photoionization threshold.

As a consequence of the unitary value of the transformation of the final and initial states (Levinson and Nikitin 1962), the principle of spectroscopic stability was proven by D'yachkov (1986) for the microfield model, that is, for an atom in a constant uniform (on the atomic scale) external field.

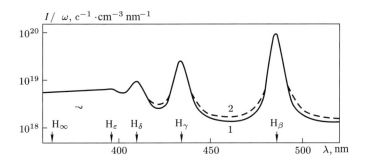

FIG. 8.11.   The radiation spectrum of a hydrogen plasma, $n_e = 9.3 \cdot 10^{16}$ cm^{-3}, $T = 1.41 \cdot 10^4$
K (D'yachkov *et al.* 1987): 1, experiment (Wiese, Kellecher, and Paquette 1972); 2, theory
(D'yachkov *et al.* 1987).

D'yachkov *et al.* (1987) have calculated the threshold spectra of hydrogen
within the microfield model without dividing them into the discrete and continuous components, which is the usual practice, and without assuming the spectroscopic stability. A spectrum is represented in the form of a single continuum
with the Stark resonances forming lines in the latter. Therefore, the spectroscopic
stability in the microfield model is derived by calculation rather than entry into
the latter. Figure 8.11 provides a comparison of experiment with theory.

8.6   Continuous spectra of strongly nonideal plasma

The effect of weak nonideality on a continuous spectrum is accounted within the
ideology of spectroscopic stability. It has been revealed by Radtke and Günter
(1986), Radtke *et al.* (1986, 1988), and Gavrilova *et al.* (1997) that no other
manifestations of nonideality are observed in hydrogen at $n_e \leq 7 \cdot 10^{17}$ cm^{-3}
(Fig. 8.12). The Coulomb density of oscillator strengths agrees well with the
"experimental" value. The latter was derived from the results of measurement
of the absorption coefficient of the Balmer continuum of a pulsed gas discharge.

In a series of studies, however, the measured value of the absorption coefficient in continuous spectra of argon and xenon plasma at frequencies above
the threshold is less than the value calculated by the formulas in Section 8.2
(Kobzev *et al.* 1977). Apparently, these effects were first revealed by Andreev
and Gavrilov (1970) and Andreev (1975) at $n_e \leq 10^{18}$ in xenon. Results of more
recent measurements of the absorption index of xenon, carried out with a three–
stage pneumatic shock tube, are shown in Fig. 8.13. They are compared with
calculation for the ideal plasma. One can see that the theory reproduces the
effect recorded in the experiment.

With the purpose of revealing the effects of plasma nonideality, a wide and
continuously varying range of parameters in argon, from $n_e \leq 10^{18}$ cm^{-3} and
$\Gamma \leq 0.3$ to extremely high parameters $n_e \leq 2 \cdot 10^{20}$ cm^{-3} and $p \sim 0.5$ GPa, where
$\Gamma = (1–6)$, was realized by Bespalov, Gryaznov, and Fortov (1979). For these parameters strong nonideality could result in qualitative changes of the radiation

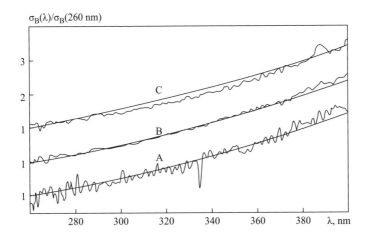

σ_B(λ)/σ_B(260 nm)

FIG. 8.12.   The density of oscillator strengths of the Balmer continuum, $\sigma_B(\lambda)$, normalized to the wavelength 260 nm (Radtke and Günter 1986). The experimental curves reflect the presence of noise in the plasma, and the theoretical curves are smooth. The plasma parameters have the following meanings: A, $T = 1.5 \cdot 10^4$ K, $p = 26$ kPa, $n_e = 5 \cdot 10^{16}$ cm^{-3}; B, $T = 2.15 \cdot 10^4$ K, $p = 340$ kPa, $n_e = 5.4 \cdot 10^{17}$ cm^{-3}; C, $T = 1.9 \cdot 10^4$ K, $p = 426$ kPa, $n_e = 7 \cdot 10^{17}$ cm^{-3}.

characteristics of the plasma. Note that the initial interpretation of these measurements was revised by the authors (Kulish et al 1995). Higher concentrations of charges were amounted by Mintsev and Zaporogets (1989) in experiments on the reflection of radiation by the shock wave front.

The coefficient of reflection of electromagnetic radiation R of an ideal heavily ionized plasma may be calculated by Frenel's formula:

$$R = \left| \frac{\sqrt{\varepsilon(\omega)} - 1}{\sqrt{\varepsilon(\omega)} + 1} \right|^2, \qquad (8.21)$$

where $\varepsilon = 1 - i4\pi\sigma/\omega$ is the plasma perittivity, and Drude's formula for the electrical conductivity:

$$\sigma = \frac{\omega_{\mathrm{p}}}{4\pi\nu} \frac{1}{1 + i\omega/\nu}, \qquad (8.22)$$

where $\omega_{\mathrm{p}} = \sqrt{4\pi e^2 n_e/m}$ is the plasma frequency and ν is the electron collision frequency.

In the limit $\nu/\omega \to 0$, the full reflection of the electromagnetic radiation at frequencies less the plasma frequency, $\omega \leq \omega_{\mathrm{p}}$, takes place. There comes a point where at a given frequency the electron concentration becomes higher than the "critical" value, $n_c = m\omega^2/4\pi e^2$. Typical dependencies of the coefficient of reflection of the plasma on the electron concentration are shown in Fig. 8.14. The

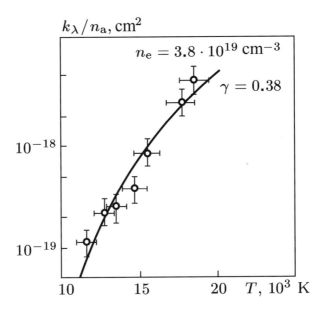

FIG. 8.13. The absorption index of xenon in the continuous spectrum counting, per single atom: points, experimental results at $\lambda = 444$ nm (Sechenov 1981); curve, results of calculation.

steepness of the dependence of reflectivity near the critical concentration is determined by the ratio ν/ω. At low collision frequencies, when a clear step in the dependence of R on n_e is observed, it is possible to determine the electron concentration by means of changing a frequency of the probing radiation. In the strongly coupled plasma the dependence of reflectivity on electron concentration has more complicated character and contains new information about the properties of matter.

The coefficient of reflection of laser radiation with a wavelength of 1.06 μm from a nonideal shock–compressed xenon plasma with electron concentration $n_e \sim (2\text{–}9){\cdot}10^{21}$ cm^{-3}, density $\rho \sim (0.5\text{–}4)$ g cm^{-3}, and temperature $T \geq 3 \cdot 10^4$ K was measured by Fortov, Mintsev, and Zaporogets (1984). These states of matter were obtained by means of compression and irreversible heating of xenon in the explosive generator of rectangular shock waves (Fig. 8.15).

Because of the significant intrinsic heat radiation of the shock–compressed plasma, the probing beam was created by a pulsed (10^{-8} s) yttrium aluminum garnet laser, which gives a high spectral radiance temperature, $T \sim 6{\cdot}10^7$ K, and small angular divergence of the light flux. The laser system was provided with a Pockels' shutter with electronic control, which permits the synchronization of the explosive generation and laser probing to 10^{-8} s. The coefficient of the plasma optical reflection has been determined determined by comparison of the intensities of incident and reflected radiation.

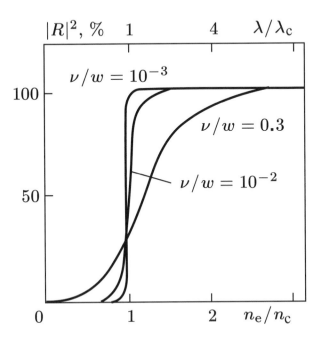

FIG. 8.14.   The coefficient of reflection of an ideal heavily ionized plasma as a function of
the electron concentration at different collision frequencies (Valuev, Mintsev, and Norman
2000).

The analysis of the spatial pattern of the ionizing shock wave has shown,
that three distinctive regions exist, which can influence the propagation of the
electromagnetic wave (Fig. 8.16). The existence of the highly heated plasma
behind the shock wave front results in intensive radiant heat exchange with the
resting gas ahead of the front of the shock wave. This gives rise to its heating
and ionization (region I, "precursor"). Estimations have shown (Valuev et $al.$
2000) that in the condition of the experiments discussed, the "precursor" does
not affect the propagation of the radiation. The structure of the shock wave front
(region II) is determined by relaxation processes in the plasma. Corresponding
estimations have shown that the width of the shock wave front d is about 10^{-5}
cm, order of value less the wavelength λ of the radiation. This gives grounds
for considering that the reflection of the laser radiation is determined by the
properties of the plasma behind the shock wave front (region III). The condition
$d \ll \lambda$ allows one to use Frenel's formula (8.21).

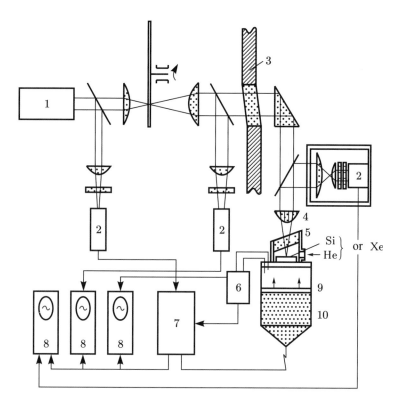

FIG. 8.15. Schematic of the apparatus for measuring the coefficient of reflection of the shock-compressed plasma (Valuev, Mintsev, and Norman 2000): 1, laser; 2, photomultiplier; 3, explosive chamber; 4, ring lens; 5, cavity; 6, impulse generator; 7, delay block; 8, oscillographs; 9, plunger; 10, explosive charge.

For determination of the thermodynamic parameters of the shock–compressed matter, the velocity of the shock wave front, and the run–up velocity of the metal plunger were fixed in each experiment by electro–contact basis method. The measured values of the velocities, together with the conservation laws were used for the determination of the pressure, density, internal energy, and ionization composition of the plasma.

The results of measurements of the reflectivity of xenon by Fortov, Mintsev, and Zaporogets (1984) and Mintsev and Zaporogets (1989) are shown in Fig. 8.17. In the experimental conditions the reflectivity amounts to high values (about 50%), close to values typical for metals. One can see that the step in the dependence of R on n_e, which is characteristic for a weakly nonideal plasma, is absent. The measurement results point to the plasma transition, as the nonideality increases, from a weakly reflecting state (ideal plasma) to a state with high reflectivity (nonideal plasma). The transition is realized in the range of $\gamma \cong 1$ and $\omega/\omega_p \cong 1$, and cannot be described within any single theoretical model. The

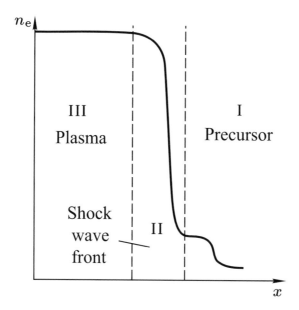

FIG. 8.16.   Scheme showing possible regions of reflection of the laser radiation

curves in Fig. 8.17 are extrapolations to the transition region of the expressions that are valid for an ideal plasma (curve 1 is the extrapolation from the region where $\gamma \ll 1$ and $\omega \ll \omega_p$) and for a strongly nonideal plasma (curve 2 is the extrapolation from the region where $\gamma \gg 1$ and $\omega \gg \omega_p$). The reflection coefficient was calculated by Atrazhev and Iakubov (1989) using Eqs (8.21) and (8.22).

Simple estimations show that the best agreement with the experiment could be attained using Eqs. (8.21) and (8.22), if we accept that $\nu/\omega \approx 2$. Such a high value of this parameter is evidence that the system under investigation is characterized by so strong an interparticle interaction that even the applicability of Drude's formula raises doubts. One can suggest that a significant reorganization of the spectrum is realized. A similar effect takes place in easily boiled metals during the metal–dielectric transition for closed nondimensional parameters of the matter (see Section 8.7).

Ng et al. (1986) measured the reflection of laser radiation from an expanding aluminum plasma formed after the arrival of a powerful laser–induced shock wave to a free surface. Recently the reflection coefficient in shock–compressed deuterium at a pressure 17–50 GPa was measured by Celliers et al. (2000) and analyzed by Collins et al. (2001). In all these experiments, high values of the reflection coefficient were also recorded.

The theory of bremsstrahlung and recombination radiation of a nonideal plasma (Longhvan Kim, Pratt, and Tseng 1985; Totsuju 1985; Artem'ev and Iakubov 1988; Kawakami et al. 1988) allows for the increasing effect made on

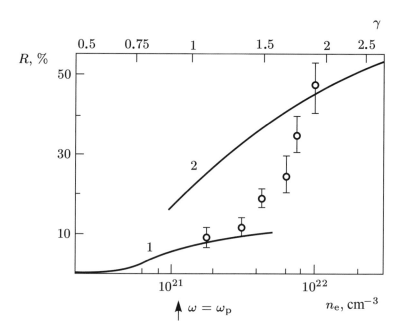

FIG. 8.17. The coefficient of reflection from a nonideal xenon plasma: experiment (Bespalov, Gryaznov, and Fortov 1979; Kulish *et al* 1995): points; theory (Atrazhev and Iakubov 1989): 1, extrapolation from the $\gamma \ll 1$ region; 2, extrapolation from the $\gamma \gg 1$ region.

radiation by the increase in the interion correlation and screening as the nonideality parameter increases. The radiation is suppressed at low frequency, where it becomes less intensive than its value calculated by Kramers' formula. This suppression of the radiation is due to the fact that the portions of electron trajectories, on which low frequencies are radiated, are not realized in nonideal plasma. Two effects are responsible for this: screening and correlation. Artem'ev and Iakubov (1988) have cited the characteristic frequency below which the continuum is suppressed,

$$\omega \leq \omega^* \gamma_{ei}^{3/2}, \tag{8.23}$$

where $\omega^* = (6\pi e^2 n_e/m)^{1/2}$ is a frequency close to the plasma frequency, and γ_{ei} is the parameter of the electron–ion interaction. It is believed that these effects show up in the experiments of Bespalov *et al.* (1979), Fortov *et al.* (1984), and Ng *et al.* (1986).

8.7 The optical properties of low–temperature metal plasma

Strong nonideality may result in qualitative reorganization of the radiation and absorption spectra. Such conditions are realized in the region of transition of some metals from the gaseous plasma state to the fluid-metal state. In the plasma of vapors of mercury and alkali metals, such phenomena occur at fairly low temperatures (close to 2000 K).

The results of measurements of frequency dependencies of the coefficients of optical absorption, as well as reflection for mercury in the "dielectric–metal" transition region, were presented in Chapter 2 (Figure 2.16, where the measurement results demonstrate the variation of the optical radiation reflection coefficient). The dependencies characteristic of a metal have been obtained at $\rho \geq 12$ g cm^{-3}. At low frequencies, the reflection coefficient R reaches approximately 90%. It also decreases with density; this is accompanied by a variation of the frequency dependence. At a density of $\rho \sim 9$ g cm^{-3}, the supercritical mercury metallizes and a maximum value is observed on the $R(\omega)$ curve. This is characteristic of matter with an energy gap. At low and high frequencies, the reflection coefficient drops (Ikezi et $al.$ 1978).

Now, the frequency dependence of the absorption coefficient for mercury, $k(\omega)$, at subcritical densities (Fig. 2.15) bears no resemblance to that characteristic of metal. First of all, one's attention is arrested by the sharp edge of the wide absorption band shifting toward the red with an increase in density. At low frequencies ($\hbar\omega \leq 1$ eV), with an increase in density and a decrease in temperature, the $k(\omega)$ relationship acquires a plateau. We now discuss these two main peculiarities of $k(\omega)$.

An optical transition between two atomic levels in rarefied gas causes absorption in the spectral line. The spectral line in gas is quite narrow. The line width and shift, as well as the frequency dependence of the absorption coefficient, $k(\omega)$, are defined by the interaction processes perturbing the high and low levels.

One can assume that the high–frequency component of the dense mercury spectrum is the edge of a strongly broadened band, which emerges upon the transitions of mercury atoms from the ground $6s$ state to exciton states. The exciton states occur as a result of evolution of the first excited states ($6p$ and $7s$) of mercury atoms upon their interaction with the environment, that is, with fluctuations in the density of atoms in the ground state (Bhatt and Rice 1979). First, these interactions serve to reduce E_{opt}, the minimum excitation energy, to the so–called optical gap. For an isolated atom, this gap is $E_{\text{exc}} = 4.8$ eV. Secondly, the absorption band broadens strongly with an increase in density. This is because as a result of the interaction with radiation, an electron passes to a greater orbit and finds itself in a cluster containing, with a fairly wide scatter, some or other number of atoms. Therefore, the density of bound electron states is defined by the interaction with its own ion and with the atoms of the medium. Following the reasoning in Section 6.3, one can conclude that in a first approximation,

$$\rho(\hbar\omega) \sim \exp[-(\hbar\omega - \Delta E_{\text{opt}})^2/\Delta E_{\text{t}}^2], \qquad (8.24)$$

where ΔE_{t} is the half–width of the absorption band. It is obvious that the ΔE_{opt} and ΔE_{t} values, which are divided by the average number of atoms with which the bound electron interacts, should be of the order of 0.1 eV (such is the scale of the energy of the electron's interaction with a mercury atom). Assuming that

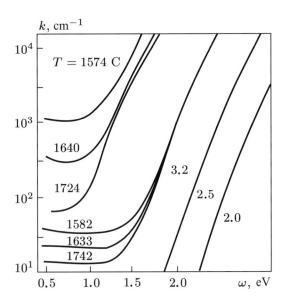

FIG. 8.18. Optical absorption coefficient for mercury (Uchtmann *et al.* 1980), $\rho = 4\,\mathrm{g\,cm^{-3}}$.

$k(\omega) \sim \rho(\hbar\omega)$, Bhatt and Rice (1979) and Popielawski *et al.* (1979) performed an evaluation by fitting Eq. (8.24) to the observed $k(\omega)$ relationship, Fig. 2.16. The ΔE_{opt} and ΔE_{t} values per single atom are estimated at 0.37 eV and 0.18 eV, respectively.

The experimental data enabled Hefner and Hensel (1982) and Overhoff, Ucht-mann, and Hensel (1976) to plot the density dependence of ΔE_{opt} (Fig. 2.13). At low densities, it can be readily extrapolated to the value of the mercury atom excitation energy E_{exc}. It is very important that the ΔE_{opt} gap decreases as the density increases, and closes at $\rho \cong \rho_{\mathrm{c}}$, that is, at the same density as the transport gap ΔE. Therefore, the high–frequency component of the spectrum has found a convincing interpretation.

We now discuss the low–frequency plateau of the absorption coefficient. For a long time, no explanation could be found for the emergence of a plateau although the behavior of the density and temperature dependencies indicated clearly that this was a new optical process taking place in a disordered, dense metallic medium. The results of additional measurements, shown in Fig. 8.18, led Uchtmann, Hensel, and Overhoff (1980) to conclude that the plateau had emerged as a result of radiation absorption by charged clusters. Roughly, the plateau in Fig. 8.18 is the new absorption band. It follows from Fig. 8.18 that $k(\omega)$ increases with an increase in density and a decrease in temperature on the plateau. The same is true of the cluster concentration (cf. Section 4.3). Naturally, this gave an absorption band in the neighborhood of the eigenfrequency of the electron bound in a cluster (Khrapak and Iakubov 1971).

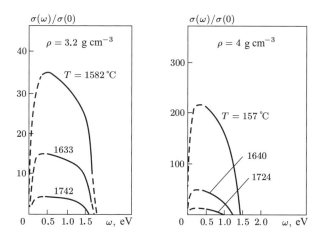

FIG. 8.19. The $\sigma(\omega)/\sigma(0)$ relationship for gaseous dense mercury at different temperatures and densities (Uchtmann *et al.* 1980).

The conclusions by Uchtmann *et al.* (1980) were directly based on the results by Lagar'kov and Sarychev (1978). The latter had calculated $\sigma(\omega)$ using the same techniques employed by them to study the static electrical conductivity $\sigma(0)$ (Chapter 6). The $\sigma(\omega)/\sigma(0)$ relationships obtained are very close to those presented in Fig. 8.19. The results by Lagar'kov and Sarychev (1978) lead one to conclude that a qualitative agreement exists between theory and experiment.

We now consider the dielectric permeability, $\varepsilon(\omega)$, of a nonideal mercury plasma in the region of the "metal–dielectric" transition. The $\varepsilon(\omega)$ relationship can be obtained by processing the experimental transmission and reflection coefficients (Verleur 1968; Hefner, Schmutzler, and Hensel 1980; Kawakami *et al* 1988; Uchtmann *et al* 1988). Verleur (1968) demonstrated the possibilities of a relatively simple approach based on an approximation of the optical properties of matter by a set of damped classical oscillators:

$$\varepsilon(\omega) = \varepsilon_{\mathrm{r}} + i\varepsilon_{\mathrm{i}} = 1 - \frac{\nu_{po}^2}{\omega(\omega + i\Gamma_0)} - \sum_{l \geq 1} \frac{f_l \nu_l^2}{\omega(\omega + i\Gamma_l) - \nu_l^2}. \qquad (8.25)$$

Here, f_l, ν_l, Γ_l denote the oscillator strength, frequency, and bandwidth of the lth oscillator, respectively. The first two terms in the right side of Eq. (8.25) correspond to the Drude model (model of almost free electrons) and must describe the observed values of transmission coefficients in the liquid–metal region. Subsequent oscillator terms will become significant when a gap occurs in the energy spectrum, a peculiarity sometimes referred to as the pseudogap. The values of the parameters ν_{po}, ν_l, Γ_l, f_l are the values selected for each point in the $\{\rho, T\}$ diagram, such that the calculated curves $R(\omega)$ should approximate the experimental curves as closely as possible. It follows from Fig. 2.16 that this can

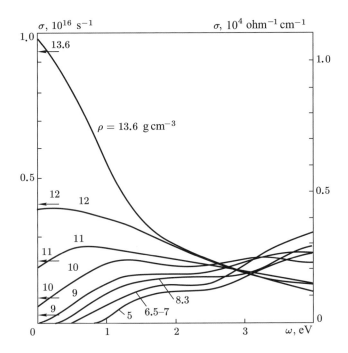

Fig. 8.20. The frequency dependence of the electrical conductivity of mercury obtained from the experimental values of $R(\omega)$ for the same $p - \rho - T$ as in Fig. 2.16.

be done to great advantage by using three resonance terms in Eq. (8.25), $l \leq 3$. Therefore, the values of ν_{po}, ν_l, Γ_l, f_l are found, and it is possible to calculate the $\varepsilon_{\mathrm{r}}(\omega)$ and $\sigma(\omega)$ frequency dependences for the same $\{\rho, T\}$ set. Note, however, that the problem of interpreting the oscillators thus derived remains open. Strictly speaking, the problem of unambiguity of fitting remains open. Still, the fitting accuracy attained appears very convincing.

Figure 8.20 shows the calculated $\sigma(\omega)$ curves. At $\rho \geq 12$ g cm^{-3}, they correspond rather well to almost free electrons (Drude model). However, even at 11 g cm^{-3}, qualitative changes are observed which point to a partial loss of metallic properties. At $\rho < 8$ g cm^{-3}, the quantity ν_{po}^2 disappears altogether and $\sigma(\omega)$ is fully described by the resonance terms. This means that the "metal–nonmetal" transition is already accomplished. The values of $\sigma(\omega)$ are directly related to those of $k(\omega)$. Indeed, it turns out that the $\sigma(\omega)$ values on the isochor of 5 g cm^{-3} presented in Fig. 8.20 agree well with those obtained from $k(\omega)$ in Fig. 2.15. Note further that the static conductivity $\sigma(0) = \nu_{po}^2/(4\pi\nu_0)$ agrees well with the results of static measurements (shown by arrows in Fig. 8.20). In this manner, the results of $R(\omega), k(\omega), \sigma(\omega)$ measurements are coordinated with each other.

Perhaps that the most interesting results were obtained at low densities in a

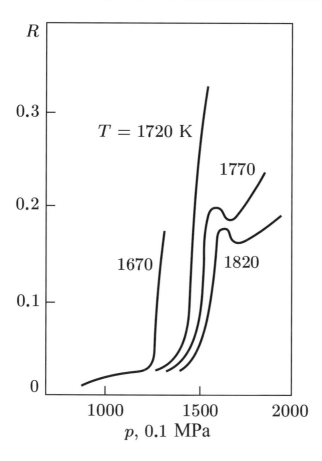

FIG. 8.21. Isotherms of the reflection coefficient for mercury at the wavelength of 650 nm (Yao, Uchtmann, and Hensel 1984).

dense plasma of mercury vapors by Hefner and Hensel (1982), Yao, Uchtmann, and Hensel (1984), and Uchtmann *et al* (1988) (see Fig. 8.21). The reflection coefficient increases anomalously at density values above 3 $g\,cm^{-3}$. The value of ε_r in this (still far from the critical) region increases accordingly, and the $\varepsilon_r(p)$ relationship passes through a maximum. This relationship does not conform to the Clausius–Mosotti formula that points to a smooth and minor rise of ε_r to 1.6. Instead, in a narrow density range, ε_r passes through a maximum to increase approximately fourfold at the density values at which the $k(\omega)$ plateau occurs. The latter leads one to conclude that the dielectric anomaly is likewise related to the transition of charged particles to clusterized states. Moreover, there are good reasons to believe that the clusters are metallized near the saturation line, an assumption made earlier from general considerations. If this is so, the matter is a disordered, two–phase medium where the liquid–metal phase is represented by charge-stabilized, finely dispersed droplets.

References

Andreev, S. I. (1975). Calculation of wall limited pulsed discharges in xenon. *JTP*, **45**, 1010–1118.

Andreev, S. I. and Gavrilov, V. E. (1970). Emissive power of xenon discharge plasma in a quartz tube. *J. Appl. Spectroscopy*, **13**, 988–991.

Antonov, E. E. and Popovich, V. I. (1983). Diagnostics of excited states of alkali metal atoms. In *Abstracts of 6 all–union conference on low–temperature plasma physics*. Leningrad, pp. 272–275.

Armstrong, B. H. (1964). Apparent positions of photoelectric edges and the merging of spectrum lines. *JQSRT*, **4**, 207–214.

Artem'ev, A. A. and Iakubov, I. T. (1988). Radiation of multi–ionized over dense plasma. In *Proceedings of the 14th international symposium on physics of ionized gases*, Konjevic, N., Tanovic, L., and Tanovic, N. (eds). University of Sarajevo, Sarajevo. Contributed papers, pp. 365–368.

Atrazhev, V. M. and Iakubov, I. T. (1989). Reflection coefficient of a short-wave radiation from a nonideal plasma boundary. In *Proceedings of the 19th international conference on phenomena in ionized gases*, Labat, J. M. (ed.). University of Beograd, Beograd. Contributed papers, Vol. 1, pp. 62–63.

Avilova, I. V., Biberman, L. M., Vorob'ev, V. S., Zamalin, V. M., Kobzev, G. A., Lagar'kov, A. N., Mnatsakanyan, A. K., and Norman, G. E. (1970). *Optical properties of hot air*. Nauka, Moscow.

Baranger, M. (1962). Spectral line broadening in plasmas. In *Atomic and molecular processes*, Bates, D. R. (ed.), 493–548. Academic Press, New York.

Batenin, V. M. and Minaev, A. V. (1975). High intensity emission sources. In *Plasma chemistry. Vol. 2*, Smirnov, B. M. (ed.), pp. 199–244. Atomizdat, Moscow.

Bespalov, V. E., Gryaznov, V. K., and Fortov, V. E. (1979). Radiation emitted by a shock–compressed high–pressure argon plasma. *JETP*, **49**, 71–75.

Bespalov, V. E., Kulish, M. I., and Fortov, V. E. (1986). Shift of spectral lines of argon in nonideal plasma. *High Temp.*, **24**, 995–997.

Bhatt, R. N. and Rice, T. M. (1979). Theory of optical absorption in expanded fluid mercury. *Phys. Rev. B*, **20**, 466–475.

Biberman, L. M., and Norman, H. E. (1967). Continuous spectra of atomic gases and plasma. *Sov. Phys. Uspekhi*, **10**, 52–90.

Bober, L. and Tankin, R. S. (1970). Investigation of equilibrium in an argon plasma. *JQSRT*, **10**, 991–1000.

Born, W. and Wolf, E. (1964). *Principles of optics*. Pergamon Press, Oxford.

Celliers, P. M., Collins, G. W., DaSilva, L. B., Gold, D. M., Cauble, R., Wallace, R. J., Foord, M. E. and Hammel, B. A. (2000). Shock–induced transformation of liquid deuterium into a metallic fluid. *Phys. Rev. Lett.*, **84**, 5564–5567.

Collins, L. A., Bickham, S. R., Kress, J. D., Mazevet, S., and Lenosky, T. J. (2001). Dynamical and optical properties of warm dense hydrogen. *Phys. Rev.*

B, **63**, 184110/1–11.

Dharma–wardana, M. W. C. and Perrot, F. (1986). Electric microfield distributions in plasmas of arbitrary degeneracy and density. *Phys. Rev. A*, **33**, 3303–3313.

D'yachkov, L. G. (1986). Oscillator–strength density observation for a spectral series in the single–particle potential approximation. *Opt. Spectrosc.*, **61**, 688–692.

D'yachkov, L. G. (1995). Optical properties of nonideal plasma. In *Transport and optical properties of nonideal plasma*, Kobzev G. A., Iakubov I. T., and Popovich M. M. (eds), pp. 177–213. Plenum Press, New York.

D'yachkov, L. G. (1996). Quasiclassical method of calculation of continuous radiation of low–temperature atomic plasma. *Opt. Spectrosc.*, **81**, 855–862.

D'yachkov, L. G. (1997). Calculation of probability of destruction of bound atom states in plasma microfields. *High Temp.*, **35**, 811–814.

D'yachkov, L. G. (1998). Approximation for the probabilities of the realization of atomic bound states in a plasma. *JQSRT*, **59**, 65–69.

D'yachkov, L. G. (2000). Continuous spectra. In *Encyclopedia of low temperature plasma. Introductory volume 1*, Fortov, V. E. (ed.), pp. 391–400. Nauka, Moscow.

D'yachkov, L. G. and Kobzev, G. A. (1981). Photo–processes with participation of unstable negative ions in atomic plasma. In *Plasma chemistry. Vol. 8*, Smirnov, B.M. (ed.), pp. 122–155. Atomizdat, Moscow.

D'yachkov, L. G., Kobzev, G. A., and Pankratov, P. M. (1987). Transformation of hydrogen line spectrum into continuous spectrum in plasma microfields. *Opt. Spectrosc.*, **63**, 250–255.

D'yachkov, L. G., Kurilenkov, Y. K., and Vitel, Y. (1998). Radiative continua of noble gas plasma. *JQSRT*, **59**, 53–64.

Ecker, G. (1972). *Theory of fully ionized plasma*. Academic Press, New York.

Fortov, B. E., Mintsev, V. B., and Zaporogets, Y. B. (1984). (unpublished).

Gavrilova, T. V., Aver'yanov, V. P., Vitel, I., D'yachkov, L. G., and Kurilenkov, Y. K. (1997). Absorption spectrum of dense hydrogen plasma in the region of Balmer series. *Opt. Spectrosc.*, **82**, 701–708.

Griem, H. R. (1964). *Plasma spectroscopy*. McGraw–Hill, New York.

Griem, H. R. (1974). *Spectral line broadening by plasmas*. Academic Press, New York.

Griem, H. R. (2000). Stark broadening of the hydrogen Balmer–α line in low and high density plasma. *Contrib. Plasma Phys.*, **40**, 46–56.

Gündel, H. (1971). Zustandssumme und Effektive Ionisierungsspannung Eines Atoms im Plasma und Möglichkeiten Ihrer Experimentellen Uberprufung. II. *Beitr. Plasma Phys.*, **11**, 1–12.

Günter, K., Hess, H., and Radtke, R. (1985). (unpublished).

Höhe, F. E. and Zimmermann, R. (1982). Oscillator strengths in dense hydrogen plasma – no "transparency window". *J. Phys. B*, **15**, 2551–2561.

Hefner, W., and Hensel, F. (1982). Dielectric anomaly and the vapor–liquid phase transition in mercury. *Phys. Rev. Lett.*, **48**, 1026–1029.

Hefner, W., Schmutzler, R. W., and Hensel, F. (1980). Optical reflectivity measurements of fluid mercury. *J. de Phys.*, **41**, C8, 62–65.

Iglesias, C. A., Lebowitx, J. L., and MacGowan, D. (1983). Electric microfield distributions in strongly coupled plasmas. *Phys. Rev. A*, **28**, 1667–1672.

Ikezi, H., Schwarzenegger, K., Simons, A., Passner, A. L., and McCall, S. L. (1978). Optical properties of expanded fluid mercury. *Phys. Rev. A*, **18**, 2494–2499.

Jones, D. W., Musiol, K., and Wiese, W. L. (1983). *Spectral line shapes. Vol. 2.* Walter de Gruyter, New York.

Kawakami, R., Mima, K., Totsuju, H., and Yokoyama, Y. (1988). Bremsstrahlung from hot, dense, partially ionized plasmas. *Phys. Rev. A*, **38**, 3618–3627.

Kettlitz, M., Radtke, R., Spanke, R., and Hitzschke, L. (1985). Shift and oscillator strength of Xe(I) lines emitted from an electric pulse discharge. *JQSRT*, **34**, 275–282.

Khrapak, A. G. and Iakubov, I. T. (1971). Electrical conductivity of nonideal low–temperature plasma and its metallization. *High Temp.*, **9**, 1050–1060.

Klein, P. and Meiners, D. (1977). Measurement of width and shift of rare gas lines emitted from a stock–tube plasma. *JQSRT*, **17**, 197–203.

Kobzev, G. A. (1984). *Optical properties of air plasma, $T = 2 \cdot 10^4 – 3 \cdot 10^6$ K.* Preprint IVTAN, No. 10134, Moscow.

Kobzev, G. A., Kurilenkov, Yu. K., and Norman H. E. (1977). Theory of the optical properties of nonideal plasma. *High Temp.*, **15**, 163–166.

Kulish, M. I., Gryaznov, V. K., Kvitov, S. V., Mintsev, V. B., Nikolaev, D. N., Ternovoi, V. Y., Filimonov, A. S., Fortov, V. E., Golubev, A. A., Sharkov, B. Y., Hoffman, D., Stockl, K., and Wetzler, H. (1995). Absorption coefficients of dense argon and xenon plasma. *High Temp.*, **33**, 966–969.

Kurilenkov, Y. K. and Filinov, V. S. (1976). Theory of microfield in nonideal plama. *High Temp.*, **14**, 783–785.

Kurilenkov, Y. K. and Filinov, V. S. (1980). Perturbation theory for microfield distribution in a nonideal plasma. *High Temp.*, **18**, 345–351.

Lagar'kov, A. N. and Sarychev, A. K. (1978). Dynamics of an electron in a random field and the conductivity of a dense mercury plasma. *High Temp.*, **16**, 773–782.

Levinson, I. B. and Nikitin, A. A. (1962). *Handbook on theoretical calculation of line intensity in atomic spectra.* Nauka, Leningrad.

Lisitsa, V. S. (1977). Stark broadening of hydrogen lines in plasmas. *Sov. Phys. Uspekhi*, **20**, 603–630.

Lisitsa, V. S. (2000). Broadening of spectral lines in plasma. In *Encyclopedia of low temperature plasma. Introductory volume 1*, Fortov, V. E. (ed.), pp. 366–376. Nauka, Moscow.

Longhvan Kim, Pratt, R. H., and Tseng, H. H. (1985). Bremsstrahlung spectra for Al, Cs, and Au atoms in high–temperature, high–density plasmas. *Phys.*

Rev. A, **32**, 1693–1702.

Mintsev, V. B. and Zaporogets, Y. B. (1989). Reflectivity of dense plasma. *Contrib. Plasma Phys.*, **29**, 493–501.

Ng, A., Parfeniuk, D., Celliers, P., DaSilva, A., More, R. M. and Lee, Y. T. (1986). Electrical conductivity of a dense plasma. *Phys. Rev. Lett.*, **57**, 1595–1598.

Overhoff, H., Uchtmann, H., and Hensel, F. (1976). Band theoretical study of the optical gap in expanded fluid mercury. *J. Phys. F*, **6**, 523–537.

Popielawski, J., Uchtmann, H., and Hensel, F. (1979). Shape of the absorption–edge in compressed fluid mercury–vapor. *Ber. Bunsenges. Phys. Chem.*, **83**, 123–127.

Radtke, R. and Günter, K. (1986). Study of Balmer spectrum of hydrogen from a high–pressure arc discharge. I. Spectroscopic measurements and plasma analysis. *Beitr. Plasma Phys.*, **26**, 143–150.

Radtke, R., Günter, K., and Spanke, R. (1986). Study of Balmer spectrum of hydrogen from a high–pressure arc discharge. II. Comparison of experiment and theory. *Beitr. Plasma Phys.*, **26**, 151–158.

Radtke, R., Serick, F., Spanke, R., and Zimmermann, R. (1988). On the Balmer recombination continuum in dense hydrogen plasma. *J. Phys. D*, **21**, 535–537.

Raizer, Y. P. (1991). *Gas discharge physics*. Springer, Berlin.

Sechenov, V. A. (1981). An investigation of the optical properties of dense xenon plasma. *Plasma Phys. Rep.*, **7**, 1172–1175.

Sechenov, V. A. and Novikov, V. N. (1987). Spectral line shift of xenon atom at high electron densities. In *Proceedings of the 18th international conference on phenomena in ionized gases*. Swansea. Contributed papers, vol. 1, pp. 242–243.

Sevastyanenko, V. G. (1985). The influence of particles interaction in low–temperature plasma on its composition and optical properties. *Contrib. Plasma Phys.*, **25**, 151–197.

Tighe, R. J. and Hooper, C. F. (1977). Low–frequency electric microfield distributions in a plasma containing multiply–charged ions: Extended calculations. *Phys. Rev. A*, **15**, 1773–1779.

Totsuju, H. (1985). Bremsstrahlung in high–density plasmas. *Phys. Rev. A*, **32**, 3005–3010.

Truong–Bach, Richou, J., Lesage, A., and Miller, M. H. (1981). Stark broadening of neutral xenon. *Phys. Rev. A*, **24**, 2550–2555.

Tsarikis, G. D. and Eidman, K. (1987). An approximate method for calculating Planck and Rosseland mean opacities in hot, dense plasmas. *JQSRT*, **38**, 353–368.

Uchtmann, H., Hensel, F., and Overhoff, H. (1980). D.c. and infrared a.c. conductivity in dense mercury vapours. *Philos. Mag. B*, **42**, 583–586.

Uchtmann, H., Bruisius, U., Yao, M., and Hensel, F. (1988). Optical properties of fluid mercury in liquid–vapor critical region. *Z. Phys. Chem.*, **156**, 151–155.

Unsold, A. (1948). Zur Berechnung der Zustandssummen für Atome und Ionen im einen teilweise ionisierten Gas. *Z. Astrophys*, **24**, 355.

Valuev, A. A., Mintsev, V. B., and Norman, H. E. (2000). Reflectivity of strongly ionized nonideal plasma. In *Encyclopedia of low temperature plasma. Introductory volume 1,* Fortov, V. E. (ed.), pp. 487–490. Nauka, Moscow

Verleur, H. W. (1968). Determination of optical constants from reflectance or transmittance measurements on bulk crystals or thin films.*J. Opt. Soc. Amer.,* **58**, 1356–1364.

Vorob'ev, V. S. (1978). Calculation of the optical properties of plasma at elevated pressures. *High Temp.,* **16**, 391–398.

Wiese, W., Smith, M., and Glennon, B. (1966). *Atomic transitions probabilities.* NBS, Washington.

Wiese, W. L., Kellecher, D. E., and Paquette, D. R. (1972). Detailed study of the Stark broadening of Balmer lines in high–density plasma. *Phys. Rev. A,* **6**, 1132–1141.

Yao, M., Uchtmann, H., and Hensel, F. (1984). 9unpublished).

Zel'dovich, Y. B. and Raizer, Y. P. (1966). *Physics of shock waves and high–temperature hydrodynamic phenomena.* Academic Press, Dover–New York.

9

METALLIZATION OF NONIDEAL PLASMAS

The behavior of a hydrogen plasma under conditions of extreme heating and compression – as well as of any other plasma produced from a dielectric substance – is of general interest for physics. The properties of such plasmas are also of substantial importance for various applications, e.g., in astrophysics and power engineering. For astrophysics, the importance follows from the fact that the cosmological abundance of hydrogen is about 90 at.%. Giant planets, like Jupiter and Saturn, consist mostly of hydrogen, heated and compressed at pressures of dozens of Mbar. Knowledge of the equation of state for hydrogen and its mixture with helium allows us to calculate the inner structure of stars and giant planets – e.g., knowing the magnitude of the metallization pressure, one can estimate the size of the metallic core and, therefore, the magnitude of the magnetic field. The equation of state and physical properties of liquid hydrogen (and its isotopes) is very important for inertial fusion. Stimulated by searching for metallic hydrogen (Wigner and Huntington 1935; Abrikosov 1954; Maksimov and Shilov 1999) in connection with its possible high–temperature superconductivity in a metastable atomic state (Ashcroft 1968), a great deal of effort has been put into the investigation of the dielectric metallization at high pressures. One of the most brilliant ideas about the possibility for hydrogen to form a metastable metallic phase at zero pressure belongs to Brovman, Kagan and Kholas (1971), who showed that strongly anisotropic string–like structures can exist. Although it is rather obvious that the transition to the conductive state should occur in any dielectric, provided the compression is strong enough (see, e.g., the quasiclassical calculations by Kirzhnits *et al.* 1975), the calculations of the transition pressure and density require the most sophisticated methods of band theory to be applied. Such calculations (in some cases supported by experimental measurements) showed that the metallization of dielectrics can be expected in a fairly broad range of pressures – from tens of kbars up to tens of Mbars, whereas the compression of metals, e.g., Ni (McMahan and Albers 1982), Li, Na (Neaton and Ashcroft 1999, 2001; Fortov *et al.* 2001), can be accompanied by the formation of quasi–dielectrics. In fact, properties of strongly compressed matter are much more diverse than one can deduce from extremely simplified quasiclassical models, and the investigations of the electronic structure of hot substances at sub-Mbar pressures is currently one of the "hot topics" in the physics of condensed matter.

Note that in nature not only the metallization of dielectrics is possible, but also the transition of metals into a dielectric state can take place. Implementing the isothermal expansion of low–boiling metals at supercritical pressures (Hensel and Franck 1968; Alekseev and Vedenov 1971), one can continuously pass from a high–conducting "metallic" state to low–conducting gaseous "dielectric" states. It is established that a metal–dielectric transition occurs in a narrow range of densities that are close to (for Cs, Rb, K), or somewhat greater (for Hg) than the density at the critical point (see Chapter 6 for details). For the majority of other metals, which constitute 80% of the elements of the periodic table, critical temperatures and pressures are extremely high and are inaccessible with static experiments. Recent results obtained with fast electric explosion of metallic conductors at supercritical pressures (DeSilva and Katsouros 1998; Saleem et al. 2001) show that the metallic conductivity probably disappears upon significant expansion (a factor of 5 to 7) of solid metals.

In the previous chapters we mostly discussed properties of plasmas created by strong heating up to the temperatures comparable to the ionization potential of atoms ($kT \sim I$). However, the plasma can also be formed at low temperatures by strong compression, when the atomic size r_a becomes comparable to the interparticle distance $n_a^{-1/3}$. The latter case is referred to as "cold" ionization or "pressure–induced" ionization. While the processes accompanying the thermal ionization are well studied (Fortov 2000), investigations of pressure–induced ionization are very complicated, since one has to deal with a "cold" ($kT \ll I$) compression of a plasma up to Mbar pressures and densities that considerably exceed the solid state values. In this regime the electron shells of atoms and molecules overlap and the typical electrical conductivity is comparable to the metallic one. Although Zeldovich and Landau (1944), Mott and Davis (1979), and Hensel and Franck (1968) showed that a metal can be distinguished from a dielectric only by deviations in the electron spectra at $T = 0$, but not by the magnitude of the conductivity, this regime is nevertheless often called "metallization". (For instance, relatively rarefied tokamak plasmas, with $n_e \sim n_i \sim 10^{14}$ cm^{-3} and $kT \sim$5–10 keV, have conductivity close to that of pure copper.) Estimates of the metallization pressure p^* obtained for various substances by employing the methods of the band theory of solids at $T = 0$ yield values in the Mbar range ($p_{H_2}^* \sim 0.3$ TPa (Wigner and Huntington 1935; Abrikosov 1954; Trubitsyn 1966; Ashcroft 1968; Brovman et al. 1971; Maksimov and Shilov 1999), $p_{Xe}^* \sim 0.15$ TPa (Ross and McMahan 1980), $p_{He}^* \sim 11$ TPa (Young et al. 1981), $p_{Ne}^* \sim 134$ TPa (Boettger 1986). Although the static experimental technique of diamond anvils provides pressures as high as $\simeq 0.5$ TPa (Maksimov and Shilov 1999), the metallization with this technique was obtained only recently, in xenon at $p^* = 150$ GPa (Goettel et al. 1089), whereas hydrogen seems to remain a dielectric at pressures $\simeq 340$ GPa.

By using the technique of strong shock waves for the compression and irreversible heating of matter, one can obtain very high pressures (currently, the record value is $p \sim 400$ TPa; Avrorin et al 1993). The pressures are limited only by the intensity of the shock wave generator, whereas for static compression the upper limit is the strength of diamond under static conditions. The viscous dissipation of the flux kinetic energy occurring in the shock wave front along with the compression leads to significant heating, thus stimulating the thermal ionization of a plasma. Kinetics and thermodynamics of such plasmas have been studied in detail both for the ideal and strongly nonideal case. In comparison to the fully developed thermal ionization, the influence of density effects on the ionization equilibrium is not very strong and is described by various models of the reduced ionization potential (see Chapter 4). The thermodynamic states obtained so far in static and dynamic experiments (Hawke *et al.* 1978; Grigor'ev *et al.* 1978; Pavlovskii *et al.* 1987; Holmes *et al.* 1995; Weir *et al.* 1996a; Da Silva *et al.* 1997; Nellis *et al.* 1998; Fortov *et al.* 1999a; Ternovoi *et al.* 1999; Mostovych *et al.* 2000; Mochalov and Kuznetsov 2001; Knudson *et al.* 2001; Belov *et al.* 2002) are shown in the phase diagram of molecular hydrogen in Fig. 9.1. Also, theoretical estimates for plasma phase transitions are given along with the corresponding critical points (Robnik and Kundt 1983; Haronska *et al.* 1987; Saumon and Chabrier 1989, 1992; Beule *et al.* 1999; Nellis 2000; Mulenko *et al.* 2001).

In order to separate the density and thermal effects of ionization, one should naturally try to suppress the effects of irreversible heating by employing a quasi–isentropic compression. For this purpose, in the work by Fortov *et al.* (2003) the compression was produced by a sequence of direct and reflected shock waves, due to the reverberation in planar and cylindrical geometry. The explosive devices of the end–face and cylindrical throwing were implemented to generate shock waves (see Chapter 3). By using processes of multiple shock compression, one can reduce the heating by an order of magnitude and increase the plasma compression by a factor of ten, as compared to what direct wave compression gives. Also, in experiments with H_2, He, Xe, Kr, and some other substances, the plasma conductivity rose by five orders of magnitude in a narrow density range, which is peculiar to the regime of "cold" ionization.

9.1 Multiple shock wave compression of condensed dielectrics

9.1.1 *Planar geometry*

A typical sketch of experiments to implement multiple shock wave compression of condensed hydrogen and inert gases in a planar geometry is shown in Fig. 9.2 (Fortov *et al.* 1999a, 2003; Ternovoi *et al.* 1999; Fortov *et al.* 2003; Mintsev *et al.* 2000; Ternovoi *et al.* 2002). Shock waves were generated by the impact of a steel impactor 2 (of 1–3 mm thickness and 30–40 mm diameter). It was accelerated by detonation products of an explosive (hexogen) 1 to velocities of 3–8 km s^{-1}, by employing the "gradient cumulation" effect (Ternovoi 1980). Explosive throwing devices developed for these experiments provided a diameter 15–30 mm of the flat part of the impactor by the moment when it hit the bottom of the experimental assembly. The absence of melting and evaporation of the impactor material as well as the absence of mechanical fracture of the impactor during acceleration was tested in a series of dedicated experiments. The transition of a shock wave from a metallic screen 3 (of 1–1.5 mm thickness) to the studied substance 4 (of the initial thickness of 1 to 5 mm) generated the

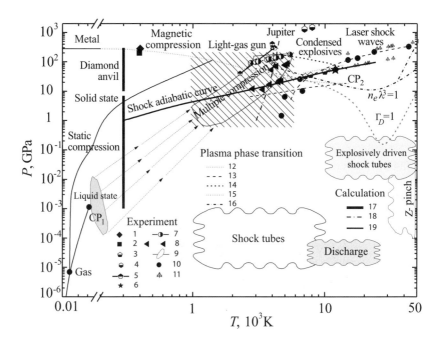

FIG. 9.1. Phase diagram of hydrogen. Experimental data: 1 and 2, magnetic compression (Hawke *et al.* 1978; Pavlovskii *et al.* 1987); 3, Z–pinch (Knudson *et al.* 2001); 4 and 5, cylindrical compression (Grigor'ev *et al.* 1978; Mochalov and Kuznetsov 2001); 6, spherical compression (Belov *et al.* 2002); 7 and 8, single and multiple compression in a light–gas gun (Holmes *et al.* 1995; Weir *et al.* 1996a; Nellis *et al.* 1998); 9, multiple shock compression (Fortov *et al.* 1999a; Ternovoi *et al.* 1999); 10 and 11, shock compression by laser (Da Silva *et al.* 1997; Mostovych *et al.* 2000). Estimates for the plasma phase transition: 12, Beule *et al.* (1999); 13, Robnik and Kundt (1983); 14, Saumon and Chabrier (1989, 1992); 15, Haronska *et al.* (1987), 16, Mulenko *et al.* (2001). Experimental and simulation data: 17, compression in diamond anvils (DA) (Maksimov and Shilov 1999); 18, Jupiter atmosphere (Nellis 2000); 19, shock adiabat for hydrogen (Da Silva *et al.* 1997).

first shock wave with pressure $p_1 = 2$–80 GPa. The wave, being reflected from a transparent sapphire window 5 (of 4–5 mm thickness and 20 mm diameter) excited a shock of secondary compression. Further re-reflection of shock waves between screen 3 and window 5 led to multiple shock compression of the sample up to maximum pressures of $p \sim 100$–200 GPa. This magnitude was determined by the velocity of the impinging impactor, its thickness, and sizes of the studied substance.

The initial states of the substances for a further multiple compression corresponded either to the gaseous region of the phase diagram, at pressures $p_0 = 5$–35 MPa and temperatures $T_0 = 77.4$–300 K, or to the liquid region at $p_0 \sim 0.1$–1 MPa and $T_0 \sim 20.4$–160 K. In the latter case, the liquefaction was performed

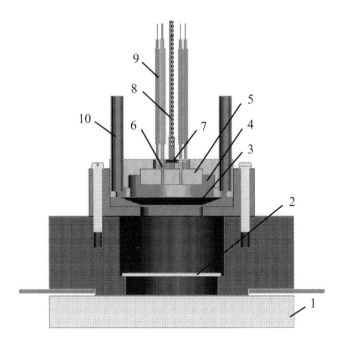

FIG. 9.2. Experimental setup for multiple shock wave compression of condensed hydrogen
and inert gases in planar geometry (Fortov *et al.* 2003). 1, explosive; 2, steel plate; 3,
bottom part of the assembly; 4, studied substance; 5, leucosapphire window; 6, iridium
electrodes; 7, shunt resistance; 8, quartz–quartz light guide; 9, coaxial electric cables; 10,
gas supply pipes.

from high–purity gases supplied to the setup through pipes 10. A two–contour
system of cooling was used: To liquefy hydrogen, the external contour was filled
with nitrogen, whereas for xenon the internal contour was filled with ethanol.
The temperature in the assembly was monitored by thermocouples and platinum
resistance thermometers.

The process of multiple compression was observed by means of fast optoelec-
tronic converters and a five–channel fiber–optic–coupled pyrometer 8 with the
time resolution of 2–5 ns. Since the shock compressed sapphire of optic window
5 retained transparency up to 20 GPa and, hence, made it possible to record the
moments of the shock wave reflections from its surface at higher pressures, and
also, since its insulating properties were at an acceptable level under compression
up to $\simeq 220$ GPa (Weir *et al.* 1996b), five to six reverberations of shock waves
could be detected by measuring the conductivity of the compressed layer and
the optical radiation. The initial stages of the compression (up to 20 GPa) were
recorded in individual experiments by using a VIZAR differential laser interferometer (Barker *et al.* 1986).

In this experimental scheme, the compression and irreversible heating of the substance were performed by a series of shock waves produced upon successive reflections from the sapphire window and the steel screen. A hydrodynamic analysis of the process suggested that, after the propagation of the first two shocks through the compressed layer, the further compression proceeded quasi–isentropically. This made it possible to achieve higher densities ($\rho/\rho_0 \sim 10$–100) in comparison to the case of a single wave compression and to reduce the final temperature, thus increasing the effects of the interparticle interaction. The shock reverberation is clearly revealed by distinct "steps" seen in oscillograms of radiation and electrical conductivity. The measured moments of the shock arrival at the plasma boundaries allow us to determine independently the thermodynamic parameters of the shock compression, p, ρ, and E, by using mass, momentum, and energy conservation (see Chapter 3). The data obtained up to pressures of 30–60 GPa for the caloric and thermal equations of state of hydrogen and helium – the latter was chosen as a reference substance – are in agreement both with the "chemical" model of nonideal plasma (Ebeling *et al.* 1991) and with the results given by the semiempirical equation of state of hydrogen (Grigor'ev *et al.* 1978; Juranek *et al.* 1999). At pressures above 60 GPa, however, no reliable data on the thermodynamics of hydrogen were obtained. In that case, the thermodynamic parameters of multiple shock compression at the final stage were calculated on the basis of 1D hydrodynamic codes that employ semiempirical equations of state for hydrogen (Grigor'ev *et al.* 1978; Juranek *et al.* 1999) and the materials used in the assembly (Bushman *et al.* 1992).

The obtained set of gas–dynamic and temperature measurements was used to determine the thermodynamic parameters of shock compression at the initial stages. Also, the measurements were used as input data (along with the velocity of the impactor) in testing 1D and 2D gas–dynamic codes, which were employed – together with semiempirical broad–range equations of state – to determine the values of pressure, density, and temperature after multiple compression. The errors in the p, ρ, and T values obtained with this method were about 5, 10, and 20%, respectively.

The electrical conductivity of the shock–compressed plasma was determined with electrical probes. An electric current through the compressed plasma was supplied via electrodes (6) oriented perpendicular to the shock front. The current flowed along the compressed sample, then arrived at the surface of steel screen (3), and left the compressed region through a grounding electrode. The electric signals were transferred by high–frequency coaxial cables (9) and recorded by multichannel digital oscilloscopes with transmission bandwidth 500 MHz. The two– or three–electrode schemes were implemented to measure the resistance. In the latter case, it was possible to suppress the cophased noises and, thus, to record moments of the shock reflection not only from the optical window but also from the screen. In order to eliminate the breakdown and arc effects during the transmission of the transport current through a plasma, the current density was maintained below 10^4 A cm^{-2}. In a dedicated series of measurements, where the current density was varied within the range of 10^3–10^4 A cm^{-2}, the current–voltage characteristics of the plasma was shown to be linear. Based on these measurements of the plasma–gap resistance, the plasma electrical conductivity was determined by employing numerical and electrostatic simulations of the corresponding electrostatic problem. The resulting accuracy of the measured plasma conductivity was estimated at 20–100%.

9.1.2 Cylindrical geometry

Another experimental setup for shock wave compression (see Fig. 9.3) has been employed by Adamskaya et al. (1987) and Urlin et al. (1992, 1997). A cylindrical explosive 1 (40/60 trotyl/hexogen alloy) with outer diameter of 30 cm was initiated over the outer surface at 640 points. This generated a highly symmetric detonation wave at the inner surface of the explosive (variations in the arrival time were within 100 ns). The arrival of the wave at the inner surface caused the centripetal motion of the cylindrical steel impactor 2 with the initial velocity $\simeq 5$ km/s. The deceleration of the impactor against the metallic surface of the chamber 3 filled with the studied gas at initial pressures of up to 70 MPa generated a converging shock wave. The intensity of the wave increased as it approached the center, because of the geometric cumulation (Zababakhin 1979). Thereafter, successive reflections of the shock wave from the center of symmetry and from the moving inner surface of the chamber caused multiple shock compression. Similar to the case of planar geometry, the compression was close to isentropic.

The evolution of the profiles of the thermodynamic parameters caused by multiple compression was determined from 1D or 2D gas–dynamic calculations. Broad–range semiempirical equations of state for the explosive, materials used in the assembly, and target plasmas were employed in these calculations. In some experiments, the process of cylindrical compression was monitored by measuring the velocity of the impactor by electrocontact and fiber–light optical methods, as well as by using two sources of hard radiation (Pavlovskii et al. 1965) emitting beams crossed at 135° angle. This made it possible to track the dynamics of compression and to test the quality of gas–dynamic calculations and provided additional boundary conditions for the simulation codes. The parameters found for the shock–compressed plasma are: for deuterium, the pressure was 1.25–1.44 TPa at temperatures of 12 500–14 000 K and densities of 2 g cm^{-3}; for xenon, the pressure and density were 200 GPa and 13 g cm^{-3}, respectively.

Brish et al. (1960) measured the electrical conductivity by implementing the classic two–point circuit diagram involving a reference resistance connected in parallel with the resistance of the studied sample. The resistance of hydrogen was determined by using two stainless steel electrodes 6 (of 2 mm diameter) placed on the axis of the setup with a gap of 6.5 mm in between. A high–capacitance capacitor was discharged through the resistance R_{sh} shunting the hydrogen sample. The decrease in the resistance of the compressed hydrogen led to a decrease in the total resistance, which in turn caused the voltage across the

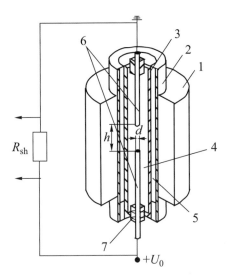

FIG. 9.3. Experimental setup for multiple shock compression of condensed hydrogen and inert gases in cylindrical geometry (Fortov *et al.* 2003). 1, explosive, 2, casing, 3, chamber (steel), 4, hydrogen (gaseous), 5, air ($p_0 = 0.1$ MPa), 6, measuring electrodes, 7, insulator.

measuring electrodes (recorded by an oscilloscope) to change. In these experiments, $R_{sh} = 3$ ohm was used. In order to determine the specific conductivity from the measured resistance values, the actual geometry of the current distribution between the electrodes was taken into account, with the evolving geometry of the electrodes and the profiles of the thermodynamic parameters of hydrogen obtained from hydrodynamic codes. The error in determining the specific conductivity was estimated at 50%.

The typical plasma parameters obtained in experiments with the planar and cylindrical geometry are shown in Table 9.1.

9.1.3 *Light–gas guns*

Metallization of hydrogen and other dielectrics was studied with shock reverberation in light–gas guns (see Chapter 3 for details). A sketch of the experimental cell used by Nellis *et al.* (1999) is shown in Fig. 9.4. A two–stage light–gas gun accelerated impactor 1 made of Al or Cu to supersound velocities. Liquid hydrogen or another liquid dielectric in sapphire anvils 6 was placed in casing 5 made of Al and cooled down to 20 K. Upon collision of the impactor with the casing a strong shock was generated, which propagated through the studied substance and compressed it. The multiple shock reflection from the sapphire anvils was utilized to decrease the temperature of the sample and provide quasi–isentropic heating to temperatures above the melting value. The setup included also electrodes 4 to measure the conductivity of the sample. In addition to hydrogen, also deuterium was used for the investigations. This is because at temperature 20 K the initial

Table 9.1

Substance	Initial state	Final state	p, GPa	ρ, g cm^{-3}	T, 10^3 K	σ, ohm^{-1}cm^{-1}
		Planar geometry				
H$_2$	$p_0 = 25.6$ MPa, $T_0 = 77.4$ K	Maximal compression	227	0.94	5.3	1600
He	$p_0 = 28$ MPa, $T_0 = 77.4$ K	Maximal compression	126	1.37	15	1080
Xe	$p_0 = 0.1$ MPa, $T_0 = 160$ K	Maximal compression	126	10	25	500
		Cylindrical geometry				
H$_2$	$p_0 = 50$ MPa, $T_0 = 293$ K	Maximal compression	1440	2.4	14	550
H$_2$	$p_0 = 70$ MPa, $T_0 = 293$ K	Maximal compression	1250	2	12.5	1100

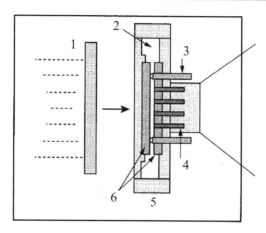

FIG. 9.4. Experimental setup for multiple shock compression of condensed hydrogen and inert gases in light–gas guns (Nellis *et al.* 1999). 1, metal impactor; 2, studied substance; 3, triggering contact; 4, measuring electrodes; 5, aluminum casing; 6, sapphire insulators.

densities of liquid H$_2$ and D$_2$ differ by a factor of two or even more, so that in the shock–compressed substance the values of the density, temperature and conductivity differ substantially as well. The maximum pressure in hydrogen achieved in light–gas guns with shock reverberation was 180 GPa. In accordance to the estimates made by the authors, the corresponding temperatures were relatively low, about 3 000 K.

9.2 Measurements of the electrical conductivity. Model of pressure–induced ionization.

The properties of a hydrogen plasma at high pressures – as well as of plasmas of the hydrogen isotopes – is of significant interest. Hydrogen is the most abundant chemical element in nature which has, in addition, the simples single-electron structure. Table 9.2 summarizes the chronology of hydrogen investigations, starting from its discovery in 1766.

In the experiments described below the multiple shock compression of hydrogen and inert gases is implemented, which makes it possible to obtain new physical information about an unexplored part of the phase diagram, which is depicted in Fig. 9.1. One can see that the region of pressures of up to 1.5 TPa and temperatures of 3000–7000 K was reached by means of dynamic compression. The achieved densities are one order of magnitude higher than those of solid hydrogen and solid inert gases at the triple point, where the mean spacing between protons, $n_a^{-1/3} \sim 1$ Å, is comparable to the typical sizes of molecules (about 0.74 Å) and even isolated atoms in the ground state.

From the physical point of view, this region is interesting because it corresponds to a strong collective interparticle interaction and – when the ionization is high – to a strong Coulomb coupling ($\gamma \sim 10$). The situation is additionally complicated by the fact that the type of statistics changes upon compression – electrons become degenerate and, instead of the temperature, the Fermi energy starts playing the role of the kinetic energy scale. All these features make a theoretical description of strongly nonideal states very complicated, preventing the employment of perturbation theory or parameter–free classical computer MC and MD methods (Zamalin *et al.* 1977) developed for Boltzmann statistics.

Experimental measurements of the electrical conductivity of shock–compressed hydrogen and inert gases are presented in Figs. 9.5–9.9, along with the results obtained on the basis of some theoretical models. Let us first point out some general features in the behavior of the electrical conductivity of strongly nonideal plasmas. The most prominent peculiarity is that, at the final stages of compression the electrical conductivity increases sharply (by three to five orders of magnitude) in a narrow range of "condensed" densities ($\rho \sim 0.3$–1 g cm^{-3} for hydrogen and $\rho \sim 8$–10 g cm^{-3} for xenon) at Mbar pressures, reaching values of about 10^2–10^3 ohm^{-1}cm^{-1}, which is typical for alkali metals. The measurements exhibit a pronounced threshold effect in the density and are therefore in qualitative contradiction with models of weakly nonideal plasmas, which predict a monotonic decrease in the plasma electrical conductivity in response to the isothermal compression (see Chapter 6).

Indeed, at low degrees of the plasma ionization the electrical conductivity is determined by the scattering of electrons on neutrals and is qualitatively described by the Lorentz formula (6.5), according to which the electrical conductivity is proportional to the concentration of free electrons. In turn, the plasma composition is governed by the Saha ionization-equilibrium equation (6.5). Thus, based on the Lorentz and Saha formulas, one can conclude that in weakly ion-

Table 9.2

1766	Cavendish — discovery of "burning gas" – hydrogen
1898	Dewar — liquid and solid H_2 – transparent (not a metal, as expected)
1927	Herzfeld — Clausius–Mossotti "dielectric catastrophe" at $0.6 \, \mathrm{g \, cm^{-3}}$
1935	Wigner and Huntington — metallization at 25 GPa
1954	Abrikosov — metallization at 250 GPa
1968	Ashcroft — high–temperature superconductivity of atomic hydrogen
1971	Kagan — metallization of H_2 at 300 GPa
1972	Kormer — explosive quasi–isentropic compression at 210–400 GPa
1978	Hawke — explosive magnetic compression at 200 GPa, 400 K, $\sigma \sim$ 1 $\mathrm{ohm^{-1}cm^{-1}}$
1980	Mao, Bell, Hemley, and Silvera — static compression in diamond anvils at 10 GPa
1983	Ross — metallization at 300–400 GPa, based on shock–compression data
1987	Pavlovskii — explosive magnetic compression at 100 GPa, 300 K, $\sigma \sim$ 100 $\mathrm{ohm^{-1}cm^{-1}}$
1990	Ashcroft — dissociation/metallization at 300 GPa
1991	Hemley — hcp nonmetallic phase
1993	Silvera, Mao, Hemley, and Ruoff — solid H_2, not a metal up to 250 GPa
1996	Nellis — light–gas gun – multiple shock compression of liquid H_2 at 140 GPa, 2600 K – "metallic" conductivity
1997	Fortov and Ternovoi — explosive multiple shock compression of gaseous and liquid H_2 at 150 GPa, 600 K, $\sigma \sim 10^3 \, \mathrm{ohm^{-1}cm^{-1}}$. Nonideal plasma – pressure–induced ionization
1997	Da Silva, Cauble et al. — laser–induced shocks at 200 GPa, 4500 K – nonideal plasma
2001	Trunin, Fortov et al. — explosion–induced spherical shocks at 100 GPa
2001	Asay, Knudson et al. — shocks in Z–pinch at 100 GPa

ized ideal plasmas the electrical conductivity under the conditions of isothermal compression scales as $\sigma \sim 1/\rho$. Nonideality, which must be taken into account under the discussed conditions, is introduced to the Saha equation via a density–dependent shift of the ionization potential, ΔI. This results in a nonthermal growth of the ionization fraction and, hence, causes the plasma electrical conductivity to increase upon isothermal compression. Then the curve representing the electrical conductivity versus density at $T = \mathrm{const.}$ exhibits a minimum, and its depth decreases with temperature. At $kT \sim I$ the minimum levels out, because the thermal ionization effects become more pronounced than effects associated with pressure–induced ionization (the latter are significant at $kT \ll I$).

As the density and/or temperature increases further, ionization processes described by the Saha formula are completed. Then, instead of the Lorentz formula, one should employ the Spitzer formula (6.8) – when the Boltzmann statistics is applicable, or relation $\sigma \sim n_e/\Lambda$ – in the case of Fermi statistics (where Λ is the

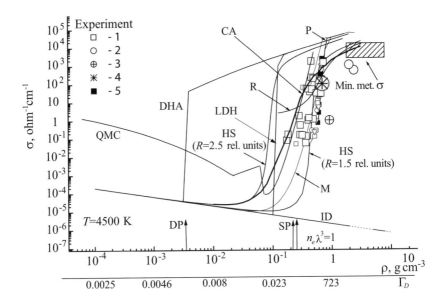

FIG. 9.5.   Electrical conductivity of hydrogen versus density. Experimental data: 1, planar geometry (Fortov *et al.* 2003); 2, cylindrical geometry (Fortov *et al.* 2003); 3 and 4,
magnetic confinement (Hawke *et al.* 1978), and (Pavlovskii *et al.* 1987), respectively; 5,
light–gas guns (Nellis *et al.* 1998, 1999; Weir *et al.* 1996a).

Coulomb logarithm). Thus, at high temperatures the exponential dependence
on the electron density is changed to a weaker, logarithmic or linear, scaling.
To estimate the conductivity in this case, one can use the so–called "minimal
metal" Regel–Ioffe conductivity (7.14), which is widely used in the theory of
simple metals and semiconductors (Seeger 1977),

$$\sigma \sim \frac{e^2}{m} n_e \frac{r_s}{v_T}, \tag{9.1}$$

where r_s is the Wigner–Seitz radius and v_T is the mean thermal velocity of
electrons. The Regel–Ioffe limit is applicable when the electron mean free path
becomes comparable to the distance between atoms.

One can see that the exponential growth of the electron density caused by the
reduction of the ionization potential due to strong interparticle interaction is the
main reason behind the sharp increase in the measured electrical conductivity.
At the same time, the increase of the frequency of electron collisions with atoms
and ions is not important and can be estimated from the standard models discussed in Chapters 6 and 7. It should be emphasized that the pressure–induced
ionization model discussed here leads to an exponential variation of the electrical
conductivity with temperature,

$$\sigma = \sigma_0 \exp(-\Delta(\rho)/2kT), \tag{9.2}$$

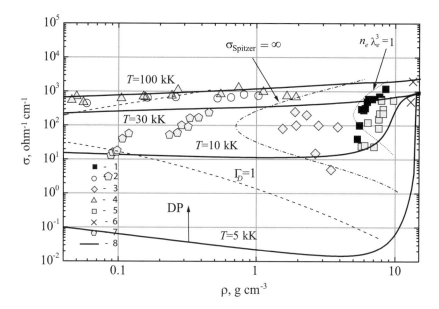

FIG. 9.6.   Electrical conductivity of xenon versus density. Experimental data: 1, Adamskaya
et al. (1987); 2, Ivanov et al. (1976); 3, Mintsev and Fortov (1979); 4, Mintsev et al.
(1980); 5, Mintsev et al. (2000); 6, Eremets et al. (2000); 7, Dudin et al. (1998). Also
shown are lines of electron degeneration ($n_e \lambda_e^3 = 1$) and the constant value of the coupling
parameter, $\Gamma = 1$, where the conductivity calculated from the Spitzer formula diverges.
Line 8 corresponds to the conductivity calculated with the model proposed by Fortov et
al. (2003).

similar to what the semiconductor thermal-excitation model yields (see, for example, Seeger 1977), where the energy gap $\Delta(\rho)$ decreases with density. This
model was used by Ternovoi et al. (1999) and Nellis et al. (1999) to analyze
experiments with light–gas guns. In the pressure range studied, the conductivity
can be well approximated with the following dependence of the energy gap on
density:

$$\Delta(\rho) = 1.22 - 62.6(\rho - 0.30), \qquad (9.3)$$

where Δ is measured in eV, ρ is in $mol\,cm^{-3}$, and the conductivity σ_0 is taken
to be constant and equal to 90 $ohm^{-1}cm^{-1}$. For a density of 0.32 $mol\,cm^{-3}$,
which corresponds to a pressure of 120 GPa, the energy gap becomes equal to
$\simeq 2600$ K. The authors suggest that the metallization of hydrogen ($\Delta = 0$)
occurs at $p = 140$ GPa and $T = 2600$ K. The pressure dependence practically
vanishes at higher p,

Thus, the experimental data on the electrical conductivity obtained by multiple shock compression at $kT \ll I$ provide a unique opportunity to make an
adequate choice of the thermodynamic model for the reduction of the ionization
potential. For example, analysis of the data in Figs. 9.5–9.9 reveals that the

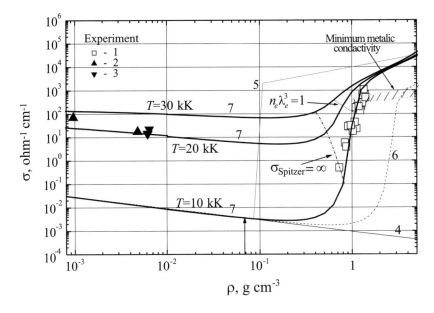

FIG. 9.7.   Electrical conductivity of helium versus density. Experimental data: 1, Ternovoi
et al. (2002); 2, Ivanov *et al.* (1976); 3, Dudin *et al.* (1998). The dotted line corresponds to
the notations in Fig. 9.6. Theoretical curves: 4, conductivity for a composition calculated
from the model of ideal plasma; 5, composition calculated from the Debye–Hückel model; 6,
composition calculated from the bounded atom model, with fixed atomic radius $r_a = 1.3a_0$,
7, Fortov *et al.* (2003).

standard Debye–Hückel model strongly overestimates the effects of the Coulomb
interaction, leading to the pressure–induced ionization at densities that are two
orders of magnitude lower than the experimental values.

9.3   Metallization of dielectrics

The "bounded atom model" by Gryaznov *et al.* (1980) can probably be considered as the most adequate theory of strongly nonideal plasmas. The model
explicitly takes into account the finite size of the phase space for the realization
of the bound states of atoms and ions. We used this theory in the previous chapters to describe the thermodynamics of the shock–compressed inert gases and
cesium. Actually, this is a combination of the Wigner–Seitz solid state model and
the plasma model of the ionization equilibrium. In the framework of this theory,
atoms and ions are treated as rigid spheres, whose thermodynamic functions are
constructed on the basis of MD and MC simulations, whereas the contribution of
the bound electrons is described by the quantum–mechanical Hartree–Fock approximation. Figure 9.10 shows the calculated energy spectrum of a compressed
hydrogen atom versus its radius r_c. In the calculation, the radial component of
the wavefunction satisfies the following boundary conditions:

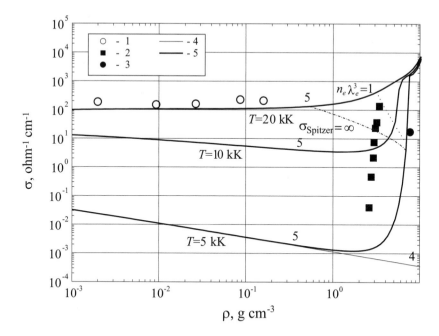

FIG. 9.8.   Electrical conductivity of argon versus density. Experimental data: 1, Ivanov *et al.* (1976); 2, Gatilov *et al.* (1985); 3, Dolotenko *et al.* (1997). The dotted line corresponds to the notation in Fig. 9.6. Theoretical curves: 4, conductivity for a composition calculated from the model of ideal plasma; 5, Fortov *et al.* (2003).

$$f_{nl}(r)|_{r=r_c} = 0, \quad \left.\frac{\partial f_{nl}(r)}{\partial r}\right|_{r=r_c} = 0. \tag{9.4}$$

In the framework of the solid–state model (Zaiman 1972), this corresponds to the upper and lower boundaries of the energy band to which the transition of the corresponding energy level of an isolated ($r_c \to \infty$) atom occurs as the result of compression. In this approach, the width of the forbidden energy gap, ΔE, can be taken equal to the difference between the upper boundary of the ground state band (curve 1s in Fig. 9.10) and the lower boundary of the first excited state band (curve 2p). One can see that ΔE decreases with density (see Fig. 9.11), and its value is in good agreement with the data of Weir *et al.* (1996b) obtained from a direct analysis of experiments on the multiple compression of hydrogen and deuterium. A modification of this model was successfully used by Gryaznov *et al.* (1982, 1998) and Gryaznov and Fortov (1987) to describe the thermodynamics of metal plasmas in the region of high and ultrahigh (up to 400 TPa) pressures.

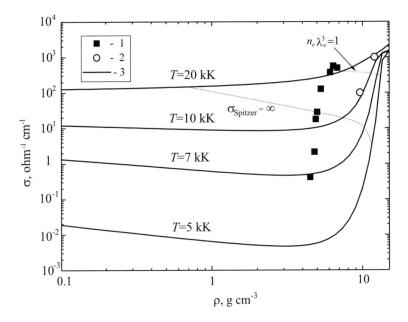

FIG. 9.9.   Electrical conductivity of argon versus density. Experimental data: 1, Glukhodedov *et al.* (1999); 2, Veeser *et al.* (1998). The dotted line corresponds to the notation in
Fig. 9.6. Theoretical curve 3, Fortov *et al.* (2003).

The parameter region investigated in the experiments is characterized by extremely complicated and diversified processes that require appropriate physical
models. First of all, as the plasma is compressed by many orders of magnitude,
the thermodynamic composition is changed and the interparticle interaction becomes strong, including the Coulomb interaction between electrons and ions, the
polarization interaction between charged and neutral particles, and the short–
range interaction between neutral particles. Since the typical interparticle distance in the plasma is comparable to the characteristic size of atoms and ions,
the phase space occupied by them becomes inaccessible to other particles. Therefore, their kinetic energy grows, which provides the corresponding contribution
to the free energy of strongly compressed disordered systems. Moreover, strong
compression causes the energy spectrum of bound states in atomic and molecular systems to change. Also, one has to take into account the changes occurring
in the continuous energy spectrum of electrons – the transition from the Boltzmann to Fermi statistics – because the degeneracy parameter (1.4) changes from
$\xi \simeq 5 \cdot 10^{-2}$ to $\xi \simeq 1.5 \cdot 10^2$ for the conditions of the experiments.

Fortov *et al.* (2003) calculated the plasma thermodynamics in the TPa pressure range as follows:

The free energy of a quasineutral mixture of electrons, ions, atoms and molecules was presented as a superposition of the ideal gas contribution and the term

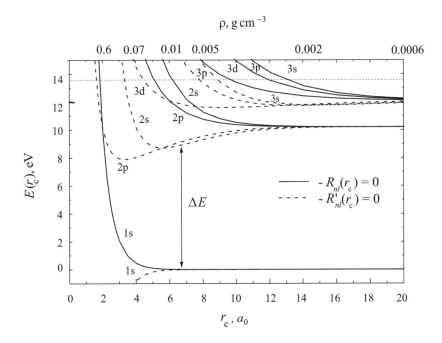

FIG. 9.10. Energy spectrum of the hydrogen atom.

responsible for the interparticle interaction. It was assumed that heavy particles (atoms, ions, molecules) obey Boltzmann statistics, whereas the electrons were treated as an ideal Fermi gas. A version of the pseudopotential model with multiple ionization was employed by Iosilevskii (1980) and Gryaznov *et al.* (2000) to include the Coulomb interaction. The principal point of this model is that the interaction of free charges at short distances deviates from the Coulomb form. This results in a noticeable positive shift not only in the potential energy, but also in the mean kinetic energy of free charges. The electron–ion pseudopotential in the Glauberman–Yukhnovskii form is given by Eq. (5.63). The contribution of the short–range repulsion of molecules, atoms, and ions is described phenomenologically within the soft–sphere approximation (Young 1977) generalized to the case of a multicomponent mixture.

The thermodynamic model of Fortov *et al.* (2003) provides the correct asymptotic behavior at low plasma densities, where it coincides with well–known theories of a rarefied plasma. In the region of extremely high densities, the applicability of the model was tested by comparing with the available experimental data.

9.3.1 *Hydrogen*

In the phase diagram of hydrogen shown in Fig. 9.1, the transition to the metallic state at low temperatures is shown in accordance with the estimates given by

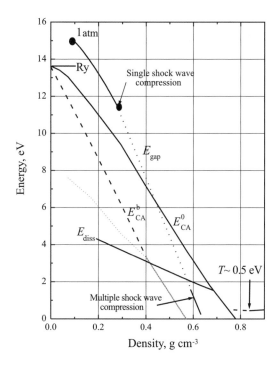

FIG. 9.11.   Energy gap of hydrogen.

Ebeling *et al.* (1991) at a pressure of about 300 GPa. The triple point, at which the metal phase coexists with solid molecular hydrogen and a molecular liquid, is predicted by Ebeling *et al.* 1991 to occur at $p = 100$ GPa and $T = 1500$ K. Both critical points lie in the molecular liquid phase. The first critical point as well as the curve of gas–liquid coexistence are well known and lie in the low–temperature region. The locations of the second critical point and the coexistence curve – which is associated with an abrupt change in the dissociation and degree of ionization and is of major interest – are not known precisely. According to the estimates of Ebeling *et al.* (1991), $T_{c2} = 16\,500$ K, $p_{c2} = 22.8$ GPa, and $\rho_{c2} = 0.13$ g cm^{-3}. Also shown in Fig. 9.1 are the estimates made by other authors (Robnik and Kundt 1983; Haronska *et al.* 1987; Saumon and Chabrier 1989, 1992; Beule *et al.* 1999; Mulenko *et al.* 2001) for the coexistence curve and the critical point for this plasma phase transition (curves 12–16).

    One can see that the parameter regions accessible in experiments with multiple shock compression in planar systems (Fortov *et al.* 1999a, 2003) (region 9) as well as in cylindrical systems (Mostovych *et al.* 2000; Knudson *et al.* 2001; Fortov *et al.* 2003) (curves 4, 5) and in experiments with light–gas guns (Saumon and Chabrier 1989, 1992) (curves 7, 8) partly overlap a rather large region of the possible plasma phase transition. The shock adiabats of singly compressed liquid hydrogen (deuterium) that were obtained in experiments with high–power lasers

(Da Silva *et al.* 1997; Mostovych *et al.* 2000) (curves 10, 11), in a high–current Z–pinch (Knudson *et al.* 2001) (curve 3), and in explosive spherical systems (Belov *et al.* 2002) (stars 6), also locate in this region, but at higher temperatures. In experiments on the isentropic compression by strong magnetic fields in explosive magnetic compression systems (Hawke *et al.* 1978; Pavlovskii *et al.* 1987), temperatures of about 700 K were realized (boxes 1, 2). Pressures of up to 300 GPa were obtained via the isothermal compression of hydrogen ($T \sim 300$ K) in diamond anvils (Maksimov and Shilov 1999) (curve 17). The regions where strong Coulomb interaction and the degeneracy of the electron component are important lie above the curves $\gamma = 1$ and $n_e \lambda_e^3 = 1$, respectively. Curve 19 represents the shock adiabat for liquid hydrogen (Da Silva *et al.* 1997), while curve 18 corresponds to parameters of the atmosphere of Jupiter (Nellis 2000). Figure 9.1 also displays the regions of typical parameters achievable with ordinary (Gaydon and Hurle 1963) and explosive (Mintsev and Fortov 1982) shock tubes in discharges and usual low–current pinches.

For hydrogen, there exists a large "monomolecular" region ($\rho \leq 0.3$ g cm^{-3}), where the thermodynamics is almost entirely determined by H$_2$–H$_2$ interaction. In the framework of the soft–sphere model (Young 1977), the parameters of H$_2$–H$_2$ interaction were chosen by Fortov *et al.* (2003) to be close to those of the rigorous "nonempirical" atom–atom approximation (Yakub 1990, 1999), and the anisotropy of the interaction was neglected. The calculations where the "soft–core" repulsion $V(r) \sim r^{-6}$ was used revealed satisfactory agreement with the molecular part of the $T = 0$ isotherm ("cold curve"), as well as with a considerable part of the shock wave experiments and results of precise MC simulations of the H$_2$+H thermodynamics (Yakub 1990, 1999).

The major issue in applying the chemical model to describe nonideality (including the case of dense hydrogen) is the correct choice of the entire set of effective potentials for the interaction between species. This concerns the interactions involving both charged and neutral particles – first of all, interactions of H$_2$–H and H–H pairs. It is important is that the effective interaction of free atoms differs drastically from the singlet (attractive) and triplet (repulsive) branches given by the rigorous theory for the total potential of H–H interaction. This is because the contribution of H–H pairs interacting via the singlet branch has been already partially taken into account in the intramolecular motion. The same is true for the effective interaction involving (free) charged particles, since the contributions of free and bound states must be consistent in the chemical model. At present, there exist serious contradictions in the suggestions given by different approaches for the form and parameters of the effective potentials. Off the monomolecular region, the major issue is the choice of parameters of the short–range repulsion in H–H and H$_2$–H pairs. We note that, according to Fortov *et al.* (2003), the appropriate choice of parameters for the H$_2$–A$^\pm$ interaction (where A$^\pm$ stands for all charged components) is equally important. One possible choice follows from the nonempirical atom–atom approximation (Yakub 1990, 1999), which leads to relatively large "eigenvolumes" of the hydrogen atom. In terms of the modified soft-sphere model (Young 1977), the obtained results correspond almost exactly to the "additive volume" approximation, with $(D_{H_2})^3 \simeq 2(D_H)^3$, where D is the atom or molecule diameter. For $\rho \leq \rho^* \sim 0.3 \, \mathrm{g\,cm}^{-3}$, the results at $T \leq 10^4$ K are in very good agreement with the precise MC calculations (Yakub 1990, 1999), whereas at $T \geq 10^4$ K they coincide with the nonanomalous part of the results obtained from quantum MC simulations (PIMC, Pierleoni *et al.* 1974). At such temperatures, the data are also in satisfactory agreement with the results of other *ab initio* approaches, including the methods of quantum MD simulations (TBMD, Collins *et al.* 1995, Lenosky *et al.* 1997) and "wave packets" (WPMD, Knaup *et al.* 1999).

Figure 9.12 presents the entire set of currently available experimental data on single shock compression of liquid deuterium. Pressures of up to 25 GPa (circles 1) were achieved by Nellis *et al.* (1983, 1998) and Weir *et al.* (1996a) with direct shock waves generated in light–gas guns. In the experiments of Da Silva *et al.* (1997) and Mostovych *et al.* (2000) on the shock generation with high–power lasers (2, 3), pressures of up to 300 GPa were achieved and an anomalously high compressibility of deuterium was discovered at pressures $p > 40$ GPa. However, more recent experiments performed by Knudson *et al.* (2001) with Z–pinches (4) and by Belov *et al.* (2002) with explosive spherical systems (5) did not confirm this anomaly up to $p \sim 70$ GPa.

The shock adiabats calculated with the SESAME code (Kerley 1972) (curve 6) do not predict this anomaly in the shock compressibility, nor does it emerge in calculations with semiempirical equations of state (Grigor'ev *et al.* 1978). Moreover, the anomaly is not expected from *ab initio* approaches, like the quantum MC method (Pierleoni *et al.* 1974) (curve 7) and the MD method (Collins *et al.* 1995, Lenosky *et al.* 1997). Ross (1998) proposed an interpolation equation of state for deuterium (curve 10), which provides a qualitative description of the experimental results obtained with lasers.

The approach of Fortov *et al.* (2003) also does not reproduce this abrupt change of the compressibility ($\sigma_{max} \equiv \rho_{max}/\rho_0 \simeq 6.5$ versus expected $\sigma_{max} \simeq 4$) in the behavior of the deuterium shock adiabat at $p \simeq 50$–200 GPa (curve 11), nor does it lead at $\rho \geq 1 \, \mathrm{g\,cm}^{-3}$ to anomalies typical of phase transitions.

The thermodynamics of compressed hydrogen (deuterium) exhibits quite different behavior if the H–H (D–D) interaction is calculated with the H–H potential (which is widely used for approximate calculations, Ross *et al.* 1983), and the H_2–H interaction is derived from standard composition rules. In terms of the modified soft sphere model of Fortov *et al.* (2003), this corresponds to a much smaller ratio of H and H_2 "eigen–volumes" ($D_H/D_{H_2} \simeq 0.4 \rightarrow 2V_H/V_{H_2} \simeq 0.13$). Such a choice of the atom "eigen–size" immediately leads to "pressure–induced dissociation" at $\rho \geq 0.3 \, \mathrm{g\,cm}^{-3}$, which is accompanied by a dip in the deuterium shock adiabat (curve 12).

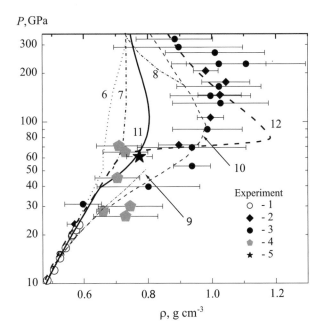

FIG. 9.12.   Shock adiabat of deuterium. Experimental data: 1, Knudson *et al.* (2001); 2,
    Pavlovskii *et al.* (1987), 3, Da Silva *et al.* (1997); 4, Mostovych *et al.* (2000), 5, Grigor'ev
    et al. (1978). Calculated curves: 6, Kerley (1972), 7, Pierleoni *et al.* (1974), 8, Ebeling *et
    al.* (1991), 9, Beule *et al.* (1999), 10, Ross (1998), 11 and 12, Fortov *et al.* (2003).

9.3.2   *Inert gases*

The phase diagram of xenon is shown in Fig. 9.13. In experiments where the
electrical conductivity of xenon was measured under the conditions of multiple
shock compression, densities up to 9.5 $\mathrm{g\,cm^{-3}}$, pressures up to 120 GPa, and
temperatures up to $(5\text{--}20)\cdot 10^3$ K were achieved. The electron number density was
as high as $3\cdot 10^{22}$ $\mathrm{cm^{-3}}$, with the ionization fraction being less than or equal to 0.5.
In the region of maximum parameter values, the plasma was degenerate ($n_e\lambda_e^3 \sim$
50) and strongly nonideal, both for the Coulomb ($\gamma \sim 10$) and interatomic
($\gamma_a \sim 1$) interactions.

    In Fig. 9.13, the phase boundaries of xenon states are depicted according
to Ebeling *et al.* (1991). Metallization of xenon under the conditions of static
compression in diamond anvils was experimentally observed by Goettel *et al.*
(1989), Eremets *et al.* (2000), and Reichlin *et al.* (1989) at densities of about
12.3 $\mathrm{g\,cm^{-3}}$ and pressures in the range 130–150 GPa, which is in agreement with
earlier calculations of Ross and McMahan (1980). In accordance with estimates
of Ebeling *et al.* (1991), the phase boundary for the metal–dielectric transition
intersects the xenon melting line at the triple point (Tp$_2$) with $p \sim 50$ GPa
$T \sim 6000$ K and terminates at the critical point C$_2$ with $p \sim 10$ GPa $T \sim$

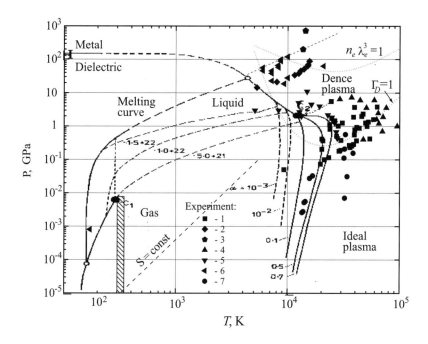

FIG. 9.13. Phase diagram of xenon. Measured equation of state: 1, shock compression of gas (Gryaznov *et al.* 1980), 2, shock compression of liquid (Keeler *et al.* 1965; Nellis *et al.* 1982; Urlin *et al.* 1992), 3, quasistatic compression (Urlin *et al.* 1992). Measured electric conductivity: 4, Mintsev *et al.* (1980); 5, Ivanov *et al.* (1976), Mintsev and Fortov (1979); 6, Mintsev *et al.* (2000). Measured optical properties: 7, Kulish *et al.* (1995).

10 000 K in the plasma region. At high temperatures, the phase transition is accompanied by a sharp change in the concentration of free electrons in a narrow range of plasma densities, which is shown in Fig. 9.13 by the curves of constant ionization fraction. Extensive experimental data from shock wave experiments, where the equation of state (Keeler *et al.* 1965; Gryaznov *et al.* 1980; Urlin *et al.* 1992; Fortov *et al.* 2001), optical properties (Radousky and Ross 1988; Urlin *et al.* 1992; Kulish *et al.* 1995), and electrical conductivity (Ivanov *et al.* 1976, Mintsev and Fortov 1979; Mintsev *et al.* 1980; Urlin *et al.* 1992; Mintsev *et al.* 2000) of the xenon plasma were measured, show no indications of any unusual feature in this parameter range. A sharp increase of the electrical conductivity observed in experiments by Mintsev *et al.* (2000) occurred at somewhat higher densities $\simeq$ 8–10 g cm^{-3}, and pressures $\sim$ 100 GPa. Note that, upon a formal interpolation of the melting curve to the parameter region of interest, some of the experimental points fall into the solid phase part.

Validity of the thermodynamic model of Fortov *et al.* (2003) was tested by comparing the results with experimental data obtained from the shock compression of liquid xenon in light–gas guns (Keeler *et al.* 1965; Nellis *et al.* 1982;

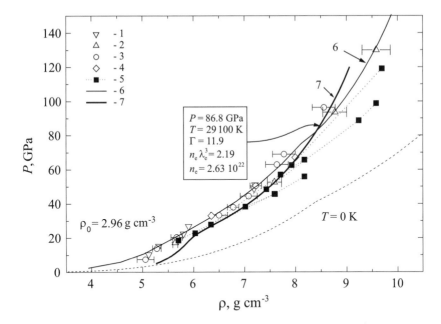

FIG. 9.14.   Shock adiabat for xenon. Experimental data: 1, Keeler *et al.* (1965); 2, Nellis
et al. (1982); 3, Urlin *et al.* (1992); 4, Radousky and Ross (1988); 5, Mintsev *et al.* (2000).
Calculated curves: 6, Urlin *et al.* (1992), 7, Fortov *et al.* (2003), dashed line represents the
"cold curve".

Radousky and Ross 1988) and explosive experiments (Urlin *et al.* 1992). Figure 9.14 shows this along with the results of experiments by Mintsev *et al.*
(2000) on multiple shock compression. In general, results of the Fortov *et al.*
model are in satisfactory agreement with experiments. The discrepancy seen at
low temperatures and pressures can be attributed to an insufficiently accurate
approximation of xenon states in the liquid phase.

The Fortov *et al.* (2003) model provides a fairly good description of the shock
adiabats also for liquid argon and krypton (see Figs. 9.15 and 9.16, where the
experimental data are borrowed from Glukhodedov *et al.* 1999 and Grigor'ev *et
al.* 1985). It is noteworthy that good agreement could also be reached for the
measured values of the brightness temperature and the speed of sound in these
substances.

The situation with the thermodynamic description of helium turns out to be
more complicated, since much fewer experimental data are available and because
the helium phase diagram is quite unusual (see Fig. 9.17) (Ebeling *et al.* 1991).
Metallization of helium at low temperatures is expected to occur at extremely
high pressures of $\simeq 1.1$ TPa. However, plasma phase transitions associated with
a sharp change in the ionization composition must take place at much lower
pressures. Indeed, the estimates performed by Ebeling *et al.* (1991) suggest that

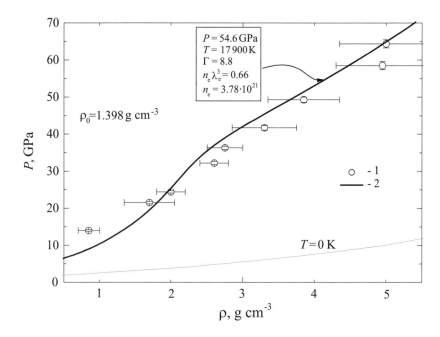

FIG. 9.15.   Shock adiabat for argon. Experimental data: 1, Grigor'ev *et al.* (1985). Calculated curves: 2, Fortov *et al.* (2003); dashed line represents the "cold curve".

the melting of solid helium due to the temperature increase will be accompanied
by the direct transition into a singly ionized plasma state (triple point Tp_3)
followed by the transition into the double ionized state (triple point Tp_4).

The curves of the plasma phase transition terminate at the critical points C_1
and C_2, yet the parameters of the first point ($p \simeq 660$ GPa and $T \simeq 35\,000$ K)
lie very close to the experimentally accessible region. In Fig. 9.17 the bullets and
closed boxes represent the parameters of helium in the incident and reflected
shock, respectively (Nellis *et al.* 1984). The shaded region corresponds to helium
states obtained by Ternovoi *et al.* (2002) in the multiple shock experiments. It
should be emphasized that the thermodynamic model of Fortov *et al.* (2003)
satisfactorily describes the data on the shock compression in the incident and
reflected waves (Nellis *et al.* 1984).

9.3.3 Oxygen

The number of dielectric materials which exhibit the transition to a state with
high "metallic" conductivity at extremely high densities increases every year.
Metallic conductivity and even superconductivity were detected by Desgreniers
et al. (1990) and Shimizu *et al.* (1998) in solid molecular oxygen at very low temperatures and pressures about 100 GPa. Recently, Bastea *et al.* 2001 performed
experiments with shock reverberation in light–gas guns and found a metallic
phase in the liquid molecular oxygen. Multiple shock compression provided a

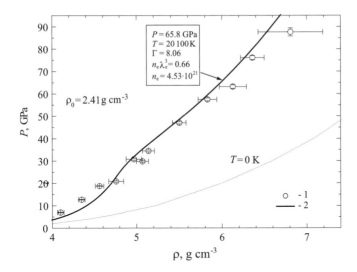

FIG. 9.16. Shock adiabat for krypton. Experimental data: 1, Glukhodedov *et al.* (1999). Calculated curves: 2, Fortov *et al.* (2003), dashed line represents the "cold curve".

density increase by a factor of four, the pressure reached a value of 190 GPa, whereas the temperature did not exceed 7000 K. One can distinguish two main regimes for the pressure (or temperature) dependence of the electrical resistivity shown in Fig. 9.18. At pressures between 30 and 100 GPa there is a rapid drop of the resistivity by six orders of magnitude. Between 100 and 200 GPa the resistivity shows little sensitivity to pressure variations. Similar to the case of hydrogen, the temperature dependence of conductivity can be approximated by Eq. (9.2) with $\sigma_0 = 1205$ ohm^{-1}cm^{-1}. By employing the Regel–Ioffe limit (9.1) for σ_0, one can estimate the magnitude of the electron density in the region of saturated conductivity. This estimate yields the effective number of conductive electrons per molecule of $\simeq 0.1$, whereas for good metals (e.g., Cu) this is $\simeq 0.5$.

The difference observed in the conductivity after single and multiple shock compression (see Fig. 9.18) is mostly related to the corresponding difference in the temperature. For a single compression, the states along the Hugoniot shock correspond to much higher temperatures. This causes dissociation of the molecules and, hence, provides an additional contribution to the conductivity due to the atomic compound. The estimates performed by Bastea *et al.* (2001) yield for $p = 40$ GPa and $T = 1200$ K after multiple shock compression, the dissociation fraction below $10^{-4}\%$, whereas at the Hugoniot adiabat at the same pressure and temperature $T = 6500$ K estimates give the dissociation fraction of about 7%.

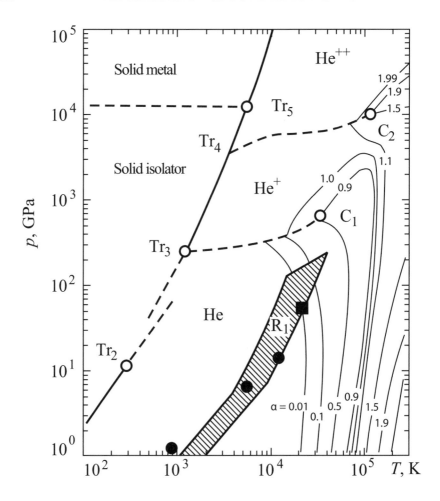

FIG. 9.17.   Phase diagram of helium. Triple points (for single and double ionization) are Tp3 and Tp4. Critical points of plasma phase transitions (for single and double ionization) are C_1, C_2. Shaded region represents the states obtained in experiments with multiple shock compression (Ternovoi et al. 2002).

9.3.4 Sulfur

Sulfur is another substance used to study the dielectric–metal transition. The possibility of the transition was shown by Le Neindre et al. (1976), by extrapolating the shift of the absorption band edge to the high pressure range. Later on, this was verified by direct measurements of the electrical resistivity of the samples under pressure. The investigations were carried out both under static (Vereschagin et al. 1974; Chhabildas and Ruoff 1977; Dunn and Bundy 1977; Evdokimova and Kuzemskaya 1978) and shock wave compression (David and Hamann 1958; Berger et al. 1960; Berg 1964; Yakushev et al. 1974; Nabatov et al. 1979). Naba-

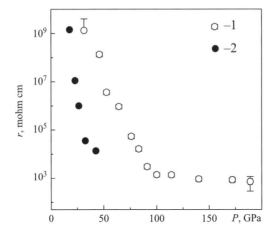

<Fig.> 9.18.   Electrical resistivity of oxygen versus pressure. 1, multiple shock compression (Bastea *et al.* 2001); 2, single shock compression (Hamilton *et al.* 1988).

tov *et al.* (1979) used an experimental setup which provided the detection of the dielectric–metal transitions in the pressure range 100–200 GPa. The sulfur sample, together with the insulating teflon pads, was compressed between two copper screens. An impactor (accelerated by explosion) generated a shock with a rectangular profile in the screen. Since both sulfur and teflon are not as hard as copper, the maximum pressure corresponding to the pressure produced by the impactor on the screen was achieved by the shock reverberation between the screens. The resistivity of the sample was measured by employing the four–point circuit diagram. The shunt resistor was made of manganin foil and placed next to the sample, in order to diminish the parasitic inductance. In a plane parallel to the resistor, the manganin pressure sensor was installed, to be used for the shunt resistance measurements. Since no information about the sulfur equation of state in the high–pressure range was available, the resistivity ρ was calculated assuming that the sample was compressed to the maximum pressure by a single shock. The shock adiabat (Chhabildas and Ruoff 1977) extrapolated up to 100 GPa was used for these estimates. The corresponding values of p and ρ are shown in Fig. 9.19 along with the data obtained by Berg (1964) at lower pressures with a single compression. One can see a fairly steep decrease of resistivity, followed by a transition to (approximately) constant value, $\rho \sim 10^{-2}$ ohm cm. The transition occurs at pressures about 19–20 GPa.

In order to analyze the qualitative influence of temperature on the resistivity, Postnov *et al.* (1986) decreased the temperature of the sample (with respect to the value corresponding to the shock adiabat) by employing multiple shock compression. For pressures below the transition value, ρ grows as the temperature decreases (by about three orders of magnitude), which is typical for semiconductors. Above the transition value, ρ weakly varies with temperature (decreases

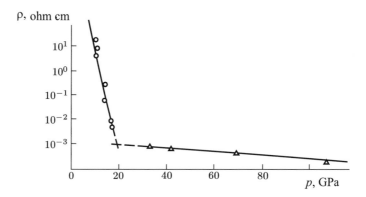

FIG. 9.19.   Resistivity of sulfur versus pressure: △, multiple shock compression (Nabatov *et al.* 1979), ○, single shock compression (Berg 1964).

less than by a factor of two), which is typical for metals. This allows us to conclude that the dielectric–metal transition in sulfur occurs at $p = 19$–20 GPa, with electrons being the major charge carriers.

9.3.5   *Fullerene*

In the crystalline state, fullerene is a semiconductor with an energy gap ΔE of about 2.1 eV and with rather narrow valence and conduction bands (about 0.5 eV each). The C_{60} molecules are mainly bound by the van der Waals forces, and the compressibility of C_{60} crystals is very high. One would expect that the valence and conduction band widths, which exponentially depend on the intermolecular distance, will rapidly increase with pressure. This process should be accompanied by a decrease in the energy gap, and, at some pressure, the crystal may transform to the metal state.

Analysis of the optical absorption spectra measured with C_{60} crystals at high pressure by Meletov *et al.* (1992) and Moshary *et al.* (1992) shows that the energy gap decreases rapidly with pressure. However, different estimates of the pressure corresponding to the transition to the metal state exhibit considerable scatter, from 20 to 70 GPa. Moreover, it is still unclear whether the transition of C_{60} crystals to the metal state is possible before the collapse of the molecules or their polymerization. The broad scatter of experimental data can be caused by several factors: First, high pressures may lead to the polymerization of C_{60} molecules. This is a rather slow process which can be present, to some extent, in all experiments and hence affect the data for $\Delta E(p)$. Second, in most of the hydrostatic experiments a pressure-transmitting medium was used (such as liquid xenon, alcohol mixtures, etc.). Because of the large diameter of C_{60} molecules, the crystal structure is characterized by the presence of big intermolecular voids, which can be easily filled with the molecules of the pressure–transmitting medium. The latter process can strongly affect the electron band structure of the samples.

One can avoid the problems mentioned above by performing specially designed experiments with shock wave compression, which can provide a fairly
smooth quasi–isentropic loading regime with minimal heating (Osip'yan *et al.*
2002). The idea of the method is as follows: The sample is placed between two
plates made of glass or fused quartz. One plate is in contact with a metal screen.
The dynamic loading occurs from the side of the screen, which is hit by a metal
impactor accelerated by explosion products to a velocity of $\simeq 2$ km s^{-1}. Because
of the anomalous compressibility of glass and fused quartz below the elastic limit
and by virtue of the thermodynamic laws governing the transition to the plastic
state in these materials, no shock wave can exist at pressures below $\simeq 12$ GPa
(Kanel' and Molodets 1976). The shock wave is smeared our as it travels from
the boundary between the quartz plate and the metal screen and transforms to a
continuous wave of isentropic compression, which makes it possible to considerably reduce the irreversible shock heating of the sample. As a result, the sample
is loaded smoothly – without a shock – under the conditions of the dynamic
experiment. Earlier, this method had been successfully used by Fortov *et al.*
(1999b, 2001) to study the "dielectrization" of lithium and sodium (see Section
9.3.7).

Using the dynamic compression method described above, the conductivity of
the crystalline fullerene samples with density of 1.67 g cm^{-3} was studied under
pressures of up to 20 GPa. It was found that, when the initial temperature was
293 K, the conductivity of the C$_{60}$ samples reached values of $\sigma \simeq 5$ ohm^{-1}cm^{-1},
whereas for samples cooled to 77 K before the loading, the conductivity ob-
served under a similar dynamic compression was almost two orders of magnitude
smaller, $\sigma \simeq 0.07$ ohm^{-1}cm^{-1}. Note that in the first case, σ increases with pressure by seven to eight orders of magnitude with respect to the initial value. This
kind of temperature dependence is typical for semiconductors and is similar to
that observed in hydrogen, xenon, and oxygen. The results obtained by Osip'yan
et al. (2002) suggest that the energy gap of C$_{60}$ crystals decreases sharply with
pressure. However, the gap does not decrease to zero, so that the samples compressed to a maximum pressure of 20 GPa remain semiconducting. Therefore,
higher pressures are necessary in order to study the possibility of metallization
in fullerene with the preserved molecular structure.

9.3.6 *Water*

Water is one of the most abundant substances on the earth and in the universe.
In particular, along with methane and ammonia it forms the main component
of the inner shells of the giant planets in the solar system, Neptune and Uranus,
occurring in the liquid phase at temperatures up to several thousand degrees
and pressures up to hundreds of GPa. It is quite possible that the electrical conductivity of highly compressed and heated water is responsible for the magnetic
field of these planets (Cavazzoni *et al.* 1999).

The physical properties of water under these extreme conditions have recently
attracted detailed study. In particular, *ab initio* calculations of its phase diagram

using MD simulations by Cavazzoni *et al.* (1999) have shown that at pressures above 30 GPa and temperatures higher than 2000 K, water should be in a superionic state characterized by anomalously high proton mobility. The question of the predicted transition of water to the metallic state under the action of high pressures and temperatures (Cavazzoni *et al.* 1999) by analogy with molecular hydrogen is also fundamentally unresolved.

At present, the only source of experimental information on the properties of liquid water at high pressures and temperatures are experiments using powerful shock waves, which have already yielded unique data on its equation of state up to pressures of around 100 GPa by using explosives as the energy source (Bakanova *et al.* 1975), up to 230 GPa in experiments with light–gas guns (Mitchell and Nellis 1982), and at pressures of approximately 100 GPa (Volkov *et al.* 1980) and 1425 GPa (Podgurets *et al.* 1972) in experiments using underground nuclear explosions.

Results of the first measurements of the electrical conductivity of water beyond a shock wave front were obtained by David and Hamann (1959, 1960) who observed a sharp increase in conductivity as the dynamic pressure increased from 2 to 13 GPa and attributed this to an increase in the degree of dissociation of the water molecules to form ions. At about the same time Brish *et al.* (1960) made independent measurements of the conductivity at a dynamic pressure of 10 GPa and obtained a value of 0.2 $ohm^{-1}cm^{-1}$ in good agreement with the data of David and Hamann (1959, 1960). Somewhat later Hamann and Linton (1966) represented refined experimental values of the conductivity which were obtained at dynamic pressures up to approximately 22 GPa. Measurements of the electrical conductivity in aqueous solutions of KCl, KOH, and HCl in the range 7–13.3 GPa by Hamann and Linton (1969) allowed them to conclude that at pressures above 15–20 GPa the water beyond the shock wave front is almost completely dissociated to form ions and has an electrical conductivity of 10 $ohm^{-1}cm^{-1}$. The ionic nature of the electrical conductivity of dynamically compressed water is also supported by its optical transparency, observed by Zel'dovich *et al.* (1961) at pressures up to 30 GPa, and also by the good agreement between the dynamic and static (Holzapfel and Franck 1966) measurements of the electrical conductivity made at similar temperatures and pressures. Finally, using the two–stage light–gas gun, Mitchell and Nellis (1982) succeeded in making accurate measurements of the electrical conductivity of water at pressures of 25–59 GPa beyond the shock wave front.

Yakushev *et al.* (2000) substantially expanded the range of pressures to measure the electrical conductivity of water. This was accomplished by using a multistage dynamic loading regime in which the sample is exposed to the action of a series of successive shock waves circulating between two plane–parallel plates having a substantially higher dynamic rigidity than that of water. As a result, the water conductivity was measured up to pressures of 130 GPa and a density of 3.2 $g\,cm^{-3}$. The ionic nature of the electrical conductivity of water under these experimental conditions was confirmed by recording the electrochemical

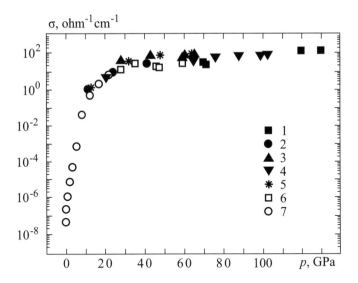

FIG. 9.20.   Electric conductivity of water versus pressure. 1–5, Yakushev *et al.* (2000); 6, Mitchell and Nellis (1982); 7, Hamann and Linton (1966).

potentials.

Figure 9.20 summarizes the experimental results on the electrical conductivity of water. It can be seen that starting from ~ 30 GPa the conductivity depends weakly on pressure and saturates at approximately 150 ohm^{-1}cm^{-1}. Following Hamann and Linton (1966), this can be attributed to the complete dissociation of water. In electrochemical experiments, galvanic cells having electrodes of various metals and water as the electrolyte were subjected to dynamic compression. The characteristics of the recorded e.m.f. of these cells, along with the fact that the measured conductivity is far from a typical metallic value, indicate that the high electrical conductivity of highly compressed water is of an ionic nature.

9.3.7   Dielectrization of metals

The conventional point of view in solid states physics is that the structural phase transitions occur in a solid as the density and pressure increase, thus causing it to form a close–packed phase with maximal coordination number. As a result, insulators become conductors and poor metals improve the metallic properties. Numerous experiments, including those discussed above, seem to support this general picture. However, modern sophisticated quantum–mechanical calculations (Neaton and Ashcroft 1999) predict much reacher and interesting behavior of matter at high pressures. For example, lithium, like other alkali metals, has long been considered as a prototype "simple" metal. At normal conditions, alkali metals have a simple bcc structure, metallic sheen and conductivity. However, theory shows that lithium under pressure transforms into an orthorhombic phase at 50 GPa. At higher pressures near 100 GPa, lithium nuclei form pairs, which results in the formation of structures similar to condensed phases of molecular hydrogen, with the electron properties being close to the properties of semiconductors with narrow energy gap. Finally, at even higher pressures, lithium reverts to a monoatomic metal.

In experiments carried out in diamond anvil cells to a pressure of 60 GPa (Struzhkin et al. 1999; Mori and Ruoff 1999) it was found that the metallic sheen of lithium disappears under compression – it becomes gray and then black (i.e., strongly absorbing) at $p = 50$ GPa. Recent X–ray diffraction studies carried out at pressures up to 50 GPa by Hanfland et al. (2000) have also revealed a sequence of structural transitions. According to these measurements, near 39 GPa lithium transforms from a high–pressure fcc phase through an intermediate rhombohedral phase to a complex bcc phase with 16 atoms per cell. Calculations performed by Hanfland et al. (2000) predicted high stability of this phase up to pressures of about 165 GPa.

In experiments by Fortov et al. (1999b, 2001), direct measurements of electrical resistivity of compressed lithium revealed anomalous behavior of its electrophysical properties. Lithium was compressed in multistep shock experiments up to a pressure of 210 GPa and density of 2.3 g cm^{-3}. The data obtained by Fortov et al. (2001) are shown in Fig. 9.21 in the form of normalized resistivity ρ/ρ_{293} as a function of density. As one can see, the resistivity increases monotonically with density for all experiments corresponding to a maximum pressure of 100 GPa at initial temperatures of 77 K and 293 K. The data obtained at higher pressures (160 and 212 GPa) yield the same dependence in the investigated density range up to 1.75 g cm^{-3}. At higher densities of 2.0–2.3 g cm^{-3}, the resistivity decreases dramatically. Lithium melts under conditions of dynamic experiment in the first or second shock at pressures below 7.3 GPa and temperatures below 530 K, depending on the intensity of the incident shock wave. The final states of dynamically compressed lithium, according to the results of 1D numerical modeling with real equation of state, correspond to the liquid state at temperatures from 955 to 2833 K. The estimated thermal contribution to the lithium resistivity is about 20–25% of the total value at the maximum density. Therefore, the main reason for the change in lithium resistivity is a decrease in the interatomic distances. One should note that the dependence of resistivity on density is similar both in the solid and liquid states. Another interesting and unusual fact is that, under conditions of dynamic experiments, liquid lithium is a poor conductor up to 160 GPa, whereas at higher pressures the resistivity is decreasing. Presumably this suggests that compressed lithium has an ordered structure which is destroyed at 160 GPa, and then lithium again becomes a "good" metal.

Recently, the existence of new phases in compressed sodium at pressures $p > 130$ GPa has been predicted by Neaton and Ashcroft (2001), based on the density functional method employed to calculate the atomic and electron structure. Similar to the case of lithium, the new phases are different from those expected for "simple" metals – they have low structural symmetry and semimetallic electron properties. The experimental confirmation of a possible transi-

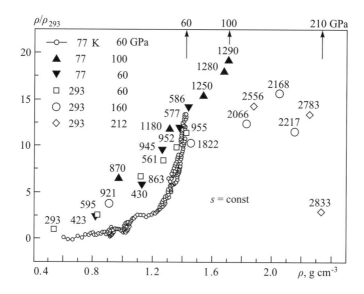

FIG. 9.21.   Resistivity of lithium versus density (Fortov *et al.* 2001). The legend of points
includes the initial temperature and the maximum pressure, and numbers near the points
show the calculated temperature.

tion of sodium into the dielectric state was reported by Fortov *et al.* (2001).

The dramatic changes in the atomic and electron structure of lithium and
sodium obtained in the quantum mechanical calculations by Neaton and Ashcroft
(1999, 2001) indicate strong electron–phonon interaction. This suggests the possibility of superconductivity at relatively high temperatures (Christensen and
Novikov 2001).

The transition to the dielectric state is predicted not only for the alkali metals. Self–consistent augmented–plane–wave calculations by McMahan and Albers
(1982) predict that nickel should transform from a metal to an insulator at a
pressure of 34 TPa, and then revert back to a metal at 51 TPa. This prediction
still requires experimental confirmation.

9.4   Ionization by pressure

The effect of pressure–induced ionization is most pronounced in the case of hydrogen. Figure 9.5 shows the data on the quasi–isentropic compression of liquid
and gaseous hydrogen in planar 1 and cylindrical 2 geometries (Fortov *et al.* 2003)
together with the results of the compression in the light–gas gun 5 (Weir *et al.*
1996a) and the results of explosive cylindrical compression in an axial magnetic
field 3 (Hawke *et al.* 1978) and 4 (Pavlovskii *et al.* 1987). Because of the small
molecular weight, the multiple shock compression of hydrogen is accompanied by
relatively weak heating – even at maximum pressures of 0.1–1 TPa, the typical
values of the temperature do not exceed $T = 10^4$ K, which favors the regime of

"cold" ionization. For hydrogen compressed to densities of $\rho \sim 0.01$–1.2 $\mathrm{g\,cm^{-3}}$ and heated to $T \sim 10^4$ K at pressures below 1.5 TPa, a wide spectrum of plasma states was realized, characterized by a fully developed ionization, $\alpha \simeq 1$, and a high electron concentration, $n_e \simeq 2 \cdot 10^{23}$ $\mathrm{cm^{-3}}$. At maximum compressions, the plasma is degenerate, $n_e \lambda_e^3 < 200$, and is strongly nonideal both with respect to Coulomb ($\gamma \sim 10$) and to interatomic ($n_a r_a^3 \sim 1$) interactions.

It is interesting that the extrapolation of the simplest plasma models to this region of strong nonideality leads to the thermodynamic instability of Debye–Hückel models (the "Coulomb collapse" indicated by the arrow DP in Fig. 9.5) and to the divergence of the Spitzer formula (arrow SP). The first of these approximations is depicted by the DHA curve in Fig. 9.5 and predicts the "pressure–induced ionization" at densities approximately two orders of magnitude lower than the experimental values. The shock compression leads to the overlap of the wavefunctions for neighboring atoms and, hence, to the percolation conductivity mechanism suggested by Likal'ter (1998, 2000), which is described in terms of the density–dependent reduction of the ionization potential (see the discussion of the percolation model in Chapter 4).

A decrease in the ionization potential with density is also predicted by the Mott model (Mott and Davis 1979), which was used by Ebeling *et al.* (1991) to construct a semiempirical broad–range model of ionization equilibrium and transport properties (curve M in Fig. 9.5) of compressed and hot matter,

$$\Delta I = -I \ln \left\{ 1 + \exp \left[-2 \left(\frac{R - a(I)}{\Delta(I)} \right) \right] \right\}. \tag{9.5}$$

The parameters a, R and Δ were chosen to reproduce experimental data on pressure–induced ionization of alkali metals. One can see that the proposed approximations provide a good qualitative description of experimental results.

By using the ring (Debye) approximation in a big canonical ensemble of statistical mechanics (curve LDH) to describe Coulomb nonideality, one can reduce the discrepancy between the theoretical and experimental results down to one order of magnitude. The remaining disagreement can be eliminated by introducing the hard–sphere model to describe the short–range repulsion of atoms and ions (curve HS) and by taking into account the compression–induced change in the energy spectrum of atoms and ions within a simplified model discussed in Section 9.3 (curve CA). An attempt to take into account the "hopping" character of the electrical conductivity in nonideal plasmas was made by Redmer *et al.* (2001). The corresponding results are marked by the symbol "R" in Fig. 9.5. The QMC curve corresponds to the calculation of the conductivity by the quantum MC method (Filinov *et al.* 2000a,b, 2001; Mulenko *et al.* 2001).

Figures 9.6–9.9 display the results of the electrical conductivity of shock–compressed Xe, Ar, Kr, and He. Similar to the case of hydrogen, at "low" temperatures one can clearly see the effect of the pressure–induced ionization occurring at higher plasma densities ~ 1–10 $\mathrm{g\,cm^{-3}}$. For multielectron atoms it is also natural to expect that, as compression is increased further, the first pressure–induced ionization will be followed by the next stages of multiple ionization. These stages should be accompanied by the emergence of new boundaries in the phase diagram. Unfortunately, experimental investigation of the regimes of multiple ionization is beyond the currently available capabilities of the explosive experimental equipment.

Along with the results of multiple ("cold") compression, the same figures show data obtained previously (Ivanov et al. 1976; Mintsev and Fortov 1979; Mintsev et al. 1980) by measuring the electrical conductivity of singly and doubly compressed plasmas. The temperature in this case is almost one order of magnitude higher, so that the effect of thermal ionization becomes dominant. The effect increases with increasing molecular weight of the studied substance and is particularly clear for xenon (see Fig. 9.6). It can be seen that upon thermal ionization at $T \simeq (4-10) \cdot 10^3$ K, a high level of the electrical conductivity of $\sim 10^3$ ohm^{-1}cm^{-1} is achieved even at low densities of $\rho \simeq 0.04-1$ g cm^{-3}, whereas the pressure–induced ionization can provide the same conductivity in cold ($T \sim 10^4$ K) matter only at extremely high densities of $\rho \sim 10$ g cm^{-3}. It can also be seen that, with increasing molecular weight of substances, the relative jump in the conductivity due to pressure–induced ionization decreases and becomes as low as two orders of magnitude for xenon. It is noteworthy that the electrical conductivity of xenon plasma measured in experiments with multiple shock compression is close to that obtained under static compression in diamond anvils (crosses in Fig. 9.6).

It is important to note that some of the models discussed here lose thermodynamic stability in the parameter regime achieved in the experiments. This might be considered as an indication of a "plasma" first–order phase transition (see Chapter 5) leading to the stratification of a strongly nonideal plasma into phases characterized by different degrees of ionization and compressibility. A sharp increase in the electrical conductivity of a dense plasma might suggest the occurrence of such a phase transition.

References

Abrikosov, A. A. Equation of state of hydrogen and high pressures. (1954). *Astron. Zh.*, **31**,112.

Adamskaya, I. A., Grigor'ev, F. V., Mikhailova, O. L., Mochalov, M. A., Sokolova, A. I., and Urlin, V. D. (1987). Quasi–isentropic compression of liquid argon at pressures up to 600 kbar. *JETP*, **66**, 366–368.

Alekseev, V. A. and Vedenov, A. A. (1971). Conductivity of dense cesium vapor. *Sov. Phys. Uspekhi*, **13**, 830–831.

Ashcroft, N. W. (1968). Metallic hydrogen: A high–temperature superconductor? *Phys. Rev. Lett.*, **21**, 1748–1749.

Avrorin, E. N., Vodolaga, B. K., Simonenko, B. A., and Fortov, V. E. (1993). Intense shock waves and extreme states of matter. *Phys. Uspekhi*, **36**, 337–364.

Bakanova, A. A., Zubarev, V. N., Sutulov, Y. N., and Trunin, R. F. (1975). Thermodynamic properties of water at high pressures and temperatures. *JETP*, **41**, 544–548.

Barker, L. M., Trucano, T. G., Wize, J. L., and Asay, J. R. (1986). (private communication).

Bastea, M., Mitchell, A. C., and Nellis, W. J. (2001). High pressure insulator–metal transition in molecular fluid oxygen. *Phys. Rev. Lett.*, **86**, 3108–3111.

Belov, S. B., Boriskov, G. V., Bykov, A. I., Il'kaev, R. I., Luk'yanov, N. B., Matveev, A. Y., Mikhailova, O. L., Selemir, V. D., Simakov, G. V., Trunin, R. F., Trusov, I. P., Urlin, V. D., Fortov, V. E., and Shuikin, A. N. (2002). Shock compression of solid deuterium *JETP Lett.*, **76**, 433–435.

Berg, U. I. Investigations of a very high pressure transducer. (1964). *Arkiv for Fysik*, **25**, 111–122.

Berger, J., Joigneau, S., and Bottet, G. (1960). Behaviour of sulphur under the action of a shock wave. *CR Acad. Sci. (Paris)*, **250**, 4331–4333.

Beule, D., Ebeling, W., Forster, A., Juranek, H., Nagel, S., Redmer, R., and Röpke, G. (1999). Equation of state for hydrogen below 10000 K: From the fluid to the plasma. *Phys. Rev. B*, **59**, 14177–14181.

Boettger, J. C. (1986). Equation of state and metallization of neon. *Phys. Rev. B*, **33**, 6788–6791.

Brish, A. A., Tarasov, M. S., and Tsukerman, V. A. (1960). Electrical conductivity of dielectrics in strong shock waves. *JETP*, **11**, 15–17.

Brovman, E. G., Kagan, Y., and Kholas, A. (1971). Structure of metallic hydrogen at zero pressure. *JETP*, **34**, 1300–1315.

Bushman, A. B., Lomonosov, I. V., and Fortov, V. E. (1992). *Equations of state of metals at high energy densities*. Joint Institute of Chemical Physics, Chernogolovka.

Cavazzoni, C., Chiarotti, G. L., Scandolo, S., Tosatti, E., Bernasconi, M., and Parrinello, M. (1999). Superionic and metallic states of water and ammonia at giant planet conditions. *Science*, **283**, 44–46.

Chhabildas, L. C. and Ruoff, A. L. (1977). The transition of sulfur to a conducting phase. *J. Chem. Phys.*, **66**, 983–985.

Christensen, N. E. and Novikov, D. L. (2001). Predicted superconductive properties of lithium under pressure. *Phys. Rev. Lett.*, **86**, 1861–1864.

Collins, L. A., Kwon, I., Kress, J. D., Troullier, N., and Lynch, D. (1995). Quantum molecular dynamics simulations of hot, dense hydrogen. *Phys. Rev. E*, **52**, 6202–6219.

Da Silva, L. B., Celliers, P., Collins, G. W., Budil, K. S., Holmes, N. C., Barbee Jr., T. W., Hammel, B. A., Kilkenny, J. D., Wallace, R. J., Ross, M., and Cauble, R. (1997). Absolute equation of state measurements on shocked liquid deuterium up to 200 GPa (2 Mbar). *Phys. Rev. Lett.*, **78**, 483–486.

David, H. G. and Hamann, S. D. (1958). Sulfur: A possible metallic form. *J. Chem. Phys.*, **28**, 1006.

David, H. G. and Hamann, S. D. (1959). The chemical effects of pressure. Part

5.–The electrical conductivity of water at high shock pressures. *Trans. Faraday Soc.*, **55**, 72–78.

David, H. G. and Hamann, S. D. (1960). The chemical effects of pressure. Part 6.–The electrical conductivity of several liquids at high shock pressures. *Trans. Faraday Soc.*, **56**, 1043–1050.

Desgreniers, S., Vohra, Y. K., and Ruoff, A. L. (1990). Optical response of very high density solid oxygen to 132 GPa. *J. Phys. Chem.*, **94**, 1117–1122.

DeSilva, A. W. and Katsouros, J. D. (1998). Electrical conductivity of dense copper and aluminum plasmas. *Phys. Rev. Lett.*, **57**, 5945–5951.

Dolotenko, M. I., Bykov, A. I., *et al.* (1997). (private communication).

Dudin, S. V., Fortov, V. E., Gryaznov, V. K., Mintsev, V. B., Shilkin, N. S., and Ushnurtsev, A. E. (1998). Investigation of shock–compressed plasma parameters by interaction with magnetic field. In *Shock compression of condensed matter. 1999*, Schmidt, S. C. (ed.), pp. 793–795. AIP Conference Proceedings 429, New York.

Dunn, K. J. and Bundy, F. P. (1977). Electrical behavior of sulfur up to 600 kbar–metallic state. *J. Chem. Phys.*, **67** 5048–5053.

Ebeling, W., Förster, A., Fortov, V., Gryaznov, V., and Polishchuk, A. (1991). *Thermophysical properties of hot dense plasmas.* Teubner, Stuttgart–Leipzig.

Eremets, M. I., Gregoryanz, E. A., Struzhkin, V. V., Mao, H., Hemley, R. J., Mulders, N., and Zimmerman, N. M. (2000). Electrical conductivity of xenon at megabar pressures. *Phys. Rev. Lett.*, **85**, 2797–2800.

Evdokimova, V. V. and Kuzemskaya, I. G. (1978). Superconductivity of sulfur at high pressure. *JETP Lett.*, **28**, 360–362.

Filinov, V. S., Bonitz, M., Ebeling, W., and Fortov, V. E. (2001). Thermodynamics of hot dense H–plasmas: Path integral Monte Carlo simulations and analytical approximations. *Plasma. Phys. Control. Fusion*, **43**, 743–759.

Filinov, V. S., Bonitz, M., and Fortov, V. E. (2000a). High–density phenomena in hydrogen plasma. *JETP Lett.*, **72**, 361–365.

Filinov, V. S., Fortov, V. E., Bonitz, M., and Kremp, D. (2000b). Pair distribution functions of dense partially ionized hydrogen. *Phys. Lett. A*, **274**, 228–235.

Fortov, V. E. (Ed.) (2000). *Encyclopedia of low temperature plasmas. Introductory volume 1.* Nauka, Moscow.

Fortov, V. E., Gryaznov, V. K., Mintsev, V. B., Ternovoi, V. Y., Iosilevski, I. L., Zhernokletov, M. V., and Mochalov, M. A. (2001). Thermophysical properties of shock–compressed argon and xenon. *Contrib. Plasma Phys.*, **41**, 215–218.

Fortov, V. E., Ternovoi, V. Y., Mintsev, V. B., Nikolaev, D. N., Pyalling, A. A., and Filimonov, A. S. (1999a). Electrical conductivity of nonideal hydrogen plasma at megabar dynamic pressures. *JETP Lett.*, **69**, 926–931.

Fortov, V. E., Ternovoi, V. Y., Zhernokletov, M. V., Mochalov, M. A., Mikhailov, A. L., Filimonov, A. S., Pyalling, A. A., Mintsev, V.B., Gryaznov, V. K., and Iosilevski, I. L. (2003). Pressure–produced ionization of nonideal plasma in a megabar range of dynamic pressures. *JETP*, **97**, 259–278.

Fortov, V. E., Yakushev, V. V., and Kagan, K. L., Lomonosov, I. V., Postnov, V. I., and Yakusheva, T. I. (1999b). Anomalous electrical conductivity of lithium under quasi–isentropic compression to 60 GPa (0.6 Mbar). Transition into a molecular phase? *JETP Lett.*, **70**, 628–632.

Fortov, V. E., Yakushev, V. V., Kagan, K. L., Lomonosov, I. V., Postnov, V. I., Yakusheva, T. I., and Kur'yanchik, A. N. (2001). Anomalous resistivity of lithium at high dynamic pressures. *JETP Lett.*, **74**, 418–421.

Gatilov, L. A., Glukhodedov, V. D., Grigor'ev, F. V., Kormer, S. B., Kuleshova, L. V., and Mochalov, M. A. (1985). Electric conductivity shock–compressed condensed argon at pressure from 20 to 70 GPa. *J. Appl. Mech. Techn. Phys.*, **26**, 99–102.

Gaydon, A. G. and Hurle, I. R. (1963). *The shock tube in high temperature chemical physics*. Chapman and Hall, London.

Glukhodedov, V. D., Kirshanov, S. I., Lebedeva, T. S., and Mochalov, M. A. (1999). Properties of shock–compressed liquid krypton at pressures of up to 90 GPa. *JETP*, **89**, 292–298.

Goettel, K. A.,Eggert, J. H., Silvera, I. F., and Moss, W. C. (1989). Optical evidence for the metallization of Xe at 132(5) GPa. *Phys. Rev. Lett.*, **62**, 665–668.

Grigor'ev, F. V., Kormer, S. B., Mikhailova, O. L., Mochalov, M. A., and Urlin, V. D. (1985). Shock compression and brightness temperature of a shock wave front in argon. *JETP*, **61**, 751–757.

Grigor'ev, F. V., Kormer, S. B., Mikhailova, O. L., Tolochhko, A. P., and Urlin, V. D. (1978). Equation of state of molecular yydrogen. Phase transition into the metallic state. *JETP*, **48**, 847–852.

Gryaznov, V. K. and Fortov, V. E. (1987). Thermodynamics of aluminum plasma at ultrahigh energy densities. *High Temp.*, **25**, 1208–1210.

Gryaznov, V. K., Fortov, V. E., Zhernokletov, M. V., Simakov, G. V., Trunin, R. F., Trusov, L. I., and Iosilevskii, I. L. (1998). Shock–wave compression of nonideal plasma of metals. *JETP*, **87**, 678–690.

Gryaznov, V. K., Iosilevskii, I. L., and Fortov, V. E. (1982). Thermodynamics of strongly compressed plasma of the megabar pressure range. *JTP Lett.*, **8**, 1378–1381.

Gryaznov, V. K., Iosilevskii, I. L., and Fortov, V. E. (2000). Thermodynamics of shock–compressed plasma in conception of chemical model. In *Shock waves and extremal states of matter*, Fortov, V. E., Al'tshuler, L. V., Trunin, R. F., and Funtikov, A. I. (eds), pp. 342–387. Nauka, Moscow.

Gryaznov, V. K., Zhernokletov, M. V., Zubarev, V. N., Iosilevskii, I. L., and Fortov, V. E. (1980). Thermodynamic properties of a nonideal argon or xenon plasma. *JETP*, **51**, 288–295.

Hamann, S. D. and Linton, M. (1966). Electrical conductivity of water in shock compression. *Trans. Faraday Soc.*, **62**, 2234–2241.

Hamann, S. D. and Linton, M. (1969). Electrical conductivities of aqueous solutions of KCl, KOH and HCl, and the ionization of water at high shock pressures. *Trans. Faraday Soc.*, **65**, 2186–2193.

Hamilton, D. C., Nellis, W. J., Mitchell, A. C., Ree, F. H., and van Thiel, M. (1988). Electrical conductivity and equation of state of shock–compressed liquid oxygen. *J. Chem. Phys.*, **88**, 5042–5050.

Hanfland, M., Syassen, K., Christensen, N. E., and Novikov, D. L. (2000). New high–pressure phases of lithium. *Nature*, **408**, 174–178.

Haronska, P., Kremp, D. *et al.* (1987). (unpublished).

Hawke, P. S., Burgers, T. J., Duerre, D. E., Huebel, J. G., Keeler, R. N., Klapper, H., and Wallace, W. C. (1978). Observation of electrical conductivity of isentropically compressed hydrogen at megabar pressures. *Phys. Rev. Lett.*, **41**, 994–997.

Hensel, F. and Franck, E. U. (1968). Metal–nonmetal transition in dense mercury vapor. *Rev. Mod. Phys.*, **40**, 697–703.

Holmes, N. S., Ross, M., and Nellis, W. J. (1995). Temperature measurements and dissociation of shock–compressed liquid deuterium and hydrogen. *Phys. Rev. B.*, **52**, 15835–15845.

Holzapfel, W. and Franck, E. U. (1966). conductivity and ionic dissociation of water at 1000 C and 100 kbar. *Ber. Bunsenges. Phys. Chem.*, **70**, 1105–1112.

Iosilevskii, I. L. (1980). Equation of state of nonideal plasma. *High Temp.*, **18**, 355–359.

Ivanov, Y. V., Mintsev, V. B., Fortov, V. E., and Dremin, A. N. (1976). Electric conductivity of nonideal plasma. *JETP*, **44**, 112–116.

Juranek, H., Redmer, R., Röpke, G., Fortov, V. E., and Pyalling, A. A., (1999). A comparative study for the equation of state of dense fluid hydrogen. *Contrib. Plasma Phys.*, **39**, 251–261.

Kanel', G. I. and Molodets, A. M. (1976). Behavior of glass k–8 under dynamic compressing and subsequent unloading. *Techn. Phys*, **46**, 398–407.

Keeler, R., van Thiel, M., and Alder, B. (1965). Corresponding states at small interatomic distances. *Physica*, **31**, 1437–1440.

Kerley, G. I. (1972). *A theoretical equation of state for deuterium. NTIS document LA–47766*. National technical information service, Springfield, VA.

Kirzhnits, D. A., Lozovic, Y. E., and Shpatakovskaya, G. V. (1975). Statistical model of matter. *Sov. Phys. Uspekhi*, **18**, 3–48.

Knaup, M., Reinhard, P. G., and Topffer, C. (1999). Wave packet molecular dynamics simulations of hydrogen near the transition to a metallic fluid. *Contrib. Plasma Phys.*, **39**, 57–60.

Knudson, M. D., Hanson, D. L., Bailey, J. B., Hall, C. A., Asay, J. R., and Anderson, W. W. (2001). Equation of state measurements in liquid deuterium to 70 GPa. *Phys. Rev. Lett.*, **87**, 225501/1–4.

Kulish, M. I., Gryaznov, V. K., Kvitov, S. V., Mintsev, V. B., Nikolaev, D. N., Ternovoi, V. Y., Filimonov, A. S., Fortov, V. E., Golubev, A. A., Sharkov, B. Y., Hoffmann, D., Stockl, K., and Wetzler, H. (1995). Absorption coefficients of dense argon and xenon plasma. *High Temp.*, **33**, 966–969.

Le Neindre, B., Suito, K., and Kawai, N. (1976). Fixed oints for pressure calibration above 10 GPa. *High Temp.–High Pres.*, **8**, 1–20.

Lenosky, T. J., Kress, J. D., Collins, L. A., and Kwon, I. (1997). Molecular–dynamics modeling of shock–compressed liquid hydrogen. *Phys. Rev.B*, **55**, 11907–11910.

Likal'ter, A. A. (1998). Is superdense fluid hydrogen a molecular metal? *JETP*, **86**, 598–601.

Likal'ter, A. A. (2000). Critical points of condensation in Coulomb systems. *Phys. Uspekhi*, **43**, 777–797.

Maksimov, E. G. and Shilov, Y. I. (1999). Hydrogen at high pressure. *Phys. Uspekhi*, **42**, 1121–1138.

McMahan, A. K. and Albers, R. C. (1982). Insulating nickel at a pressure of 34 TPa. *Phys. Rev. Lett.*, **49**, 1198–1201.

Meletov, K. P., Dolganov, V. K., Zharikov, O. V., Kremenskaya, I. N., and Ossipyan, Y. A. (1992). Absorption spectra of crystalline fullerite 60 at pressures up to 19 GPa. *J. Phys. I France*, **2**, 2097–2105.

Mintsev, V. B. and Fortov, V. E. (1979). Electrical conductivity of xenon under supercritical conditions. *JETP Lett.*, **30**, 375–378.

Mintsev, V. B. and Fortov, V. E. (1982). Explosive shock tubes. *High Temp.*, **20**, 745–764.

Mintsev, V. B., Fortov, V. E., and Gryaznov, V. K. (1980). Electrical conductivity of a high–temperature nonideal plasma. *JETP*, **52**, 59–63.

Mintsev, V. B., Ternovoi, V. Y., Gryaznov, V. K., Pyalling, A. A., Fortov, V. E., Iosilevskii, I. L. (2000). Electrical conductivity of shock–compressed xenon. In *Shock compression of condensed matter – 1999*, Furnish, M. D., Chhabildas, L. C., and Hixson, R.S. (eds), pp. 987–990. AIP Press, New York.

Mitchell, A. C. and Nellis, W. J. (1982). Equation of state and electrical conductivity of water and ammonia shocked to the 100 GPa (1 Mbar) pressure range. *J. Chem. Phys.* **76**, 6273–6281.

Mochalov, M. A. and Kuznetsov, O. N. (2001). Measurement of conductivity of gaseous hydrogen compressed quasi–isentropically to pressure 1500 GPa. *Abstracts of "III Khariton's subject readings"*, Sarov, p. 108

Mori, Y. and Ruoff, A. L. (1999). Lithium becomes a molecular solid at high pressure. *Bull. Am. Phys. Soc.*, **44**, 1489.

Moshary, F., Chen, N. H., and Silvera, I. F. (1992). Gap reduction and the collapse of solid C60 to a new phase of carbon under pressure. *Phys. Rev. Lett.*, **69**, 466–469.

Mostovych, A. N., Chan, Y., Lehecha, T., Schmitt, A., and Sethian, J. D. (2000). Reflected shock experiments on the equation–of–state properties of liquid deuterium at 100–600 GPa (1–6 Mbar). *Phys. Rev. Lett.*, **85**, 3870–3873.

Mott, N. F. and Davis, E. A. (1979). *Electron processes in noncrystalline materials*. Clarendon Press, Oxford.

Mulenko, I. A., Olejnikova, E. N., Khomkin, A. L., Filinov, V. S., Bonitz, M., and Fortov, V. E. (2001). Phase transition in dense low–temperature molecular gases. *Phys. Lett. A*, **289**, 141–146.

Nabatov, S. S., Dremin, A. N., Postnov, V. I., and Yakushev, V. V. (1979). Measurement of the electrical conductivity of sulfur under superhigh dynamic pressures. *JETP Lett.*, **29**, 369–372.

Neaton, J. B. and Ashcroft, N. W. (1999). Pairing in dense lithium. *Nature*, **400**, 141–144.

Neaton, J. B. and Ashcroft, N. W. (2001). On the constitution of sodium at higher densities. *Phys. Rev. Lett.*, **86**, 2830–2833.

Nellis, W. J. (2000). Metallization of fluid hydrogen at 140 GPa (1.4 Mbar): Implications for Jupiter. *Plan. Space Sci.*, **48**, 671–677.

Nellis, W., van Thiel, M., and Mitchel, A. (1982). Shock compression of liquid xenon to 130 GPa (1.3 Mbar). *Phys. Rev. Lett.*, **48**, 816–818.

Nellis, W. J., Mitchell, A. C., van Thiel, M., Devine, G. J., Trainor, R. J., and Brown, N. (1983). Equation–of–state data for molecular hydrogen and deuterium at shock pressures in the range 2–76 GPa (20–760 kbar). *J. Chem. Phys.*, **79**, 1480–1486.

Nellis, W. J., Holmes, N. C., Mitchell, A. C., Trainor, R. J. , Governo, G. K., Ross, M., and Young, D. A. (1984). Shock compression of liquid helium to 56 GPa (560 kbar). *Phys. Rev. Lett.*, **53**, 1248–1251.

Nellis, W. J., Weir, S. T., and Mitchell, A. C. (1998). *Rev. High Press. Sci. Technol.*, **7**, 870–872.

Nellis, W. J., Weir, S. T., and Mitchell, A. C. (1999). Minimum metallic conductivity of fluid hydrogen at 140 GPa (1.4 Mbar). *Phys. Rev. B*, **59**, 3434–3449.

Osip'yan, Y. A., Fortov, V. E., Kagan, K. L., Kveder, V. V., Kulakov, V. I., Kur'yanchik, A. N., Nikolaev, R. K., Postnov, V. I., and Sidorov, N. S. (2002). Conductivity of C_{60} fulerene crystals under dynamic compression up to 200 kbar. *JETP Lett.*, **75**, 563–565.

Pavlovskii, A. I., Boriskov, G. V., Bykov, A. I., Dolotenko, M. I., Egorov, N. I., Karpikov, A. A., Kolokolchikov, N. P., and Mamyshev, V. I. (1987). Isentropic solid hydrogen compression bt ultrahigh magnetic field pressure in megabar range. In *Megagauss technology and pulsed power applications*, Fowler, C. M., Caird, R. S., and Erickson, D. J. (eds), pp. 255–262. Plenum Press, New York and London.

Pavlovskii, A. I., Kuleshov, G. D., Sklizkov, G. V., Zusin, Y. A., and Gerasimov, A. I. (1965). High–current iron–free betatron. *DAN USSR*, **160**, 68–70.

Pierleoni, C., Ceperley, D. M., Bernu, B., and Magro, W. R. (1974). Equation of state of the hydrogen plasma by path integral Monte Carlo simulation. *Phys. Rev. Lett.*, **73**, 2145–2149.

Podurets, M. A., Popov, L. V., Trunin, R. F., Simakov, G. V., and Moiseev, B. N. (1972). Compression of water by strong shock waves. *JETP*, **35**, 375–376.

Postnov, V. I., Anan'eva, L. A., Dremin, A. N., Nabatov, S. S., and Yakushev, V. V. (1986). Electric conductivity and compressibility of sulfur at shock compression. *Comb., Expl., Shock Waves*, **22**, No. 4, 106–109.

Radousky, H. B. and Ross, M. (1988). Shock temperature measurements in high density fluid xenon. *Phys. Lett. A*, **129**, 43–46.

Redmer, R., Röpke, G., Kuhlbrodt, S., and Reinholz, H. (2001). Metal–nonmetal transition in dense hydrogen. *Contrib. Plasma Phys.*, **41**, 163–166.

Reichlin, R., Brister, K. E., McMahan, A.K., Ross, M., Martin, S., Vohra, Y. K., Ruoff, A. L. (1989). Evidence for the insulator–metal transition in xenon from optical, X–ray, and band–structure studies to 170 GPa. *Phys. Rev. Lett.*, **62**, 669–672.

Robnik, M. and Kundt, W. (1983). Hydrogen at high pressures and temperatures. *Astronom. Astrophys*, **120**, 227–233.

Ross, M. (1998). Linear–mixing model for shock–compressed liquid deuterium. *Phys. Rev. B*, **58**, 669–677.

Ross, M. and McMahan, A. K. (1980). Condensed xenon at high pressure. *Phys. Rev. B*, **21**, 1658–1664.

Ross, M., Ree, F., and Young, D. (1983). The equation of state of molecular hydrogen at very high density. *J. Chem. Phys.*, **79**, 1487–1494.

Saleem, S., Haun, J., and Kunze, H.–J. (2001). Electrical conductivity measurements of strongly coupled W plasmas. *Phys. Rev. E*, **64**, 056403/1–6.

Saumon, D., and Chabrier, G. (1989). Fluid hydrogen at high density: The plasma phase transition. *Phys. Rev. Lett.*, **62**, 2397–2400.

Saumon, D., and Chabrier, G. (1992). Fluid hydrogen at high density: Pressure ionization. *Phys. Rev. A*, **46**, 2084–2100.

Seeger, K. (1977). *Semiconductor physics*. Springer, Wien–New York.

Shimizu, K., Suhara, K., Ikumo, M., Eremets, M. I., and Amaya, K. (1998). Superconductivity in oxygen. *Nature*, **393**, 767–769.

Struzhkin, V. V., Hemley, R. J., and Mao, H. K. (1999). Compression of Li to 120 GPa. *Bull. Am. Phys. Soc.*, **44**, 1489.

Ternovoi, V. Y. (1980). (private communication).

Ternovoi, V. Y., Filimonov, A. S., Fortov, V. E., Kvitov, S. V., Nikolaev, D. N., and Pyalling, A. A. (1999). Thermodynamic properties and electrical conductivity of hydrogen under multiple shock compression to 150 GPa. *Physica B*, **265**, 6–11.

Ternovoi, V. Y., Filimonov, A. S., Pyalling, A. A., Mintsev, V. B., and Fortov, V. E. (2002). Thermophysical properties of helium under multiple shock compression. In *Shock compression of condensed matter. 2001*, Furnish, M. D., Thadhani, N. N., and Horie, Y. (eds), pp. 107–110. AIP Press, New York.

Trubitsyn, V. P. (1966). Phase transition in hydrogen crystals. *Sov. Phys. Solid State*, **8**, 862–865.

Urlin, V. D., Mochalov, M. A., and Mikhailova, O. L. (1992). Liquid xenon study under shock and quasi–isentropic compression. *High Press. Res.*, **8**, 595–605.

Urlin, V. D., Mochalov, M. A., and Mikhailova, O. L. (1997). Quasi–isentropic compression of liquid argon up to 500 GPa. *JETP*, **84**, 1145–1148.

Veeser, L. I., Ekdah, C. A., Oona, H. *et al.* (1998). Isentropic compression of argon. In *Abstracts of the 8th international conference on megagauss magnetic field generation and related topics,* Tallahasse, p. 239.

Vereschagin, L. F., Yakovlev, E. N., Vinogradov, B. V., and Sakun, V. P. (1974). Transition of Al_2O_3, NaCl, and S into the conducting state. *JETP Lett.*, **20**, 246–247.

Volkov, L. P., Voloshin, N. P., Mangasarov, R. A., Simonenko, V. A., Sin'ko, G. V., and Sorokin, V. L. (1980). Shock compressibility of water at pressure of ~ 1 Mbar. *JETP Lett.*, **31**, 513–515.

Weir, S. T., Mitchell, A. C., and Nellis, W. J. (1996a). Metallization of fluid molecular hydrogen at 140 GPa. *Phys. Rev. Lett.*, **76**, 1860–1863.

Weir, S. T., Mitchell, A. C., and Nellis, W. J. (1996b). Electrical resistivity of single–crystal Al_2O_3 shock–compressed in the pressure range 91–220 GPa (0.91–2.20 Mbar). *J. Appl. Phys.* , **80**, 1522–1525.

Wigner, E. and Huntington, H. B. (1935). On the possibility of a metallic modification of hydrogen. *J. Chem. Phys.*, **3**,764–770.

Yakub, E. S. (1990). Equation of state of shock–compressed liquid hydrogen. *High Temp.*, **28**, 490–496.

Yakub, E. S. (1999). Diatomic fluids at high pressures and temperatures: nonempirical approach. *Physica*, **265**, 31–38.

Yakushev, V. V., Nabatov, S. S., and Yakusheva, O. V. (1974). Physical properties and transformation of acrylonitrile at high dynamic pressures. *Comb., Expl., Shock Waves*, **10**, 583–594.

Yakushev, V. V., Postnov, V. I., Fortov, V. E., and Yakusheva, O. V. (2000). Electrical conductivity of water during quasi–isentropic compression to 130 GPa. *JETP*, **90**, 617–622.

Young, D. (1977). A soft–sphere model for liquid metals. *LLNL report UCRL–52352*, University of California.

Young, D. A., McMahan, A. K., and Ross M. (1981). Equation of state and melting curve of helium to very high pressure. *Phys. Rev. B*, **24**, 5119–5127.

Zababakhin, E. I. (1979). Unbounded cumulation effects. *Mechanics in USSR for 50 years*. Nauka, Moscow.

Zaiman, J. M. (1972). *Principles of theory of solids*. Cambridge University Press, London.

Zamalin, V. M., Norman, G. E., and Filinov, V. S. (1977). *Method Monte Carlo in statistical thermodynamic*. Nauka, Moscow.

Zel'dovich, Y. B. and Landau, L. D. (1944). On correlation between liquid and gaseous states of metals. *Acta Phys.–Chim. USSR*, **18**, 194–198.

Zel'dovich, Y. B., Kormer, S. B., Sinitsyn, M. V., and Yushko, K. B. (1961). Investigation of optical properties of transparent substances at superhigh pressures. *DAN USSR*, **138**, 1333–1336.

10

NONNEUTRAL PLASMAS

Nonneutral plasmas are plasmas consisting of particles with a single sign of charge. These can be pure electron or positive ion plasmas (the latter may include several species), positron plasmas, as well as electron–antiproton plasmas. Nonneutral plasmas provide some unique research opportunities that are not available with "conventional" quasineutral plasmas (Davidson 1974; Davidson 1990; Dubin and O'Neil 1999). Due to strong the repulsion between like–charged particles, external fields are required to confine nonneutral plasmas. The confinement, which is usually provided by static electric and magnetic fields, can last a very long time – a few hours or even days. Since recombination cannot occur, nonneutral plasmas can be cooled to ultracryogenic temperatures (< 1 mK) where the kinetic energy of ions is much smaller than the energy of the mutual electrostatic interaction, and therefore the formation of liquid– and crystal–like states is possible.

10.1   Confinement of nonneutral plasmas

Nonneutral plasmas can be confined by employing various electric and magnetic traps. Electrons can be localized directly on the surface of liquid helium, forming a two–dimensional system where the Wigner crystallization is observed. By using laser cooling of ions, strongly coupled nonneutral plasmas can be obtained and studied in Penning and Paul traps. Recently, the remarkable example of plasma condensation, the "crystalline beams", was discovered by cooling the ions in a storage ring. Let us discuss briefly these methods of confinement.

10.1.1   Electrons on a surface of liquid He

The interaction of a free electron with atoms of a liquid has two contributions: The long–range attraction due to the polarization and the short–range repulsion due to the exchange interaction. The former one causes the potential energy of the electron to decrease (as compared with the vacuum value) whereas the latter one increases the energy. Hence, both the sign and the magnitude of the mean potential energy in a dielectric medium, V_0 (viz., the bottom of the conduction band for the electron embedded into the liquid), is determined by the competition of the polarization attraction and the exchange repulsion. For atoms or molecules with small polarizability, like He, Ne, H_2, the exchange repulsion prevails and, hence, $V_0 > 0$. This implies that an electron with a kinetic energy smaller than V_0 cannot penetrate into the liquid. However, being outside the liquid, the electron is attracted to the surface by the image force. The corresponding potential is

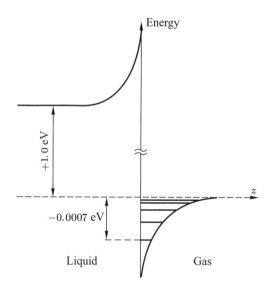

FIG. 10.1. Illustration of the electron energy spectrum near the surface of liquid He (Shikin and Monarkha 1989).

$$V(z) = -\frac{Q}{z}, \quad Q = \frac{e^2(\varepsilon - 1)}{4(\varepsilon + 1)}, \tag{10.1}$$

where ε is the dielectric permittivity of the liquid and the z–axis is pointed perpendicular to the surface. In the liquid ($z < 0$) we have $V(z) \simeq V_0$. Thus, we have a one–dimensional potential well (see Fig. 10.1) where the electron can be trapped in the direction perpendicular to the surface, whereas along the surface it can move freely.

For liquid He, the potential barrier, $V_0 \sim 1$ eV, is large compared to the electron binding energy. Therefore, with sufficient accuracy one can set $V_0 = \infty$ and thus employ $\Psi|_{z=0} = 0$ as the boundary condition for the electron wavefunction. In this case, it is easy to see that the problem of the energy spectrum of the trapped electron is reduced to that of the hydrogen atom (with the substitution $e^2 \to Q$),

$$E_n(\mathbf{k}) = \frac{\hbar^2 k^2}{2m} + \frac{mQ^2}{2\hbar^2 n^2}, \quad n = 1, 2, \ldots, \tag{10.2}$$

where $\mathbf{k}$ is the two–dimensional wavevector of the electron along the surface. Due to the low polarizability of the liquid He the binding energy of the electron in the ground state is low ($E_1(0) \simeq 8$ K $\ll V_0$) and, hence, the assumption $V_0 = \infty$ is well justified. Electrons are localized at $\sim 10^{-6}$ cm from the He surface and, therefore, the particular structure of $V(z)$ at shorter (atomic) distances is not important. The properties of electrons localized on the surface of liquid He are discussed in more detail by Cole (1974), Shikin (1977), and Shikin and Monarkha (1989).

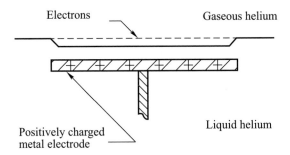

FIG. 10.2. Sketch of the experiment with electrons localized near the surface of liquid He.

There exists a simple method to create localized electrons on the surface of a liquid He (Shikin and Monarkha 1989). The source of electrons (e.g., corona discharges or radioactive isotopes) is located above the liquid surface, whereas the positively biased planar electrode is embedded into the liquid, close to the surface (see Fig. 10.2). The equilibrium number density of electrons on the surface, n_s, is determined by the condition that the total electric field is equal to zero above the surface, which yields $E_\perp = 2\pi e n_s$. By varying the magnitude of the bias on the electrode one can easily change n_s in a very broad range, from 10^5 to 10^9 cm^{-2}. The electron subsystem is not degenerate – even for $n_s = 10^9$ cm^{-2} the Fermi energy is small enough, $\varepsilon_F/k = \pi\hbar^2 n_s/km \leq 10^{-2}$ K. Depending on the values of $E_\perp$ and T, the coupling parameter of the two–dimensional electron subsystem, $\gamma = \pi^{1/2} e^2 n_s^{1/2}/kT$, can be changed from 0 up to 10^2, thus covering the regions of an ideal gas, electron liquid and the Wigner crystal. The upper limit of γ is determined by two factors: the instability of the charged surface at high n_s (and, hence, at large $E_\perp$), and degeneration of electrons at low temperatures.

10.1.2 Penning trap

The simple example of the Penning trap is shown in Fig. 10.3. The trap was proposed by Penning (1936) and since then it has been used extensively to confine charged particles. A conductive cylinder is divided into three sections: The central one is grounded, whereas the other two sections are biased positively (here, we suppose the ions to be charged positively). The external magnetic field **B** is parallel to the cylinder axis. The ions collect themselves in the region of the central grounded section, where the radial confinement is due to the magnetic field and the axial one is electrostatic. The radial confinement is associated with the azimuthal rotation of ions, which induces the radial Lorenz force balancing the centrifugal, electrostatic, and pressure forces pointed radially outwards.

Usually the size of the ion cloud is small compared to the trap dimensions, so that the potential well of the confinement is almost a parabolic one. The equilibrium shape of the cloud trapped in such a confinement is a rotational ellipsoid. The cloud rotates due to the pondermotive force, so that the analysis

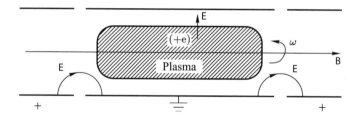

FIG. 10.3.   Sketch of the Penning trap (Penning 1936; Dubin and O'Neil 1999).

should pe performed in the rotating reference frame. Thus, the effective potential for charged particles confined in the Penning trap and rotating with frequency ω is (Dubin and O'Neil 1999)

$$\Phi(r, z) = \frac{M_i \omega_z^2}{2}(z^2 + \beta r^2),\qquad(10.3)$$

where the parameters β and ω_z are

$$\beta = \frac{\omega(\Omega_c - \omega)}{\omega_z^2} - \frac{1}{2}, \quad \omega_z^2 = \frac{2kZ_iU}{M_i}.\qquad(10.4)$$

Here M_i is the ion mass, U is the potential difference, ω_z is the eigenfrequency of the axial particle oscillations, k is a geometrical factor, and $\Omega_c = eB/M_ic$ is the ion cyclotron frequency. The parameter β determines the symmetry of $\Phi(\mathbf{r})$. For $\beta = 1$, the potential is spherically symmetric. For $\beta > 1$, the cloud is stretched along z–axis, and for $\beta < 1$ it is compressed.

The possibilities of the experimental investigation of strongly coupled non-neutral plasmas improved dramatically due to development of Doppler laser cooling (Chu 1999; Cohen–Tannoudji 1998; Phillips 1998). The main idea of laser cooling is easy to understand (Wineland et al. 1985; Dubin and O'Neil 1999). The laser beam which has a frequency slightly below the frequency of one of the electron transitions, is directed through the plasma (see Fig. 10.4). For ions having the velocity component directed opposite to the direction of the beam propagation, viz., $\mathbf{k} \cdot \mathbf{v} < 0$ (where $\mathbf{k}$ is the wavevector of the laser beam and $\mathbf{v}$ is the ion velocity), the transition energy is shifted towards resonance due to the Doppler effect. Hence, these ions absorb more effectively than those moving in the direction of the beam. After the photon absorption, the ion velocity is decreased by $\Delta\mathbf{v} = -\hbar\mathbf{k}/M_i$. The subsequent spontaneous reemission is spherically symmetric and, therefore, its contribution to the ion momentum is equal on average to zero. Thus, the net effect of the absorption and reemission processes is cooling of the plasma.

The original idea of laser cooling was proposed by Hänsch and Schawlow (1975), and Wineland and Dehmelt (1975). Three years later it had been implemented in experiments with ions of Mg$^+$ by Wineland et al. (1978) and Ba$^+$ by Neuhauser et al. (1978). The lower temperature limit (Doppler limit) which can

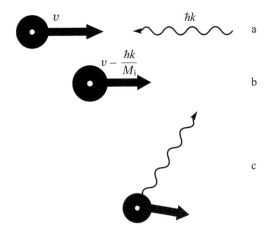

FIG. 10.4.   The sequence of processes that lead to the decrease of mean kinetic energy of
ions during laser cooling: An ion moving with velocity **v** interacts with a photon having
momentum $\hbar\mathbf{k}$ (a); the ion absorbs the photon and slows down by $\hbar\mathbf{k}/M_i$ (b); reemission of
the photon in an arbitrary direction causes the ion velocity to decrease on average (Phillips
1998).

be achieved with the laser cooling is of the order of $\hbar\Gamma$ (Letokhov *et al.* 1977,
and Wieman *et al.* 1999), where Γ is the rate of spontaneous reemission from the
excited state. This temperature is determined by the balance between the laser
cooling and heating processes caused random phonon absorption and emission
(Phillips 1998). The Doppler limit is typically about a few tenths of a mK for the
allowed dipole transitions. By now, there have been a few methods developed to
cool atoms and ions to temperatures much below the Doppler limit (Chu 1999;
Cohen–Tannoudji 1998; Phillips 1998; Wieman *et al.* 1999). These are, e.g., sub–
Doppler laser cooling and evaporative cooling (Wieman *et al.* 1999). The latter
is analogous to the cooling of hot water in a tea–cup, due to the evaporation
of more energetic molecules. After the evaporation, the energy of the remaining
ions and atoms is redistributed due to collisions and the resulting temperature
is decreased. Evaporative cooling allows one to decrease the temperature down
to 50 nK or even lower (Wieman *et al.* 1999).

One of the most remarkable recent achievements of the laser cooling technique is the experimental observation of macroscopic quantum systems – the
Bose–Einstein condensates predicted by Einstein in 1924. Examples of Bose–
Einstein condensation are, e.g., superfluidity and superconductivity. The condensation of rarefied atomic gas was first achieved by the cooling of Rb atoms
confined in a magneto-optical trap. The temperature in the experiment was low
enough ($\sim$ 200 nK) to ensure the overlapping of the de Broglie wave envelopes
of individual atoms (Anderson 1995). Under these conditions most of the atoms
condense in the ground state with zero momentum. Pre–cooling to tempera-

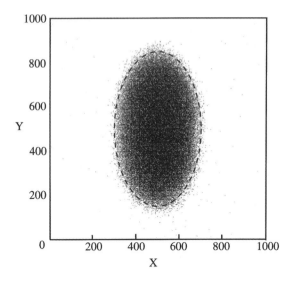

FIG. 10.5.   Side view of the plasma cloud confined in an asymmetric Penning trap and consisting of $\sim 8 \cdot 10^4$ ions $^9\text{Be}^+$ (Dubin and O'Neil 1999).

tures ~ 10 K was achieved with Doppler laser cooling, and then the evaporation method was employed. The condensate can be visualized both in real space, due to a sudden increase of the atom density in the middle of the trap, where the confining potential has a minimum, and in momentum space, because of the steep peak in the velocity distribution at zero value. The first observations of the Bose–Einstein condensates triggered enormous theoretical and experimental activity. Many theoretical hypotheses have been confirmed – among these, e.g., the shape of the condensate wavefunction and its dependence on the mutual atom interaction, the temperature dependence of the condensed atom fraction and the thermal capacity, etc.

Let us focus again on the properties of nonneutral plasmas. The shape of the plasma cloud can be determined experimentally by measuring light scattering or the ion fluorescence (Dubin and O'Neil 1999; Bollinger *et al.* 2000). Figure 10.5 shows the side view of a small plasma cloud ($\sim 8 \cdot 10^4$ of Be^+ ions) in the Penning trap. Ions are crystallized into a bcc lattice. This can be seen both with the analysis of Bragg scattering and by direct observation of the luminescence (the latter employs the stroboscopic effect, which allows us to "freeze" the rotation of the cloud at frequency $\sim 10^2$ kHz).

10.1.3   *Linear Paul trap*

The Paul trap is a quadrupole consisting of four parallel cylindrical electrodes, as shown in Fig. 10.6 (Raizen *et al.* 1992). Two diagonal electrodes are grounded, and a r.f. voltage is applied to two other electrodes. Each cylinder consists of two or three segments, so that a positive potential at edges of the trap together with

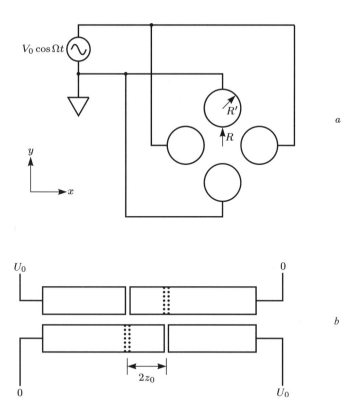

FIG. 10.6.   Schematics (a) and side view (b) of the Paul trap (Raizen *et al.* 1992).

the r.f. field provides the axial confinement. Similar to the case of the Penning trap, the potential in the central region can be well approximated by a parabolic profile, Eq. (10.3). Now, the parameter β is given by the following expression:

$$\beta = \frac{\omega_r^2}{\omega_z^2} - \frac{1}{2}, \tag{10.5}$$

where

$$\omega_r^2 = \frac{Z_i^2 U_{\text{rf}}^2}{2 M_i^2 r_0^4 \Omega^2}, \qquad \omega_z^2 = \frac{2k Z_i U_{\text{dc}}}{M_i}. \tag{10.6}$$

Here U_{dc} is the magnitude of the positive bias, U_{rf} and Ω are the amplitude and frequency of the r.f. voltage, respectively, r_0 is the distance measured from the central axis of the trap to the electrode surface, and k is some constant coefficient which takes into account the particular geometry. In Paul traps one can obtain long quasicrystalline structures (see the discussion in Section 10.2).

10.1.4   *Storage ring*

Figure 10.7 shows the ion storage ring PALLAS which was employed to investigate strongly coupled OCP plasmas (Schätz *et al.* 2001; Schramm *et al.* 2002).

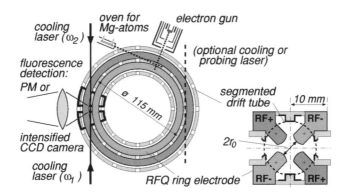

FIG. 10.7.   Axial and radial cross–section of the r.f. quadrupole storage ring PAL-
LAS (Schramm *et al.* 2002).

Basically, the storage ring is a rf quadrupole trap which is very similar to the
Paul trap. Sixteen segmented drift tubes enable the manipulation of ions along
the orbits. The radial confinement is provided by the r.f. field with amplitude
$U_{rf} = 200$ V and frequency $\Omega = 2\pi \cdot 6.3$ MHz applied between the quadrupole
electrodes. In the storage ring, the ions are confined in a parabolic potential well,

$$\Phi(r) = \frac{M_i \omega_r^2}{2} r^2, \tag{10.7}$$

where ω_r is given by Eq. (10.6) with $r_0 = 2.5$ mm.

In order to accumulate ions in the storage ring, a weak collimated beam of
^{24}Mg atoms is used. The atoms are ionized by a focused electron beam. Periodically, the trapped ions cross the oppositely directed laser beams, as shown
in Fig. 10.7. Both lasers are tuned to frequencies close to the $3s^2 S_{1/2} - 3p^2 P_{3/2}$
transition in the ^{24}Mg$^+$ ion, which corresponds to the wavelength 280 nm and
provides the conventional Doppler laser cooling. The resonance ion fluorescence
is detected by a photomultiplier and a CCD matrix.

After the accumulation of a sufficient number of ions with velocities ~ 1000
m s^{-1}, the resonant light pressure of the (first) laser beam directed along the ion
motion is employed to accelerate the ensemble. The frequency ω_1 grows continuously, until the ions occur in resonance with the (second) decelerating beam.
A fixed value of frequency ω_2 is chosen to provide the necessary stationary velocity of the ion beam (in experiments by Schätz *et al.* 2001 the velocity was
2800 m s^{-1}). The longitudinal velocity distribution in the ion beam is substantially narrowed, because of the dispersive nature of the force produced by the
laser beams. The decay of the ion motion in the transverse direction is provided
by the mutual Coulomb collisions, and also because of the small noncollinearity
of the ion and laser beams.

Apparently, the first reported crystallization of ions in a storage ring was
observed by Dement'ev *et al.* (1980) and Budker *et al.* (1976) in experiments with

a proton beam cooled by electrons. Recently, the phase transition of a "gaseous" ion beam into a one–dimensional "crystallized" beam was also observed in the storage ring PALLAS (Schätz *et al.* 2001; Schramm *et al.* 2002). The obtained results are discussed in the next section.

10.2 Strong coupling and Wigner crystallization

In the limit of high densities and low temperatures, the electrons behave very much like a degenerate ideal Fermi gas. This is because the mean energy of the Coulomb interaction $(V \sim r_s^{-1})$ does not increase with the density as fast as the Fermi energy $(\varepsilon_F \sim r_s^{-2})$. In metals V and ε_F are about the same, so that the electron subsystem is a weakly nonideal degenerate plasma. In accordance with the prediction made by Wigner 1934, as the density decreases the mean kinetic (Fermi) energy can become much smaller than the mean potential energy. Then the electron localization and the formation of lattices becomes energetically favorable. The electrons perform small oscillations around the equilibrium (note that usually the zero–point vibrations can be neglected). The melting of a Wigner crystal can be achieved only by increasing the density, since the temperature is small compared to the Fermi energy. In the classical limit, electron crystallization is also possible. The properties of classical electron crystals are well described in the framework of the OCP model (see discussion in Chapter 5).

For electrons localized on the surface of liquid He, it is rather easy to achieve the conditions necessary for crystallization (Shikin and Monarkha 1989). For typical surface densities of electrons, $n_s = 10^8$–10^9 cm^{-2}, the corresponding Fermi energy $(\varepsilon_F/k < 10^{-2}$ K) is much smaller than both the potential energy and the temperature. The surface electrons in experiments obey classical statistics, so that the strength of the mutual interaction between the electrons is characterized by the coupling parameter $\gamma = \pi^{1/2}e^2 n_s^{1/2}/kT$. The first experimental evidence of Wigner crystallization was presented by Grimes and Adams 1979, who observed a series of resonances in the r.f. energy absorption during the excitation of electrons on the surface of liquid He (see Fig. 10.8). These resonances only appear when the temperature is below a certain critical value T which is a function of the electron density, $T \sim n_s^{1/2}$. The explanation of these results was given by Fisher *et al.* (1979): It turned out that at low temperatures electrons form a triangular lattice with the period $d = 2^{1/2}3^{-1/4}n_s^{-1/2} \geq 2 \cdot 10^{-5}$ cm. Each electron causes surface distortion, so that the motion in the external electric field drives capillary waves (ripplons). The electron–ripplon interaction induces coupling of the electron and ripplon oscillations and, hence, causes resonance absorption of the r.f. radiation when the capillary–wave wavenumber is equal to a reciprocal vector of the electron lattice.

The Wigner crystal melts and the resonances disappear as the temperature increases. With good accuracy, the coupling parameter remains constant on the melting line, $\gamma = 137 \pm 15$ (see Fig. 10.9). Similar to the melting of solids, electron lattices start melting at high temperatures because the formation of dislocations becomes thermodynamically favorable, which eventually causes destruction

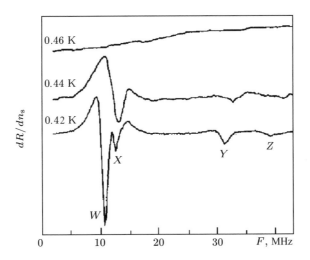

FIG. 10.8.   Resonance absorption of the electromagnetic waves by a Wigner crystal (Grimes and Adams 1979). The resonance emerges at $T = 0.457$ K, when the two–dimensional electron system crystallizes.

of the lattices. This dislocational mechanism of melting is confirmed by the values of the melting temperature and the shear modulus obtained from the MD simulations by Morf (1979). The corresponding coupling parameter on the melting line is $120 < \gamma < 140$.

Let us now discuss the properties of nonideal plasmas where the ions are strongly correlated, so that liquid–like (short–range order) and crystalline (long–range order) structures have been predicted and observed with the increase of the coupling parameter γ. The properties of such ordered structures can vary dramatically. They are determined not only by the value of γ, but also by the parameters of the trap and the size of the structure. Based on the spatial scales of the plasma, one can distinguish three different regimes (Dubin and O'Neil 1999): macroscopic plasmas, mesoscopic plasmas, and Coulomb clusters. The size of the macroscopic plasmas is so large that the surface effects have practically no influence on the average physical properties. The microscopic structure inside such plasmas coincides with that of infinite homogeneous plasmas. For the strongly correlated case, when ions form crystalline structures, the bulk properties prevail over the surface properties when the number of ions is fairly large, $N_i \geq 10^5$. However, when the correlation is not very strong ($\gamma \leq 10$), the surface effects can be neglected at $N_i \geq 10^3$ (Dubin and O'Neil 1999). This is because the correlation length in a liquid phase is of the order of one or two interparticle distances, and the surface effects do not penetrate beyond this length. When

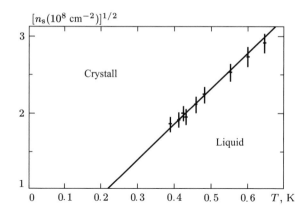

FIG. 10.9.   Phase diagram of a two–dimensional electron Wigner crystal localized on the surface of liquid He (Grimes and Adams 1979). The linear fit corresponds to a constant value of the coupling parameter $\gamma = 137$.

the number of particles is low enough, so that the plasma shape and size play an important role, but on the other hand is large enough $(N_i \geq 10^2)$ in order to employ major methods of statistical mechanics, the plasma is referred to as a mesoscopic plasma. And finally, when the number of particles is very low $(N_i \leq 10)$, the ions form, at low temperatures, simple geometrical configurations – the Coulomb clusters. The structure of clusters is determined by the confining fields and the number of particles. In early experiments with strongly coupled nonneutral plasmas confined in the Penning and Paul traps, the number of ions was relatively low, $10^2 < N_i < 10^4$ (Bollinger and Wineland 1984; Gilbert et al. 1988; Raizen et al. 1992). Therefore, most of the numerical simulations have been performed for a mesoscopic plasma confined in a parabolic potential well. Figure 10.10 shows the results obtained with Monte Carlo simulations for a cloud of 400 ions confined in the Penning trap (Dubin 1996). The angular velocity of the cloud rotation, ω, was tuned to provide the spherical symmetry of the effective potential Φ (see Eq. (10.3) with $\beta = 1$). For a weak correlation, $\gamma \sim 1$, the ion density slowly falls off and approaches zero at the cloud surface. As γ increases, the density decrease near the surface becomes steeper, approaching a step function shown by the dashed line. In addition, the emerging oscillatory structure suggests local ordering (the spatial scale characterizing decay of the oscillations is a measure of the correlation length). The formation of the oscillatory structure can be considered as a precursor of crystallization.

Similar behavior was observed in the one–component plasma model by Ichimaru et al. (1987). By increasing γ, the oscillation amplitude grows until the minimum ion density between the neighboring peaks becomes equal to zero, so that the ion cloud is separated into a set of concentric shells. The distance between the consecutive shells as well as the interparticle distance within each shell is close to the Wigner–Seitz radius r_s. Thus, the number of ions in the shell is

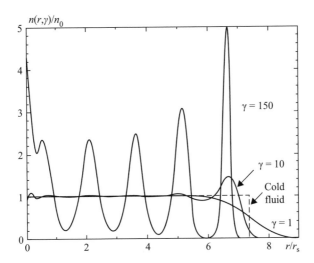

FIG. 10.10.   Equilibrium ion density versus the radius in a spherically symmetric cloud
of nonneutral plasma containing 400 ions at different values of the coupling parameter
γ (Dubin 1996). The dashed line corresponds to the low–temperature limit of the mean
field theory with $n_i(r) = \overline{n}_i$ and $r_s = (4\pi \overline{n}_i/3)^{-1/3}$.

proportional to its surface, $N_i^{2/3}$, and the number of shells is proportional to
$N_i^{1/3}$.

For $\beta \neq 1$ the cloud is a spheroid consisting of concentric shells. At $\beta \to \infty$
the cylindrical concentric shells are formed (Rahman and Schiffer 1986), whereas
for $\beta \to 0$ ions arrange themselves into a two–dimensional lattice in the xy–plane
(Rafac et al. 1991; Bedanov and Peeters 1994).

Totsuji et al. (2002) have investigated with MD simulations the nonneutral
plasma containing a fairly large number of ions, $5 \cdot 10^3 < N_i < 1.2 \cdot 10^5$, and
confined in a spherically symmetric parabolic well. It turns out that the shell
structure is energetically favorable only for $N_i < 10^4$, whereas for larger number
of particles the formation of the bcc bulk lattice surrounded by a few spherical
shells near the surface provides the energy minimum (see Fig. 10.11).

In experiments by Gilbert et al. (1988) and Bollinger et al. (2000), the shell
structure and the bcc lattice have been observed for $N_i = 1.5 \cdot 10^4$ and $N_i = 2 \cdot 10^5$, respectively. A laser cooled plasma of Be$^+$ ions at temperature ~ 10 mK
was studied in Penning traps. For typical ion density $n_i \simeq 4 \cdot 10^8$ cm^{-3}, this
corresponds to the coupling parameter $\gamma > 200$.

In experiments by Schätz et al. 2001, three laser beams crossing the plasma
cloud at different angles were used. The beams induced luminescence, which allowed them to obtain a three–dimensional visualization of the cloud. Figure 10.12
shows the image of the structure which contains $1.5 \cdot 10^4$ ions and consists of
11 shells. Both the number of shells and the distances between them agree very

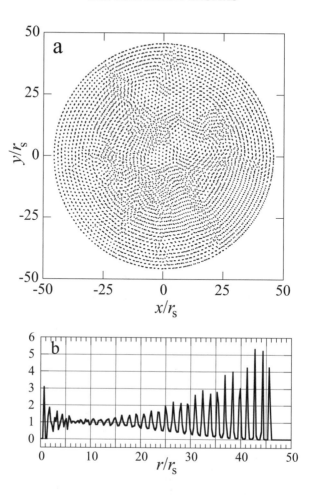

FIG. 10.11.   Cross–section of the bcc lattice containing 120 032 ions (Totsuji *et al.* 2002).
The equatorial slice at $|z| < 1.19 r_s$ (a) and the radial distribution normalized to the mean
ion density (b) are shown.

well with the results of numerical simulations. Wineland *et al.* (1985), used the
Bragg scattering of the cooling laser light along with the direct observation of
the luminescence. Figure 10.13 shows the diffraction image of a spherically symmetric plasma cloud of about $7.5 \cdot 10^5$ Be$^+$ ions. The image was obtained by
employing Bragg scattering from different crystalline planes. The light spots in
Fig. 10.13 represent very well the rectangular reciprocal lattice, which clearly
suggests the formation of the bcc bulk lattice with (110) plane perpendicular to
the laser beam.

Ordered crystalline structures have also been observed in Paul traps by
Raizen *et al.* (1992), Drewsen *et al.* (1998), and Hornekaer *et al.* (2001). The

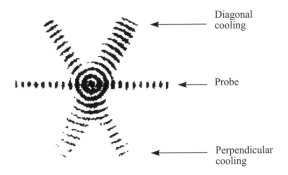

FIG. 10.12.   Experimentally observed ordered structure containing about $1.5 \cdot 10^4$ ions and consisting of 11 shells plus the central string. The image is obtained by employing three crossing laser beams (Gilbert *et al.* 1988).

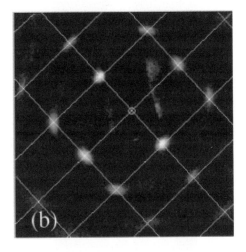

FIG. 10.13.   Bragg scattering image of an ionic crystal in the Penning trap, as obtained by video strobing at the frequency of the crystal rotation (Bollinger *et al.* 2000).

simplest examples are the string–like crystals. By changing the parameters of the trap (e.g., the amplitude of the r.f. field or the potential difference between the electrodes) one can obtain two– or three–dimensional structures. Figure 10.14 shows a zigzag–like structure formed by 11 ions (10 $^{199}Hg^+$ ions and one unidentified ion which did not produce fluorescence). The ions do not form a helix but remain in a plane, because of the weak azimuthal asymmetry of the confinement. Well–ordered structures containing more than 10^5 Mg^+ ions have been investigated by Drewsen *et al.* (1998). The structures are stretched significantly along the trap axis and have up to 10 concentric cylindrical shells around the central string (see Fig. 10.15). The transitions associated with the appearance of new

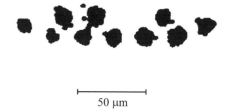

50 μm

FIG. 10.14.   Image of a zigzag crystal containing 10 ^{199}Hg$^+$ ions and one unidentified ion which does not fluoresce (Raizen *et al.* 1992).

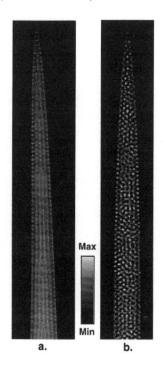

Max

Min

a.   b.

FIG. 10.15.   Experimental image (a) and results of the MD simulations (b) of a stretched crystal containing 3500 ions (Drewsen *et al.* 1998).

shells along the axis can be seen, which is fully in agreement with the results of the MD simulations performed in the same work. Drewsen *et al.* 1998 studied the structural transitions occurring in strongly coupled nonneutral plasmas with an increase of the temperature (i.e., decrease of γ). As γ decreases the shell structure becomes less pronounced, and at the lowest achieved value $\gamma \simeq 4$ the shells practically disappear. Interesting investigations of structural properties of two–component ionic crystals have been performed by Hornekaer *et al.* (2001), where ^{24}Mg$^+$ and ^{40}Ca$^+$ ions were used. Equations (10.3), (10.5), and (10.6)

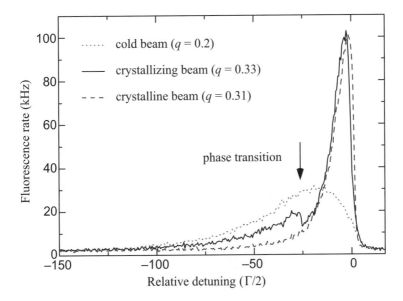

FIG. 10.16. Fluorescence intensity of the ion beam versus the detuning frequency of the cooling lasers $\Delta\omega$ (frequency is in units of the line half–width $\Gamma = 2\pi \cdot 42.7$ MHz). The beam contains $1.8 \cdot 10^4$ ions Mg^+. The arrow indicates inhomogeneity of the curve at $q = 0.33$ suggesting the transition to a crystalline state (Schätz et al. 2001).

suggest that the axial confinement in the Paul trap does not depend on the ion mass, whereas in the radial direction the confinement is mass–dependent. Therefore, the lighter Mg^+ ions concentrate in the vicinity of the trap axis and form an inner structure consisting of cylindrical shells, which resembles very much the structures observed in one–component crystals. The outer part of the cloud consists of Ca^+ ions and has a spheroidal shape which is practically unaffected by the Mg^+ ions – only a few shells close to the axis acquire the cylindrical form. Discussion of earlier work on the centrifugal separation of ions (mostly in Penning traps) is presented in the review by Dubin and O'Neil (1999).

The phase transition of gaseous–like ion beams into one–, two–, or three– dimensional crystalline structures has been recently observed in the r.f. quadrupole storage ring PALLAS (Schätz et al. 2001; Schramm et al. 2001, 2002); the setup is described in section 10.1.4. Typical behavior of a cold ion beam of about $1.8 \cdot 10^4$ particles is shown in Fig. 10.16 for different values of the stability parameter, $q = 2eU_{rf}/m\Omega^2 r_0^2$ (Schätz et al. 2001). The fluorescence grows as the detuning of the cooling lasers, $\Delta\omega = \omega_2 - \omega_1$, decreases. This growth, however, is limited because in close proximity of the resonance, the light absorption can be effective only for one of two photons having almost the same frequencies and moving in the opposite directions. The velocity distribution of ions broadens and, therefore, the fluorescence decreases as the resonance is approached. The increase of the stability parameter q (or the confining potential) causes a noticeable change of the ion beam properties. For $q = 0.33$ one can see an abrupt decrease in the fluorescence intensity at a certain value of $\Delta\omega$ (indicated by the arrow in the figure), followed by a narrow asymmetric peak. Such behavior can be interpreted as a result of the balance between the laser cooling and the heating of ions in a r.f. field. The efficiency of the heating rapidly falls off as the ions become more ordered. The slight decrease of the stability parameter ($q = 0.31$) and, hence a decrease of the r.f. heating, causes the phase transition to occur at larger values of $\Delta\omega$. Because of the noise, this makes detection of the transition very difficult. Another indication of the phase transition is a sudden decrease of the beam diameter. For the experimental conditions corresponding to Fig. 10.16, the formation of the rotating ion string (viz., one–dimensional ion crystal) with the interparticle distance $a \simeq 20$ μm was observed. The longitudinal temperature measured in the reference frame moving together with the beam was about 3 mK, which corresponds to the coupling parameter $\gamma > 500$. By changing the parameters of the plasma and the trap (first of all, the number of ions), two–dimensional zigzag and three–dimensional helix structures have been obtained and investigated by Schramm et al. (2002). It was shown that the stability range of the crystalline structures formed in the storage ring is substantially smaller than that obtained in stationary ionic crystals and in the MD simulations.

10.3   Melting of mesoscopic crystals

The nonneutral plasmas form crystalline states at sufficiently low temperatures. As the temperature grows the crystal melts, yet keeping the correlation and the short–range order, and when the temperature is high enough the plasma is in a disordered gaseous state. Both the numerical simulations and experiments suggest that the stable crystalline state of the macroscopic plasma is the bcc lattice, and melting occurs at $\gamma \simeq 173$. In mesoscopic plasmas, different types of ordered structures can be formed. As discussed in the previous section, ions confined in a parabolic potential well (typical for ion traps) form clouds with a shell structure and a well–defined surface. In the surface layer and in each shell ions arrange themselves into a triangle lattice. As the number of ions in the cloud increases, the transition to a bcc lattice is observed in the bulk region, which is typical of infinite plasmas.

The melting of such finite–size crystalline structures has been studied by Schiffer (2002) with MD simulations. In finite systems the transitions between ordered and disordered states do not occur abruptly – they develop continuously in a certain range of parameters (e.g., temperature or density). Figure 10.17 shows two outer spherical shells of the cloud containing 10^4 ions at two different temperatures. Also, the radial density distribution and the pair correlation function $g(r)$ are shown, as obtained from simulations at three different temperatures corresponding to the crystalline, liquid, and gaseous states.

The potential energy and the heat capacity per particle versus the temperature are shown in Fig. 10.18. One can see the well–defined melting transition,

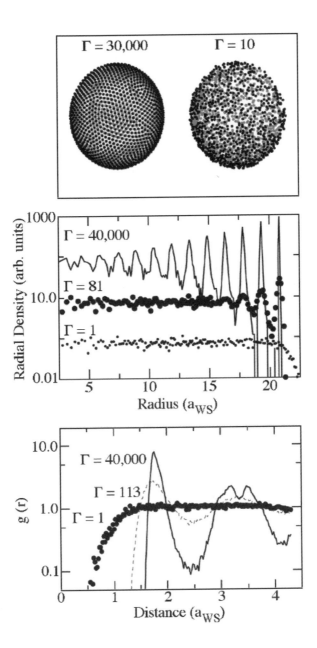

FIG. 10.17. The upper figure shows the image of two outer shells containing 10^4 ions. Ions of the outmost shell are in black, the next shell is in gray. The middle figure shows the radial distribution of the ion density, and the lower one, the pair correlation function $g(r)$ (the length is in units of the Wigner–Seitz radius) (Schiffer 2002).

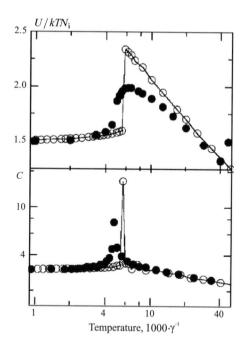

FIG. 10.18. Total energy U and thermal capacity C per particle for an ensemble of charges (Schiffer 2002). The calculations are for an infinite system (o) and for systems containing 10^4 ions (•). Temperature is in units of γ^{-1}.

both in a finite system of 10^4 ions and in an infinite Coulomb system, although in the former case the melting temperature is somewhat smaller and the transition itself is smeared out. Based on the analysis of the diffusion coefficients performed by Schiffer (2002), a significant increase in the ion diffusion within each shell, as compared with the diffusion between the neighboring shells, has been revealed in the vicinity of the melting transition.

Schiffer (2002) also addressed the question: Which factor plays the more important role in the decrease of the melting temperature observed in mesoscopic crystals – the finite size or different types of ordered structures? Figures 10.19 and 10.20 show the heat capacity and the melting temperature calculated for a system of 100, 1000 and 10 000 ions, as well as for an infinite system. The height of the heat capacity peak decreases and its maximum shifts towards smaller temperatures as the number of ions decreases. The obtained results exhibit a smooth dependence on the size, regardless of the particular crystalline structure. Figure 10.20 shows the melting temperature, T_m, versus the number of ions in the outmost shell, N_s, normalized to the total ion number. One can see that $\Delta T_m = T_m(\infty) - T(N) \propto N_s$. Hence, one can claim that the decrease of the melting temperature is almost solely due to the finiteness of the system.

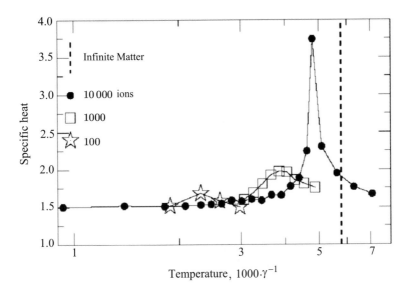

FIG. 10.19. Thermal capacity of nonideal plasmas near the melting point (Schiffer 2002). The calculations are for 100 ($\star$), 1000 ($\square$), and 10000 ($\bullet$) ions as well as for an infinite system (dashed line).

10.4 Coulomb clusters

In this section we discuss the properties of small ($N_i \sim 10$) systems of charged particles. At sufficiently low temperatures they form ordered structures called "Coulomb clusters". Although the properties of such clusters and mesoscopic crystalline systems are similar, the Coulomb clusters exhibit a number of quite interesting features. Below we discuss the clusters trapped in a parabolic potential well, since this confinement is typical for most experiments and numerical (MD and MC) simulations. The equilibrium configurations of particles are determined by the minimum of the potential energy. Taking into account Eq. (10.3), the energy is

$$U = \sum_{i>j} \frac{e^2}{|\mathbf{r}_i - \mathbf{r}_j|} + \frac{M\omega_z^2}{2} \sum_i (z_i^2 + \beta\, r_i^2). \tag{10.8}$$

By introducing dimensionless variables,

$$r \to (e^2/M\omega_z^2)^{1/2} r, \quad U \to \left(\frac{e^4 M\omega_z^2}{2}\right)^{1/3} U, \tag{10.9}$$

one can reduce the potential energy to the form which depends only on the number of particles and the asymmetry parameter β. In a spherically symmetric trap, $\beta = 1$, we have

$$U = \sum_{i>j} \frac{1}{|\mathbf{r}_i - \mathbf{r}_j|} + \sum_i r_i^2. \tag{10.10}$$

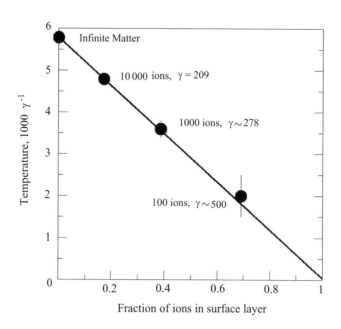

FIG. 10.20. Temperature corresponding to the maximum of the thermal conductivity, versus the fraction of ions in the outmost shell, N_s/N_i (Schiffer 2002).

Figure 10.21 shows equilibrium configurations of clusters consisting of different numbers of particles and confined in a spherically symmetric potential well. At a glance, one can think that all ions in clusters with $N_i \leq 12$ lie on a sphere. In fact, however, this is true only for "symmetric" configurations with $N_i = 2$, 3, 4, 6, 8, 12, whereas for $N_i = 5$, 7, 9, 10, 11 the radial positions of ions are not the same (Dubin and O'Neil 1999). Starting from $N = 13$, the second shell is formed. As the number of ions increases further, the concentric shell structure is forming, as described in the previous sections.

Tsuruta and Ichimaru (1993) calculated the potential energy of Coulomb clusters with $N_i \leq 60$. The interaction energy per particle, $U/(N_i e^2/r_s)$, decreases as the number of particles grows (see Fig. 10.22). This is because the role of surface effects becomes weaker. One can see several minima in the dependence U versus N_i; the deepest ones correspond to $N_i = 6$, 12, and 38. Following nuclear physics terminology, Tsuruta and Ichimaru (1993) refereed to these N_i as "magic numbers" for spherical Coulomb clusters. Note that the "magic" clusters have the most–symmetric configurations corresponding to an octahedron for $N_i = 6$, an icosahedron for $N_i = 12$, and a two–block structure consisting of an octahedron inside a phase–centered icosahedron for $N_i = 38$ (Tsuruta and Ichimaru 1993).

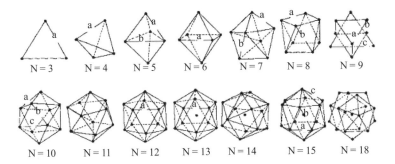

FIG. 10.21. Configurations of Coulomb clusters in a spherically symmetric parabolic potential well (Rafac *et al.* 1991).

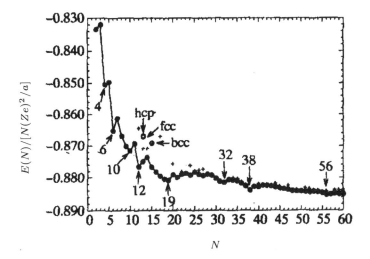

FIG. 10.22. Dimensionless energy of the electrostatic interaction for N–particle clusters (Tsuruta and Ichimaru 1993).

In anisotropic confinement the equilibrium configuration is determined by the value of β. For $\beta \gg 1$, the ion cloud is stretched along the z–axis, forming one–dimensional ordered structures. For $\beta \ll 1$ ions form two–dimensional structures in the xy–plane. The ground state of two–dimensional Coulomb clusters is well studied (Lozovik 1987; Lozovik and Mandelshtam 1990, 1992; Rafac *et al.* 1991; Bedanov and Peeters 1994). For $N_i \leq 5$ the ions lie on a circle, for $6 \leq N_i \leq 8$ one of the ions is located in the center, and for $N_i = 9$ two ions are inside the ring. For $N_i = 15$ the second (inner) ring is completed (5, 10) and for $N_i = 16$ one ion again appears in the center (1, 5, 10). The two–dimensional ring structures are completely analogous to the three–dimensional shell structures in spheroidal nonneutral plasmas.

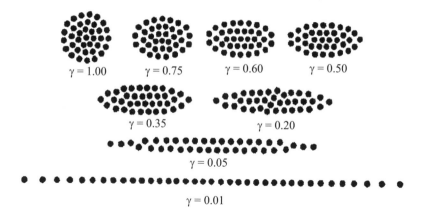

F IG . 10.23.   Configurations of a two–dimensional cluster containing 37 ions for different
values of the anisotropy parameter γ (Lozovik and Rakoch 1999).

The two–dimensional confinement can be anisotropic as well. Figure 10.23
shows different configurations of a two–dimensional Coulomb cluster of 37 ions,
corresponding to different values of the anisotropy parameter (Lozovik and Rakoch 1998, 1999). Here, the parameter γ determines the anisotropy of the trap,

$$U_{\text{ext}} = \gamma \sum_i x_i^2 + (2 - \gamma) \sum_i y_i^2, \qquad (10.11)$$

so that $\gamma = 1$ corresponds to circular symmetry. As γ increases the number of
rings decreases and finally the cluster is stretched into a one–dimensional string
with inhomogeneous interparticle distance.

As the temperature increases, the ion oscillations around the equilibrium
increase as well and eventually the cluster melts (Lozovik 1987; Bedanov and
Peeters 1994; Lozovik and Rakoch 1998, 1999). The melting occurs in two stages,
both in two– and three–dimensional cases. At the first stage, when the temperature is relatively low, the orientational melting starts – the transition from the
steady configuration to the state with enhanced rotational oscillations of the
neighboring (rings) shells, yet the ions within each shell remain stable. At the
second stage the radial order disappears as well. As the number of ions in a
cluster increases, the first stage becomes less pronounced and the orientational
melting can only be observed in a few shells near the surface.

References

Anderson, M. H., Ensher, J. R., Matthews, M. R., Wieman, C. E., and Cornell,
    E. A. (1995). Observation of Bose–Einstein condensation in a dilute atomic
    vapor. *Science*, **269**, 198–201.
Bedanov, V. M. and Peeters, F. M. (1994). Ordering and phase transitions of
    charge particles in a classical finite two–dimensional system. *Phys. Rev. B*, **49**,
    2667–2676.

Bollinger, J. J. and Wineland, D. J. (1984). Strongly coupled nonneutral ion plasma. *Phys. Rev. Lett.*, **53**, 348–351.

Bollinger, J. J., Mitchell, T. B., Huang, X.-P., Itano, W. M., Tan, J. N., Jelenkovic, B. M., and Wineland, D. J. (2000). Crystalline order in laser–cooled, nonneutral ion plasmas. *Phys. Plasmas*, **7**, 7–13.

Budker, G. I., Dikanskiy, N. S., Kudelaynen, V. I., Meshkov, I. N., Parchomchuk, V. V., Pestrikov, D. V., Skrinsky, A. N., and Sukhina, B. N. (1976). Experimental studies of electron cooling. *Part. Accel.*, **7**, 197–211.

Chu, S. (1999). The manipulation of neutral particles. *Rev. Mod. Phys.*, **70**, 685–706.

Cohen–Tannoudji, C. N. (1998). Manipulations atoms with photons. *Rev. Mod. Phys.*, **70**, 707–719.

Cole, M. W. (1974). Electronic surface states of liquid helium. *Rev. Mod. Phys.*, **46**, 451–464.

Davidson, R. C. (1974). *Theory of nonneutral plasmas*. Benjamin, Reading.

Davidson, R. C. (1990). *Physics of nonneutral plasmas*. Addison–Wesley, Redwood City.

Dement'ev, E. N., Dikanskiy, N. S., Medvedko, A. S., Parkhomchuk, V. V., and Pestrikov, D. V. (1980). Measuring of the proton beam thermal noises on NAP–M storage. *Soviet Phys.–JTP*, **50**, 1717–1721.

Drewsen, M., Brodersen, C., Hornekaer, L., Hangst, J. S., and Schifffer, J. P. (1998). Large ion crystals in a linear Paul trap. *Phys. Rev. Lett.*, **81**, 2878–2881.

Dubin, D. H. E. (1996). Effect of correlations on the thermal equilibrium and normal modes of a nonneutral plasma. *Phys. Rev. E*, **53**, 5268–5290.

Dubin, D. H. E. and O'Neil, T.M. (1999). Trapped nonneutral plasmas, liquids, and crystals (the thermal equilibrium states). *Rev. Mod. Phys.*, **71**, 87–172.

Fisher, D. S., Halperin, B. I., and Platzman, P. M. (1979). Phononripplon coupling and the two–dimensional electron solid on a liquid–helium surface. *Phys. Rev. Lett.*, **42**, 798–801.

Gilbert, S. L., Bollinger, J. J., and Wineland, D. J. (1988). Shell–structure phase of magnetically confined strongly coupled plasmas. *Phys. Rev. Lett.*, **60**, 2022–2025.

Grimes, C. C. and Adams, G. (1979). Evidence of liquid–to–crystal phase transition in a classical two–dimensional sheet of electrons. *Phys. Rev. Lett.*, **42**, 795–798.

Hänsch, T. and Schawlow, A. (1975). Cooling of gases by laser radiation. *Opt. Commun.*, **13**, 68–69.

Hornekaer, L., Kjaergaard, N., Thommesen, A.M., and Drewsen, M. (2001). Structural properties of two–component Coulomb crystals in linear Paul traps. *Phys. Rev. Lett.*, **86**, 1994–1997.

Ichimaru, S., Iyetomi, H., and Tanaka, S. (1987). Statistical physics of dense plasmas: Thermodynamics, transport coefficients and dynamic correlations. *Phys. Rep.*, **149**, 91–205.

Letokhov, V. S., Minogin, V. G., and Pavlik, B. D. (1977). Cooling and capture of atoms and molecules by resonant light field. *JETP*, **45**, 698–705.

Lozovik, Y. E. (1987). Ion and electron clusters. *Phys.–Uspekhi*, **30**, 912–913.

Lozovik, Y. E. and Mandelshtam, V. A. (1990). Coulomb clusters in a trap. *Phys. Lett. A*, **145**, 269–271.

Lozovik, Y. E. and Mandelshtam, V. A. (1992). Classical and quantum melting of a Coulomb cluster in a trap. *Phys. Lett. A*, **165**, 469–472.

Lozovik, Y. E. and Rakoch, E. A. (1998). Energy barriers, structure, and two-stage melting of microclusters of vortices. *Phys. Lett. A*, **240**, 311–321.

Lozovik, Y. E. and Rakoch, E. A. (1999). Structure, melting, and potential barriers in mesoscopic clusters of repulsive particles. *JETP*, **89**, 1089–1102.

Morf, R. H. (1979). Temperature dependence of the shear modulus and melting of two–dimensional electron solid. *Phys. Rev. Lett.*, **43**, 931–935.

Neuhauser, W., Hohenstatt, M., Toschek, P., and Dehmelt, H. (1978). Optical–sideband cooling of visible atom cloud confined in parabolic well. *Phys. Rev. Lett.*, **41**, 233–236.

Penning, F. M. (1936). The spark discharge in low pressure between coaxial cylinders in an axial magnet field. *Physika*, **3**, 873–894.

Phillips, W. D. (1998). Laser cooling and trapping of neutral atoms. *Rev. Mod. Phys.*, **70**, 721–741.

Rafac, R., Schiffer, J. P., Hangst, J. S., Dubin, D. H. E., and Wales, D. J. (1991). Stable configurations of confined cold ionic systems. *Proc. Natl. Acad. Sci. USA.*, **88**, 483–486.

Rahman, A. and Schiffer, J. P. (1986). Structure of a one–component plasma in an external field: A molecular dynamic study of particle arrangement in a heavy–ion storage ring. *Phys. Rev. Lett.*, **57**, 1133–1136.

Raizen, M. G., Gilligan, J. M., Bergquist,W. M., Itano, W. M., and Wineland, D. J. (1992). Ionic crystals in linear Paul trap. *Phys. Rev. A*, **45**, 6493–6501.

Schätz, T., Schramm, U., and Habs, D. (2001). Crystalline ion beams. *Nature*, **412**, 717–720.

Schiffer, J. P. (2002). Melting of crystalline confined plasmas. *Phys. Rev. Lett.*, **88**, 205003/1–4.

Schramm, U., Schätz, T., and Habs, D. (2001). Bunched crystalline ion beams. *Phys. Rev. Lett.*, **87**, 184801/1–4.

Schramm, U., Schätz, T., and Habs, D. (2002). Three–dimensional crystalline ion beams. *Phys. Rev. E*, **66**, 036501/1–9.

Shikin, V. B. (1977). Mobility of charges in liquid, solid, and gaseous helium.*Sov. Phys. Uspekhi*, **20**, 226–248.

Shikin, V. B. and Monarkha, Y. P. (1989). *Two–dimensional charged systems in helium (in Russian)*. Nauka, Moscow

Totsuji, H., Kishimoto, T., Totsuji, C., and Tsuruta, K. (2002). Competition between two forms of ordering in finite Coulomb clusters. *Phys. Rev. Lett.*, **88**, 125002/1–4.

Tsuruta, K. and Ichimaru, S. (1993). Binding energy, micrustructure, and shell model of Coulomb clusters. *Phys. Rev. A*, **48**, 1339–1344.

Wieman, C. E., Pritchard, D. E., and Wineland, D. J. (1999). Atom cooling, trapping, and quantum manipulation. *Rev. Mod. Phys.*, **71**, S253–S262.

Wigner, E. (1934). On the interaction of electrons in metals. *Phys. Rev.*, **46**, 1002–1011.

Wineland, D. and Dehmelt, H. (1975). Proposed $10^{14}\Delta\nu < \nu$ laser fluorescence spectroscopy on Tl$^+$ mono–ion oscillator III. *Bull. Am. Phys. Soc.*, **20**, 637.

Wineland, D., Drullinger, R. and Walls, F. (1978). Radiation–pressure cooling of bound resonant absorbers. *Phys. Rev. Lett.*, **40**, 1639–1642.

Wineland, D. J., Bollinger, J. J., Itano, W. M., and Prestage, J. D. (1985). Angular–momentum of trapped atomic particles. *J. Opt. Soc. Am. B*, **2**, 1721–1729.

11

DUSTY PLASMAS

11.1 Introduction

Dusty plasmas constitute ionized gases containing charged particles of condensed matter. Other terms used for such systems are "complex plasmas", "colloidal plasmas", and "plasmas with a condensed disperse phase".

The term "complex plasmas" is currently widely used in the scientific literature to designate dusty plasmas specially designed to study the properties of the dust component. Of particular interest is the situation in which the dust particles are strongly coupled (and form liquid–like and crystal–like structures) and investigation of plasma processes at the kinetic level is possible. The main attention here is given to such systems. At the same time, dusty plasmas in space and the atmosphere, chemical plasmas with growing particles, etc., are hardly considered here. In light of this we will not distinguish between dusty and complex plasmas and will mostly use the term "dusty plasma" as generally accepted.

Dust and dusty plasmas are quite natural in space. They are present in planetary rings, comet tails, and interplanetary and interstellar clouds (Goertz 1989; Northrop 1992; Tsytovich 1997; Bliokh *et al.* 1995). Complex plasmas are found in the vicinity of artificial satellites and space stations (Whipple 1981; Robinson and Coakley 1992) and in thermonuclear facilities with magnetic confinement (Tsytovich and Winter 1998; Winter and Gebauer 1999; Winter 2000). The properties of dusty plasmas are very actively investigated in the laboratory, too. Dust particles are not only deliberately introduced into a plasma but can also form and grow due to different physical processes. The widespread presence of dusty plasmas, combined with a number of their unique (relative simplicity of producing, observing, and controlling the parameters, feasibility of taking measurements at a kinetic level) and unusual (openness of dusty plasma systems, inconstancy of the dust particle charge, high dissipativity, self–organization and formation of ordered structures) properties, make dusty plasmas extraordinarily attractive and interesting objects for investigations.

Dust particles immersed in a plasma acquire an electric charge and constitute an additional charged component. However, the properties of dusty plasmas are much more diverse than that of the usual multicomponent plasmas consisting of electrons and different types of ions. The dust particles are recombination centers for plasma electrons and ions and sometimes sources of electrons (thermo-, photo-, and secondary electron emission). This implies that the dust component can significantly influence the plasma ionization balance. In addition, the dust particle charge is not fixed, but is determined by the surrounding plasma parame-

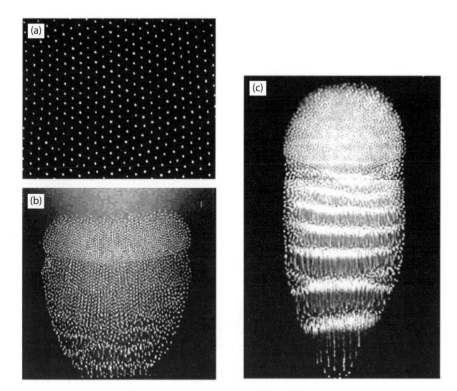

FIG. 11.1. Typical images of the ordered structures of dust particles in r.f. and d.c. discharges. (a) A horizontal cross section of hexagonal structure (shown is the area 6.1×4.2
mm^2 containing 392 particles of 6.9 μm in diameter) formed in the sheath region of an r.f.
discharge (Morfill and Thomas 1996). (b) Vertical cross–section of the ordered structure
in a stratified d.c. discharge. In the lower region of the structure, the vertical oscillation
of dust particle density is present, in the central region high ordering appears, and in the
periphery of the upper region the particles experience convective motion. (c) Self–excited
dust acoustic wave in a d.c. gas discharge (Fortov $et\ al.$ 2000) (neutral gas pressure $p = 0.2$
torr, wave frequency $\omega \sim 60$ s^{-1}, wavenumber $k \sim 60$ cm^{-1}, and propagation velocity
$v_{ph} \sim 1$ cm s^{-1}).

ters, and can vary both spatially and temporarily. Moreover, charge fluctuations
are present even for constant plasma parameters because of the stochastic nature
of the charging process.

Due to the large charges carried by the dust particles, the electrostatic energy of interaction between them (proportional to the product of charges of
interacting particles) is high. For this reason, strong coupling in the dust subsystem can be achieved much more easily than in the electron–ion subsystem,
whilst typically the dust concentration is much lower than that of electrons and
ions. Hence, the appearance of short–range order and even crystallization in the

dust subsystem become achievable. The first experimental realization of the ordered (quasicrystal–like) structures of charged microparticles was implemented by Wuerker et al. (1959) with the help of a modified Paul trap (Paul and Raether 1955). The possibility of dust subsystem crystallization in a nonequilibrium gas discharge plasma was predicted by Ikezi (1986). Experimentally, the ordered structures of dust particles were observed almost a decade later, first near the sheath edge of r.f. discharges (Chu and I 1994; Thomas et al. 1994; Hayashi and Tachibana 1994; Melzer et al. 1994), where the strong vertical electric field can compensate for gravity and makes dust particle levitation possible. Figure 11.1(a) shows a horizontal cross–section of a typical ordered (quasicrystal–like) structure found in the sheath region of an r.f. discharge (Morfill and Thomas 1996). In the vertical direction, the particles are also ordered and arranged one beneath the other to form chain–like structures. Later on, ordered structures of macroparticles were observed in a thermal dusty plasma at atmospheric pressure (Fortov et al. 1996b,c; Nefedov et al. 1997), in the positive column of a d.c. glow discharge (Fortov et al. 1996a; Nefedov et al. 2000), and in a nuclear–induced dusty plasma (Fortov et al. 1999a,b). Figure 11.1(b) displays an example of the ordered structure of dust particles in a d.c. glow discharge. In the lower region of the structure, self–excited nonlinear oscillations of the dust particle density are present; most of the central region is occupied by a highly ordered regular structure, and in the upper region the particles experience convective motion (Fortov et al. 1999a). Dust component crystallization and phase transitions in different types of dusty plasmas constitute a wide area of current research.

The presence of the dust component is essential for the collective processes in dusty plasmas. Dust not only modifies the wave spectrum in comparison with dust–free plasma, but can also lead to the appearance of new modes and new mechanisms of damping and instability. The presence of dust introduces new characteristic spatial and temporal scales. For example, the dust plasma frequency is several orders of magnitude smaller than the ion plasma frequency, because the dust particles are extremely massive compared to ions. This leads to the emergence of a new very–low–frequency mode – the dust acoustic wave, as it is called, which represents the oscillations of particles against the quasi–equilibrium background of electrons and ions (in some sense it is analogous to the ion acoustic wave, which represents ion oscillations in a gas against the equilibrium background of electrons). The characteristic frequency range of this mode is $10–100 \text{ s}^{-1}$, which makes it particularly interesting from the experimental point of view. An illustration is presented in Fig. 11.1(c) which shows a typical pattern of dust acoustic waves. These waves were self–excited under certain conditions in a d.c. discharge plasma (Fortov et al. 2000), revealing the existence of some mechanisms of dust acoustic wave instability.

Such properties as relative simplicity of production, observation, and control, as well as fast relaxation and response to external disturbances make dusty plasmas not only attractive objects for investigation, but also very effective instruments for studying the properties of strongly coupled plasma, fundamental properties of crystals, etc. Especially important is that the dust particles can be usually seen with the naked eye or easily visualized with the help of a simple optical technique. This allows us, in principle, to perform measurements at the kinetic level, yielding the dust particle distribution function in coordinate and momentum phase space, $f_d(\mathbf{r}, \mathbf{p}, t)$. The detailed investigation of phase transitions, particle transport, low–frequency waves, etc. at the kinetic level is thus possible. In addition, the diagnostics of dust particles and surrounding plasmas is considerably simplified.

Despite the fact that dusty plasmas in the laboratory were first discovered by Langmuir (1924) in the 1920s, active investigation of them began only a few decades ago, mainly in connection to such applications as rocket-fuel combustion products, solid-fuel-fired magnetohydrodynamic generators, and dusty clouds in the atmosphere (Zhukhovitskii *et al.* 1984; Yakubov and Khrapak 1989; Sodha and Guha 1971; Soo 1990). In the late 1980s, interest moved to such issues as dust charging, propagation of electromagnetic waves and their damping and instability, mostly in connection to space dusty plasmas (Goertz 1989; Havnes *et al.* 1987; Pilipp *et al.* 1987). Growth in interest in the field in the early 1990s was mostly connected with wide utilization of plasma deposition and etching technologies in microelectronics, as well as with the production of thin films and nanoparticles (Selwyn *et al.* 1993, 1996; Bouchoule 1999; Kersten *et al.* 2001). This interest was caused by the fact that the presence of the dust particles in a processing plasma not only leads to the contamination of a semiconductor element surface and, hence, to the increased yield of defect elements, but can also perturb the plasma in an often unpredictable way. To prevent or lower the role of these negative impacts, it is necessary to understand the processes involved in the formation and growth of dust particles in a gas discharge plasma, mechanisms of dust transport, and the influence of dust on plasma parameters. Finally, in the mid–1990s, crystal–like structures of dust particles in different types of dusty plasmas were discovered experimentally (Chu and I 1994; Thomas *et al.* 1994; Hayashi and Tachibana 1994; Melzer *et al.* 1994; Fortov *et al.* 1996a,b,c, 1999a,b; Nefedov *et al.* 1997, 2000). These findings led to tremendous growth in the investigation of dusty plasmas, which is still in progress (we note in this context that active investigations of Wigner crystallization of ions in different types of ion traps (Dubin and O'Neil 1999), and electrons on the surface of liquid helium (Shikin 1989) are also being performed).

Among the present–day directions of dusty plasma investigations we can distinguish the following:

- formation of ordered structures, including crystallization and phase transitions in the dust subsystem;

- elementary processes in dusty plasmas: charging of dust for different plasma and particle parameters, interactions between the particles, external forces acting on the particles;

- linear and nonlinear waves in dusty plasmas (solitons, shock waves, Mach cones), their dynamics, damping, and instability.

Of exceptional significance are the experiments under microgravity conditions, first conducted in 1998 on–board the "Mir" space station, and later (in the beginning of 2001) started on–board the International Space Station (the PKE–Nefedov experiment). In the absence of gravity, the influence of strong electric fields (which are necessary on the ground to levitate particles) and strong plasma anisotropies caused by these fields is substantially reduced.

Investigations into dusty plasmas are nowadays a rapidly growing field of research, which includes such aspects as fundamental problems in the physics of plasmas, hydrodynamics, kinetics of phase transitions, nonlinear physics, solid state physics, and several applications (e.g., in nanotechnology, plasma technology, new materials). More and more research groups from different countries are taking up the field. The number of scientific publications appearing every year is increasing constantly. Taking this into account, the focus of this chapter is first and foremost on the aspects that are required for understanding the basic properties of dusty plasmas and are most frequently needed in research. From an immense amount of material we tried to select those experimental and theoretical results which, in our opinion, illustrate the most important achievements in the investigation of dusty plasmas, reported over the last decade.

11.2 Elementary processes in dusty plasmas

11.2.1 *Charging of dust particles in plasmas (theory)*

In this section different processes leading to the charging of dust particles immersed in plasmas are considered. Expressions for the ion and electron fluxes to (from) the particle surface, caused by different processes (collection of plasma electrons and ions, secondary, thermionic and photoelectric emission of electrons from the particle surface), are given. Problems such as the stationary surface potential, kinetics of charging, changing of plasma charge composition in response to the presence of the dust, as well as dust particle charge fluctuations due to the stochastic nature of the charging process, are considered. More detailed examination of charging processes can be found in Tsytovich *et al.* (2002) and Morfill *et al.* (2003). Here we mostly focus on the processes which are important for the problems addressed in this book. Below in this section we consider spherical dust particles as being of radius a.

11.2.1.1 *Charging in gas discharge plasmas.* In a nonequilibrium plasma of low–pressure gas discharges the ions, atoms, and macroscopic charged particles typically remain "cold", whilst the electron energies are relatively high. In the absence of emission processes, the charge of a dust particle is negative. This is connected to the fact that the electron and ion fluxes are directed to the surface of an uncharged particle. Similar to the theory of electric probes (Chung *et al.* 1975; Allen 1992) it is customary to assume that electrons and ions recombine on the particle surface, and neutral particles appearing in the process of recombination either remain on the surface or return to the plasma (in the first case, the charging can be accompanied by particle growth). Due to higher mobility of the electrons

their flux exceeds considerably that of the ions, and the neutral particle begins
to charge negatively. The emerging negative charge on the particle leads to the
repulsion of the electrons and the attraction of the ions. The charge grows (in
absolute magnitude) until the electron and ion fluxes are balanced. On longer
timescales, the charge is turned practically constant and experiences only small
fluctuations around its equilibrium value.

The stationary surface potential of the dust particle (connected to the particle
charge) is defined (with the accuracy of a coefficient on the order of unity) as
$\varphi_s = -T_e/e$, where T_e is the electron temperature (in this chapter temperature is
in energy units). Physically, this can be explained by the requirement that in the
stationary state most of the electrons should not have kinetic energies sufficient
to overcome the potential barrier between the particle surface and surrounding
plasma.

11.2.1.2 *Orbit motion limited approximation.* For a quantitative description
of particle charging in gas discharge plasmas, probe theory is generally adopted.
One of the most frequently used approaches is the orbit motion limited (OML)
theory (Chung *et al.* 1975; Allen 1992; Goree 1994). This approach allows one to
determine the cross–sections for electron and ion collection by the dust particle
only from the laws of conservation of energy and angular momentum. Usually,
the conditions of applicability of the OML theory are formulated as (Goree 1994)

$$a \ll \lambda_D \ll \ell_{i(e)}, \tag{11.1}$$

where λ_D is the plasma screening length (the corresponding Debye radius), and
$\ell_{i(e)}$ is the mean free path of the ions (electrons). It is also assumed that the
dust particle is isolated in the sense that other dust particles do not affect the
motion of electrons and ions in its vicinity.

Assuming that the electrons and the ions are collected if their trajectories
cross or graze the particle surface, the corresponding velocity–dependent cross–
sections are given by

$$\sigma_e(v) = \begin{cases} \pi a^2 \left(1 + \dfrac{2e\varphi_s}{m_e v^2}\right), & \dfrac{2e\varphi_s}{m_e v^2} > -1, \\ 0, & \dfrac{2e\varphi_s}{m_e v^2} < -1, \end{cases} \tag{11.2}$$

$$\sigma_i(v) = \pi a^2 \left(1 - \dfrac{2e\varphi_s}{m_i v^2}\right), \tag{11.3}$$

where $m_{e(i)}$ is the electron (ion) mass, and v denotes the velocity of the electrons
and ions relative to the dust particle. Here, the surface potential φ_s of the dust
particle is negative ($\varphi_s < 0$), and the ions are singly charged. We note that
within OML theory $\sigma_e(v)$ and $\sigma_i(v)$ are independent of the exact form of the
electrostatic potential around the dust particle. Limitations to this approach are
considered below.

Electron and ion fluxes to the particle surface are determined by the integration of the corresponding cross–sections with velocity distribution functions $f_{e(i)}(v)$:

$$I_{e(i)} = n_{e(i)} \int v\sigma_{e(i)}(v) f_{e(i)}(v) d^3 v, \qquad (11.4)$$

where $n_{e(i)}$ is the electron (ion) number density. For the Maxwellian velocity distribution of plasma particles, viz.

$$f_{e(i)}(v) = (2\pi v_{T_{e(i)}}^2)^{-3/2} \exp\left(-\frac{v^2}{2v_{T_{e(i)}}^2}\right), \qquad (11.5)$$

where

$$v_{T_{e(i)}} = \sqrt{\frac{T_{e(i)}}{m_{e(i)}}} \qquad (11.6)$$

is the electron (ion) thermal velocity, the integration in Eq. (11.4) performed with the use of formulas (11.2) and (11.3) gives

$$I_e = \sqrt{8\pi} a^2 n_e v_{T_e} \exp\left(\frac{e\varphi_s}{T_e}\right), \qquad (11.7)$$

$$I_i = \sqrt{8\pi} a^2 n_i v_{T_i} \left(1 - \frac{e\varphi_s}{T_i}\right). \qquad (11.8)$$

The stationary potential of the dust particle surface (floating potential) is determined by the balance of electron and ion fluxes collected by the particle:

$$I_e = I_i. \qquad (11.9)$$

It is convenient to introduce the following dimensionless parameters which are widely used throughout this chapter:

$$z = \frac{|Z_d|e^2}{aT_e}, \qquad \tau = \frac{T_e}{T_i}, \qquad \mu = \frac{m_e}{m_i}. \qquad (11.10)$$

Here, z is the absolute magnitude of the particle charge in units of aT_e/e^2, while τ and μ are the electron–to–ion temperature and mass ratios, respectively. We note that sometimes τ is used to denote the inverse ratio T_i/T_e (see, e.g., Tsytovich et al. 2002; Morfill et al. 2003). Typically, in gas discharge plasmas $\tau \gg 1$ ($\tau \sim 10-100$), $z \sim 1$, and, of course, $\mu \ll 1$. It is also assumed that the particle charge and surface potential are connected through the expression $Z_d e = a\varphi_s$, where Z_d is the dust particle charge number (namely, the charge expressed in units of the elementary charge). This connection (like for a charged sphere in a vacuum) follows from a solution of the linearized problem concerning a potential distribution around a spherical macroparticle in Boltzmann plasmas, when $a \ll \lambda_D$. This is usually a good approximation, whilst, in principle, there may

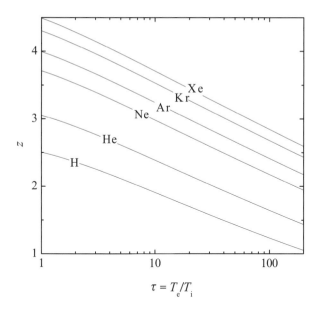

FIG. 11.2.   Dimensionless charge $z = \frac{|Z_d|e^2}{aT_e}$ of an isolated spherical particle as a function of electron–to–ion temperature ratio $\tau = T_e/T_i$ for different types of isotropic plasmas.

be some deviations due to strongly nonlinear screening and/or nonequilibrium distribution of the electrons and ions around the dust particle.

In dimensionless quantities (11.10), the flux balance equation (11.9) with expressions (11.7) and (11.8) taken into account can be rewritten in the form

$$\exp(-z) = \frac{n_i}{n_e} \left(\frac{\mu}{\tau}\right)^{1/2} (1 + z\tau). \qquad (11.11)$$

For the isolated particle, the quasineutrality condition reduces to $n_i = n_e$. In this case, the dimensionless surface potential z depends on two parameters only – the electron–to–ion temperature ratio, and the gas type (electron–to–ion mass ratio). In Fig. 11.2, the values of z are presented for different gases (H, He, Ne, Ar, Kr, Xe) as functions of τ. The particle potential decreases with increasing $\tau = T_e/T_i$, and builds up with increasing gas atomic mass. For typical values of $\tau \sim 10 - 100$, the dimensionless potential ranges over $z \sim 2 - 4$. For a particle with $a \sim 1$ μm and $T_e \sim 1$ eV, the characteristic charge number is $(1-3) \cdot 10^3$.

Dusty plasmas are often subject to external electric fields, which occurs, for example, near walls or electrodes of gas discharges. In this case, the ion drift relative to the stationary dust component substantially affects the particle charging. Using instead of Eq. (11.5) the shifted Maxwellian velocity distribution for ions, viz.

$$f_i(\mathbf{v}) = (2\pi v_{T_i}^2)^{-3/2} \exp\left[\frac{(\mathbf{v} - \mathbf{u})^2}{2v_{T_i}^2}\right], \qquad (11.12)$$

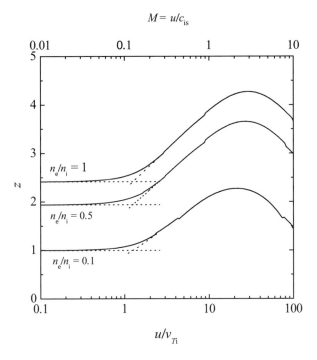

$$M = u/c_{\mathrm{is}}$$

FIG. 11.3.   Dimensionless charge $z = \frac{|Z_{\mathrm{d}}|e^2}{aT_{\mathrm{e}}}$ of an isolated spherical particle as a function of the ratio $u/v_{T_{\mathrm{i}}}$ of the ion drift to ion thermal velocity (or Mach number $M = u/c_{\mathrm{i}}$, where $c_{\mathrm{i}} = \sqrt{T_{\mathrm{e}}/m_{\mathrm{i}}}$ is the ion speed of sound) for a plasma with ion drift – the calculations were done for three values of the ratio of electron–to–ion number density and correspond to an argon plasma with $\tau = 100$.

(where $\mathbf{u}$ is the average ion drift velocity), and the cross section (11.3), we get for the ion flux the following expression (Whipple 1981; Uglov and Gnedovets 1991; Kilgore *et al.* 1994) (remembering that $\varphi_{\mathrm{s}} < 0$):

$$I_{\mathrm{i}} = \sqrt{2\pi} a^2 n_{\mathrm{i}} v_{T_{\mathrm{i}}} \left\{ \sqrt{\frac{\pi}{2}} \frac{1 + u^2/v_{T_{\mathrm{i}}}^2 - 2e\varphi_{\mathrm{s}}/T_{\mathrm{i}}}{u/v_{T_{\mathrm{i}}}} \mathrm{erf}\left(\frac{u}{\sqrt{2}v_{T_{\mathrm{i}}}} \right) \right.$$
$$\left. + \exp\left(-\frac{u^2}{2v_{T_{\mathrm{i}}}^2} \right) \right\}. \tag{11.13}$$

Correction to expression (11.13) for $u \ll v_{T_{\mathrm{i}}}$ is a second–order infinitesimal, $\Delta I_{\mathrm{i}} \sim O(u^2/v_{T_{\mathrm{i}}}^2)$, whilst for $u \gg v_{T_{\mathrm{i}}}$ expression (11.13) reduces to

$$I_{\mathrm{i}} = \pi a^2 n_{\mathrm{i}} u \left(1 - \frac{2e\varphi_{\mathrm{s}}}{m_{\mathrm{i}} u^2} \right). \tag{11.14}$$

The dimensionless surface potential of the dust particle, calculated from the flux balance condition (11.9) with the help of formulas (11.7) and (11.13), is shown in Fig. 11.3 as a function of the ion drift velocity for three values of n_e/n_i. The calculation corresponds to an argon plasma with $\tau = 100$. The figure demonstrates that the dimensionless charge is practically constant for $u \leq v_{T_i}$, then increases with increasing drift velocity, reaching a maximum at $u \sim c_i = \sqrt{T_e/m_i}$ (where c_i is the ion sound velocity), after which it decreases. Comparison of the exact expression (11.13) with approximations (11.8) and (11.14) (shown by dashed lines) indicates that some difference exists only in a narrow region near $u \sim v_{T_i}$. Numerical simulation (Lapenta 1999) supports the applicability of the expressions presented for describing the dust particle charging in plasmas with drifting ions.

So far, we have considered the charging of an individual particle in plasmas. In reality, however, the dust concentration can be high. In this case, an increase in dust particle concentration leads to a decrease in their surface potential (and hence charge) in the absolute magnitude due to a reduction in the electron concentration compared to that of the ions ($n_e < n_i$). To illustrate this phenomenon, we show a simplified consideration which gives a correct qualitative picture of the physics involved. Namely, assuming that conditions (11.1) are satisfied we use expressions (11.7) and (11.8) for the electron and ion fluxes, respectively, taking into account the contribution of the dust component to the plasma charge composition. The latter leads to the following quasineutrality condition

$$n_e = n_i + Z_d n_d. \tag{11.15}$$

The equilibrium surface potential is then determined (instead of formula 11.11) by

$$\exp(-z) = \left(\frac{\mu}{\tau}\right)^{1/2} (1 + z\tau)(1 + P), \tag{11.16}$$

where the dimensionless parameter

$$P = \frac{|Z_d|n_d}{n_e} \tag{11.17}$$

determines the ratio of the charge residing on the dust component to that on the electron component (the so-called Havnes parameter) (Havnes et al. 1987). The particle charge tends to the charge of an isolated particle, when $P \ll 1$ (see Fig. 11.2), whilst for $P > 1$ it is reduced considerably in absolute magnitude. Notice that sometimes instead of expression (11.17) the quantity $aT_e e^{-2}(n_d/n_e)$ is used, which differs from P by the factor $1/z$.

11.2.1.3 Applicability of the orbit motion limited approach.
Above, the inequalities (11.1) were employed to determine the conditions of applicability of OML. Here we define more precisely the latter conditions and point out the physical mechanisms which can make the OML approach inapplicable.

First, we must remember that the expressions for electron and ion collection cross-sections (11.2) and (11.3) do not depend on the exact form of the potential distribution around the dust particle. However, OML is applicable only if the potential satisfies certain conditions. The point is that the motion of the ions approaching the dust particle is determined by the effective interaction potential U_{eff}, which in addition to the attractive electrostatic potential $U(r)$ between a positive ion and negatively charged particle contains a component associated with the centrifugal repulsion due to ion angular momentum conservation. The effective potential normalized on the ion initial kinetic energy $E = m_i v_i^2 / 2$ is given by

$$U_{\text{eff}}(r, \rho) = \frac{\rho^2}{r^2} + \frac{U(r)}{E}, \tag{11.18}$$

where ρ is the impact parameter, and $U(r) < 0$ for attraction. For a given ρ, the distance r_0 at which $U_{\text{eff}}(r_0, \rho) = 1$ corresponds to the distance of the closest approach between the ion and the dust particle. The ion is collected when $r_0 \leq a$, whilst for $r_0 > a$ it experiences elastic scattering by the particle potential, but does not reach its surface. Inserting $r = a$ and $U(a) = e\varphi_s$ into Eq. (11.18) we find the maximum impact parameter for ion collection:

$$\rho_c^{\text{OML}} = a \sqrt{1 - \frac{2e\varphi_s}{m_i v_i^2}}. \tag{11.19}$$

Taking into account that $\sigma = \int_0^{\rho_c} 2\pi\rho d\rho$ we immediately get the cross–section (11.3). However, the equation $U_{\text{eff}}(r, \rho) = 1$ does not necessarily have only one root. It can be shown that the solution is unique only when $|U(r)|$ decreases more slowly than $1/r^2$ (Allen *et al.* 2000). In reality, however, $|U(r)| \propto 1/r$ close to the particle, and $|U(r)| \propto 1/r^2$ far from it, and at intermediate distances the potential can decrease faster. In this case, the equation $U_{\text{eff}}(r, \rho) = 1$ can have multiple roots (the distance r_0 of the closest approach is given by the largest one). This means that a potential barrier for ions moving to the dust particle emerges: some ions are reflected at $r_0 > \lambda_D > a$ (see below) and, hence, cannot reach the particle surface.

As a useful example let us consider the screened Coulomb (Debye–Hückel or Yukawa) interaction potential between the ions and the dust particle, $U(r) = -(U_0/r)\exp(-r/\lambda_D)$, where $U_0 = e|\varphi_s|a\exp(a/\lambda_D)$. Using the normalized distance $\tilde{r} = r/\lambda_D$ it is easy to show that the behavior of the effective potential U_{eff} is governed by two dimensionless parameters: $\beta = U_0/2E\lambda_D$ and $\tilde{\rho} = \rho/\lambda_D$. Curves of the effective potential for two values of β and different values of $\tilde{\rho}$ are displayed in Fig. 11.4. The potential barrier is absent at $\beta = 10$, whilst at $\beta = 20$ the existence of the barrier leads to an abrupt jump in the distance of closest approach from $r_0/\lambda_D \sim 0.7$ to $r_0/\lambda_D \sim 2.6$ for $\rho \sim 3.8\lambda_D$. It is possible to show that when the potential barrier exists, the distance of closest approach cannot be shorter than $\sim 1.62\lambda_D$ (Khrapak *et al.* 2003a).

For the Debye–Hückel (Yukawa) potential, *necessary condition* for the existence of the potential barrier reduces to $\beta > \beta_{\text{cr}} \approx 13.2$ (Kilgore *et al.* 1993). The barrier emerges for ions with the impact parameters $\tilde{\rho} \geq \tilde{\rho}_* \sim \ln\beta + 1 - 1/(2\ln\beta)$

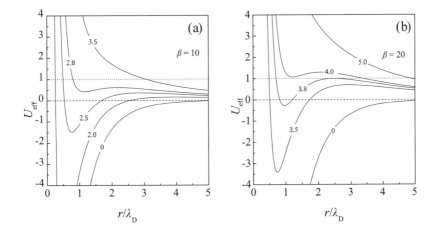

FIG. 11.4. Curves of the effective potential for the radial ion motion in the dust particle electric field (for the screened Coulomb interaction potential) for two values of the scattering parameter β (see text) and different impact parameters ρ (indicated in the figures). The potential barrier is absent at $\beta = 10$ and present at $\beta = 20$.

(Khrapak *et al.* 2003a). The OML approach is applicable only for ions with $\rho_c^{OML} \leq \rho_*$, because the ions with $\rho > \rho_*$ are reflected from the barrier at $r_0 > \lambda_D > a$ (see Fig. 11.4b) and are not collected. If this applicability condition is not satisfied for a significant fraction of the ions, then the OML approximation overestimates the ion flux to the dust particle surface, i.e., underestimates the absolute value of the stationary particle charge. At very large β (slow ions), OML fails because $\rho_c^{OML} \propto \sqrt{\beta}$, whilst $\rho_* \sim \ln \beta$ for $\beta \gg 1$. For example, for the Maxwellian ion velocity distribution there are always ions for which OML is not applicable (Allen *et al.* 2000). However, the inequality $\rho_c^{OML}(v_{T_i}) \ll \rho_*(v_{T_i})$ guarantees that the corrections are small and can be ignored. Equating the impact parameters $\rho_c^{OML}(v_{T_i}) = \rho_*(v_{T_i})$ and taking into account that $\beta(v_{T_i}) = (z\tau a/\lambda_D) \exp(a/\lambda_D)$ we can determine the values of the ratio a/λ_D (for a given value of $z\tau$) for which the OML approach starts to be inapplicable. As an example we give the following numbers: when $z\tau$ increases from 50 to 500 (typical values in gas discharge plasmas), the critical value of the ratio a/λ_D decreases from ~ 0.4 to ~ 0.2; at the same time, $\beta(v_{T_i})$ increases from ~ 30 to ~ 100.

Note that for a repulsive interaction potential the barrier of the effective potential does not show itself. That means that the difficulty considered above is absent when describing the collection of the electrons within the OML approach.

Taking ion collisions into consideration can considerably narrow the regime of OML applicability as well. We note that in a weakly ionized gas discharge plasma the dominant type of collisions is of course that of ion–neutral collisions, which are mostly determined by the resonant charge exchange mechanism. Trajectories of the ions approaching the dust particle from the unperturbed plasma are assumed to be unaffected by collisions. This at least requires the inequality $\ell_i \gg \lambda_D$ to be satisfied — for distances $r \gg \lambda_D$, the dust particle does not perturb the plasma. However, this inequality does not guarantee that the OML approach is applicable. Indeed, following the work of Zobnin et $al.$ (2000) let us consider how many charge exchange collisions occur within a sphere of radius ϵ surrounding the dust particle. If $|U(\epsilon)|/T_n \geq 1$, the majority of newly created (due to charge exchange collisions) ions having a kinetic energy determined by the neutral gas energy spectrum insufficient for returning to the unperturbed plasma, are falling on the dust particle surface. The ion flux associated with this process can be roughly estimated as $I_{i,coll} = 4\pi\epsilon^3 n_i \nu_{in}/3$, where $\nu_{in} \sim v_i \sigma_{in} n_n$ is the frequency of charge exchange collisions, σ_{in} is the corresponding cross–section, and n_n is the density of neutral atoms. Comparing $I_{i,coll}$ with the ion flux given by the OML approach (11.8) we find a condition allowing us to ignore the effect of ion–neutral collisions on the dust particle charging. This condition has the form $I_{i,OML} \gg I_{i,coll}$ and can be much stricter than the right–hand side of Eq. (11.1). Numerical simulation (Zobnin et $al.$ 2000) shows that collisions can considerably increase the ion flux compared to that given by OML even when ℓ_i is larger than λ_D. This effect leads to a decrease of the particle charge in the absolute magnitude and can be quite important under typical dusty plasma conditions, as demonstrated recently experimentally (Ratynskaia et $al.$ 2004).

Finally, the existence of extremes (minimum and maximum) in the curve of the effective potential can lead to the appearance of "trapped ions" when charge exchange collisions are present. Trapped ions are those that move in closed orbits around the dust particle. Trapped ions can affect both the particle charging and the particle charge screening by the surrounding plasma. Evidently, this effect can be substantial even for the almost collisionless regime for ions, when $\ell_i > \lambda_D$. However, this question is not fully understood. Hence, we only give reference to original works (Goree 1992; Kilgore et $al.$ 1993; Lampe et $al.$ 2001a) where the effect of trapped ions was discussed.

11.2.1.4 $Charging$ in a $drift$–$diffusion$ $regime.$ When the condition $\ell_{i(e)} \ll \lambda_D$ is satisfied, the OML approach is no longer applicable. This situation is often called the drift–diffusion regime of charging. As before, the dust particle charge (potential) is determined by the balance of the electron and ion fluxes to its surface, which in the case considered can be written as

$$I_i = 4\pi r^2 \left[n_i \mu_i \frac{d\varphi}{dr} + D_i \frac{dn_i}{dr} \right], \tag{11.20}$$

$$I_e = -4\pi r^2 \left[n_e \mu_e \frac{d\varphi}{dr} - D_e \frac{dn_e}{dr} \right], \tag{11.21}$$

where $\varphi(r)$ is the electrostatic potential around the dust particle, and $\mu_{i(e)}$ and $D_{i(e)}$ are the mobility and the diffusion coefficient of the ions (electrons), respectively. Equations (11.20) and (11.21) are supplemented by the Poisson equation

$$\nabla^2 \varphi = -4\pi e(n_{\rm i} - n_{\rm e}) \qquad (11.22)$$

and corresponding boundary conditions. It is common to set the following boundary conditions on the collecting surface:

$$\varphi(a) = \varphi_{\rm s}\,, \qquad\qquad \varphi(\infty) = 0\,;$$

$$(11.23)$$

$$n_{\rm i}(a) = n_{\rm e}(a) = 0\,, \qquad n_{\rm i}(\infty) = n_{\rm e}(\infty) = n_0\,.$$

We note that due to the fact that the ion–neutral collision cross section is usually much larger than the cross–section for electron–neutral collisions, the situation in which $\ell_{\rm i} \ll \lambda_{\rm D} \ll \ell_{\rm e}$ can be realized. In this case, the electron flux is described by the OML approach as before, and only for the ion flux should the drift–diffusion approximation be used.

A number of works were dedicated to solving the formulated or similar problems (e.g., when accounting for ionization and recombination processes in the vicinity of a collecting body, modifications of boundary conditions, etc.) starting from electric probe theory to the investigation of dust particle charging in plasmas. Since this limit is still quite rarely realized in dusty plasma experiments, we give here only original references (Su and Lam 1963; Smirnov 2000; Pal' et al. 2002; Tsytovich et al. 2002) where some solutions have been obtained and their applicability conditions discussed.

11.2.1.5 *Different charging mechanisms.* The collection of ions and electrons from a plasma is not the only possible mechanism for charging dust particles. In particular, the electrons can be emitted from the particle surface due to thermionic, photoelectric, and secondary electron emission processes. These processes are of special importance for dust charging in the working body of solid-fuel MHD generators and rocket engines (Zhukhovitskii et al. 1984; Yakubov and Khrapak 1989; Sodha and Guha 1971; Soo 1990), in the upper atmosphere, in space (Whipple 1981; Bliokh et al. 1995; Mendis 2002), and in some laboratory experiments, for instance, in thermal plasmas (Fortov et al. 1996b,c; Nefedov et al. 1997) or in plasma induced by UV irradiation (Fortov et al. 1998), with photoelectric charging of dust particles (Sickafoose et al. 2000), with electron beams (Walch et al. 1995), etc. Emission of electrons increases the dust particle charge and, under certain conditions, the particles can even reach a positive charge, in contrast to the situation discussed previously. Moreover, due to the emission processes, the existence of a two–component system consisting of dust particles and the electrons emitted by them is principally possible. In such a case, the equilibrium potential (charge) of the dust particle is determined by the balance of the fluxes that are collected by the particle surface and emitted from it, and the quasineutrality condition takes the form

$$Z_{\rm d} n_{\rm d} = n_{\rm e}\,. \qquad (11.24)$$

Such a system serves as the simplest model for investigating different processes associated with emission charging of dust particles (Zhukhovitskii et al. 1984;

Yakubov and Khrapak 1989; Khrapak *et al.* 1999). Let us briefly consider each of the emission processes listed above.

Thermionic emission. The flux of the emitted thermal electrons increases with increasing dust particle surface temperature and depends on the sign of the particle charge. This is because the emitted electrons should not only be capable of escaping from the particle surface but also possess enough energy to overcome the potential barrier (for a positively charged particle) between the surface and quasineutral plasma. For an equilibrium plasma characterized by a temperature T, it is common to use the following expressions for the flux of thermoelectrons (Sodha and Guha 1971):

$$I_{\text{th}} = \frac{(4\pi aT)^2 m_{\text{e}}}{h^3} \exp\left(-\frac{W}{T}\right) \times \begin{cases} 1, & \varphi_{\text{s}} < 0, \\ \left(1 + \frac{e\varphi_{\text{s}}}{T}\right)\exp\left(-\frac{e\varphi_{\text{s}}}{T}\right), & \varphi_{\text{s}} > 0. \end{cases} \quad (11.25)$$

Values of the work function W of thermoelectrons for different metals and semi-conductors lie typically within the ranges from 2 to 5 eV. In the case of dielectric particles, where free electrons appear due to ionization, thermionic emission can-not play a significant role because the particles usually melt before the thermionic emission makes a substantial contribution to the electron flux. In the case of negatively charged particles, the electric field is directed in such a way that it accelerates the electrons from the particle surface. In this case, some increase in emission current can be expected due to reduction of the work function by the effect of the field (Schottky effect). As usual, the equilibrium particle charge ($\propto \varphi_{\text{s}}$) can be found from the balance of the plasma particle fluxes to/from its surface.

Photoelectric emission. Dust particles can be positively charged due to pho-toelectric emission, when irradiated in a buffer gas by a flux of photons with energies exceeding the work function of photoelectrons from the particle surface (Rosenberg and Mendis 1995; Rosenberg *et al.* 1996). The characteristic value of the work functions for most materials does not exceed 6 eV, and hence photons with energies ≤ 12 eV can charge dust particles without ionizing a buffer gas. The flux of emitted electrons depends on the properties of the irradiation source, particle material, and the sign of the particle charge in the following way (Goree 1994; Rosenberg *et al.* 1999):

$$I_{\text{pe}} = 4\pi a^2 Y J \begin{cases} 1, & \varphi_{\text{s}} < 0, \\ \exp\left(-\frac{e\varphi_{\text{s}}}{T_{\text{pe}}}\right), & \varphi_{\text{s}} > 0, \end{cases} \quad (11.26)$$

where J is the photon flux density, and Y is the quantum yield for the particle material. It is also assumed that radiation is isotropic, the efficiency of radiation absorption is close to unity, which occurs when the particle size is larger than the radiation wavelength, and the photoelectrons possess a Maxwellian velocity distribution with temperature T_{pe}. The last of these lies in most cases within the ranges from 1 to 2 eV. It is noteworthy that the quantum yield strongly depends not only on the particle material, but also on the direction of the incident radiation. The quantum yield is very low just above the threshold, and for the most interesting regime of vacuum ultraviolet it can reach a value of one photoelectron per several photons. Therefore, the photoelectric emission mechanism of particle charging can be especially important in space.

Secondary electron emission. The flux I_{se} of secondary electrons is connected to that of primary electrons, I_e, through the secondary emission coefficient δ which determines the number of emitted electrons per incident electron: $I_{se} = \delta I_e$. The coefficient δ depends both on the energy E of primary electrons and on the dust particle material. The dependence $\delta(E)$ turns out to be practically universal for different materials, if δ is normalized on the maximum yield δ_m of electrons, and E is normalized on the value E_m of energy at which this maximum is reached. The corresponding expressions for the case of monoenergetic electrons can be found in Whipple (1981) and Goree (1994). The values of the parameters δ_m and E_m for some materials given by Whipple (1981) lie within the ranges: $\delta_m \sim (1-4)$, and $E_m \sim (0.2-0.4)$ keV. For the case of Maxwellian–distributed electrons, the expression for δ was given, for instance, by Goree (1994). Note that the number of secondary electrons, which can reach the surface of the material and escape from it, decreases exponentially from the surface to the bulk of the material. This means that the yield of secondary electrons is mostly associated with the thin near–surface layer.

11.2.1.6 *Kinetics of dust particle charging.* The kinetic equation for dust particle charging in plasmas is written as follows

$$\frac{dZ_d}{dt} = \sum_j I_j = I, \qquad (11.27)$$

where the summation is made over all the fluxes I_j of charged particles collected or emitted by the dust particle, taken with the corresponding sign. The stationary dust particle charge is determined from the condition $dZ_d/dt = 0$. Let us consider particle charging in the absence of emission processes. In so doing, we use the standard equations (11.7) and (11.8) for the electron and ion fluxes to the negatively charged spherical isolated particle, derived within the OML approach. Introducing dimensionless time

$$t^* = \frac{\omega_{pi}}{\sqrt{2\pi}} \left(\frac{a}{\lambda_{Di}} \right) t,$$

where $\lambda_{Di} = \sqrt{T_i/4\pi e^2 n_i}$ is the ionic Debye radius, and $\omega_{pi} = v_{T_i}/\lambda_{Di}$ is the ion plasma frequency, we get instead of Eq. (11.27) the following equation

$$\frac{dz}{dt^*} = \frac{1}{\sqrt{\mu\tau}} \left[\exp(-z) - \left(\frac{\mu}{\tau} \right)^{1/2} (1 + \tau z) \right]. \qquad (11.28)$$

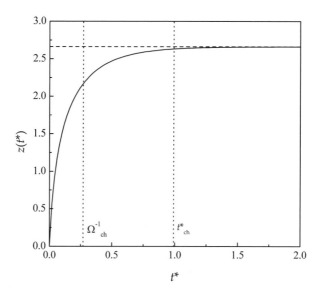

FIG. 11.5. Dimensionless charge $z = \frac{|Z_\mathrm{d}|e^2}{aT_\mathrm{e}}$ of an isolated spherical particle as a function of dimensionless time $t^* = (\omega_\mathrm{pi}/\sqrt{2\pi})(a/\lambda_\mathrm{Di})t$ for an argon plasma with $\tau = 50$. The particle is initially uncharged, the horizontal dashed line corresponds to the stationary value of the particle charge, the vertical dashed lines show two possibilities for determining the characteristic charging time (for details see text).

Combined with the initial condition $z(t^* = 0) = 0$, this equation allows us to determine the stationary value of the particle charge $z = z(\tau, \mu)$ for $t^* \to \infty$, and the characteristic time of charging, $t^*_\mathrm{ch}(\tau, \mu)$, from the uncharged state. Notice that the condition of charge stationarity: $dZ_d/dt^* = 0$ coincides with Eq. (11.11) for $n_\mathrm{e} = n_\mathrm{i}$. In Fig. 11.5, the solution to equation (11.28) with the initial condition $z(t^* = 0) = 0$ is shown for an argon plasma at $\tau = 50$. The characteristic charging time can be determined as, for example, $t^*_\mathrm{ch} = |Z_\mathrm{d}|/I_0$, where $I_0 = I_\mathrm{e0} = I_\mathrm{i0}$ are the electron and ion fluxes in a stationary state. However, it is useful to introduce the following definition which is extensively utilized in describing different processes in dusty plasmas. Let us define the charging frequency Ω_ch (inverse charging time) as the relaxation frequency for a small charge deviation from the stationary value:

$$\Omega_\mathrm{ch} = \left.\frac{dI}{dZ_\mathrm{d}}\right|_{Z_\mathrm{d0}}, \tag{11.29}$$

where the derivative should be evaluated for the stationary value of the particle charge. In the case considered we get, using Eq.s (11.7) and (11.8), the following expression

$$\Omega_\mathrm{ch} = I_0 \frac{e^2}{aT_\mathrm{e}} \frac{1 + \tau + z\tau}{1 + z\tau}. \tag{11.30}$$

The values of Ω_{ch}^{-1} and t_{ch}^{*} are shown by the vertical dashed lines in Fig. 11.5. The thermionic and photoelectric charging mechanisms are considered by Khrapak *et al.* (1999) and Khrapak and Morfill (2001) for the simplest system consisting of dust particles and electrons emitted from their surfaces. We note that dust particle charging is usually a fast process with the characteristic time scale $t_{ch}^{-1} \sim \omega_{pi}(a/\lambda_{Di})$. Therefore, the real time for the charge to achieve its stationary value can be determined by other processes proceeding in the system. For example, when cold particles are injected into a hot plasma, in which the thermionic emission plays a considerable role, then the charging time may, in principle, correspond to the time that is required for the particle surface to reach the temperature of the surrounding plasma.

11.2.1.7 *Charge composition of dusty plasmas.* The presence of dust particles in plasmas often leads to considerable changes in plasma charge composition. The point is that the dust particles are the ionization and recombination centers for the plasma electrons and ions. Particles that emit electrons and are turned positively charged may increase the electron concentration in the plasmas. Conversely, when the particles absorb electrons from the plasma they become negatively charged and reduce the number of free electrons. The quasineutrality condition in dusty plasma, which is expressed by Eq. (11.15), allows us to formulate the condition indicating when the presence of a dust component drastically influences the charge composition of the plasma. This condition is given by the inequality $|Z_d|n_d/n_e \equiv P \geq 1$. In the absence of emission processes, electrons and ions recombine on the dust particles. The frequency of plasma loss is determined then by the formula

$$\nu_{Le(i)} = I_{e(i)} \frac{n_d}{n_{e(i)}}, \tag{11.31}$$

where $I_{e(i)}$ is the flux of electrons (ions) absorbed by the dust particle surface. For considerable dust concentrations, the frequency of electron and ion losses on the particles can exceed the recombination frequency in the dust–free plasma (volume recombination and/or plasma losses in the walls of a discharge camera). In this case, the existence conditions for the plasma can change, because an increase in the recombination frequency should be compensated for by a corresponding increase in the ionization frequency (Lipaev *et al.* 1997). When the particles emit electrons, they serve as ionization sources as well. The particle contribution to the ionization is characterized by the frequency $\nu_{Ie(i)}$ equal to the flux of emitted electrons (ions). In the limiting case, emitting particles put in a nonionized gas completely determine the charge composition of the plasma, playing the roles of sources and sinks for the electrons. The two–component system of dust particles and the electrons emitted by them is characterized then by the quasineutrality condition (11.24).

11.2.1.8 *Dust particle charge fluctuations.* In Eqs. (11.7), (11.8), (11.25), and (11.26) for the electron and ion fluxes, as well as in the charging kinetic equation

(11.27), the discreteness of the electrostatic charges is ignored. In other words, the particle charge is treated as a continuous variable. However, the charging currents represent in reality sequences of events bound to electron and ion absorption or emission by the dust particle surface. These sequences and time intervals between the successive acts of absorption and emission are random. As a result, the particle charge can fluctuate around its average value. Several studies in recent years have addressed the problem of small charge fluctuations that arise from the random nature of the charging process (Cui and Goree 1994; Matsoukas and Russel 1995, 1997; Matsoukas et al. 1996; Khrapak et al. 1999; Vaulina et al. 1999a). In particular, the case of laboratory gas discharge plasma, where dust is charged by collecting electrons and ions, was mainly considered within the framework of the OML approach. Nevertheless, in the work of Khrapak et al. (1999), several different charging mechanisms, including thermionic and photoelectronic emission processes, were also considered. The key results of these investigations can be formulated as follows. Independently of the charging mechanism, random charge fluctuation can often be described as a stationary, Gaussian and Markovian process, or the Ornstein–Uhlenbeck (1930) process. This process was originally adopted to describe the stochastic behavior of the velocity of a Brownian particle. In our case, it describes the behavior of the randomly fluctuating deviation of a particle charge from its average value: $Z_1(t) = Z_d(t) - Z_0$, where $Z_0 = \langle Z_d(t) \rangle$ is the average particle charge determined by flux equality condition (11.9). Let us summarize the main properties of charge fluctuations:

(1) The charge fluctuation amplitude has zero average:

$$\langle Z_1(t) \rangle = 0; \tag{11.32}$$

(2) Its temporal autocorrelation function has an exponential form

$$\langle Z_1(t) Z_1(t') \rangle = \langle Z_1^2 \rangle \exp\left(-\Omega_{\mathrm{ch}} |t - t'|\right), \tag{11.33}$$

where Ω_{ch} is the charging frequency given by Eq. (11.29);

(3) The average of the squared amplitude of the charge fluctuation is proportional (similar to many random processes) to the absolute magnitude of the average charge:

$$\langle Z_1^2 \rangle = \gamma |Z_0|, \tag{11.34}$$

where γ is a coefficient of proportionality. The analytical expressions for γ can be found in Khrapak et al. (1999) for several different charging mechanisms. Within the OML approach for electron and ion collection, one finds

$$\gamma = \frac{1 + z\tau}{z(1 + \tau + z\tau)}. \tag{11.35}$$

For typical dusty plasma parameters in gas discharges $\tau \sim 10^2$ and $z \sim 3$ giving $\gamma \sim 0.3$;

(4) The process defined as $Y(t) = \int_0^t Z_1(x)\mathrm{d}x$ for $t \geq 0$ is Gaussian but neither stationary nor Markovian. With the help of formula (11.33), the following equation

$$\langle Y(t)^2 \rangle = \frac{2\langle Z_1^2 \rangle}{\Omega_{\mathrm{ch}}^2} \left[\Omega_{\mathrm{ch}} t + \exp(-\Omega_{\mathrm{ch}} t) - 1 \right] \qquad (11.36)$$

can be obtained. Usually, it is enough to use these properties for investigating the influence of charge fluctuations on dynamic processes in dusty plasmas. In particular, the following work can be mentioned: dust particle "heating" (in terms of the kinetic energy) in an external electric field due to charge fluctuations was investigated by Vaulina et al. (1999a,b, 2000) and Quinn and Goree (2000); instabilities of dust particle oscillations due to charge fluctuations were considered by Morfill et al. (1999a) and Ivlev et al. (2000a), and dust diffusion across a magnetic field due to random charge fluctuations was studied by Khrapak and Morfill (2002) with applications to astrophysical plasmas.

11.2.1.9 *Experimental determination of dust particle charge.* In previous sections most attention was given to theoretical concepts of dust particle charging in plasmas. Experimental examination of dust particle charges is of extreme importance, especially in cases where the plasma parameters are unknown or cannot be determined with sufficient accuracy. Here we point out several original papers describing the experimental determination of particle charges under various conditions. In the work of Walch et al. (1995), the charging of particles of various materials and diameters from 30 to 120 μm by thermal and monoenergetic suprathermal electrons was experimentally investigated. For conditions in which the charging was dominated by suprathermal electrons, the particles were charged to a potential proportional to the electron energy and the charge magnitude proportional to the particle radius, in agreement with theoretical models. When the electron energy reached a threshold value (different for various materials), from which the secondary electron emission became important, a sharp decrease in the particle potential and charge was found. Sickafoose et al. (2000) studied photoelectric emission charging of dust particles with diameters of ~ 100 μm. Conducting particles acquired a positive floating potential and charge, both increasing linearly with the decreasing work function of the photoelectrons. The behavior of dust particles charged by solar radiation in microgravity conditions on–board the orbital station "Mir" was investigated by Fortov et al.(1997a, 1998) and Nefedov et al. (2000). An analysis of particle dynamics after UV irradiation, reported by Fortov et al. (1998), revealed that the particles with mean radius 37.5 μm were charged to approximately $10^4 e$. In the work of Barkan et al. (1994), probe measurements were performed in order to investigate the influence of dust concentration on particle charge. In accordance with the theoretical predictions, a significant reduction in the charge was found when the dust concentration increased. Measurements of the particle charge in a stratified d.c. discharge plasma were performed by Fortov et al. (2001b). In this work, aperiodic oscillations of an isolated particle excited by a focused laser beam were studied. A nonlinear dependence of the particle charge on the particle size was observed. This nonlinear dependence was later attributed by Samarian and Vladimirov (2002) to the dependence of plasma parameters on particle size: The particles with different radii were levitating in different regions of striations characterized by different plasma parameters, i.e., different ion and electron densities and electron temperature. This caused the particle charge to deviate from the linear dependence on the particle size. An experimental determination of particle charge in a bulk d.c. discharge plasma, covering a wide range of neutral gas pressures, was reported by Ratynskaia *et al.* (2004). The charge was obtained by two independent methods: One based on an analysis of the particle motion in a stable particle flow and another on an analysis of the transition of the flow to an unstable regime. Molecular dynamics simulations of the particle charging for conditions similar to those of the experiment were also performed. Results of both experimental methods and the simulations demonstrated good agreement. The charge obtained was several times smaller than predicted by the collisionless OML theory, and thus the results serve as an experimental indication that ion–neutral collisions significantly affect particle charging. Other methods of charge determination based on an analysis of interparticle collisions and particle oscillations will be considered in Sections 11.2.4 and 11.4.1, respectively.

11.2.2   *Electrostatic potential around a dust particle*

The distribution of the electrostatic potential $\varphi(r)$ around an isolated spherical dust particle of charge Z_d in an isotropic plasma satisfies the Poisson equation (11.22) with the boundary conditions $\varphi(\infty) = 0$ and $\varphi(a) = \varphi_s$. The potential is connected to the particle charge through the relationship

$$\left.\frac{d\varphi}{dr}\right|_{r=a} = -\frac{Z_d e}{a^2}. \tag{11.37}$$

In a plasma with a Boltzmann distribution of electrons and ions, where the condition $|e\varphi_s/T_{e(i)}| < 1$ is satisfied, the right–hand side of Eq. (11.22) can be linearized, which yields

$$\varphi(r) = \varphi_s(a/r)\exp\left(-\frac{r-a}{\lambda_D}\right), \tag{11.38}$$

where $\lambda_D^{-2} = \lambda_{De}^{-2} + \lambda_{Di}^{-2}$ in the case under consideration. The surface potential is connected to the charge through the formula $\varphi_s = Z_d e/a(1 + a/\lambda_D)$. For the potential distribution in the case $a \ll \lambda_D$, we can use the following expression

$$\varphi(r) = (Z_d e/r)\exp(-r/\lambda_D). \tag{11.39}$$

The potential (11.39) is the screened Coulomb potential which is often applied to describe the electrostatic interaction between the particles in dusty plasmas. In different physical systems this form of the potential is also known as the Debye–Hückel or Yukawa potential. If the surface potential is not small compared to the temperatures of electrons and/or ions, then at sufficiently large distances from the particle surface one can still use Eq. (11.39) in which, however, the surface potential φ_s should be replaced by some effective potential φ_{eff}, with $|\varphi_{eff}| < |\varphi_s|$. The effective potential can be calculated numerically for known plasma parameters (Dubin 1995; Nefedov *et al.* 1998).

The equilibrium case considered above is uncommon for dusty plasmas. In particular, if the charging is determined by electron and ion collection as well as their recombination on the particle surface, then plasma is continuously absorbed on the particles. This means that some external ionization sources, which continuously supply energy to the plasma, are required for its occurrence. Hence, the system considered is open. In this case, the equilibrium (Boltzmann) distributions of electrons and ions around the particle are disrupted – the electron and ion fluxes directed from the particle surface to the plasma are absent. Thus, formula (11.39) is, strictly speaking, inapplicable independently of whether the Boltzmann distributions can be expanded or not in terms of the potential (in dusty plasma this expansion is usually not possible because typically $|e\varphi_s/T_e| \geq 1$, and $|e\varphi_s/T_i| \gg 1$). Let us consider first the asymptotic behavior of the potential at large (compared to the Debye radius) distances. To calculate the electrostatic potential around a relatively small ($a \ll \lambda_D$) dust particle within the framework of the OML approach, we follow Tsytovich (1997) and represent the ion velocity distribution function in the form

$$f_i(\mathbf{v}) = \begin{cases} f_0(v), & \theta > \theta_*, \\ 0, & \theta \leq \theta_*, \end{cases} \tag{11.40}$$

where $f_0(v)$ is a Maxwellian distribution function, and θ is the angle between the $\mathbf{v}$ and $\mathbf{r}$ vectors. The angle θ_* defines the solid angle in the velocity space where the ions (moving away from the particle) are absent due to absorption. At large distances from the particle, the angle θ_* is small and can be determined using the OML approach:

$$\sin^2 \theta_* \approx \frac{a^2}{r^2}\left(1 + \frac{2e|\varphi_s|}{m_i v^2}\right). \tag{11.41}$$

Assuming $\varphi_s \approx Z_d e/a$ we arrive at the following asymptotic behavior of the electrostatic potential around the particle (see, e.g., Al'pert *et al.* 1965; Tsytovich 1997; Khrapak *et al.* 2001):

$$\frac{e\varphi(r)}{T_e} \approx -\frac{1+2z\tau}{4(1+\tau)}\frac{a^2}{r^2}, \tag{11.42}$$

which holds at sufficiently large distances from the particles, when $r \gg a\sqrt{z\tau}$ and $r > \lambda_D \ln(\lambda_D/a)$. The first of these inequalities ensures that θ_* is small, while the second means that the screened Coulomb potential (11.39) is small compared to the asymptotic value (11.42). We note that the dependence $\varphi(r) \propto r^{-2}$ at large distances from an absorbing body in plasmas is well known from probe theory (see, e.g., Al'pert et $al.$ 1965; Chung et $al.$ 1975). In typical gas discharge plasmas $\tau \gg 1$ and $z \sim 1$, so that Eq. (11.42) can be considerably simplified:

$$\varphi(r) = \frac{Z_{\mathrm{d}}ea}{2r^2}. \tag{11.43}$$

At smaller distances from the particle, the electrostatic potential can also differ from the screened Coulomb potential (Al'pert et $al.$ 1965). The question of how significant this difference might be is not fully understood: there exist only several numerical calculations for a limited set of plasma parameters. In particular, Daugherty et $al.$ (1992) and Lipaev et $al.$ (1997) calculated numerically from a self–consistent solution for the Poisson–Vlasov equation the electrostatic potential around a spherical particle in an isotropic plasma in a collisionless regime for the ions and electrons. The main results of these works can be formulated as follows: for not too large distances from a relatively small ($a \ll \lambda_{\mathrm{D}}$) particle, the electrostatic potential can be approximated with reasonable accuracy by the potential of the form (11.39) with the screening length λ_{L} close to the ionic Debye radius: $\lambda_{\mathrm{L}} \sim \lambda_{\mathrm{D}} \sim \lambda_{\mathrm{Di}}$; for larger particles, formula (11.39) is still applicable, but with the screening length increasing with the particle size and reaching values close to the electron Debye length λ_{De} or even larger. For still larger distances, the potential asymptotically tends to the r^{-2} dependence in accordance with expression (11.43).

Finally, as shown by Lampe et $al.$ (2001a), under certain conditions the existence of trapped ions can lead to even better agreement between the potential calculated in a self–consistent way and the screened Coulomb potential for distances up to several Debye radii (at larger distances $\varphi(r) \sim r^{-2}$, as discussed above).

So far, we have assumed that the plasma is isotropic. Often, especially in laboratory experiments, ions are drifting with a nonzero velocity $\mathbf{u}$ relative to the dust particles at rest. A test particle immersed in such a plasma creates a perturbed region of flow downstream from the particle – a wake. The electrostatic potential created by a point–like charge at rest is defined in the linear approximation (see Alexandrov et $al.$ 1978) as

$$\varphi(\mathbf{r}) = \frac{Z_{\mathrm{d}}e}{2\pi^2} \int d\mathbf{k} \frac{\exp(i\mathbf{k} \cdot \mathbf{r})}{k_i k_j \varepsilon_{ij}(\mathbf{k}, 0)}, \tag{11.44}$$

where $\varepsilon_{ij}(\mathbf{k}, \omega)$ is the permittivity tensor of the plasma, and $\mathbf{k}$ is the wavevector. Using a certain model for $k_i k_j \varepsilon_{ij}(\mathbf{k}, \omega)$ one can, in principle, calculate the distribution $\varphi(\mathbf{r})$ (at least for some limiting cases), which is anisotropic in this case. This was done by Nambu et $al.$ (1995); Vladimirov and Nambu (1995); Vladimirov and Ishihara (1996); Ishihara and Vladimirov (1997); Xie et $al.$ (1999); Lemons et $al.$ (2000); and Lapenta (2000). The potential distribution can also be obtained from numerical modeling (Melandsøand Goree 1995; Lemons et $al.$ 2000; Lampe et $al.$ 2000; Maiorov et $al.$ 2001; Winske 2001; Lapenta 2002; Vladimirov et $al.$

2003). Physically, generation of electrostatic wakes in dusty plasmas is analogous to the generation of electromagnetic waves by a particle at rest, which is placed in a moving medium (Bolotovskii and Stolyarov 1992; Ginzburg 1996). When dealing with supersonic ion drift velocities $u > (T_e/m_i)^{1/2}$, which are evident in the sheath region of r.f. discharges, the analogy with the Vavilov–Cherenkov effect can be useful. The qualitative description of the effect under consideration is the following. In the directions upstream from and perpendicular to the ion flow, the potential has the form of a screened Coulomb potential; downstream from the flow (within a certain solid angle) the potential has a periodically oscillating character with decay. As shown by numerical modeling, the shape of the wake potential is sensitive to ion collisions (ion–neutral collisions) (Hou *et al.* 2003) and the electron-to-ion temperature ratio which governs Landau damping (Lampe *et al.* 2001b). In typical situations, these mechanisms can effectively reduce the wake structure to one oscillation; at larger distances, the wake structure is smeared. The shape of the wake potential depends on the value of the Mach number $M = u/c_i$, however, the wake itself appears both in supersonic and subsonic regimes of the ion drift. In this context, we mention the work of (Lampe *et al.* 2000) where some examples of the wake structures calculated numerically for different plasma conditions are presented. The effects of finite particle size and asymmetry of the charge distribution over its surface are considered by Hou *et al.* (2001) and Ishihara *et al.* (2000). The wake effect is usually invoked for explaining the vertical ordering of the dust particles (chain formation) in ground–based experiments: due to the opposite sign of the potential in the wake, there may appear an attractive force between the particles (of the same sign of charge) situated along the ion flow.

Recently, another effect was also pointed out, which might play some role in the vertical ordering of dust particles along the ion flow. This effect is connected with a distortion of the ion velocity field by the upstream particle and the appearance of a horizontal component of the force, caused by ion momentum transfer in absorbing and Coulomb collisions with the downstream particle (Lampe *et al.* 2000; Lapenta 2002). This force – the ion drag force – brings the downstream particle back to the axis with the origin at the upstream particle position and parallel to the ion flow. Numerical modeling of the ion velocity field in the wake showed that for certain conditions the force associated with the perturbation of ion velocities prevails over the electrostatic one (Lapenta 2002). We note once again that both effects considered are sensitive to ion–neutral collisions. First, the collisions reduce the ion directed velocity in an external electric field (and, hence, plasma anisotropy). Second, they limit the perturbed plasma region (both the potential and the ion velocity field) around a probe particle to a length scale on the order of the ion free path. Therefore, the mechanisms considered can be effective only at sufficiently low pressures.

Note that in this section we were dealing with the potential distribution around an "isolated" particle. Such a consideration is justified when the dust component does not play the role of a real plasma component and the interparticle separation significantly exceeds the Debye radius. In the opposite situation, the dust component can also contribute to the plasma screening of a test charge. Equation (11.22) should also be modified to take into account the charge variation in response to the plasma perturbation. We will not consider this situation here; the details can be found in the recent review by Morfill *et al.* (2003).

11.2.3   *Main forces acting on dust particles in plasmas*

The main forces acting on dust particles in plasmas can be conveniently divided into two groups: the first one includes the forces which do not depend on the particle charge (force of gravity, neutral drag force, thermophoretic force), while the second one includes forces which depend directly on the particle charge (electrostatic force and the ion drag force).

11.2.3.1   *Force of gravity.* The gravitational force is determined by the expression

$$F_\mathrm{g} = m_\mathrm{d} g, \qquad\qquad (11.45)$$

where g is the gravitational acceleration. The gravitational force is proportional to the particle volume, viz. $F_\mathrm{g} \sim a^3$.

11.2.3.2   *Neutral drag force.* When the particle is moving, it experiences a force of resistance from the surrounding medium. In the case of weakly ionized plasma, the main contribution to this resistance force comes from the neutral component. This resistance force – the neutral drag force – is in most cases proportional to the particle velocity, because the latter is usually much smaller than the thermal velocity of neutral atoms or molecules. One should distinguish between two regimes depending on the value of the Knudsen number, Kn, which is the ratio of the atomic or molecular free path to the particle size: $\mathrm{Kn} = \ell_\mathrm{n}/a$. It is common to call the regime for which $\mathrm{Kn} \ll 1$ the hydrodynamic regime. In this limit, the resistance force is given by the Stokes expression (see, e.g. Landau and Lifshitz 1988)

$$F_\mathrm{n} = -6\pi\eta a u, \qquad\qquad (11.46)$$

where η is the neutral gas viscosity, and u is the particle velocity relative to the neutral gas. The minus sign means that the force is directed opposite to the vector of relative velocity. In the opposite limiting case of $\mathrm{Kn} \gg 1$, which is often called the free molecular regime, and for sufficiently small relative velocities ($u \ll v_{T_\mathrm{n}}$), the neutral drag force can be written as (Lifshitz and Pitaevskii 1979; Epstein 1924)

$$F_\mathrm{n} = -\frac{8\sqrt{2\pi}}{3}\delta a^2 n_\mathrm{n} T_\mathrm{n}\frac{u}{v_{T_\mathrm{n}}}, \qquad\qquad (11.47)$$

where n_n and T_n are the concentration and temperature of the neutrals, respectively, and δ is a coefficient on the order of unity that depends on the exact processes proceeding on the particle surface. For example, $\delta = 1$ for the case of complete collection or specular reflection of neutrals from the particle surface at collisions, and $\delta = 1 + \pi/8$ for the case of diffuse scattering with full accommodation. For high relative velocities $(u \gg v_{T_n})$, the neutral drag force is proportional to the velocity squared (see, for example, Draine and Salpeter 1979; Nitter 1996):

$$F_n = -\pi a^2 n_n m_n u^2, \qquad (11.48)$$

where m_n is the mass of neutrals. It should be noted that these expressions were originally derived for uncharged particles in a neutral gas. In this way, the polarization interaction, which is associated with a nonuniform electric field in the vicinity of the dust particle, was not taken into account. However, the radius of the polarization interaction turns out to be much smaller than the particle size, so that in most cases the corrections can be ignored.

For most dusty plasmas, formula (11.47) is applicable. We note that it can be rewritten in the form $F_n = -m_d \nu_{dn} u$, where ν_{dn} stands for the effective momentum transfer frequency of dust–neutral collisions.

11.2.3.3 Thermophoretic force.

If a temperature gradient is present in a neutral gas, then the particle experiences a force directed opposite to this gradient, i.e., in the direction of lower temperatures. This is associated with the fact that the larger momentum is transferred from the neutrals coming from the higher temperature region. In the case of full accommodation, this force (called the thermophoretic force) can be expressed as (Talbot et al. 1980)

$$F_{th} = -\frac{4\sqrt{2\pi}}{15} \frac{a^2}{v_{T_n}} \kappa_n \nabla T_n, \qquad (11.49)$$

where κ_n is the thermal conductivity coefficient of the gas. The simplest estimation of this coefficient gives $\kappa_n \approx C\bar{v}/\sigma_{tr}$ (Raizer 1991), where C is a numerical factor on the order of unity, so that $C = 5/6$ for atoms, and $C = 7/6 (9/6)$ for diatomic molecules with unexcited (excited) vibrational states, respectively; $\bar{v} = \sqrt{8T_n/\pi m_n}$, and σ_{tr} is the transport scattering cross–section for gas atoms or molecules. Inserting the value of $C = 5/6$ for the atomic gas into Eq. (11.49) we get

$$F_{th} \approx -\frac{16}{9} \frac{a^2}{\sigma_{tr}} \nabla T_n. \qquad (11.50)$$

This estimate shows that the thermophoretic force depends on the particle radius, gas type (through σ_{tr}), and temperature gradient, but does not depend on the neutral gas pressure and temperature. For particles of about 1 μm radius and mass density ~ 1 g cm^{-3} in an argon plasma, the thermophoretic force is comparable to the force of gravity at temperature gradients $\nabla T_n \sim 10$ K cm^{-1}. Note that the expression (11.49) was derived for an unbounded system. If the dust particle is situated near the electrode or the walls of a discharge camera, then one should take into account corrections associated with the accommodation of neutrals when they collide with the electrode or wall surfaces. These corrections basically change the numerical factor in formula (11.49) (Havnes et al. 1994).

Experimental investigation of the effect of thermophoretic force on the behavior of dust particles in gas discharge plasmas was performed by Jellum et $al.$ (1991); Balabanov et $al.$ (2001); and Rothermel et $al.$ (2002). In these works, it is shown that the thermophoretic force can be used for particle levitation and controlled action on the ordered structures of dust particles in the quasineutral bulk of a plasma.

11.2.3.4 $Electrostatic$ $force.$ If an electric field E is present in plasmas, it acts on a charged particle. For conducting particles, the electrostatic force is given by (Daugherty et $al.$ 1993)

$$\mathbf{F}_{\mathrm{el}} = Z_d e \mathbf{E} \left[1 + \frac{(a/\lambda_{\mathrm{D}})^2}{3(1 + a/\lambda_{\mathrm{D}})} \right]. \tag{11.51}$$

An effective electric field can be introduced as follows:

$$\mathbf{E}_{\mathrm{eff}} = \mathbf{E} \left[1 + \frac{(a/\lambda_{\mathrm{D}})^2}{3(1 + a/\lambda_{\mathrm{D}})} \right],$$

then $\mathbf{F}_{\mathrm{el}} = Z_d e \mathbf{E}_{\mathrm{eff}}$. An increase in E_{eff} compared to E is connected to plasma polarization in the vicinity of the dust particle, which is induced by the external electric field. Plasma polarization induces a dipole moment $\mathbf{p} \approx a^3 \mathbf{E}_{\mathrm{eff}}$ on the dust particle directed along the field. If the electric field is nonuniform, then such a dipole experiences a force

$$\mathbf{F}_{\mathrm{dp}} = (\mathbf{p}\nabla)\mathbf{E}. \tag{11.52}$$

In typical dusty plasmas the inequality $a \ll \lambda_{\mathrm{D}}$ is valid, so that the electrostatic force is $\mathbf{F}_{\mathrm{el}} \approx Z_d e \mathbf{E}$, and the dipole moment is $\mathbf{p} \approx a^3 \mathbf{E}$, as for a conducting sphere in a vacuum. The dipole moment is usually very small and one can typically ignore the force $\mathbf{F}_{\mathrm{dp}}$, in contrast to $\mathbf{F}_{\mathrm{el}}$. Let us mention the work Hamaguchi and Farouki (1994a,b) where the action of the electrostatic force on the dust particle in a nonuniform plasma with ion flows was calculated. It was shown that there appears an additional component of the force proportional to the gradient of the plasma charge density or to the gradient of the corresponding Debye radius (which is dependent on the ion flow velocity), and acting in the direction of decreasing Debye length. In typical dusty plasma conditions ($a \ll \lambda_{\mathrm{D}}$), this component is smaller than F_{el} by approximately a factor of a/λ_{D}.

11.2.3.5 Ion $drag$ $force.$ If there exists a drift of ions (electrons) relative to the dust particle, there appears a force associated with the momentum transfer from the plasma to the dust particle. Due to the larger ion mass, the effect associated with the ions typically dominates; see, however, Khrapak and Morfill (2004). The movements of ions relative to dust particles can be associated with the presence of an external electric field or with the (thermal) motion of the particles relative to the stationary background of ions. The force considered is called the ion drag force. The ion drag force is connected with two processes: momentum transfer from the ions that are collected by the particle, and momentum transfer from the ions that are elastically scattered in the electric field of the particle.

The fact that the ion drag force can be important for various processes in dusty plasmas was ascertained even before the active laboratory investigation of dusty plasmas started (Morfill and Grün 1979; Northrop and Birmingham 1990; and Barnes et al. 1992). Presently, it is considered to be established that ion drag affects (or even determines) the location and configuration of the dust structures in laboratory plasma facilities (Barnes et al. 1992), is responsible for dust structure (e.g., clusters) rotation in the presence of a magnetic field (Konopka et al. 2000b; Kaw et al. 2002; Ishihara et al. 2002), affects the properties of low–frequency waves in dusty plasmas (D'Angelo 1998; Khrapak and Yaroshenko 2003), causes the formation of a "void" (a region free of dust particles) in the central part of an r.f. discharge in experiments under microgravity conditions (Morfill et al. 1999b; Goree et al. 1999; Tsytovich et al. 2001b; Khrapak et al. 2002; Tsytovich 2001), and governs the diffusion and mobility of weakly interacting Brownian particles in strongly ionized plasmas (Trigger 2003a, 2003b). All this indicates the importance of a correct estimation of the ion drag force and its dependence on plasma parameters. However, a self–consistent model describing all possible situations has not yet been constructed. The complications here are caused by the necessity of taking into account ion–neutral collisions in the vicinity of the dust particle, the effect of neighboring particles, nonlinear coupling of the ions to the particles, the necessity of knowing the exact distribution of the electrostatic potential around the particle, and other problems which cannot be considered to be fully solved. Rather, there exist several approaches which can be utilized under certain conditions.

So far, most of the results have been obtained for binary collision (BC) approximations, i.e., for the case of collisionless ions with $\ell_i \gg r_{int}$, and "isolated" dust particles with $\Delta \gg r_{int}$, where r_{int} is the characteristic radius of interaction between the ion and the dust particle, and Δ is the mean interparticle distance. This situation will be considered below. The general expression for the ion drag force is written as (see, for example, Lifshitz and Pitaevskii 1979)

$$\mathbf{F}_i = m_i n_i \int \mathbf{v} f_i(\mathbf{v}) \sigma_i^{tr}(v) v d\mathbf{v}, \qquad (11.53)$$

where $f_i(\mathbf{v})$ is the ion velocity distribution function, and $\sigma_i^{tr}(v)$ is the momentum transfer cross–section for ion collisions with the dust particle. In weakly ionized plasmas, it is often reasonable to use a shifted Maxwellian distribution (11.12) for ion velocities. For subthermal ion drifts, it can be expanded as follows

$$f_i(\mathbf{v}) \simeq f_{i0}(v)\left[1 + \frac{\mathbf{u}\mathbf{v}}{v_{T_i}^2}\right]. \qquad (11.54)$$

Here, u is the drift velocity of ions relative to the dust particle ($u < v_{T_i}$), and $f_{i0}(v)$ is the isotropic Maxwellian function (11.5) for ions. Thus, to calculate the ion drag force it is necessary to determine the momentum transfer cross–section $\sigma_i^{tr}(v)$ for ion–particle collisions.

First, let us study the case of a point–like particle. In this case, the momentum transfer cross–section is defined in the following way (Landau and Lifshitz 1998):

$$\sigma_i^{tr}(v) = \int_0^\infty \left[1 - \cos\chi(\rho, v)\right] 2\pi\rho d\rho. \tag{11.55}$$

Here, ρ is the impact parameter, χ is the scattering angle, and $\chi(\rho, v) = \left|\pi - 2\varphi_0(\rho, v)\right|$, where

$$\varphi_0 = \int_{r_0}^\infty \frac{\rho dr}{r^2 \sqrt{1 - U_{eff}(r, \rho, v)}}, \tag{11.56}$$

and the effective potential U_{eff} is determined from Eq. (11.18). The momentum transfer cross–section for ion collisions with a massive dust particle of charge Z_d can be obtained either by self–consistent calculation of the potential distribution around a sufficiently small particle ($a \ll \lambda_D$) and subsequent integration in expressions (11.55) and (11.56), or by assuming a certain form of the interaction potential $U(r)$ a priori with subsequent evaluation of the same integrals.

The results of different calculations of the momentum transfer cross–section are presented in Fig. 11.6(a). It is useful to express the dependence of the cross–section on the ion velocity through the dimensionless scattering parameter β introduced in Section 11.2.1.3. In the case considered, it can be written down as

$$\beta(v) \approx \frac{|Z_d|e^2}{m_i v^2 \lambda_D}, $$

where v denotes the relative velocity over which the averaging should be performed. A self–consistent determination of the cross section with preliminary numerical calculation of the potential distribution around a small dust particle was reported by Kilgore et al. (1993). Numerical computations for the attractive screened Coulomb potential of the interaction were performed by Hahn et al. (1971) and Khrapak et al. (2003a, 2004). As can be seen from Fig. 11.6(a), these numerical results demonstrate good agreement, which should be attributed to the fact that for $a \ll \lambda_D$ the potential distribution calculated self–consistently can be quite well approximated by the screened Coulomb potential within a few screening lengths from the particle (Daugherty et al. 1992); longer distances do not contribute considerably to the momentum transfer.

The standard theory of Coulomb collisions of charged particles in plasmas, which is based on a pure Coulomb interaction potential with cutoff at impact parameters exceeding the plasma screening length, gives the following result for the momentum transfer cross section:

$$\frac{\sigma_i^{tr}}{\lambda_D^2} = 4\pi\beta^2\Lambda_C, \tag{11.57}$$

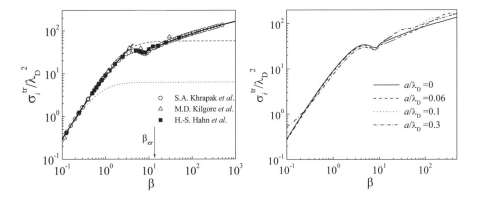

FIG. 11.6. Momentum transfer cross–section (normalized on the square of screening length λ_D^2) versus the scattering parameter β: (a) for a point–like particle; the symbols correspond to the results of numerical calculations (Hahn *et al.* 1971) – ■, (Kilgore *et al.* 1993) – $\triangle$, (Khrapak *et al.* 2003a, 2004) – $\bigcirc$; dotted line corresponds to the standard Coulomb scattering theory (Eq. (11.57); dashed line corresponds to the approximation (11.58) proposed in (Kilgore *et al.* 1993); solid curves correspond to the analytical expressions proposed in (Khrapak *et al.* 2002) for $\beta < 5$ and (Khrapak *et al.* 2003a) for $\beta > \beta_{cr}$; and the arrow indicates the value of $\beta = \beta_{cr} \simeq 13.2$; (b) for a finite size particle; the numerical results are shown for different values of a/λ_D, indicated in the figure.

where $\Lambda_C = \ln(1 + 1/\beta^2)/2$ is the Coulomb logarithm (Lifshitz and Pitaevskii 1979). Figure 11.6(a) shows that the standard Coulomb scattering theory significantly underestimates the cross–section above $\beta \sim 1$, i.e., for sufficiently slow ions. This is not surprising, because the theory of Coulomb collisions can be grounded only for $\beta \ll 1$. In this case, the interaction radius – the Coulomb radius $r_C = |Z_d|e^2/(m_i v^2)$ – is small compared to the screening length λ_D. The ratio of the momentum transfer from collisions with $r_C < \rho < \lambda_D$ to that with $\rho < r_C$ is approximately equal to the Coulomb logarithm $\Lambda_C \approx \ln(1/\beta) \gg 1$. The relative contribution from collisions with $\rho > \lambda_D$ is also logarithmically small ($\sim \Lambda_C^{-1}$) because of the screening (Khrapak *et al.* 2004). Therefore, the momentum transfer is mostly associated with the scattering in the bare Coulomb potential (for $r \ll \lambda_D$, the screening can be neglected), and the Coulomb scattering theory adequately (with logarithmic accuracy) determines the momentum transfer cross section for $\beta \ll 1$. We note that the dominant contribution to the momentum transfer in this case is due to small-angle scattering (the deflection is strong only if $\rho < r_C$).

In a common electron–ion plasma, the condition $\beta \ll 1$ is usually very well satisfied (Khrapak *et al.* 2002, 2003a), excluding the case of a strongly nonideal plasma, for which the pair-collision approximation considered here is meaningless anyway. However, for dusty plasmas the particle charge is large ($|Z_d| \gg 1$) and the scattering parameter for ion–particle collisions is not necessarily small.

For a typical particle size of 1 μm, the scattering parameter is on the order of $\beta \sim 10$. In this case, the interaction range exceeds the screening length, it is not possible to neglect impact parameters larger than the screening length, and the main contribution to the momentum transfer is due to large-angle scattering. The theory of Coulomb scattering fails under these conditions (see Fig. 11.6(a)). Therefore, all the calculations of the ion drag force, which use expression (11.57), are also incorrect in this case.

Kilgore et al. (1993) performed a fit of the obtained numerical data to the analytical form of Eq. (11.57). Based on this fit the following analytical expression was proposed:

$$\frac{\sigma_i^{tr}}{\lambda_D^2} = C_1 \beta^2 \ln\left(1 + \frac{C_2}{\beta^2}\right) \tag{11.58}$$

with $C_1 \approx 3.748$ and $C_2 \approx 15.33$ determined from the best agreement of Eq. (11.58) with the calculated results. As follows from Fig. 11.6(a), this approximation lies much closer to the numerical data compared to the standard result (11.57). However, this fit is physically ungrounded and, as a consequence, demonstrates improper asymptotic behavior for large enough β ($\beta > \beta_{cr}$).

Analytical expressions for the momentum transfer cross–section were recently obtained by Khrapak et al. (2002, 2003a) for an attractive screened Coulomb interaction potential. In these works, different approaches were used. This is connected to the fact that at $\beta = \beta_{cr} \approx 13.2$ the behavior of ion trajectories in the vicinity of the dust particle changes qualitatively due to the appearance of the barrier in the effective potential (see Section 11.2.1.3).

Khrapak et al. (2002) proposed for the case $\beta < \beta_{cr}$ to take into account the ions (with impact parameters $\rho > \lambda_D$) approaching the particle surface closer than the screening length ($r_0 \leq \lambda_D$). This basically leads to the following modification of the Coulomb logarithm

$$\tilde{\Lambda}_C = \ln \frac{1 + \beta}{\beta + a/\lambda_D}. \tag{11.59}$$

In the limit of point–like particles, $\tilde{\Lambda}_C = \ln(1 + 1/\beta)$ and when $\beta \ll 1$ expression (11.59) coincides with the result of the standard Coulomb scattering theory. However, for $\beta \gtrsim 1$ there is a considerable difference. Figure 11.6(a) shows that the approach proposed precisely describes the numerical results up to $\beta \approx 5$.

The case of $\beta > \beta_{cr}$ was considered by Khrapak et al. (2003a) and the authors arrived at the following analytical expression for the cross–section

$$\sigma_i^{tr} = A\pi\rho_*^2(\beta) + B\lambda_D^2(1 + 2\ln^{-1}\beta), \tag{11.60}$$

where $A \approx 0.81$, $B \approx 6.4$, and

$$\rho_*(\beta) \approx \lambda_D\left(\ln\beta + 1 - \frac{1}{2\ln\beta}\right).$$

Figure 11.6(a) demonstrates good agreement between Eq. (11.60) and numerical results.

When taking into account the finite size of the dust particles, then in addition to the parameter β describing the scattering from a point–like particle, there appears a second parameter a/λ_D. The physics of the ion momentum transfer changes as follows: the ions with $\rho < \rho_c$ experience inelastic collisions with the particle (collection), while the ions with $\rho > \rho_c$ are scattered in the electric field of the particle. The total momentum transfer cross–section for the case of a finite–sized particle is the sum of collection (absorption) and scattering cross–sections: $\sigma_i^{tr} = \sigma_i^c + \sigma_i^s$. Assuming that at absorbing collisions the ions transfer their initial momentum (i.e., ignoring the possible physical processes on the particle surface), we find that the collection formally corresponds to the scattering angle $\chi = \pi/2$, yielding

$$\sigma_i^c = \int_0^{\rho_c} 2\pi\rho d\rho = \pi\rho_c^2.$$

The scattering part σ_i^s of the cross–section is given now by Eq. (11.55), with the lower (zero) limit of integration replaced by ρ_c. When OML approach is valid, the impact parameter corresponding to collection, $\rho_c = \rho_c^{OML}$, is determined by formula (11.19), and the collection cross–section is given by expression (11.3). If $\rho_c^{OML} > \rho_*$, then OML is no longer applicable. In this case, for $a < \lambda_D$ we have $\rho_c = \rho_*$ and $\sigma_i^c = \pi\rho_*^2$.

The influence of the finite particle size on the total momentum transfer cross–section is discussed in detail by Khrapak et al. (2003a, 2004). Figure 11.6(b) shows the dependence of the total cross–section $\sigma_i^{tr}(\beta)$ for several values of a/λ_D. The main result is that the cross–section for a finite–sized particle does not differ dramatically from that of a point–like particle even in the region $\beta \gg 1$, where the main contribution to the momentum transfer is due to ion collection (Khrapak et al. 2003a).

When the dependence of the momentum transfer cross–section on β (i.e., on the velocity) is known, it is possible to calculate the ion drag force F_i by performing the integration in formula (11.53). For subthermal drifts one should also use formula (11.54). When $\beta(v_{T_i}) \lesssim 5$, one can use the expression

$$F_i = -\frac{8\sqrt{2\pi}}{3}a^2 n_i m_i v_{T_i} u\left(1 + \frac{1}{2}z\tau + \frac{1}{4}z^2\tau^2\Pi\right), \qquad (11.61)$$

derived by Khrapak et al. (2002). Here, Π is the modified Coulomb logarithm integrated with the ion velocity distribution function:

$$\Pi = 2\int_0^\infty \exp(-x)\ln\left(\frac{2\lambda_D/a + z\tau}{2x + z\tau}\right)dx.$$

The last term in parentheses in Eq. (11.61) corresponds to elastic scattering and is dominant. In (Khrapak et al. 2003a), the estimate

$$F_i \sim -\pi\rho_*^2(v_{T_i})n_i m_i v_{T_i} u \qquad (11.62)$$

was submitted for $\beta(v_{T_i}) > \beta_{cr}$. Note that in this case $F_i \propto T_i^{3/2} m_i^{1/2}$ but depends only logarithmically on a, n_i, and τ.

In the case of highly anisotropic plasma, $u \gg v_{T_i}$, the integration in expression (11.53) can be simplified with the substitution $f_i(\mathbf{v}) \sim \delta(\mathbf{v} - \mathbf{u})$ into the corresponding integrand. Notice that in this limiting case the value of β is substantially lower than in an isotropic plasma, first due to $u \gg v_{T_i}$ and, second, due to the fact that the Debye length is determined by electrons rather than by ions: $\lambda_D \sim \lambda_{De} \gg \lambda_{Di}$. Hence, the cross–section (11.57) with the modified Coulomb logarithm (11.59) is applicable for estimations of F_i. Moreover, in the case $\beta(u) \ll 1$, the modification is even unnecessary as discussed above. Neglecting the weak logarithmic dependences we get $F_i \sim u^{-2}$ for this case, i.e., the ion drag force decreases with the relative velocity. Finally, in the limiting case of very high relative velocity only the geometrical particle size matters, and then (Nitter 1996)

$$F_i = -\pi a^2 m_i n_i u^2, \tag{11.63}$$

i.e., F_i increases again with the relative velocity. The qualitative description of this nonmonotone behavior of F_i as a function of the relative velocity was given by Goree et al. (1999).

Let us briefly discuss the question concerning which value of λ_D should be used in the expressions encountered in this section. In the case of isotropic plasma, λ_D is the linearized Debye radius:

$$\lambda_D \sim (\lambda_{De}^{-2} + \lambda_{Di}^{-2})^{-1/2},$$

which is close to λ_{Di} for $T_e \gg T_i$. Only for relatively large particles ($a \gtrsim \lambda_{Di}$) does an approximation of numerically calculated potential by the Debye–Hückel form yield a larger value for the effective screening length. In highly anisotropic plasmas, $u \gg v_{T_i}$, ions hardly screen the particle charge, and λ_D tends to λ_{De}, while in this case the potential distribution itself (and correspondingly the screening length) is anisotropic, i.e., depends on the direction (see, e.g., Lampe et al. 2001b). The choice $\lambda_D = \lambda_{De}$ for the experiments performed in a central part of an r.f. discharge (Zafiu et al. 2002, 2003), where the ambipolar electric field is weak and the ion drift is (sub)thermal, was criticized by Khrapak et al. (2003b). The employment of the results obtained for the case $\beta > 1$ and considered above, expressions (11.59) and (11.60) with $\lambda_D \simeq \lambda_{Di}$, would be more consistent.

We have considered the results which the binary collision approach yields for the ion drag force. An alternative way to obtain the force is the linear kinetic approach. Instead of deriving single ion trajectories, one solves the Poisson equation, coupled to the kinetic equation for ions, and obtains the self consistent electrostatic potential around the grain. The polarization electric field at the origin of the grain then gives the force acting on the grain. When the polarization is due to the ion flow, the corresponding force is the ion drag force. As long as the "linear dielectric response" (LR) approximation is applicable the whole problem is basically reduced to the calculation of the appropriate plasma response function (permittivity). The LR formalism can account self–consistently for ion–neutral collisions and the effects of potential anisotropies caused by the ion flow, but is applicable only when the coupling between the ions and the grain is weak ($\beta \ll 1$). This approach was recently used by Ivlev *et al.* (2004, 2005) to study the effect of ion–neutral collisions on the ion drag force. It was shown that ion–neutral collisions can increase the force in the regime of subthermal ion drift. Note that the BC and LR approaches are not really competitive but rather *complementary*: the BC approach is more appropriate to describe scattering of subthermal collisionless ions which are strongly coupled to the particle, while the LR approach is applicable for suprathermal ions and/or small submicron particles (weak ion–particle coupling) and highly collisional cases.

Finally, we point out a few problems which require further elaboration. These are, first and foremost, the influence of the exact potential distribution around the dust particle on the ion drag force, consistently taking ion–neutral collisions into consideration in the case of nonlinear ion–particle coupling ($\beta > 1$), and an investigation of the situation in which the characteristic radius of the ion–particle interaction exceeds the interparticle distance.

11.2.4 *Interaction between dust particles in plasmas*

The potential of interaction between dust particles differs from the Coulomb interaction potential between charged particles in a vacuum. As will be shown below, the potential of interaction between the dust particles is, generally speaking, determined not only by electrostatic interactions between the particles. Variability of the particle charges as well as a number of collective effects, some of them discussed below, are responsible for these differences. "Collective interactions" can also lead to the attraction of similarly charged particles, which can occur for a certain critical dust concentration (for a detailed consideration see Tsytovich *et al.* 2002). For lower dust concentrations, the interparticle interaction, screening, and charging can be calculated in the approximation of "isolated" particles. In this case, the electrostatic interaction between the particles is assessed when the electrostatic potential distribution $\varphi(r)$ in a plasma surrounding a test particle is known. The absolute value of the electrostatic force acting on a particle with a fixed charge Z_d and located at a distance r from the test particle can be presented in the form $F_{de} = -dU_{el}(r)/dr$, where

$$U_{el}(r) = Z_d e \varphi(r). \tag{11.64}$$

Thus, it is necessary to know the distribution $\varphi(r)$ of the potential in plasmas. As was previously shown, the potential of an isolated spherical particle in an isotropic plasma is purely Coulombic at small distances $r \ll \lambda_D$; for $r \sim \lambda_D$, the screening is important and the Debye–Hückel form can often be used; finally, at distances exceeding several screening lengths the potential has an inverse power–law asymptote. Hence, up to distances considerably exceeding λ_D it is reasonable to use a screened Coulomb type of the potential

$$U_{el}(r) = \frac{Z_d^2 e^2}{r} \exp\left(-\frac{r}{\lambda_D}\right). \tag{11.65}$$

For larger distances $r > \lambda_D \ln(\lambda_D/a) \approx (3-5)\lambda_D$, a long–range repulsion takes place, which is associated with the anisotropy of the plasma velocity in the vicinity of a collecting particle. According to expressions (11.43) and (11.64), the asymptotic behavior of the interaction potential is given by

$$U_{el} \approx \frac{Z_d^2 e^2 a}{2r^2}. \qquad (11.66)$$

Estimate (11.66) holds for distances not exceeding the free path of ions in collisions with dust particles or neutrals.

Different additional mechanisms governing attraction and repulsion between the dust particles can exist as a consequence of the openness of dusty plasma systems, the openness caused by continuous exchange of energy and matter between the particles and surrounding plasmas. Shortly after the first experimental discovery of dusty plasma crystallization, the possibility of the attraction between the two particles of like charge signs was pointed out by Tsytovich (1994). In this work, the attraction was attributed to a decrease in particle charges when the distance between them is decreasing. Later on, it was shown that this result is incorrect due to the incompleteness of the energy analysis, and it was also indicated that the formation of dusty plasma crystals does not necessarily mean the existence of attraction between the dust particles, because in most of the experiments the particles are confined by external forces (Hamaguchi 1997). Nevertheless, the work of Tsytovich (1994) stimulated the investigation of supplementary (to electrostatic) mechanisms of interparticle interaction. Some of them are considered below.

The continuous flow of plasma electrons and ions on the surface of a dust particle leads to a drag experienced by neighboring particles. This can result in an effective attractive force between the particles, which is called the ion shadowing force. The magnitude of this force is mainly determined by the ion component due to larger ion masses. Such an attractive mechanism was first considered by Ignatov (1996) and Tsytovich et $al.$ (1996), and later on in the work of Lampe et $al.$ (2000) and Khrapak et $al.$ (2001). Note that the ion shadowing force basically represents the ion drag force in the ion flow directed to the surface of a test particle. Strictly speaking, the ion shadowing force is not pairwise, since the interaction between several particles (more than two) depends on their mutual arrangement.

Additional attraction or repulsion mechanisms can be associated not only with the ion component but also with neutrals, if, once scattered from the particle surface, they leave the surface with an energy spectrum different from that determined by the temperature of the neutrals. This can happen if the particle surface temperature T_s is different from the neutral gas temperature T_n and full or partial accommodation takes place. The temperature of the particle surface is governed by the balance of various processes, such as radiative cooling, exchange of energy with neutrals, and recombination of electrons and ions on the surface (Daugherty et $al.$ 1993). In the case $T_s \neq T_n$, there exist net fluxes of energy and momentum between gas and particles. Hence, if two particles are located sufficiently close to each other, an anisotropy in momentum fluxes on the particles will also exert a shadowing force between them, which in this case is associated with the neutral component. This effect was first considered by Tsytovich et al (1998).

For both neutral and ion shadowing effects, the corresponding potentials U_{ns} and U_{is} scale similarly with distance, namely, $U_{ns}(r) \sim U_{is}(r) \propto 1/r$. Hence, at large distances the shadowing interaction will overcome the long–range electrostatic repulsion (11.66). The existence of attraction makes the formation of dust molecules (an association of two or more particles coupled by long–range attraction) possible. The theoretical examination of the conditions of molecular formation can be found in (Tsytovich 1997; Khrapak et $al.$ 2001). For conditions of an isotropic plasma, however, the formation of dust molecules has not yet been experimentally established. This can be first of all connected to the fact that rather large particles are needed for the substantial shadowing effect. In ground–based experiments, such particles can levitate only in the sheath regions of discharges, where the electric field is strong enough to balance gravity. In these regions, the effects of plasma anisotropy are of primary importance. The ions moving towards the cathode with superthermal velocities contribute here practically nothing to screening. In addition, the focusing of ions occurs downstream from the particle – the so–called wake is formed. This leads to differences in the interaction in the planes parallel and perpendicular to the ion flow. Along the flow, the electrostatic potential has a damped periodical structure in which attraction between particles is possible.

11.2.5 *Experimental determination of the interaction potential*

Determination of the interaction potential constitutes a delicate experimental problem. Only a few such experiments have been performed (Konopka et $al.$ 1997, 2000a; Takahashi et $al.$ 1998; Melzer et $al.$ 1999). An elegant method based on an analysis of elastic collisions between the two particles was proposed by Konopka et $al.$ (1997, 2000a). In another method, laser radiation is used to manipulate particles (Takahashi et $al.$ 1998; Melzer et $al.$ 1999). Let us describe these methods separately.

In the collision method (Konopka et $al.$ 2000a), particles of radius $a \approx 4.5$ μm and an argon plasma at pressure $p = 2.7$ Pa were utilized. The particles are introduced into an r.f. discharge through a small hole in the glass window built into the upper electrode and are levitated above the lower electrode, where the electric field compensates for gravitational force. To confine the particles horizontally, a ring is placed on the lower electrode, which introduces a horizontal parabolic confining potential. The manipulation of the particles and activation of elastic collisions between them is performed with the use of a horizontal electric probe introduced into the discharge chamber. During the collision, the particle trajectories are determined by the confining potential and the interparticle interaction potential which is a function of the interparticle spacing. An analysis

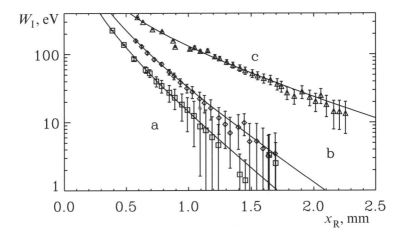

FIG. 11.7.   Potential energy W_{I} of interaction between two particles versus the distance
x_R between them (Konopka *et al.* 2000a). Measurements were taken at $p = 2.7$ Pa and
different r.f. peak–to–peak voltages U_{p}. Symbols correspond to experimental results, solid
lines show their fit to a screened Coulomb potential $W_{\mathrm{I}}(x_{\mathrm{R}}) = (Q_{\mathrm{eff}}^2/x_{\mathrm{R}})\exp(-x_{\mathrm{R}}/\lambda_{\mathrm{D}})$
leading to the following effective particle charge Q_{eff} and a screening length λ_{D} (T_{e} is
the measured electron temperature): 1 – $|Q_{\mathrm{eff}}| = 13\,900e$, $\lambda_{\mathrm{D}} = 0.34$ mm, $T_{\mathrm{e}} = 2.0$ eV,
$U_{\mathrm{p}} = 233$ V; 2 – $|Q_{\mathrm{eff}}| = 16\,500e$, $\lambda_{\mathrm{D}} = 0.40$ mm, $T_{\mathrm{e}} = 2.2$ eV, $U_{\mathrm{p}} = 145$ V; 3 –
$|Q_{\mathrm{eff}}| = 16\,800e$, $\lambda_{\mathrm{D}} = 0.90$ mm, $T_{\mathrm{e}} = 2.8$ eV, $U_{\mathrm{p}} = 64$ V. Note that the screening length
determined from the experiment is closer to the Debye radius for electrons than for ions.
This is in qualitative agreement with the conception that the ion velocity is close to the
ion sound velocity $c_{\mathrm{i}} = \sqrt{T_{\mathrm{e}}/m_{\mathrm{i}}}$ in the (collisionless) sheath and ions do not contribute to
the particle charge screening.

of recorded trajectories of the dust particles during collisions yields the coordinates and velocities of both the particles during collision. Then, the form of the
interaction potential can be reconstructed from the equation of motion. Application of this method (Konopka *et al.* 2000a; Morfill *et al.* 1999c) showed that for
low discharge powers and pressures the interaction potential coincided with the
screened Coulomb potential (11.39) within experimental uncertainties. This is
illustrated in Fig. 11.7. The role of other interaction mechanisms is insignificant
for the given plasma conditions, which, however, does not exclude the possibility
of their existence (Khrapak *et al.* 2001). The measurements also allow the determination of the particle effective charge and plasma screening length because
these two parameters determine the form of the Coulomb screened potential. A
detailed discussion of these experiments, the assumptions made in their theoretical interpretation, and some proposals for using this technique in experimental
investigations can be found in (Morfill *et al.* 2003).

A method based on laser manipulation of the dust particles was employed to
study the interaction between the particles in the direction of the ion flow (Taka-

hashi *et al.* 1998). Its modification for two particles was described by Melzer *et al.* (1999), which is discussed in some detail below. The essence of the experiment was the following. Two particles of different masses are used: the first one having a radius $a \approx 1.7$ μm, and the second one being a cluster of two such particles sticking together. The particles are introduced into a plasma of an r.f. discharge in helium at a pressure of $p \sim 50 - 200$ Pa. Because of the different charge-to-mass ratios, the particles are levitated at different vertical equilibrium positions in the anisotropic sheath region above the lower electrode – more massive particles levitate closer to the electrode. Due to the inhomogeneous electric field, the force balance (determined by the electrostatic and gravity forces) fixes their vertical positions. Both particles, meanwhile, are free to move in the horizontal plane. The first observation was the following. For sufficiently low pressures, the particles tend to form a bound state, in which the lower particle is vertically aligned to the upper one. Note that similar structures (vertically aligned chains) are quite common for many experimental investigations of multilayer dust crystal formation in gravity conditions. With increasing pressure, the bound state can be destroyed, and the particle separation in the horizontal plane is limited only by a very weak horizontal confinement due to a specially concave electrode. The existence of these states indicates some mechanisms of attraction and repulsion between the particles, which change each other in response to changes in plasma parameters. It is also found that the effect exhibits hysteresis (dependent on the pressure).

Next, to prove that the aligned bound state is due to attraction between the particles rather than its being forced by an external confinement, the particles were manipulated by laser radiation. The laser beam is focused either on the upper or the lower particle, causing its motion. It was found that when the upper particle is pushed by the laser beam, then the lower particle follows its motion and the bound state is not destroyed. This behavior proves that the lower particle is subject to an attractive horizontal force mediated by the upper particle. If the lower particle is pushed by the laser beam, then the upper particle's response is much weaker and the bound state can be easily destroyed. Hence, the interaction between the particles is asymmetric. It is clear that the attraction between the particles situated along the ion flow can be attributed to the wake effect. However, the question of whether it has an electrostatic nature or is associated with momentum transfer from the ions scattered by the upper particle (Lapenta 2002) still needs to be investigated.

11.2.6 *Formation and growth of dust particles*

In laboratory conditions, the dust particles are usually introduced into a plasma deliberately. On the other hand, they may, in principle, be self-formed in plasmas. There are several possible sources of dust particles. First is condensation leading to the appearance of solid particles or droplets. This process is typical for expanding plasmas, e.g., adiabatic plasma expansion in a vacuum or expansion of plasma in the channel of an MHD generator (Zhukhovitskii *et al.* 1984; Yakubov and Khrapak 1989). In chemically reacting mixtures, the dust particles may appear due to chemical reactions (Perrin and Hollenstein 1999). Finally, erosion of the electrodes and walls of a discharge chamber also leads to the appearance of macroparticles (Bouchoule 1999). In plasmas, the particles may grow. One of the reasons is the surface recombination of ions, which leads to a permanent sedimentation of the material on the particle surface. The agglomeration of dust particles may also become valid. In the work of Perrin and Hollenstein (1999), one of the possible scenarios of particle formation and growth was considered. It includes four stages: first, primary clusters are formed; once they have grown to a critical size, heterogeneous condensation occurs; at the next stage, the processes of coagulation and agglomeration are dominant, and at the last stage, the condensation of monomers on the isolated multiply charged particles turns out to be most important. On the whole, the processes of dust particle formation and growth in plasmas are not fully understood and require further investigation. The importance of this problem is largely connected to the needs of plasma technologies in, for example, the production of nanoparticles and thin films, and material processing.

11.3 Strongly coupled dusty plasmas and phase transitions

11.3.1 *Theoretical approaches*

11.3.1.1 *Strong coupling of dusty plasmas.* The conditions which can be realized in dusty plasmas are quite diverse and depend on relations among their characteristic parameters. One of the fundamental characteristics of a many–particle interacting system is the coupling parameter γ defined as the ratio of the potential energy of interaction between neighboring particles to their kinetic energy. For the Coulomb interaction between charged particles one finds

$$\gamma = \frac{Z^2 e^2}{T \Delta}, \tag{11.67}$$

where $\Delta = n_{\rm d}^{-1/3}$ characterizes the average interparticle spacing, and T characterizes their kinetic energy. For the plasma electrons and ions one obtains

$$\gamma_{\rm e(i)} = \frac{e^2 n_{\rm e(i)}^{1/3}}{T_{\rm e(i)}} \tag{11.68}$$

(ions are assumed to be singly charged). The system is commonly called strongly coupled when $\gamma \gtrsim 1$. It is well known that charges in plasmas are screened. Hence, in dusty plasmas, in addition to the average interparticle spacings, the Debye screening radii of each species and the dust particle radii appear as characteristic scales of length. In conditions typical of dusty plasma experiments, the number of electrons (ions) $N_{\rm e(i)}^{\rm D}$ in the electron (ion) Debye sphere is large: $N_{\rm e(i)}^{\rm D} = n_{\rm e(i)} \lambda_{\rm De(i)}^3 \gg 1$, and hence electron and ion species themselves are weakly coupled (ideal), because $\gamma_{\rm e(i)} \sim (N_{\rm e(i)}^{\rm D})^{-2/3} \ll 1$.

The situation with the dust component is qualitatively different. As before, for $N_d^D \gg 1$, the dust subsystem is weakly coupled; in this case, the dust appears as an additional plasma component which introduces new spatial and temporal scales in the system. The dust particles contribute to screening, the effective screening length now being

$$\lambda_D^{-2} = \lambda_{De}^{-2} + \lambda_{Di}^{-2} + \lambda_{Dd}^{-2} \tag{11.69}$$

(the particle charges are assumed fixed here). In the opposite case of $N_d^D \ll 1$, the dust subsystem is not always strongly coupled, because the screening can be determined only by electrons and ions. The interparticle distance can be shorter than the dust Debye radius, but the particles are not necessarily strongly interacting, being screened by the electron–ion background.

Most theories developed thus far to describe the properties of dusty plasmas employ the following model: negatively charged particles are confined within the plasma volume due to some confining force (usually of electrostatic character) and interact between themselves via the isotropic screened Coulomb (Debye–Hückel or Yukawa) repulsive potential (11.65) where the screening is governed by plasma electrons and ions. This model gives a simplified picture of dusty plasma behavior and is unsuited to some experiments, especially when plasma anisotropy plays a considerable role. Moreover, this model does not take into account variations of particle charges, long–range interactions, the exact form of the confining potential, etc. However, it has proved useful in providing qualitative results which are supported by experiments, and hence it may be considered as the basis on which one can construct more realistic models intended to represent actual dusty plasmas under various conditions.

11.3.1.2 *Phase diagram of Debye–Hückel systems.* Besides complex plasmas, particles interacting with a Debye–Hückel potential have been extensively studied in different physical systems ranging from elementary particles to colloidal suspensions. Not surprisingly, their phase diagrams have received considerable attention. Various numerical methods, usually Monte Carlo (MC) or molecular dynamics (MD) simulations, have been employed (Kremer *et al.* 1986; Robbins *et al.* 1988; Stevens and Robbins 1993; Hamaguchi *et al.* 1997).

In the case considered, the static properties of the system are completely determined by two independent dimensionless parameters. The first one measures the effective system temperature and is defined as $\tilde{T} = T/m_d \omega_E^2 \Delta^2$, where ω_E is the Einstein frequency of the crystalline structure oscillations. Since ω_E depends on the crystal structure, the fcc Einstein frequency is commonly used for definiteness. The other is the so–called structure (lattice) parameter

$$\kappa = \frac{\Delta}{\lambda_D}. \tag{11.70}$$

This choice of parameters comes historically from the theories of colloidal solutions. On the other hand, the ordering parameter commonly used for complex plasmas is the Coulomb coupling parameter in the form (11.67), namely

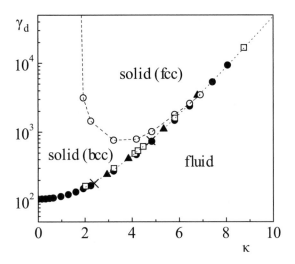

FIG. 11.8. Phase diagram of Debye–Hückel systems, obtained from numerical modeling. Open circles correspond to the bcc–fcc phase boundary (Hamaguchi *et al.* 1997). The fluid–solid phase boundary is marked by triangles (Meijer and Frenkel 1991), squares (Stevens and Robbins 1993), and solid circles (Hamaguchi *et al.* 1997). The crosses correspond to jumps in the diffusion constant, observed in the simulations of dissipative Debye–Hückel systems (Vaulina and Khrapak 2001; Vaulina *et al.* 2002). The dashed line is the fit to the numerical data judged by eye.

$\gamma_{\mathrm{d}} = Z_{\mathrm{d}}^2 e^2 / \Delta T_{\mathrm{d}}$. This is because early investigations of plasma crystallization focused on one–component plasmas (OCP), which can be considered as a limiting case for $\lambda_{\mathrm{D}} \to \infty$ or $\kappa \to 0$ of the Debye–Hückel model (Dubin and O'Neil 1999; Ishimaru 1982, 1992; Zamalin *et al.* 1977). The reduced temperature and the Coulomb coupling parameter are related by the formula

$$\tilde{T} = \frac{\omega_{\mathrm{pd}}^2}{4\pi\gamma_{\mathrm{d}}\omega_{\mathrm{E}}^2},$$

where $\omega_{\mathrm{pd}} = (4\pi Z_{\mathrm{d}}^2 e^2 n_{\mathrm{d}}/m_{\mathrm{d}})^{1/2}$ is the dust plasma frequency. The ratio $\omega_{\mathrm{E}}/\omega_{\mathrm{pd}}$ is a known function of κ (Robbins *et al.* 1988; Ohta and Hamaguchi 2000a). In Fig. 11.8, the phase diagram of the Debye–Hückel system, summarizing available numerical results, is drawn in the $(\kappa, \gamma_{\mathrm{d}})$–plane. Three phases were found, depending on the values of coupling and structure parameters. For very strong coupling, $\gamma_{\mathrm{d}} > \gamma_{\mathrm{M}}$, where γ_{M} denotes the value of γ_{d} in the melting curve, there are solid fcc and bcc phases and a liquid phase for $\gamma_{\mathrm{d}} < \gamma_{\mathrm{M}}$. The bcc phase is stable at small κ, while fcc is stable at larger κ. The "triple point" is at $\kappa \approx 6.90$ and $\gamma_{\mathrm{d}} \approx 3.47 \cdot 10^3$ (Hamaguchi *et al.* 1997).

Of particular interest for plasma crystallization experiments is the form of the melting (crystallization) curve $\gamma_{\mathrm{M}} = \gamma_{\mathrm{M}}(\kappa)$ (Ikezi 1986; Robbins *et al.* 1988;

Meijer and Frenkel 1991; Stevens and Robbins 1993; Hamaguchi *et al.* 1997; Ohta and Hamaguchi 2000a; Vaulina and Khrapak 2000). Results with OCP systems ($\kappa = 0$) indicate that the crystallization proceeds at $\gamma_\mathrm{d} = \gamma_\mathrm{OCP} \approx 106$ (or ≈ 172 if the Wigner–Seitz radius $(4\pi n_\mathrm{d}/3)^{-1/3}$ is used as the length unit instead of Δ) (Ichimaru 1982; Dubin 1990; Farouki and Hamaguchi 1993). The corresponding reduced temperature is $\tilde{T}_\mathrm{M} \approx 0.0022$. Several assumptions have been made in the literature concerning the analytical dependence of γ_M on κ. Starting from the OCP limit, Ikezi (1986) assumed that one simply has to take screening into account through $U(\Delta)/T = \gamma_\mathrm{OCP}$, so that according to formula (11.65) one arrives at

$$\gamma_\mathrm{M} = \gamma_\mathrm{OCP} \exp(\kappa). \tag{11.71}$$

Another simple argument is the following. The dimensionless temperature is proportional to the mean square of the particle oscillation amplitude in the quasi–harmonic approximation and, according to the Lindemann criterion, should be approximately constant along the melting curve. As the numerical values of the Einstein frequencies for the bcc and fcc lattices differ by less than 1% in the region where the bcc lattice is stable, one may assume that

$$\tilde{T}_\mathrm{M} \approx \tilde{T}_\mathrm{OCP}. \tag{11.72}$$

To reach a better agreement with the calculated results, Stevens and Robbins (1993) used a linear fit to their numerical data instead of relationship (11.72). They resorted to the expression

$$\tilde{T}_\mathrm{M} \approx \tilde{T}_\mathrm{OCP}(1 + 0.1\kappa). \tag{11.73}$$

Finally, Vaulina and Khrapak (2000) proposed that the characteristic dust lattice wave (DLW) frequency (see Section 11.4) be used instead of ω_E in the determination of $\tilde{T}$. Using the Lindemann criterion with this normalization, the dependence of γ_M on κ becomes

$$\gamma_\mathrm{M} = \gamma_\mathrm{OCP} \frac{\exp(\kappa)}{1 + \kappa + \kappa^2/2}. \tag{11.74}$$

The arguments used in deriving expressions (11.71)–(11.74) are not sufficiently rigorous. They can, therefore, be considered only as phenomenological melting conditions.

Note that although all the expressions (11.71)–(11.74) give the same (correct) result at $\kappa = 0$, they demonstrate different dependencies of γ_M on κ. As shown by Vaulina *et al.* (2002), expression (11.71), widely used in the literature, is in poor agreement with simulation results. Expression (11.72) provides somewhat better agreement. Expressions (11.73) and (11.74) yield the best agreement. Good agreement of formula (11.73) with the results of numerical simulations is not surprising – it is reached by using a linear fit to the numerical results. On the other hand, the functional form of Eq. (11.74) is simple and ensures better agreement with numerical experiments in the regime mostly relevant to complex plasmas experiments, namely $\kappa < 5$. Thus, it is convenient to introduce a modified coupling parameter

$$\gamma^* = \gamma_d \left(1 + \kappa + \kappa^2/2 \right) \exp(-\kappa), \tag{11.75}$$

of which the value $\gamma_M^* \approx 106$ uniquely determines the location of the melting (crystallization) curve in the phase diagram.

11.3.1.3 *Crystallization criteria.* From a practical point of view, a simple criterion is often required, which allows us to judge whether the system under consideration is in a crystalline or liquid state. Different phenomenological criteria for the crystallization (melting) of systems of interacting particles exist, which are often independent of the exact form of interaction potential between the particles. Some of them are convenient for dusty plasmas. Best known is the Lindemann (1910) criterion, according to which melting of the crystalline structure occurs when the ratio of the root–mean–square particle displacement to the mean interparticle distance reaches a value of ~ 0.15. Notice that this value can vary somewhat for various physical systems, mainly due to different procedures used to determine the average interparticle spacing. Another criterion is the value of the first maximum of the structural factor in the liquid state (Hansen and Verlet 1969), which reaches a value of ~ 2.85 in the crystallization curve. There also exists a simple crystallization criterion expressed in terms of the pair correlation function, the ratio of the minimum to the maximum of which should be approximately equal to 0.2 under crystallization. A simple dynamic crystallization criterion, similar in spirit to the Lindemann criterion, was proposed by Löwen *et al.* (1993). According to this criterion, crystallization occurs when the diffusion constant reduces to a value of ~ 0.1 compared to the diffusion constant for noninteracting particles. Later on, it was noted that this criterion holds for both 2D and 3D systems (Löwen 1996).

11.3.1.4 *Dynamics of Debye–Hückel systems.* The motion of dust particles in a dusty plasma with not too low a pressure can be considered as Brownian motion – modified, however, by the interaction between the particles themselves. The question to be answered is, therefore, to what extent does this interaction affect the particle dynamics. The dynamical properties of the dust component are fully determined by three dimensionless parameters, as can be clearly seen by normalizing the equations of motion of particles to a dimensionless form (Vaulina and Khrapak 2001). These are the parameters γ_d and κ, introduced above, while the third one is an appropriate measure of system dissipativity. Following Vaulina *et al.* (2002) let us define this parameter in the form

$$\theta = \frac{\nu_{dn}}{\omega_d} \tag{11.76}$$

and call it the *dynamic parameter.* Here, ν_{dn} stands for the damping rate associated with particle–neutral collisions (the neutral component typically determines dissipation because the degree of plasma ionization is very low in experiments, $\alpha \sim 10^{-6} - 10^{-7}$), and ω_d is some characteristic frequency associated with the charged dust component. In principle, the dust–plasma frequency or the Einstein frequency may be used for normalization. However, as will be clear from the following, it is most convenient to use for this purpose the dust lattice wave frequency.

One of the most fundamental quantities characterizing the dynamic behavior of the system is the single-particle diffusion coefficient. For 3D diffusion it is determined as

$$D(t) = \frac{\langle [\mathbf{r}(t) - \mathbf{r}(0)]^2 \rangle}{6t}, \tag{11.77}$$

where $\mathbf{r}(t)$ is the particle trajectory, and $\langle \ldots \rangle$ denotes ensemble averaging. The diffusion constant is then $D_L = \lim_{t \to \infty} D(t)$. The limit $t \to \infty$ is understood in the sense that the time t is longer than any other microscopic time in the system $(\nu_{dn}^{-1}, \omega_d^{-1})$, but shorter than the characteristic diffusion time to a distance on the order of the system size or time scale for significant changes of dusty plasma parameters in the experiment. Due to the interaction between the particles, the value of D_L is smaller than the bare Brownian diffusion constant for noninteracting particles: $D_0 = T_d/m_d\nu_{dn}$, where T_d is the temperature characterizing the chaotic (thermal) velocities $v_{T_d} = \sqrt{T_d/m_d}$ of the dust particles. In the limiting case of a crystalline structure, D_L tends to zero, as the displacement of particles located at the lattice sites is limited. Therefore, the ratio D_L/D_0 for dissipative systems appears as a natural quantity reflecting the nature and strength of the interaction potential.

Diffusion in Debye–Hückel systems has been studied using numerical modeling by Rosenberg and Thirumalai (1986); Kremer et al. (1987); Robbins et al. (1988); Löwen et al. (1993); Ohta and Hamaguchi (2000a); Vaulina and Khrapak (2001); and Vaulina et al. (2002). The problem of self–diffusion in nondissipative systems ($\theta = 0$) was considered by Rosenberg and Thirumalai (1986); Robbins et al. (1988); and Ohta and Hamaguchi (2000a). In the context of colloidal solutions, in which dissipation is many orders of magnitude higher than in dusty plasmas, the diffusion was considered by Kremer et al. (1987) and Löwen et al. (1993). In particular, the problems of subdiffusive behavior of the time–dependent diffusion coefficient (11.77) (Kremer et al. 1987) and the value of the diffusion constant at the liquid–solid phase boundary (Löwen et al. 1993) were addressed.

A systematic study of diffusion in dissipative Debye–Hückel systems by means of Brownian dynamics simulations for parameters typical of isotropic gas discharge plasmas was performed in (Vaulina and Khrapak 2001; Vaulina et al. 2002; Vaulina and Vladimirov 2002). In these works, interaction with the medium is modeled by the Langevin force consisting of two terms: one of which describes systematic friction, while the other is the random force describing the stochastic action of the medium. The friction is usually due to the neutral component, while the random component of the force is either associated with individual collisions with neutral atoms or molecules, or can have another origin: plasma

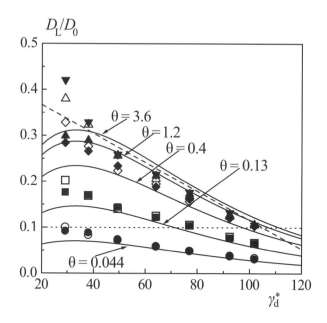

FIG. 11.9. The ratio D_L/D_0 for strongly interacting Debye–Hückel systems as a function
of the effective coupling parameter γ_d^*. The results of numerical simulations (Vaulina and
Khrapak 2001; Vaulina *et al.* 2002) for different values of the dynamic parameter θ are
shown. Solid symbols correspond to $\kappa = 2.42$, and open symbols correspond to $\kappa = 4.84$.
The values of θ are 0.044 (circles), 0.13 (squares), 0.4 (diamonds), 1.2 (triangles), 3.6
(inverted triangles) (only for $\kappa = 2.42$). The dashed line corresponds to the approximation
applicable for $\theta \gg \theta_{cr}$. The solid lines represent analytical expression (11.78) for different
values of the dynamic parameter (indicated in the figure). The horizontal dotted line
corresponds to $D_L/D_0 = 0.1$.

electric field microfluctuations, charge fluctuations, etc. Independently of its nature, the random component can be in the first approximation described as a
delta-correlated Gaussian white noise whose amplitude determines the dust temperature T_d. The equations of motion incorporating the interaction between the
particles and the particle interaction with the medium are solved numerically in
3D with periodic boundary conditions. The diffusion coefficient is determined by
appropriate averaging over particle trajectories.

The main results are summarized in Fig. 11.9, which shows the behavior of
the ratio D_L/D_0 as a function of the modified coupling parameter γ^*. As the
coupling increases, the ratio D_L/D_0 decreases and at some point it experiences
an abrupt jump, decreasing by several orders of magnitude in the very narrow
range $102 \le \gamma^* \le 106$. This jump takes place at $\gamma^* \approx \gamma_M^*$, thus indicating the
first–order phase transition (crystallization). In addition, Fig. 11.9 demonstrates
that the ratio D_L/D_0 is completely determined by the parameters γ^* and θ, and

it is not separately dependent on κ. Moreover, if the dissipation is sufficiently high, $\theta > \theta_{\mathrm{cr}}$, then the ratio D_{L}/D_0 becomes practically independent of θ as well. On the other hand, for $\theta \ll \theta_{\mathrm{cr}}$, numerical results tend towards those obtained for nondissipative systems by Ohta and Hamaguchi (2000a). In the intermediate case, the ratio D_{L}/D_0 depends on θ, as indicated above. The phenomenological expression describing this dependence was proposed by Vaulina and Vladimirov (2002), and it has the form

$$\frac{D_{\mathrm{L}}}{D_0} = \frac{\theta \gamma^* \exp(-3\gamma^*/\gamma_{\mathrm{M}}^*)}{6(1 + 2\pi\theta)}. \tag{11.78}$$

This phenomenological expression demonstrates reasonable agreement with numerical data for $\gamma^* \gtrsim 30$, as follows from Fig. 11.9. Using this expression we can also submit an estimate: $\theta_{\mathrm{cr}} \sim 1/2\pi \approx 0.2$. Finally, the dynamic criterion of crystallization, requiring that $D_{\mathrm{L}}/D_0 \approx 0.1$ in the liquid phase near the crystallization curve, turns out to be applicable to strongly dissipative systems, but is not confirmed for weakly dissipative systems $(\theta \ll \theta_{\mathrm{cr}})$.

The above–described results can, in principle, be used for dusty plasma diagnostics. However, the direct application of the results of 3D modeling of Debye–Hückel systems to the analysis of experiments on dusty plasmas under microgravity conditions and of laboratory experiments with fine powders is largely limited for a number of reasons (e.g., plasma anisotropy, long–range interactions including shadowing effects, the effect of external forces and/or boundary conditions, particle charge variations, etc.).

11.3.2   *Experimental investigation of phase transitions in dusty plasmas*

11.3.2.1   *Dusty plasma crystals in an r.f. discharge.*   Since the discovery of dusty plasma crystals in 1994, the investigation of phase transitions in different types of dusty plasmas has been actively performed in dozens of laboratories. Phase transitions in the dust component are studied in capacitive and inductive discharges, glow d.c. discharge, thermal plasmas of combustion products of different fuels, as well as nuclear– and photo–induced plasmas. Below in this section we will discuss those experiments in which strong coupling of the dust component, leading to phase transitions to ordered states, is revealed in the most striking way.

Dusty plasma crystals were discovered in a capacitively coupled low–pressure r.f. discharge in inert gases almost simultaneously in several laboratories (Chu and I 1994; Thomas *et al.* 1994; Hayashi and Tachibana 1994; Melzer *et al.* 1994). The schematic of a typical experimental setup is shown in Fig. 11.10. The experimental apparatus includes a lower electrode 5–10 cm in diameter capacitively coupled to an r.f. generator $(f = 13.56 \text{ MHz})$ and an upper grounded electrode. The electrodes are placed into a vacuum chamber. Micron-sized dust particles fill a container and can be introduced into the discharge through a metallic grid. The electrode separation is typically 3–10 cm. The particle visualization is performed with the use of laser illumination. The laser beam is transformed into a

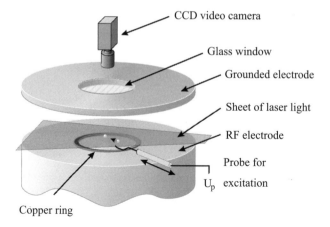

CCD video camera

Glass window

Grounded electrode

Sheet of laser light

RF electrode

Probe for

U_p excitation

Copper ring

FIG. 11.10. Schematic of experimental setup used to investigate the formation of ordered structures in an r.f. discharge.

sheet approximately 100 μm thick using a lens and illuminates the particles in the horizontal or vertical plane. The particles introduced into the plasma become (highly) charged negatively and levitate in the sheath above the lower electrode, which, on the average, acts as a cathode over a period of applied r.f. voltage. To confine the particles horizontally, a metallic ring with an inner diameter of 3–6 cm and a height of 1–3 mm is placed on the lower electrode. Sometimes, instead of a ring, a special curved electrode is used. The light scattered by the particles is recorded by a video camera.

Dust particles (of not too large a size) usually levitate in the sheath above the lower electrode due to the balance of the main forces, which include the electrostatic force (directed upwards for a negatively charged particle), the force of gravity (directed downwards), and the ion drag force (also directed downwards – in the direction of the time–averaged electric field). Under certain conditions, the particles form an ordered structure consisting of several fairly extended horizontal layers. The number of particles in such a structure can reach $\sim 10^3$–10^5, while the number of layers can vary from one to a few dozen (Zuzic *et al.* 2000). Figure 11.1a illustrates an image of the horizontal cross–section of the ordered structure (fragment of one layer) consisting of monodispersed polymer ($\rho = 1.5$ $\mathrm{g\,cm^{-3}}$) particles ~ 6.9 μm in diameter (Morfill and Thomas 1996). The structure displayed is highly ordered and exhibits a hexagonal lattice. The particles are usually settled strictly below each other in different horizontal layers, forming a cubic lattice between layers. Such an arrangement can be a consequence of ion focusing downstream from the particles – the wake effect.

It is difficult to obtain real three–dimensional structures in ground–based experiments in r.f. discharges. In these conditions, dusty plasma crystals have substantially a two–dimensional, or more exactly so–called $2\frac{1}{2}$–dimensional, character. This is directly connected to the action of gravity. To support particles against gravity, strong electric fields are required. Such strong fields can be established only in plasma sheaths which are characterized by a high degree of anisotropy and suprathermal (or even supersonic) ion flows. External forces acting on the dust particles are comparable to the interparticle forces. Hence, the dust systems are strongly compressed, inhomogeneous, and anisotropic in the vertical direction. In these conditions, comparison with theoretical results obtained for 3D Debye–Hückel systems is incorrect.

For reference, we summarize typical plasma and particle parameters in these experiments. Particle diameters usually lie in the range from 1 to 30 μm. The neutral (typically inert) gas pressure ranges from 0.01 to 1 torr. Plasma concentration in the bulk of a discharge is $n_e \sim n_i \sim 10^8 - 10^{10}$ cm^{-3}, the electron temperature is $T_e \sim 1$–5 eV, and ions are typically assumed to be in equilibrium with neutrals (usually at room temperature), so that $T_i \sim T_n \sim 0.03$ eV. Let us also emphasize some properties of sheaths where the particles levitate. The sheath is a region of a positive space charge, so that the ion concentration exceeds that of electrons even in the absence of dust. The ion velocity at the edge of the collisionless sheath satisfies the Bohm criterion $u \geq c_i = \sqrt{T_e/m_i}$. For the collisional sheath, the ion directed velocity can be lower than the velocity of sound, but the ions are still typically suprathermal: $u \gg v_{T_i}$. The characteristic spatial scale of the particle charge screening in this case is given by the electron Debye radius λ_{De}, because fast ions do not contribute considerably to screening. Finally, it should be recalled that the potential of interaction between the particles is highly anisotropic. In the direction perpendicular to the ion flow, the interparticle interaction is determined as before by a screened Coulomb potential (with a screening length close to λ_{De}), while in the direction parallel to the flow the effects caused by ion focusing downstream from the particles become important (see Section 11.2.4).

Structural properties of strongly coupled dusty plasmas were investigated by Chu and I 1994; Thomas et al. (1994); Hayashi and Tachibana (1994); Melzer et al. (1994); Quinn et al. (1996); Hayashi (1999); and Zuzic et al. (2000). Crystalline structures such as bcc, fcc, and hcp, as well as their coexistence, were found for certain plasma and particle parameters. For the quantitative analysis of the ordered structures of particles observed in experiment, it is common to use the following three characteristics (Quinn et al. 1996): the pair correlation function $g(r)$, the bound–orientational correlation function $g_6(r)$, and the structure factor $S(k)$. The pair correlation function $g(r)$ represents the probability of finding two particles separated by a distance r. For the case of an ideal crystal, $g(r)$ is a series of δ–functions (peaks) whose positions and heights depend on crystal structure type (for the gas state $g(r) \equiv 1$). The pair correlation function measures the translational order in the structure of interacting particles. The bound–orientational correlation function for two–dimensional systems is defined in terms of the nearest–neighbor bond angles with respect to an arbitrary axis. For a perfect hexagonal structure $g_6(r) \equiv 1$, while for other phases $g_6(r)$ decays with distance. The bond–orientational correlation function measures orientational order in the structure. Finally, the static structure factor measures order in the structure, similar to $g(r)$. It is defined as

$$S(k) = \frac{1}{N}\left\langle \sum_{i,j} \exp\left[i\mathbf{k}\cdot(\mathbf{r}_i - \mathbf{r}_j)\right]\right\rangle,$$

where $\mathbf{r}_i$ and $\mathbf{r}_j$ are the positions of particles i and j, and $\langle\ldots\rangle$ denotes ensemble averaging. Note that the structure factor is connected to the pair correlation function via the Fourier transform (see, e.g., Ichimaru 1982):

$$S(k) = 1 + n_d \int d\mathbf{r}\left[g(r) - 1\right]\exp(-i\mathbf{k}\cdot\mathbf{r}).$$

These characteristics are applied for the analysis of static properties of the highly ordered dust particle structure by Quinn *et al.* (1996). In the same work, a comparison between the obtained quantitative results and predictions of the 2D melting theory developed by Kosterlitz, Thouless, Halperin, Nelson, and Young (KTHNY theory) is drawn.

As already noted in the Introduction, dusty plasmas possess a number of unique properties which make these systems extremely attractive for investigation of different collective processes, including phase transitions. In particular, relatively short temporal scales for relaxation and response to external perturbations, as well as simplicity in observation, allow us not only to study static structure characteristics, but also to investigate the dynamics of phase transitions in detail (Thomas and Morfill 1996a,b; Morfill *et al.* 1999c; Melzer *et al.* 1996). Usually, phase transitions from crystal–like to liquid–like or gas–like states are experimentally investigated. Melting of a crystalline lattice can be initiated either by a decrease in the neutral gas pressure or by an increase in the discharge power. This can be attributed to the fact that the plasma parameters change under these conditions in such a way that the coupling in the system decreases. The decrease in coupling is first of all connected, at least in experiments with the lowering of the neutral pressure, with a significant increase in the kinetic energy of the dust particles. For example, the initial temperature which was close to the neutral gas temperature ($T_d \sim T_n \sim 0.03$ eV) in highly ordered crystal–like structures increased with the lowering of pressure and structure melting up to ~ 5 eV (Thomas and Morfill 1996a) or even ~ 50 eV (Melzer *at al.* 1996) at minimal pressures examined in these works. This "anomalous heating" of the dust component in plasmas indicates some source of energy which is effectively transferred into the kinetic energy of the dust particles. Dissipation of the kinetic energy in turn is determined by collisions with neutrals. With a decrease in pressure the dissipation goes down, which leads to the "heating" of the particles. Several possible mechanisms of anomalous heating were considered in the literature: stochastic fluctuation of the particle charge and energy gain in an external electric field (Morfill and Thomas 1996; Vaulina *et al.* 1999a,b, 2000; Quinn and Goree 2000); heating due to ion focusing and its attendant anisotropy in the

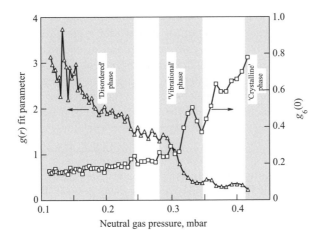

FIG. 11.11. The evolution of translational ($\triangle$) and orientational ($\square$) order parameters during the phase transition as a function of neutral gas pressure (Thomas and Morfill 1996a). Three regions are shaded: the "crystalline" phase at ~ 0.42; the "vibrational" phase at ~ 0.32 mbar; and the "disordered" phase at ~ 0.24 mbar. The intermediate "flow and floe" state (right light–colored region) occurs at ~ 0.36 mbar.

interaction of particles from different lattice planes (Schweigert *et al.* 1998); spatial particle charge variations caused by plasma nonuniformity (Zhakhovskii *et al.* 1997); and current–driven instability of liquid–like dusty plasma due to ion drift (Joyce *et al.* 2002). However, the exact origin of the heating has still not been established. Hence, further detailed experimental investigation and comparison of the results with different theoretical models are required.

Let us discuss in more detail the process of dusty plasma crystal melting induced by reducing pressure. We focus mainly on the results of a classical experiment (Thomas and Morfill 1996a) in which the melting of a "flat" plasma crystal (with a thickness of only a few lattice planes) formed by monodispersed melamine–formaldehyde particles 6.9 μm in diameter was investigated. The melting in this experiment was initiated by the continuous lowering of pressure in krypton plasma from $p = 42$ Pa, at which the stable ordered state of dust particles with thermal energy $T_d \sim T_n$ existed, to $p = 22$ Pa, for which the system lost any order and formed a "gaseous" state. The pair correlation function, bond–orientational correlation function, and kinetic energy of the dust system were measured while lowering the pressure.

From these analyses, the four "states" during the melting transition were identified. The first, the "crystalline" state is characterized by the conservation of the crystalline lattice in the horizontal plane and chain formation in the vertical direction as the pressure weakly reduced from the initial value. The particles in the lattice experience thermal oscillations ($T_d \sim T_n$) and highly occasional large–amplitude nonthermal oscillations – mostly in the vicinity of lattice defects.

The second, the "flow and floe" state, is characterized by the coexistence of islands of ordered crystalline structure (floes) and systematic directed particle motion (flows). In this state, the translational and orientational orders decrease significantly and occasional vertical particle migrations to other lattice planes are possible. Particle thermal motion still corresponds to room temperature, $v_{T_d} \sim$ 0.2 mm s^{-1}, while directed flow velocities are typically half of this. The third, the "vibrational" state, is characterized by some increase in orientational order and diminishing flow regions. However, isotropic particle vibrations with increasing amplitudes appear. Kinetic energy and vertical migrations of particles increase and the translational order continues to decrease. Finally, in the fourth, the "disordered" state, there is no translational or orientational order. The particles migrate freely both in the horizontal and vertical planes. The particle kinetic energies are hundreds of times greater than the neutral temperature ($T_d \sim 4.4$ eV).

Figure 11.11 shows the quantitative results for correlation functions $g(r)$ and $g_6(r)$ during melting. Three regions are shaded: "crystalline", "disordered" and "vibrational". In the last region, the correlation function $g_6(0)$ characterizing orientational order has a local maximum. The left light–colored region, which corresponds to the monotone decrease in translational order, separates the "disordered" state from the region of local increase in the orientational order in the "vibrational" state.

In concluding, we note once again that, strictly speaking, theoretical concepts of 2D and 3D melting are not applicable to describing these experiments because the systems are essentially $2\frac{1}{2}$–dimensional. From this point of view, the experimental investigation of monolayer crystal melting, as well as experiments on phase transitions in 3D systems under microgravity conditions, might be valuable.

11.3.2.2 Ordered structures of dust particles in a d.c. discharge.

A d.c. gas discharge can also be used for the formation of ordered structures in dusty plasmas (Fortov et al. 1996a, 1997a; Lipaev et al. 1997; Nefedov et al. 2000). The sketch of the typical experimental setup is given in Fig. 11.12. A discharge is typically created in a vertically positioned cylindrical tube. The particles introduced into the plasma can levitate in the regions where the external forces are balanced. They are illuminated by laser light and their positions are registered by a video camera. Typical conditions in the discharge are the following: a pressure in the range 0.1–5 torr, and a discharge current of $\sim 0.1-10$ mA. The ordered structures are usually observed in standing striations of the positive column of the glow discharge but can also be observed in an electric double layer formed in the transition region from the narrow cathode part of the positive column to the wide anode part, or in a specially organized multielectrode system having three or more electrodes at different potentials, etc., that is, in the regions where the electric field can be strong enough to levitate the particles. For these regions, some sheath properties considered above in the context of an r.f. discharge (e.g.,

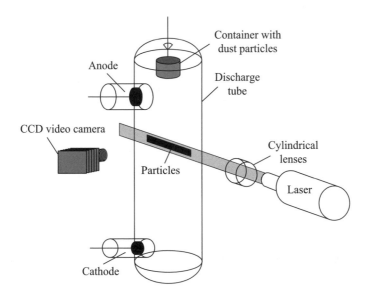

FIG. 11.12. Schematic of the experimental setup for studying the formation of ordered structures in a d.c. gas discharge.

plasma anisotropy, ion drift) are also relevant.

Most of the experiments were performed in standing striations of glow discharges. In the positive column of a low–pressure discharge, loss of electron energy in elastic collisions is small and the electron distribution function is formed under the action of the electric field and inelastic collisions. This can lead to the appearance of striations – that is, regions of spatial periodicity of the plasma parameters with the characteristic scale of the order of a few centimeters (Golubovskii et al. 1994; Golubovskii and Nisimov 1995, 1996). The electron concentration, their energy distribution, and the electric field are highly nonuniform along the striation length. The electric field is relatively strong (around 10–15 V cm^{-1} at a maximum) at the head of striation – that is, a region occupying 25–30% of the total striation length, and relatively weak (around 1 V cm^{-1}) outside this region. The maximum value of the electron concentration is shifted relative to the maximum strength of the electric field in the direction of the anode. The electron energy distribution is substantially bimodal, with the head of the striation being dominated by the second maximum whose center lies near the excitation energy of neutral gas atoms. Due to the high floating potential of the walls of the discharge tube, the striations exhibit a substantially two–dimensional character: the center–wall potential difference at the head of the striation reaches 20–30 V. Thus, an electrostatic trap is formed at the head of each striation, which in the case of vertical orientation is capable of confining particles with high enough charge and low enough mass from falling into the cathode positioned lower, while the strong radial field prevents particle sedimen-

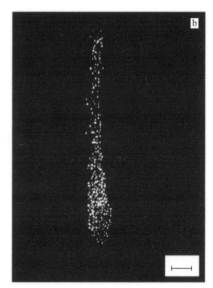

FIG. 11.13.   Video images of structures comprising charged microspheres of borosilicate
    glass: (a) in two neighboring striations (pressure around 0.5 torr, discharge current of 0.5
    mA); (b) after their coalescence (pressure around 0.4 torr, discharge current of 0.4 mA).
    The scale of length in the figure corresponds to 3 mm.

tation on the tube walls.

The process of structure formation proceeds routinely as follows: after being
injected into the plasma of the positive column, the charged particles fall past
their equilibrium position and then, over the course of several seconds, emerge
and form a regular structure which is preserved sufficiently long (until the end
of the observation) provided that the discharge parameters are unchanged. The
simultaneous existence of ordered structures in several neighboring striations
can be observed. Figure 11.13a demonstrates an image of two dust structures
formed by hollow thin-walled microspheres made of borosilicate glass with a
diameter of 50–63 μm in two neighboring striations. Figure 11.13b shows their
coalescence into one rather extended formation, occurring due to varying the
discharge parameters (Lipaev *et al.* 1997). This figure indicates the possibility
of forming structures much more extended in the vertical direction than in r.f.
discharges. In fact, the three–dimensional quasicrystalline dusty structures were
obtained in the work of Lipaev *et al.* (1997) for the first time.

In d.c. discharges, the transition of quasicrystalline structures to fluid and
gaseous states is also observed, similar to dusty plasma in r.f. discharges. This
occurs either by lowering the pressure or increasing the discharge current. For
example, for the structure comprising Al_2O_3 particles of diameter 3–5 μm at a
pressure of 0.3 torr and a current of 0.4 mA (estimated electron number density

$n_e = 10^8$ cm^{-3}, and temperature $T_e \sim 4$ eV), the correlation function reveals long–range order with four well-pronounced peaks (Lipaev et al. 1997). When the discharge current is increased by almost one order of magnitude to 3.9 mA ($n_e = 8 \cdot 10^8$ cm^{-3}), the structure "melts" and the correlation function reveals only short–range order. We note that during this "phase transition" the interparticle distance, equal to 250 μm, remains approximately constant.

Similar to dusty crystals in r.f. discharge plasmas, the effect of "anomalous heating" of particles was observed in d.c. plasmas. Under certain conditions, the particles acquired very high kinetic energies of up to 50 eV (Fortov et al. 1997a). This heating causes dust crystal melting, observed when the plasma parameters are changed.

Under certain discharge conditions, an increase in the number of small-sized particles gives rise to the formation of complex structures where different regions coexist (see Fig. 11.1b): the high ordering region ("plasma crystals"), and regions of convective and oscillatory motion of particles ("dusty plasma liquids"). Usually, in the lower part of the structure, the particles oscillate in the vertical direction (dust density waves) at a frequency of 25–30 Hz and a wavelength of about 1 mm, the mean interparticle distance being 200 μm. Such self–excited oscillations can correspond to the instability of the dust acoustic wave, the possible mechanisms of which are considered in Section 11.4. Most of the central region of the structure is occupied by a crystal–like structure with a pronounced chain–like configuration. At the periphery of the upper part of the structure, the particles undergo convective motion whose intensity decreases towards the center of the structure. This complex picture is apparently associated with a peculiar distribution of plasma parameters and forces acting on the dust particles within a striation.

11.3.2.3 *Ordered structures in a thermal plasma.* Thermal plasma constitutes a low–temperature plasma characterized by equal temperatures of the electron, ion, and neutral components. The existence of liquid or solid small–sized particles in such a plasma can significantly affect its electrophysical properties. Effects associated with the presence of particles were already observed in early experiments focused on studying plasma of hydrocarbon flames (see, e.g., Yakubov and Khrapak 1989).

Experimental investigations of the ordered dust structure formation in a thermal plasma were performed in a quasi–laminar weakly ionized plasma flow at temperatures of 1700–2200 K and atmospheric pressure (Fortov et al. 1996b,d, 1997b,c). A plasma source formed an extended and uniform volume (about 30 cm^3) of quasineutral thermal plasma, where the particles of CeO_2 were introduced. The main plasma components are the charged particles, electrons, and singly charged Na^+ ions. The electron density varies from 10^9 to 10^{12} cm^{-3}. The dust particles are charged by background electron and ion fluxes and via thermionic emission. Due to the dominant role of the latter mechanism, the particle charge is positive and equal to $10^2 - 10^3$ elementary charges. Here, in contrast to gas discharges in laboratory conditions, the plasma is isotropic and quite uniform because gravity does not play any important role.

The relatively large volume and uniformity of the plasma allow us to measure gas and particle properties by different probe and optical methods to find the characteristic plasma parameters. Different probe methods were employed to measure the densities of alkali metal ions and electrons. The gas temperature and the concentration of alkali atoms were measured by the generalized reversal technique and full absorption technique, respectively. In order to determine the mean (Sauter) diameter D_{32} and the particle concentration n_d in the plasma flow, the aperture transparency method was used. The spatial dust structure is analyzed using the pair correlation function $g(r)$ measured with a time–of–flight laser counter. The measurement volume of the counter is formed by focusing an argon laser ($\lambda = 0.488$ μm) beam in a certain region of the plasma flow. The radiation scattered by individual particles as they traverse the measurement volume is transduced to electric signals by a photodetector. The received signals are then processed in order to calculate the pair correlation function $g(r)$.

The results of measuring spatial dust structures were compared with experimental data obtained for an aerosol flow at room temperature. In the last case, only air with CeO_2 particles was injected into the inner region of the burner. Such a system models a plasma with a random (chaotic) spatial distribution of dust particles – a "gaseous plasma". Comparison of the pair correlation functions for CeO_2 particles with diameter $D_{32} \approx 1.8$ μm in aerosol flow at temperature $T_g \approx 300$ K and in a plasma at temperature $T_g = 2170$ K and particle concentration $n_d = 2.0 \cdot 10^6$ cm^{-3} showed (Nefedov et al. 1997) that they are almost identical. This means that a weakly ordered "gaseous" structure is produced. At a lower temperature $T_g = 1700$ K and significantly higher particle number density $n_d = 5.0 \cdot 10^7$ cm^{-3}, the pair correlation function $g(r)$ demonstrates short–range order which is characteristic of a liquid–like structure (Nefedov et al. 1997). In such a plasma, the measurements give for the ion number density $n_i \sim 10^9$ cm^{-3}, which is about one order of magnitude lower than the electron number density $n_e \sim 5 \cdot 10^{10}$ cm^{-3}. The particle charge evaluated from the quasineutrality condition $Z_d n_d = n_e$ is positive and equals approximately 10^3 elementary charges with an accuracy of a factor of 2. Then the ratio of the average interparticle separation to the electron Debye length is $\kappa \sim 2$, and the modified coupling parameter lies in the range $\gamma^* \sim 50-200$. These values are close to the crystallization conditions for the isotropic three–dimensional Debye–Hückel model. The fact that the structures found in the experiment are relatively weakly ordered is explained by a significant nonstationarity of the system: the time interval between particle injection into the plasma and correlation function measurements is not enough for the ordered structures to form completely. This evidence was supported by MD simulation of the dynamics of the ordered dust–particle structure formation in the conditions of the experiment (Khodataev et al. 1998; Nefedov et al. 1999).

Samaryan et al. (2000) reported experiments on determining the parameters of a plasma with a condensed disperse phase (CDP) of combustion products of synthetic solid fuels with various compositions. In most of the experiments, the coupling parameter γ was smaller than unity and ordering of CDP particles was absent. The main reason preventing the formation of ordered structures is the existence of a substantial amount of alkali–metal impurities in fuels. This leads to an increase in the degree of plasma ionization and, hence, to a decrease in the plasma screening length (Debye radius). The ordered structures of CDP particles were only observed for samples of aluminum–coated solid fuels with a low content of alkali metals. In this case, the coupling parameter reached values of 10–30, corresponding to structures with a liquid–like order. An analysis of the experimental pair correlation functions revealed a short–range order for sufficiently high particle concentrations.

11.3.2.4 *Ordered structures in nuclear–induced plasmas.* The nuclear–induced plasma is produced by nuclear–reaction products which create ion–electron pairs passing through a medium, as well as excited atoms and molecules in their tracks. In terms of physical characteristics, the nuclear–induced plasma of inert gases differs significantly from thermal and gas discharge plasmas. At the relatively low intensities of a radioactive source, typical of laboratory conditions, this plasma has a distinct track structure. The tracks are randomly distributed in space.

The experiments were performed in an ionization chamber which is placed in a hermetical transparent cell (Fortov *et al.* 1999c). Either β–particles (decay products of ^{141}Ce or α-particles and fission fragments (decay products of ^{252}Cf) were used as ionizing particles (Fortov *et al.* 1999c; Vladimirov *et al.* 2001a). The energies of reaction products, their free paths in air and dust particle material, as well as the number of secondary electrons emitted in collisions of the ionizing particle with a dust particle, are given in Table 11.1.

In a plasma of atmospheric air at an external electric field strength of 20 V cm^{-1}, the levitating particles form liquid–like structures – the pair correlation function has one maximum. At stronger electric fields, the dust particles move in closed trajectories which form a torus with the axis aligned with that of the cylindrical chamber.

In an inert gas, sufficiently dense dust clouds with sharp boundaries exist, and

Table 11.1 *Major characteristics of ionizing particles.*

	β–particle	Average fusion fragment	α–particle
Average kinetic energy, keV	138	$9 \cdot 10^4$	$6 \cdot 10^3$
Total range in solid matter (CeO_2), μm	58	5.5	20
Total range in air ($p = 1$ atm), cm	5.6	2.3	4.7
Number of secondary electrons per particle	≈ 5	≈ 250	≈ 10

FIG. 11.14.   (a) Dust structures of micron–size Zn particles in a nuclear–induced plasma
in neon at pressure 0.75 atm and electric field strength 30 V cm^{-1}. (b) Rotating dust
structures of micron–size Zn particles in neon; the agglomeration of particles is visible.

the dust particles form a liquid–like structure inside these clouds (Fig. 11.14(a)).
In a nonuniform electric field, the particles form a rotating dust structure in
which agglomeration of small particles into coarse fragments proceeds in the
course of time, Fig. 11.14(b) (Vladimirov *et al.* 2001b).

In the experiments, particles of Zn and CeO_2 were used. The particle radius
was estimated experimentally from the particle steady–state falling velocity after
the removal of the electric field and was found to be 1.4 µm. The electric charge
was inferred from the equilibrium condition for slowly moving levitating particles.
The particle charge depended on the particle radius and lay in the range from
400 to 1000 elementary charges.

For the theoretical calculation of the charge on a spherical particle in the
nuclear–induced plasma, nontraditional approaches are used because this plasma
is spatially and temporarily inhomogeneous. The charging process is considered
after the track plasma has separated into two clouds: one of electrons, and the
other of ions. When the electron or the ion cloud meets a dust particle on its drift
to the corresponding electrode, then this particle acquires some charge from the
cloud. A statistical treatment of these charging events constitutes the essence of
the mathematical model posed for calculating the particle charge. The electron
current to the particle is determined by the electron collection cross–section in
the collisionless limit. To obtain the ion current, the diffusion approximation
is used. In the simulation, the code first generates the event of creation, the
direction of emission from the source, and the type of ionizing particle (alpha
particle or a fission fragment) (Fortov *et al.* 2001a). The simulation shows that
the particle charge fluctuates with time. On the one hand, the dust particle
acquires a charge in the electron attachment process; on the other hand, its
charge decreases substantially in the less frequent events of interaction with the
ions. The contribution from alpha particles leads to a complicated dependence
of the charge on time. The characteristic time scale for changing the charge

is 10^{-3}–10^{-2} s. This time is short enough, and hence the particle interaction with external fields can be expressed in terms of effective constant charge. A time-averaged charge calculated from this model is in most cases close to the values obtained for levitating particles from the balance between gravity and electrostatic forces. The charge fluctuations with amplitudes comparable to the charge itself can be one of the reasons preventing the formation of highly ordered structures of dust particles.

When a particle gets into the track region, the cascade electrons having a mean energy of ~ 100 eV could cause charges which are sufficient for dust system crystallization. To investigate the charging process, a numerical model based on a system of equations describing two–dimensional space–time track evolution has been developed. The system of equations included the kinetic equation for electrons, the continuity equation for heavy components (ions, atoms, etc.), the Poisson equation, and equations describing chains of plasma–chemical reactions. It follows from the calculations that a particle can collect no more than 10 electrons from one track. This means that the influence of many tracks is required in order to produce the large charge on the particle.

In order for the charging rate in the track regions to prevail over that due to drift flows, an ionizing flux of a value of approximately 10^{13} cm^{-2} s^{-1} is required. Such a flux can be obtained in the beam of a charged particle accelerator. Experiments using a circular continuous beam with an aperture of 15 mm, current of 1 μA, and proton energy of 2 MeV were performed. In the absence of an external electric field, the stratification (i.e., formation of regions free of dust particles) of the dust component in neon was observed.

11.3.3 *Dust clusters in plasmas*

Dust clusters in plasmas constitute ordered systems of a finite number of dust particles interacting via the pairwise repulsive Debye–Hückel potential and confined by external forces (e.g., of electrostatic nature). Such systems are sometimes also called Coulomb or Yukawa clusters. The difference between dust clusters and dust crystals is conditional: both systems in fact consist of a finite number of particles. The term "dust clusters" is usually reserved for systems with number of particles $N \lesssim 10^2 - 10^3$, while for larger formations the term "dust crystals" can be used. A more precise definition of clusters would be the ratio of the number of particles in the outer shell to the total number of particles in the system. For crystals, this ratio should be small. Similar systems are met, for instance, in the usual singly charged plasmas in Penning or Paul traps (Dubin and O'Neil 1999; Gilbert *et al.* 1988), where the vacuum chamber is filled with the ions, as well as in colloidal solutions (Grier and Murray 1994). The distinctions between systems are due to different types of interaction potential and different forms of confining potential.

Historically, clusters consisting of repulsive particles in an external confining potential were first investigated with the use of numerical modeling (mostly by Monte Carlo and molecular dynamics methods). Taking into account the pos-

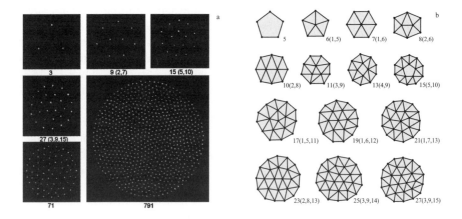

FIG. 11.15.   Video images of experimentally found (Juan *et al.* 1998) (a) typical dust
    cluster structures consisting of different number of particles (the scales are not the same
    for the pictures, typical interparticle spacing is between 0.3 and 0.7 mm); (b) typical shell
    configurations of several dust clusters composed of different numbers of particles.

sibility of applying the simulation results to dust clusters, we mention here the
works of Candido *et al.* (1998); Lai and I (1999); Totsuji (2001); Totsuji *et al.*
(2001); and Astrakharchik *et al.* (1999a,b). Most simulations were performed for
two–dimensional clusters in an external harmonic (parabolic) potential. Such a
configuration is usually realized in ground–based experiments with dusty plasma
in gas discharges. The simulations show that for a relatively small number of
particles in the cluster the "shell structure" is formed with the number of particles N_j in the jth shell $(\sum_j N_j = N)$. At zero temperature, unique equilibrium
configuration $(N_1, N_2, N_3, \ldots)$ exists for a given particle number N. Such configurations form an analog to Mendeleyev's Periodic Table, the structure of which
depends on the shape of the interaction potential, confining potential, and their
relative strengths. At finite temperatures, metastable states with energies close
to the ground state can also be realized.

    The first experimental investigation of dust clusters was reported by Juan
et al. (1998). The experiment was performed in the sheath of an r.f. discharge.
A hollow coaxial cylinder 3 cm in diameter and 1.5 cm in height was put on
the bottom electrode to confine the dust particles. Clusters with a number of
particles from a few up to 791 were investigated. A photo images of typical cluster
structures with different numbers of particles are exemplified in Fig. 11.15(a).
Figure 11.15(b) shows a series of typical shell configurations observed for dust
clusters. For a large number of particles, the inner particles form a quasi–uniform
hexagonal structure, while near the outer boundary particles form several circular
shells. The mean interparticle separation increases up to about 10% from the
center to the cluster boundary.

In the works of Klindworth *et al.* (2000) and Ishihara *et al.* (2002), the rotation of dust clusters around their symmetry axis was effected. In the first of the cited works, the cluster rotation was caused by the light pressure of the laser beam; not only could the cluster rotate as a whole, but also intershell rotation could be excited. In the second work, the cluster rotation was initiated by the presence of the magnetic field parallel to the cluster axis of symmetry. The magnetic field distorts the ion trajectories, with the result that the collisions of ions with the dust particles cause the cluster rotation.

In the work of Melzer *et al.* (2001), the oscillations of particles in clusters consisting of three, four, and seven particles were investigated. The application of the results obtained to dusty plasma diagnostics was also discussed (determination of the particle charge and plasma screening length). In this context we also mention the work of Amiranashvili *et al.* (2001) in which the stability conditions and analytical expressions for the frequencies of the basic modes were found for the simplest clusters consisting of two, three, and four particles and an arbitrary form of interaction potential between them.

11.4   Oscillations, waves, and instabilities in dusty plasmas

In this section, the wave processes in dusty plasmas are described. In selecting the material, priority was given to phenomena for which both theoretical and experimental results exist and those used for dusty plasma diagnostics in the laboratory. Our main attention is concentrated on linear waves in dusty plasmas without magnetic fields, although there also exists an extensive literature on nonlinear waves and waves in the presence of magnetic fields. We focus on the effects which are directly associated with the presence of the dust component. Often they have no analogs in the usual multicomponent plasmas. We start with the problem of individual particle oscillations in sheaths of gas discharges. Then an introduction to the theory of linear waves in dusty plasmas is given and some mechanisms of their instability and damping are considered. Finally, some experimental results are discussed, including generation of waves as a tool for dusty plasma diagnostics.

11.4.1   *Oscillations of individual particles in a sheath region of gas discharges*

In most of the ground–based experiments, negatively charged dust particles can only levitate in regions of sufficiently strong electric fields, where the electric force compensates for gravity. This occurs, for example, in the sheath of an r.f. discharge, where the electric field averaged over the oscillation period is directed along the gravity force; due to the large mass, neither the dust particles nor the ions respond to the r.f. field ($f \sim 14$ MHz). This is also true in regard to the cathode sheath of a d.c. discharge. The electric field in these regions increases almost linearly towards the electrode (Tomme *et al.* 2000) and reaches sufficiently high values on its surface. Variations in the particle charge are connected to ion acceleration under the action of the electric field and to an increase in the ratio $n_i/n_e > 1$ as the electrode is approached (see Fig. 11.3). Usually, the charge first somewhat decreases (increases in the absolute magnitude), reaches a minimum, and then increases and can even reach positive values close to the electrode. Examples of numerical calculations of the dependence of the particle surface potential on the distance from the electrode in collisionless and collisional sheaths of r.f. and d.c. discharges can be found in (Nitter 1996) for a set of plasma parameters. For not too heavy particles, there usually exists a stable particle equilibrium position in the sheath. Let us assign the vertical coordinate $z = 0$ to this position. The force balance condition gives

$$m_d g + Z_{d0} E_0 = 0, \tag{11.79}$$

where $Z_{d0} = Z_d(0)$ and $E_0 = E(0)$. For small displacements of a particle around its equilibrium position (linear oscillations), the potential energy can be expressed in the form $W(z) \approx m_d \Omega_v^2 z^2 / 2$, where

$$\Omega_v = -\frac{\partial}{\partial z} \left[\frac{Z_d(z) E(z)}{m_d} \right]_{z=0} \tag{11.80}$$

characterizes the frequency of vertical oscillations. As follows from relationship (11.80), Ω_v depends on the derivative of the electrostatic force at the equilibrium position and on the particle mass. Due to the relatively large mass of the dust particles, the value of Ω_v is not very high and typically lies in the range 1–100 s^{-1}. Hence, it is convenient to use low–frequency excitations for determining the resonance frequency of vertical oscillations of particles, which can be expressed through the plasma and particle parameters.

As a simple example we refer to a harmonic excitation of particle oscillations. The equation for small amplitude oscillations is written as

$$\ddot{z} + \nu_{dn} \dot{z} + \Omega_v^2 z = \frac{F_{ex}(t)}{m_d}, \tag{11.81}$$

where $F_{ex} = f_0 \cos \omega t$, and f_0 is the amplitude of an external force. Equation (11.81) constitutes an equation of forced oscillations in the presence of friction (see, e.g., Landau and Lifshitz 1988). Its solution for stationary oscillations has the form $z(t) = A(\omega) \cos(\omega t + \delta)$, where $A(\omega)$ is the amplitude of forced oscillations, viz.

$$A(\omega) = \frac{f_0}{m_d \sqrt{(\Omega_v^2 - \omega^2)^2 + \nu_{dn}^2 \omega^2}}, \tag{11.82}$$

δ is the phase shift, and $\tan \delta = \nu_{dn} \omega / (\omega^2 - \Omega_v^2)$. The amplitude grows when ω approaches Ω_v and reaches the maximum at $\omega = (\Omega_v^2 - \nu_{dn}^2/2)^{1/2}$. The value of ω corresponding to the maximum of the amplitude for $\nu_{dn} \ll \Omega_v$ is different from Ω_v only by a second–order correction. Hence, changing ω and measuring the oscillation amplitude (11.82), it is possible to determine Ω_v and ν_{dn}. Such measurements were performed for the first time by Melzer et al. (1994) and were later used to determine the particle charge by Trottenberg et al. (1995); Zuzic

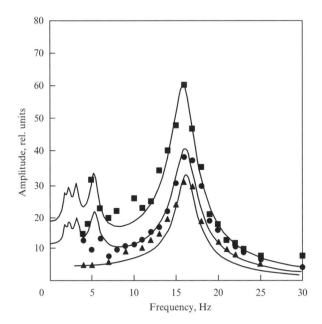

FIG. 11.16. The dependence of the amplitude of vertical oscillations of a dust particle in
the sheath of an r.f. discharge on the excitation frequency (Homann *et al.* 1999). Symbols
correspond to experimental findings, and curves to theoretical results. Different excitation
methods were used: electrode excitation with sine wave (▲), electrode excitations with
rectangular wave (●), excitation with radiation pressure from a chopped laser beam, analogous to rectangular wave (■). All the methods yield the same resonant frequency of ∼ 17
Hz, from the value of which the particle charge can be estimated.

et al. (1996); Homann *et al.* (1999); Piel and Melzer (2002); and Piel (2003).
Excitation is caused either by applying an additional low–frequency signal to
the electrode (in a modified variant, to a small probe inserted in a plasma in
the vicinity of the levitated particle) or by affecting the particle with the aid of
laser radiation (Zuzic *et al.* 1996). The exciting force is not necessarily harmonic,
but this does not influence the essential properties of the oscillations. Typical
experimental findings are presented in Fig. 11.16.

The main difficulty in estimating the particle charge from experimental results is to establish the connection between Ω_v and Z_d. It is often supposed that
the dependence of the particle charge on the vertical coordinate is much weaker
than that of the electric field, i.e., in the first approximation it is believed that
$Z_d \approx Z_{d0} = \text{const.}$, and therefore $\Omega_v^2 \approx -Z_{d0}E(0)/m_d$. The value of the derivative E' is often practically constant over the sheath and is linked through the
Poisson equation with the electron and ion number densities:

$$E'(z) = 4\pi e\big[n_i(z) - n_e(z)\big].$$

The concentrations n_i and n_e are determined by extrapolating the results of probe measurements in the bulk plasma into the sheath region, using different theoretical models. Probe measurements in the sheath are not practical because of difficulties in their interpretation. Therefore, the described method is not very accurate. However, it is widely used in experiments because of its simplicity and the absence of other methods exhibiting essentially higher accuracy.

By increasing the amplitude of the excitation force it is possible to excite nonlinear oscillations. The experiments and their theoretical interpretation are described by Ivlev et al. (2000b) and Zafiu et al. (2001). An example of a numerical model, which takes into account the dependencies of particle charge, electric field, and external force amplitude on the vertical coordinate, as well as the location of an excitation source with respect to the dust particle and force balance in the sheath, is given by Wang et al. (2002). The characteristic property of nonlinear oscillations is the hysteretic dependence of their amplitude on the excitation frequency. We also note that excitation of nonlinear oscillations allows us, in principle, to investigate the electric field and/or dust particle charge distributions along the sheath.

Vertical oscillations of particles in the sheath regions of gas discharges can be caused not only by the application of an external force, but also can be self–excited due to effects specific to dusty plasmas. For example, a drastic increase in the amplitude of vertical oscillations was examined experimentally under certain conditions (e.g., with lowering pressure) in the sheath of r.f. and d.c. discharges (Nunomura et al. 1999; Samarian et al. 2001). Different theoretical aspects of vertical oscillations were considered by Vaulina et al. (1999a,b); Ivlev et al. (2000a); and Sorasio et al. (2002). Below, we will ignore the simplest effect associated with the particle Brownian motion which is caused by the stochastic character of particle collisions with neutral atoms. This effect is always present and is additive to other effects. Its value (oscillation amplitude) is typically quite small from a practical point of view.

One of the effects leading to vertical oscillations is associated with random particle charge variations caused by the stochastic nature of particle charging (Vaulina et al. 1999a,b, 2000). To describe this effect, the right–hand side of Eq. (11.81) should be presented in the form $F_{ex} = eE_0 \delta Z_d$, where δZ_d is the random deviation of the particle charge from its average value, which is a random function of time. The oscillation amplitude will also be a random function of time. However, using the properties of random charge fluctuations (11.32)–(11.35), it is possible to derive its root–mean–square value. It can be easily shown that for typical conditions $\Omega_{ch} \gg \Omega_v \gg \nu_{dn}$, the mean square amplitude of oscillations associated with the random charge variations can be estimated as

$$\langle \Delta z^2 \rangle \approx \frac{g^2 \gamma^2}{|Z_d| \Omega_{ch} \Omega_v^2 \nu_{dn}}, \tag{11.83}$$

and increases with decreasing neutral gas pressure. Notice also that since $\Omega_{ch} \propto a$, $\nu_{dn} \propto a^{-1}$, $|Z_d| \propto a$, and in a first approximation $\Omega_v \propto |Z_d|/m_d \propto a^{-2}$, we have $\langle \Delta z^2 \rangle \propto a^3$, i.e., the mean square oscillation amplitude is proportional to the particle mass.

Another effect, which is also associated with the variability of the particle charge, results from the finite charging time. The qualitative theory of this effect was proposed to describe the following experimental evidence by Nunomura *et al.* (1999). When decreasing the neutral pressure and/or plasma density below some critical value (for the pressure p, the corresponding value is ~ 3 mtorr), self–excitation of oscillations (their amplitude grows with time) of the dust particles levitating in the sheath region of a d.c. discharge occurs. The time scale of the amplitude growth is ~ 10 s. In the final state either saturation takes place (the amplitude reaches a constant value determined, e.g., by nonlinear effects) or, if the amplitude is too large, the particle leaves the plasma reaching the electrode surface. This behavior testifies to the existence of an instability mechanism for vertical oscillations, which is effective under low neutral gas pressures. A physical interpretation of the phenomenon was proposed by Nunomura *et al.* (1999). According to the authors, due to the finite charging time, the charge of the oscillating particle experiences some delay with respect to its "equilibrium" value $Z_{\mathrm{d}}^{(\mathrm{eq})}(z)$ corresponding to infinitely slow particle motion. The particle motion is not merely potential in this case. If in the equilibrium position $Z_{\mathrm{d}}'(0) < 0$, then for a particle moving along the electric field (downwards) the absolute value of the charge $|Z_{\mathrm{d}}(z)|$ is smaller than the equilibrium value $|Z_{\mathrm{d}}^{(\mathrm{eq})}(z)|$. When the particle is moving in the opposite direction (upwards), the inverse inequality is satisfied. Thus, over the oscillation period the particle acquires energy from the electric field. The oscillations become unstable when the energy gain is higher than the energy dissipation due to friction.

Quantitative interpretation of this mechanism was formulated by Ivlev *et al.* (2000a). The right–hand side of Eq. (11.83) can be written in the considered case as $F_{\mathrm{ex}} = eE_0\delta Z_{\mathrm{d}}$, where $\delta Z_{\mathrm{d}} = Z_{\mathrm{d}}(z,t) - Z_{\mathrm{d}}^{(\mathrm{eq})}(z)$, with the equilibrium value $Z_{\mathrm{d}}^{(\mathrm{eq})}(z) = Z_{\mathrm{d}0} + zZ_{\mathrm{d}}'(0)$. The dynamics of charging is given by Eq. (11.27) which in this case takes the form

$$\frac{\partial \delta Z_{\mathrm{d}}}{\partial t} + \dot{z}\frac{\partial Z_{\mathrm{d}}}{\partial z} = -\Omega_{\mathrm{ch}}\delta Z_{\mathrm{d}}. \tag{11.84}$$

Assuming harmonic oscillations, namely $z, \delta Z_{\mathrm{d}} \propto \exp(-i\omega t)$, we get from Eq. (11.81), with the help of Eq. (11.84), the following relationship

$$(\omega_{\mathrm{ch}} - i\omega)(\omega^2 + i\nu_{\mathrm{dn}}\omega - \Omega_{\mathrm{v}}^2) = -i\omega\frac{eE_0}{m_{\mathrm{d}}}Z_{\mathrm{d}}'(0). \tag{11.85}$$

When the conditions $\Omega_{\mathrm{ch}} \gg \Omega_{\mathrm{v}} \gg \nu_{\mathrm{dn}}$ are satisfied, we have for weakly unstable (weakly damped) oscillations, $|\mathrm{Re}\,\omega| \gg |\mathrm{Im}\,\omega|$, the dependencies

$$\mathrm{Re}\,\omega \approx \Omega_{\mathrm{v}}, \qquad 2\,\mathrm{Im}\,\omega \approx -\nu_{\mathrm{dn}} - \frac{eE_0}{m_{\mathrm{d}}\Omega_{\mathrm{ch}}}Z_{\mathrm{d}}'(0). \tag{11.86}$$

The condition $\text{Im}\,\omega > 0$ corresponds to the unstable solution. Hence, the necessary condition of the instability is $Z'_\text{d}(0) < 0$, while the sufficient condition is formulated as

$$\left|\frac{Z'_\text{d}(0)}{Z_{\text{d}0}}\right| > \frac{\nu_\text{dn}\Omega_\text{ch}}{g}$$

(here the force balance condition is used). This instability mechanism was proposed to explain the increase in the amplitude of vertical oscillations with decreasing pressure in experiment (Nunomura *et al.* 1999). Some other possible instability mechanisms were proposed in the experimental work (Samarian *et al.* 2001) dealing with the investigation of instabilities of vertical oscillations in the sheath of an r.f. discharge.

11.4.2 *Linear waves and instabilities in weakly coupled dusty plasmas*

Almost simultaneously with the experimental discovery of crystallization in complex plasmas, new low–frequency waves (dust acoustic waves) and instabilities leading to self–excitation of such traveling waves were observed (Chu *et al.* 1994; D'Angelo 1995; Barkan *et al.* 1995). This stimulated interest in the theoretical investigation of wave processes in dusty plasmas. The most consistent approach to this problem, requiring solution of the kinetic equations for dusty plasmas, faces a number of difficulties associated with the appearance of a new degree of freedom (the dust particle charge) and the necessity of correctly evaluating collision integrals for different types of elastic and inelastic collisions. Attempts to construct a kinetic theory were undertaken in the works of Tsytovich and de Angelis (1999, 2000, 2001, 2002) and Tsytovich *et al.* (2001a). However, this problem cannot be considered as fully solved. Therefore, below we mostly concentrate on the simpler and clear hydrodynamic description of dusty plasmas, although in Section 11.4.2.3, which is dedicated to wave damping and instabilities, we use some of the simplest results from the kinetic theory of the conventional multicomponent plasmas.

11.4.2.1 *Basic equations.*   Let us formulate a system of equations for describing longitudinal waves in a uniform weakly coupled dusty plasma in the absence of external forces acting on the dust particles. For the dust component, the fluid continuity and momentum equations have the form

$$\frac{\partial n_\text{d}}{\partial t} + \frac{\partial}{\partial x}(n_\text{d}v_\text{d}) = 0, \tag{11.87}$$

$$\frac{\partial v_\text{d}}{\partial t} + v_\text{d}\frac{\partial v_\text{d}}{\partial t} = -\frac{eZ_\text{d}}{m_\text{d}}\frac{\partial \phi}{\partial x} - \frac{1}{m_\text{d}n_\text{d}}\frac{\partial p_\text{d}}{\partial x} - \sum_{j=e,i,n}\nu_{\text{d}j}(v_\text{d} - v_j). \tag{11.88}$$

The electric force in the field of the wave, the pressure of the dust component, and the momentum transfer in dust–electron, dust–ion, and dust–neutral collisions characterized by the effective frequencies ν_de, ν_di, and ν_dn, respectively, are taken into account on the right–hand side of Eq. (11.88). It should also be taken into account that, in contrast to ν_{dn}, the frequencies ν_{de} and ν_{di} characterize both direct collisions (collection) of electrons and ions with dust particles and their elastic (Coulomb) scattering by the particle potential. The equations for electrons and ions take the form

$$\frac{\partial n_\alpha}{\partial t} + \frac{\partial}{\partial x}(n_\alpha v_\alpha) = Q_{I\alpha} - Q_{L\alpha}, \tag{11.89}$$

$$\frac{\partial v_\alpha}{\partial t} + v_\alpha \frac{\partial v_\alpha}{\partial x} = -\frac{e_\alpha}{m_\alpha}\frac{\partial \phi}{\partial x} - \frac{1}{m_\alpha n_\alpha}\frac{\partial p_\alpha}{\partial x} - \frac{v_\alpha}{n_\alpha}\tilde{Q}_{L\alpha} - \sum_{j\neq\alpha(j=e,i,d,n)} \nu_{\alpha j}(v_\alpha - v_j),$$

$$\tag{11.90}$$

where the subscript $\alpha = e, i$ for electrons and ions, respectively, and $e_e = -e$, $e_i = e$. On the right–hand side of equation (11.89), the terms $Q_{I\alpha}$ and $Q_{L\alpha}$ describe the production and the loss of electrons and ions. Production can be associated with plasma ionization and, sometimes, with processes on the dust particle surface, i.e., thermionic, photoelectric, and secondary electron emission. The losses can be due to volume recombination, losses to the walls of a discharge chamber, collection by the dust particles, etc. In the unperturbed state $Q_{I\alpha0} = Q_{L\alpha0}$. The first two terms on the right–hand side of Eq. (11.90) describe the action of the electric field of the wave and the pressure of the corresponding component. The third one describes only the momentum loss due to "external" processes (recombination in plasma, losses to the walls), assuming that newly created particles (due to ionization) are initially at rest – the losses to the dust particles are accounted for by the next term. Finally, the fourth term describes the momentum transfer in collisions between different components, including absorbing collisions of electrons and ions with the dust particle. We note that from the momentum conservation law it follows that $n_\alpha m_\alpha \nu_{\alpha\beta} = n_\beta m_\beta \nu_{\beta\alpha}$.

Equations (11.87)–(11.90) should be supplemented by an equation of state, which in the simplest case can be written down as

$$p_j = \text{const.} \cdot n_j^{\gamma_j}. \tag{11.91}$$

The power index $\gamma_j = 1$ corresponds to isothermal perturbations of the jth component, while $\gamma_j = 5/3$ for adiabatic perturbations. In the general case, Eq. (11.91) can be considered as an equation of state of the polytrophic type.

A characteristic peculiarity of waves in dusty plasmas is that the dust particle charge is not fixed. The particle charge is determined by the local parameters of the surrounding plasma (ion and electron concentrations and velocities, plasma potential) and hence follows their perturbed behavior when waves are propagating. Therefore, the system of equations (11.87)–(11.91) should be supplemented with the charge equation. Similar to the kinetic equation (11.27), we have for the moving particles:

$$\frac{\partial Z_d}{\partial t} + v_d \frac{\partial Z_d}{\partial x} = \sum_j I_j. \tag{11.92}$$

The system is closed by the Poisson equation

$$\frac{\partial^2 \phi}{\partial x^2} = -4\pi e[n_{\mathrm{i}} - n_{\mathrm{e}} + Z_{\mathrm{d}} n_{\mathrm{d}}]. \tag{11.93}$$

The system of equations (11.87)–(11.93) can be used for investigating waves in a weakly coupled dusty plasma within thehydrodynamic approximation. As can be seen from the derived equations, dusty plasmas are characterized by a wider diversity of processes than the usual multicomponent plasma. A detailed analysis of the equations is rather complicated. Moreover, it is necessary to make certain assumptions about the nature of electron and ion sources and losses, depending on the problem under consideration. Some theoretical results on waves in dusty plasmas and their instability (stability) were obtained in original works by D'Angelo (1997, 1998); Ivlev *et al.* (1999); Ivlev and Morfill (2000); Ostrikov *et al.* (2000a); Khrapak and Morfill (2001); and Wang *et al.* (2001a). The input equations in these works are similar to the system (11.87)–(11.93). However, various assumptions about dominant processes of electron and ion production and loss, elastic and inelastic collision frequencies, etc. were made. Below we consider only several important limiting cases of linear waves in dusty plasmas, as well as their instability and damping in the simplest interpretation.

11.4.2.2 *Ion–acoustic and dust–acoustic waves.* As we have already mentioned, the presence of the dust component not only modifies the charge composition of a plasma, but also introduces new characteristic spatial and temporal scales in the system of interest. Hence, it is natural to expect not only modifications to the wave modes that exist without the dust, but also the appearance of new ones. Below, we will consider a modification of the dispersion relation for ion–acoustic waves and the appearance of the new low–frequency mode – the dust–acoustic wave. The latter is especially important from the point of view of experimental investigations into dusty plasmas because its characteristic frequency range, $\omega \sim 1 - 100$ s^{-1}, is easily accessible for processing and analysis.

Let us linearize the system of equations (11.87)–(11.93), assuming small perturbations of plasma parameters $n_j = n_{j0} + n_{j1}$, $v_j = v_{j0} + v_{j1}$ ($j = e, i, d$), and $\phi = \phi_1$ proportional to $\exp(-i\omega t + ikx)$. For simplicity, we also ignore all collisions, which is only possible for short wavelengths, when characteristic free paths are longer than the characteristic scale of the problem (wavelength). In this case, we should, strictly speaking, assume the constant charge on the dust particles, because charge variations are associated with inelastic collisions of electrons and ions with the dust particles, which are being ignored.

This idealized model is extremely simplified, compared to the conditions which are realized in most experiments and, therefore, it is inapplicable for their description. Nevertheless, it is useful to consider such a model because it can serve as the basis for investigating the effect of different processes which often determine qualitatively the final result.

The quasineutrality condition in the unperturbed state has the form (11.15). The Poisson equation (11.93) can be rewritten then as

$$-k^2 \phi_1 = -4\pi e[n_{\mathrm{i}1} - n_{\mathrm{e}1} + Z_{\mathrm{d}} n_{\mathrm{d}1}]. \tag{11.94}$$

Expressing, with the help of Eqs. (11.87)–(11.91), the perturbations of different component densities through the wave potential ϕ_1, we get the dispersion relation

$$\varepsilon(k,\omega) = 1 + \chi_e + \chi_i + \chi_d = 0, \tag{11.95}$$

where

$$\chi_e = \frac{4\pi e n_{e1}}{k^2 \phi_1}, \qquad \chi_i = -\frac{4\pi e n_{i1}}{k^2 \phi_1}, \qquad \chi_d = -\frac{4\pi Z_d e n_{d1}}{k^2 \phi_1}$$

are the susceptibilities of the corresponding components. More generally, when the dust–particle charge variations are taken into account, χ_d should be written in the form

$$\chi_d = -\frac{4\pi e}{k^2 \phi_1}(Z_{d0} n_{d1} + Z_{d1} n_{d0})$$

In the simplest approximation considered we have

$$\chi_j = -\frac{\omega_{pj}^2}{(\omega - k v_{0j})^2 - \gamma_j k^2 v_{T_j}^2}, \tag{11.96}$$

where $j = e, i, d$, i.e., the same dispersion relation as for a multicomponent collisionless plasma in the hydrodynamic approximation. A peculiarity of dusty plasmas in this simple approach only reduces to the strong asymmetry of the charge–to–mass ratios for different components:

$$\frac{e}{m_e} : \frac{e}{m_i} : \frac{|Z_d|e}{m_d} \sim 1 : 10^{-5} : 10^{-13}.$$

In this case, the inequalities $\omega_{pe} \gg \omega_{pi} \gg \omega_{pd}$ are well satisfied.

In the absence of directed drifts of plasma components ($v_{j0} = 0$), and for high frequencies $\omega \gg k v_{T_e} \gg k v_{T_i} \gg k v_{T_d}$, the dispersion relation takes the form

$$1 - \sum_{j=e,i,d} \frac{\omega_{pj}^2}{\omega^2} = 0. \tag{11.97}$$

According to the previous consideration, the presence of dust does not affect the spectrum of high–frequency Langmuir waves; in this case, $\omega \approx \omega_{pe}$.

At lower frequencies $k v_{T_e} \gg \omega \gg k v_{T_i} \gg k v_{T_d}$, we have $\chi_e \approx 1/(k^2 \lambda_{De}^2)$ ($\gamma_e = 1$) and $\chi_i \approx -\omega_{pi}^2/\omega^2$, so that the dispersion relation acquires the form

$$\omega^2 \approx \frac{\omega_{pi}^2 k^2 \lambda_{De}^2}{1 + k^2 \lambda_{De}^2}, \tag{11.98}$$

implying that $\omega \approx \omega_{pi}$ for $k \lambda_{De} \gg 1$. In the opposite limiting case, $k \lambda_{De} \ll 1$, the ion–acoustic waves are initiated: $\omega \approx k C_{IA}$, with the velocity of sound

$$C_{IA} = \left(\frac{T_e}{m_i} \frac{n_{i0}}{n_{e0}}\right)^{1/2}. \tag{11.99}$$

This expression differs from the standard expression for the ion–acoustic wave velocity in a nonequilibrium plasma, $T_e \gg T_i$ (see, e.g., Aleksandrov et al. 1978), by the factor $\sqrt{n_{i0}/n_{e0}}$. The condition for the ion–acoustic wave existence, $T_e \gg T_i$, following from $C_{IA} \gg v_{T_i}$ is weaker in this case because the existence of the negatively charged dust component implies $n_i/n_e > 1$. Therefore, the influence of charged dust particles on the ion–acoustic wave spectrum reduces to the appearance of the dependence of the velocity of wave propagation on the dust particle charge and density. The influence of the dust component on the ion–acoustic wave dispersion was first considered by Shukla and Silin (1992).

For even lower frequencies

$$kv_{T_e} \gg kv_{T_i} \gg \omega \gg kv_{T_d},$$

we have $\chi_{e(i)} \approx 1/(k^2\lambda^2_{De(i)})$ ($\gamma_e = 1, \gamma_i = 1$), $\chi_d = -\omega^2_{pd}/\omega^2$, and using the notation $\lambda_D^{-2} = \lambda_{De}^{-2} + \lambda_{Di}^{-2}$ we arrive at the dispersion relation in the form

$$\omega^2 \approx \frac{\omega^2_{pd}k^2\lambda_D^2}{1 + k^2\lambda_D^2}. \tag{11.100}$$

In the limit $k\lambda_D \gg 1$, Eq. (11.100) gives $\omega \approx \omega_{pd}$; in the opposite limit, $\omega \approx kC_{DA}$. This mode, being nonexistent in the absence of the dust particles, is called the dust–acoustic wave (DAW) and was first considered by Rao et al. (1990). The dust–acoustic wave velocity is given by

$$C_{DA} = \omega_{pd}\lambda_D = \sqrt{\frac{|Z_d|T_i}{m_d}}\sqrt{\frac{P\tau}{1 + \tau + P\tau}}. \tag{11.101}$$

The condition $C_{DA} \gg v_{T_d}$ of the dust–acoustic wave existence can be fulfilled (for not too small P) even when $T_d > T_i$, due to the large value of particle charge Z_d. Thus, the dust–acoustic wave is considerably different from the ion–acoustic wave in usual electron–ion plasmas, which can only exist when the temperature of the light component considerably exceeds the temperature of the heavy component: $T_e \gg T_i$. It will be recalled that in deriving the dispersion relations the collisions were ignored. For this reason, the dust–acoustic wave (11.101) is sometimes called the short-wavelength DAW (Tsytovich 1997).

11.4.2.3 *Damping and instabilities of waves in dusty plasmas.* Let us briefly consider several mechanisms of damping and instability of electrostatic waves in plasmas, caused by the presence of the dust component. We mostly limit ourselves to the low–frequency mode – the dust–acoustic wave. If the time evolution of waves is of interest, then the wave frequency can be presented in the form $\omega = \omega_r + i\gamma$, where $\omega_r = \text{Re}\,\omega$ and $\gamma = \text{Im}\,\omega$ are the real and imaginary parts of the frequency, respectively. In our designations $\gamma < 0$ corresponds to wave damping, and $\gamma > 0$ to wave growth (instability). In the case of weak damping (growth), the following expression for the damping decrement (instability increment) can be applied (Aleksandrov et al. 1978):

$$\gamma = -\frac{\operatorname{Im}\varepsilon(\omega, k)}{\partial\operatorname{Re}\varepsilon(\omega, k)/\partial\omega}. \tag{11.102}$$

This relation will be utilized below.

One of the mechanisms leading to wave energy dissipation corresponds to collisions. The collisions with neutrals dominate in a weakly ionized plasma. Inclusion of these collisions leads to the following modification of dispersion relation (11.100):

$$\omega(\omega + i\nu_{\mathrm{dn}}) \approx \frac{\omega_{\mathrm{pd}}^2 k^2\lambda_{\mathrm{D}}^2}{1 + k^2\lambda_{\mathrm{D}}^2}. \tag{11.103}$$

The damping decrement in the case $\omega \gg \nu_{\mathrm{dn}}$ is given by

$$\gamma = -\frac{\nu_{\mathrm{dn}}}{2}. \tag{11.104}$$

Another source of wave damping is, similarly to the common electron–ion plasma, the Landau damping mechanism which follows from the kinetic description of waves in plasma. The use of the simplest approach employing the solution of Vlasov equations for electrons, ions, and dust particles with a fixed charge in collisionless plasma leads to the following expression for the susceptibilities of the corresponding components:

$$\chi_j = \frac{1}{k^2\lambda_{\mathrm{D}j}^2}\left[1 + F\left(\frac{\omega}{\sqrt{2}kv_{T_j}}\right)\right], \tag{11.105}$$

where $F(x)$ is the so–called plasma dispersion function having the asymptotes (see, e.g., Lifshitz and Pitaevskii 1979)

$$F(x) = \begin{cases} -1 - \frac{1}{2}x^2 - \frac{3}{4}x^4 + i\sqrt{\pi}x\exp(-x^2), & x \gg 1 \\ -2x^2 + i\sqrt{\pi}x, & x \ll 1, \end{cases} \tag{11.106}$$

where x is a real argument. For the dust–acoustic waves under conditions $kv_{T_{\mathrm{d}}} \ll \omega \ll kv_{T_{\mathrm{i}}}, kv_{T_{\mathrm{e}}}$, it follows from formulas (11.105) and (11.106) that

$$\chi_{\mathrm{e(i)}} \approx \frac{1}{k^2\lambda_{\mathrm{De(i)}}^2}\left[1 + i\sqrt{\frac{\pi}{2}}\frac{\omega}{kv_{T_{\mathrm{e(i)}}}}\right],$$

$$\chi_{\mathrm{d}} \approx -\frac{\omega_{\mathrm{pd}}^2}{\omega^2}\left[1 - i\sqrt{\frac{\pi}{2}}\left(\frac{\omega}{kv_{T_{\mathrm{d}}}}\right)^3\exp\left(-\frac{\omega^2}{2k^2v_{T_{\mathrm{d}}}^2}\right)\right].$$

The dispersion relation is arrived at by substituting these expressions into Eq. (11.95). Let us consider the contribution to damping, which arises due to different components. The damping on dust particles yields

$$\gamma = -\omega\sqrt{\frac{\pi}{8}}\left(\frac{\omega}{kv_{T_{\mathrm{d}}}}\right)^3\exp\left(-\frac{\omega^2}{2k^2v_{T_{\mathrm{d}}}^2}\right). \tag{11.107}$$

The condition of weak damping requires a large exponential factor: $\omega^2/2k^2v_{T_{\mathrm{d}}}^2 \gg 1$, which is almost identical to the condition $C_{\mathrm{DA}} \gg v_{T_{\mathrm{d}}}$, and, as could be expected, determines the validity range of the hydrodynamic approach to dust–acoustic waves. The damping decrement of the dust–acoustic wave due to ions is described by the following expression

$$\gamma = -\omega\sqrt{\frac{\pi}{8}}\left(\frac{\omega_{\mathrm{pi}}}{\omega_{\mathrm{pd}}}\right)^2\left(\frac{C_{\mathrm{DA}}}{v_{T_{\mathrm{i}}}}\right)^3. \tag{11.108}$$

It is usually small because

$$\left|\frac{\gamma}{\omega_{\mathrm{r}}}\right| \sim \sqrt{\frac{|Z_{\mathrm{d}}|m_{\mathrm{i}}}{m_{\mathrm{d}}}\frac{P}{P+1}}$$

for $\tau \gg 1$ and by virtue of the condition $|Z_{\mathrm{d}}|m_{\mathrm{i}}/m_{\mathrm{d}} \ll 1$. The contribution from electrons to damping can always be ignored because $\operatorname{Im}\chi_{\mathrm{e}}/\operatorname{Im}\chi_{\mathrm{i}} = \sqrt{\mu}\tau^{-3/2}n_{\mathrm{e}}/n_{\mathrm{i}} \ll 1$.

A similar analysis can be performed when the plasma components possess directed velocities. In this case, the substitution $\omega \to \omega - ku_0$ should be made in the argument of the dispersion function F. Some particular cases will be considered below. The inclusion of collisions leads to the substitution of $\omega + i\nu_j$ for ω, where ν_j is the effective collision frequency of the jth component. In a weakly ionized plasma, ν_j is often determined by collisions with neutrals. It should also be taken into account that inclusion of collisions will change Eq. (11.105) (see, e.g., Aleksandrov et al. 1978). Moreover, the asymptotes of the dispersion function F for an imaginary argument are in some cases different from formula (11.106) (Fried and Conte 1961). However, we will not discuss this situation in detail here.

Let us now examine an effect specific to dusty plasmas – that is, dust particle charge variations. It was noted above that the dust particle charge shows itself as a new degree of freedom of the dust component, and its variations should be taken into account when describing waves. For the sake of definiteness, let us consider a plasma in which the charging is determined by only electron and ion collection, which in turn can be described within the OML approach. Assuming $n_{\mathrm{e(i)}} = n_{\mathrm{e(i)}0}+n_{\mathrm{e(i)}1}$, $Z_{\mathrm{d}} = Z_{\mathrm{d}0}+Z_{\mathrm{d}1}$, and $v_{j0} = 0$ ($j = i, e, d$) (isotropic plasma) we derive from Eq. (11.92), using the expressions (11.7) and (11.8) for electron and ion fluxes, the following equation for the particle charge perturbations:

$$\frac{\partial Z_{\mathrm{d}1}}{\partial t} + \Omega_{\mathrm{ch}}Z_{\mathrm{d}1} = I_{\mathrm{e}0}\left(\frac{n_{\mathrm{i}1}}{n_{\mathrm{i}0}} - \frac{n_{\mathrm{e}1}}{n_{\mathrm{e}0}}\right), \tag{11.109}$$

where $I_{\mathrm{e}0} = I_{\mathrm{i}0}$ is the unperturbed flux of electrons or ions to the dust particle surface, and the charging frequency is defined by expression (11.30). Equation (11.109) is also applicable for $v_{j0} \ll v_{T_j}$ ($j = e, i$), because the corrections to the fluxes are on the order of $O\left[(v_{j0}/v_{T_j})^2\right]$, as was noted in Section 11.2.1. For low–frequency dust–acoustic waves we have $n_{\mathrm{i}1}/n_{\mathrm{i}0} = -e\phi_1/T_{\mathrm{i}}$, $n_{\mathrm{e}1}/n_{\mathrm{e}0} = e\phi_1/T_{\mathrm{e}}$, so that linearization of Eq. (11.109) yields

$$Z_{d1} = -i\frac{I_{e0}(1 + T_e/T_i)}{\omega + i\Omega_{ch}}\frac{e\phi_1}{T_e}, \tag{11.110}$$

and the dispersion relation accounting for charge variations acquires the form

$$1 + \frac{1}{k^2\lambda_D^2} - \frac{\omega_{pd}^2}{\omega^2} + \frac{\Phi}{k^2\lambda_{De}^2} = 0, \tag{11.111}$$

where the last term is connected to charge variations, and

$$\Phi = i\frac{I_{e0}(1 + T_e/T_i)}{\omega + i\Omega_{ch}}\frac{n_{d0}}{n_{e0}}.$$

Typically, the condition $\Omega_{ch} \gg \omega$ is satisfied and therefore the effect considered mostly contributes to the real part of the dispersion relation. In this case, the expression for Φ is considerably simplified:

$$\Phi = \frac{P(1 + \tau)(1 + z\tau)}{z(1 + \tau + z\tau)},$$

and the dispersion relation (11.111) can be rewritten in the form

$$1 + \frac{1}{k^2\lambda_D^2}\left[1 + \frac{P(1 + \tau)(1 + z\tau)}{z(1 + \tau + z\tau)(1 + \tau + P\tau)}\right] - \frac{\omega_{pd}^2}{\omega^2} = 0. \tag{11.112}$$

As follows from this expression, the effect of particle charge variations reduces the phase velocity of the dust–acoustic wave. However, for reasonable dusty plasma parameters this effect is relatively weak. Charge variations also affect the imaginary part of the dispersion relation. From Eq. (11.111) and the expression for Φ it follows that they lead to wave damping. An expression for the damping decrement can be found, for example, in (Fortov et al. 2000). However, two circumstances have to be taken into account. First, to self–consistently determine the imaginary part γ, it is necessary to account for at least the collisions leading to the dust particle charging. One needs to keep the corresponding terms in the continuity and momentum equations for the electron and ion components, and in the momentum equation for the dust component. Moreover, in the continuity equations for electrons and ions it is necessary to take into account processes which compensate for the ion and electron recombination losses on dust particles. We will not consider details here, but instead give some references to original works in which the effect of charge variations on different wave mode propagations is considered using certain assumptions (Melandsø et al. 1993; Varma et al. 1993; Jana et al. 1993; Bhatt and Pandey 1994; Vladimirov 1994; Bhatt 1997; Tsytovich 1997; Annou 1998; Vladimirov et al. 1998; Ostrikov et al. 2000b; Ivlev and Morfill 2000; Fortov et al. 2000; Khrapak and Morfill 2001; Wang et al. 2001a). Second, in most of the laboratory facilities the main effect which influences the imaginary part of the dispersion relation is connected to collisions of dust particles (or ions for high–frequency modes, e.g., ion–acoustic waves) with neutrals.

An important example of wave instability in dusty plasmas is the streaming instability which appears due to relative drift between the ions and the stationary (on the average) dust component. This situation is common for most of the experiments on dusty plasmas, although the ion drift velocity u_0 may differ substantially depending on the concrete conditions – from $u_0 \ll v_{T_i}$ in the bulk quasineutral plasma to $u_0 \sim C_{AI} \gg v_{T_i}$ in the collisionless sheath near an electrode. Electrons can also possess a directed velocity but their motion can usually be ignored. This type of instability was analyzed by Rosenberg (1993, 1996, 2002); D'Angelo and Merlino (1996); Kaw and Singh (1997); Molotkov et $al.$ (1999); and Mamun and Shukla (2000).

Let us consider some of the results obtained for the low–frequency limit, when $kv_{T_d} \ll \omega \ll kv_{T_i}, kv_{T_e}$. Within the framework of the simplest kinetic treatment for collisionless plasma we have $\chi_e = 1/k^2\lambda_{De}^2$ and $\chi_d = -\omega_{pd}^2/\omega^2$. The ion susceptibility can be calculated using formula (11.105) by substituting $\omega \to \omega - ku_0$. In the limit $|\omega - ku_0| \ll kv_{T_i}$, this yields

$$\chi_i = \frac{1}{k^2\lambda_{Di}^2}\left[1 + i\sqrt{\frac{\pi}{2}}\frac{\omega - ku_0}{kv_{T_i}}\right]. \tag{11.113}$$

From the dispersion relation (11.95) it follows that the waves are unstable when $u_0 > \omega/k$. The instability increment of the considered dust–acoustic mode can be written for $u_0 \gg \omega/k$ in the form

$$\gamma = \omega_{pd}\sqrt{\frac{\pi}{8}}\frac{u_0}{v_{T_i}}\left(\frac{\lambda_D}{\lambda_{Di}}\right)^2\frac{k\lambda_D}{(1 + k^2\lambda_D^2)^{3/2}}. \tag{11.114}$$

As a function of the wavenumber k, it reaches a maximum at $k\lambda_D = 1/\sqrt{2}$. Here, $\omega_r = \omega_{pd}/\sqrt{3}$ and

$$\gamma_{max} = \frac{\omega_{pd}}{3}\sqrt{\frac{\pi}{6}}\left(\frac{\lambda_D}{\lambda_{Di}}\right)^2\frac{u_0}{v_{T_i}}$$

determine the frequency characteristics of the most unstable mode (Fortov et $al.$ 2000). We note that the condition $|\omega - ku_0| \ll kv_{T_i}$ used above implies subthermal ion drift, $u_0 \ll v_{T_i}$, which does not contradict the instability condition $u_0 > C_{DA}$ for dust sound, because $C_{DA} \sim v_{T_i}\sqrt{|Z_d|m_i/m_d} \ll v_{T_i}$.

In the case of suprathermal ion drift, $u_0 \gg v_{T_i}$, ignoring the exponentially small term we get from expressions (11.105) and (11.106): $\chi_i \approx -\omega_{pi}^2(\omega - ku_0)^{-2}$. Notice that the same expression for the ion susceptibility follows from formula (11.96). For low–frequency ($\omega \ll ku_0$) long–wave ($k\lambda_{De} \ll 1$) oscillations, the dispersion relation (11.95) allows for unstable solutions. Thus, for the ion drift velocity close to the ion–sound wave velocity $u_0 \sim \omega_{pi}\lambda_{De} = C_{IA}$, the wave frequency is given by

$$\omega \approx \frac{1 + i\sqrt{3}}{2}\left(\frac{\omega_{pd}}{2\omega_{pi}}\right)^{2/3}ku_0. \tag{11.115}$$

For smaller but still suprathermal drift velocities, there are two purely imaginary solutions, one of which is unstable, viz.

$$\omega \approx \pm i\left(\frac{\omega_{\mathrm{pd}}}{\omega_{\mathrm{pi}}}\right) k u_0. \tag{11.116}$$

Taking into account ion–neutral collisions one can use the hydrodynamic approach, which yields for the ion susceptibility

$$\chi_{\mathrm{i}} = -\frac{\omega_{\mathrm{pi}}^2}{(\omega - k u_0)(\omega - k u_0 + i\nu_{\mathrm{in}}) - k^2 v_{T_{\mathrm{i}}}^2}. \tag{11.117}$$

The ion susceptibility for $(\omega - k u_0)^2$, $\nu_{\mathrm{in}}|\omega - k u_0| \ll k^2 v_{T_{\mathrm{i}}}^2$ has the form (D'Angelo and Merlino 1996)

$$\chi_{\mathrm{i}} = \frac{1}{k^2 \lambda_{\mathrm{Di}}^2}\left[1 + i\nu_{\mathrm{in}}\frac{\omega - k u_0}{k^2 v_{T_{\mathrm{i}}}^2}\right]. \tag{11.118}$$

The instability condition is, as previously, $u_0 > \omega/k$, while the instability increment for $u_0 \gg \omega/k$ is determined by the expression

$$\gamma = \frac{1}{2}\omega_{\mathrm{pd}}\left(\frac{\lambda_{\mathrm{D}}}{\lambda_{\mathrm{Di}}}\right)^3\left(\frac{\nu_{\mathrm{in}}}{\omega_{\mathrm{pi}}}\right)(1 + k^2\lambda_{\mathrm{D}}^2)^{-3/2}\frac{u_0}{v_{T_{\mathrm{i}}}} \tag{11.119}$$

and passes a maximum for $k\lambda_{\mathrm{D}} \ll 1$. We note, however, that in the case $\gamma_{\mathrm{in}} \ll k v_{T_{\mathrm{i}}}$, the hydrodynamic approach ignores the dominant term associated with the Landau damping – that is, the second term in brackets in Eq. (11.113). The case of $u_0 \gg v_{T_{\mathrm{i}}}$, with the ion–neutral and dust–neutral collisions (which always stabilize the instabilities considered) taken into account in a way similar to relation (11.103), was studied by Joyce et al. (2002). It is shown that low–frequency waves can be stable or unstable depending on the relation between $\nu_{\mathrm{in}}/\omega_{\mathrm{pi}}$ and $\nu_{\mathrm{dn}}/\omega_{\mathrm{pd}}$ for a fixed drift velocity u_0. Joyce et al. (2002) assumed that this instability could be a reason for the dust component "heating" examined experimentally when lowering the neutral gas pressure.

11.4.3   Waves in strongly coupled dusty plasmas

As was already noted, in dusty plasmas it is relatively easy to achieve the situation in which the dust component is in a (strongly) nonideal state. In this case, the interaction and correlations between the dust particles can significantly influence the wave spectrum. Different theoretical approaches have been proposed to describe waves in strongly coupled dusty plasmas. We will dwell on some of them below.

The system of dust particles in the regime $1 \ll \gamma_{\mathrm{d}} \ll \gamma_{\mathrm{M}}$ is in a "liquid–like" state, and its wave properties are similar to those of a viscous liquid. For a theoretical description of these waves, the method of "generalized hydrodynamics" was employed by Kaw and Sen (1998); Kaw (2001); and Xie and Yu (2000). In

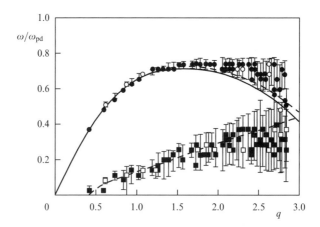

<small>FIG. 11.17.</small> The dispersion relation for the longitudinal and transverse waves in a strongly coupled Debye–Hückel system at $\gamma_{\mathrm{d}} \simeq 130$ and $\kappa \simeq 1.6$ (Ohta and Hamaguchi 2000b). The frequency is normalized to the plasma–dust frequency ω_{pd}, and the wavelength is normalized to the characteristic interparticle distance $q = ka_{\mathrm{WS}}$, where $a_{\mathrm{WS}} = (4\pi n_{\mathrm{d}}/3)^{-1/3}$ is the Wigner–Seitz radius. Symbols correspond to MD simulations. Circles are for the longitudinal mode, squares for the transverse mode (solid for $N = 800$ particles, and open for $N = 250$ particles). The curves correspond to comparison of the simulation results with theoretical models. The upper solid (dashed) curve fits the dispersion relation for the longitudinal mode, based on the quasilocalized charge approximation (generalized hydrodynamics). The lower dashed curve depicts the dispersion of the transverse mode in the approximation of generalized hydrodynamics (obtained by adjusting the parameters of the model to get better agreement with the numerical results).

the work of Murillo (1998, 2000a), the kinetic approach accounting for correlations between dust particles was applied. Rosenberg and Kalman (1997) and Kalman *et al.* (2000) obtained the dispersion relation using the quasilocalized charge approximation, originally developed to describe systems with long–range Coulomb interaction (Kalman and Golden 1990). Numerical modeling of dusty plasmas can serve as the criterion of applicability of different theoretical models. Usually, advantage is taken of MD simulations within the framework of the Debye–Hückel model (Winske *et al.* 1999; Ohta and Hamaguchi 2000b).

We point out the main properties of waves in strongly coupled dusty plasmas, which are reproduced in most of the models. First, in addition to the longitudinal mode, the transverse mode also appears, which exhibits an acoustic–like dispersion $\omega \propto k$ (the group velocity is constant) and can exist for not too long wavelengths, $k > k_{\mathrm{cr}}$ (Ohta and Hamaguchi 2000b; Murillo 2000b; Kaw 2001). Longitudinal waves have an acoustic–like dispersion in the long-wavelength limit: $k\Delta \ll 1$. For short enough wavelengths, however, the acoustic character of the dispersion is violated and the group velocity can even change its sign, $\partial\omega/\partial k < 0$, i.e., the dependence $\omega(k)$ has a maximum, $\omega_{\max}$, the value of which can be significantly smaller than ω_{pd} and typically decreases with increasing $\kappa = \Delta/\lambda_{\mathrm{D}}$. An example of the dispersion relation is shown in Fig. 11.17 which presents the results of numerical modeling along with their comparison with analytical results obtained using different theoretical models (Ohta and Hamaguchi 2000b). Notice that the dust–neutral collisions can reduce the effect of strong coupling in the dust system. The difference between strongly and weakly coupled dusty plasmas practically disappears if $\nu_{\mathrm{dn}}/\omega_{\mathrm{pd}} \sim 1$ (Rosenberg and Kalman 1997; Kaw 2001).

For stronger interparticle interactions, when $\gamma > \gamma_{\mathrm{M}}$, the particles form a crystalline structure, and the wave spectrum becomes similar to that in solids. There exist one longitudinal and two transverse modes which are usually called dust lattice waves (DLW). To describe waves in dust crystals, the screened Coulomb potential of the interaction between particles is usually employed. The electrons and ions are included in this model in an indirect way: they provide screening of the Coulomb interaction. The simplest one–dimensional model – a chain of equidistant particles – was first considered by Melandsø (1996) in connection with longitudinal waves. For linear waves, when the particle displacements from equilibrium positions are much smaller than the mean interparticle distance, the dispersion relation assumes the form

$$\omega^2 = \frac{2}{\pi}\omega_{\mathrm{pd}}^2 \sum_{n=1}^{\infty} \frac{\exp(-\kappa n)}{n^3}\left(1 + \kappa n + \frac{\kappa^2 n^2}{2}\right)\sin^2\frac{nk\Delta}{2}, \qquad (11.120)$$

where the summation is taken over all particles in the chain. In the limit of large values of κ ($\Delta \gg \lambda_{\mathrm{D}}$), it is sufficient to take into consideration only the first term, yielding

$$\omega \approx \sqrt{\frac{2}{\pi}}\omega_{\mathrm{pd}}\left(1 + \kappa + \frac{\kappa^2}{2}\right)^{1/2}\exp\left(-\frac{\kappa}{2}\right)\sin\frac{k\Delta}{2}. \qquad (11.121)$$

Acoustic–like dispersion takes place at $k\Delta \ll 1$: $\omega = C_{\mathrm{DL}}k$. Expression (11.121) plays an important role in the physics of dusty plasmas because oscillations in the one–dimensional chain of strongly interacting particles can be relatively easily realized in experiments. Moreover, relation (11.121) is convenient for simple estimations of the characteristic wave frequency in ordered dusty plasma structures. Accounting for dust particle collisions with neutral particles reduces, as ever, to the substitution of $\omega(\omega + i\nu_{\mathrm{dn}})$ for ω^2 in the left–hand side of dispersion relation (11.120).

In a similar way, dispersion relations for longitudinal waves in 2D dust crystals can be obtained (Homann et al. 1998). Dispersion relations for longitudinal and transverse waves in two–dimensional hexagonal lattices of dust particles and 3D bcc and fcc lattices were derived by Wang et al. (2001b)

Vertical oscillations in a horizontal chain of strongly interacting dust particles (transverse oscillations) in a sheath were considered by Vladimirov et al. (1997).

The dispersion relation for this mode, assuming the screened Coulomb interaction potential and taking into account only the interaction between neighboring particles, takes the form

$$\omega^2 = \Omega_{\mathrm{v}}^2 - \frac{\omega_{\mathrm{pd}}^2}{\pi} \exp(-\kappa)(1 + \kappa) \sin^2 \frac{k\Delta}{2}. \tag{11.122}$$

Replacement of the characteristic frequency Ω_{v} of vertical confinement by the horizontal one Ω_{h} gives the dispersion relation for the second transverse (horizontal) mode. Modifications of the dispersion relations (11.121) and (11.122) for longitudinal and transverse modes in a one–dimensional chain of dust particles due to ion focusing downstream from the particles (in the ion flow directed to an electrode) were considered by Ivlev and Morfill (2001).

11.4.4 *Experimental investigation of wave phenomena in dusty plasmas*

11.4.4.1 *Dust–acoustic waves.* Low–frequency fluctuations of the dust density were first observed in an experiment on dusty plasma crystallization (Chu *et al.* 1994). The experiment was performed in an r.f. magnetron discharge. At pressures of a few hundred mtorr, dust particles formed highly ordered structures. When lowering the pressure, self–excited fluctuations of plasma and dust density at a frequency $f \sim 12$ Hz were observed if the particle density exceeded some critical value. Later on, this phenomenon was interpreted as the first observation of dust–acoustic waves in the laboratory (D'Angelo 1995). In addition, low–frequency modes and possible mechanisms corresponding to these modes (including dust–acoustic waves) in an r.f. discharge were considered by Praburam and Goree (1996).

A spontaneous excitation of low–frequency perturbations of the dust density was observed in the plasma of a Q-machine by Barkan *et al.* (1995). The waves propagated with velocities $v_{\mathrm{ph}} \sim 9$ cm s^{-1}, wavelengths $\lambda \sim 0.6$ cm, and, hence, frequencies $f \sim 15$ Hz. The excitation of these waves was interpreted to be the result of the streaming instability of the dust–acoustic waves, associated with the presence of a constant electric field $E \sim 1$ V cm^{-1} in the plasma, leading to ion drift relative to the dust component.

In experimental investigations of dust–acoustic waves, a d.c. discharge is often used. For example, the facility shown schematically in Fig. 11.18(a) was employed in experiments (Thompson *et al.* 1997; Merlino *et al.* 1998). A discharge was formed in nitrogen gas at a pressure $p \sim 100$ mtorr by applying a potential to an anode disk 3 cm in diameter located in the center of the discharge chamber. A longitudinal magnetic field of about 100 G provided radial plasma confinement. If the discharge current was sufficiently high (> 1 mA), dust–acoustic waves appeared spontaneously, similar to earlier experiments (Barkan *et al.* 1995). To investigate the properties of the waves in more detail, a low–frequency sinusoidal modulation with frequencies in the range 6–30 Hz was applied to the anode. An example of the observed waves is shown in Fig. 11.18(b) for eight different excitation frequencies. Assuming that the wave frequency is determined by the

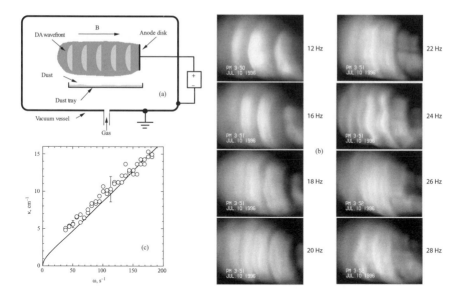

FIG. 11.18.   (a) Schematic diagram of the experimental facility used to investigate the dust–acoustic waves in a d.c. gas discharge (Merlino *et al.* 1998). (b) Video images of typical wave structures in dusty plasma for eight different values of the excitation frequency (shown to the right of the images). (c) Experimentally measured dust–acoustic wave dispersion relation (circles). The solid curve is computed from the theoretical dispersion relation which accounts for the dust–neutral collisions; see Eq. (11.103).

external excitation frequency, the dispersion relation – the $k(\omega)$ dependence in this case – can be obtained by measuring the wavelength. The results are shown in Fig. 11.18(c). The observed waves exhibit linear dispersion $\omega \propto k$ and propagate with a velocity $v_{\mathrm{ph}} \sim 12 \mathrm{~cm\,s}^{-1}$. This value is in reasonable agreement with the theoretical expression (11.101) for the dust–acoustic wave velocity (Thompson *et al.* 1997). One can conclude, therefore, that the observed waves correspond to the dust–acoustic waves.

A spontaneous excitation of low–frequency dust–acoustic waves was also observed in the positive column of a d.c. discharge (Molotkov *et al.* 1999; Fortov *et al.* 2000) in the experimental apparatus shown schematically in Fig. 11.12. By reducing the neutral gas pressure and/or the discharge current, as well as by adding new particles to the discharge, a longitudinal dust–density wave propagating in the direction of the electric field (downward) was excited. A typical wave pattern is shown in Fig. 11.1(c). The characteristic parameters of waves are: frequency $\omega \sim 60 \mathrm{~s}^{-1}$, wavenumber $k \sim 60 \mathrm{~cm}^{-1}$, and velocity of wave propagation $v_{\mathrm{ph}} \sim 1 \mathrm{~cm\,s}^{-1}$. As shown by Fortov *et al.* (2000), the ion–dust streaming instability alone cannot be responsible for exciting the waves observed because damping effects should stabilize the dust–acoustic waves. In the same work, a new instability mechanism was proposed, the combination of which with the streaming instability allows us to describe the linear stage of spontaneous wave excitation. The proposed new mechanism can be qualitatively explained in the following way. The dust particles levitate in the plasma due to the balance of electric and gravity forces. When waves propagate, the dust charge experiences variations in response to electron and ion density perturbations. When there is a certain correlation between the dust charge phases and velocity perturbations, the dust particles can gain energy from the discharge electric field, which leads to wave amplification. The quantitative results obtained by Fortov *et al.* (2000) are in reasonable agreement with experimental observations. A similar mechanism was used for explaining wave instability in an inductively coupled r.f. discharge (Zobnin *et al.* 2002).

We note that when analyzing experimental findings the theory of dust–acoustic waves in weakly coupled plasmas is often used, while in experiments the dust particles can interact strongly. As already noted in the previous section, this can be relevant in the case of strongly dissipative systems, when $\nu_{dn} \sim \omega_{pd}$ (see, e.g., Pieper and Goree 1996). In addition, density perturbations of the dust component often have large enough amplitudes, i.e., the waves are essentially nonlinear. Nevertheless, in most of the works the linear theory is used, which may also be relevant in some cases, for instance, when describing the instability threshold.

11.4.4.2 *Waves in dust crystals.* Excitation of waves in strongly ordered structures of dust particles (dust crystals) is often employed as a diagnostic tool in studying dusty plasmas. The main parameters entering into the dispersion relation (within the Debye–Hückel model) are the particle charge Z_d, and the ratio of the interparticle distance to the screening length – the lattice parameter κ. Other parameters are either known in advance (e.g., particle mass) or can be easily determined in the experiment (e.g., mean interparticle distance). Hence, Z_d and κ can be estimated by comparing an experimental dispersion relation with a relevant theoretical model.

The excitation of waves in a one–dimensional chain of particles, formed in the sheath of an r.f. discharge, is described by Peters *et al.* (1996). The excitation was effected by applying a periodic potential on a small probe placed near one of the chain ends. In a modified experiment (Homann *et al.* 1997), the edge particle was excited by a modulated laser beam focused on it. In this experiment, the lattice parameter is determined as $\kappa \approx 1.6 \pm 0.6$; the charge was measured independently by the vertical resonance method. We also note an experiment (Misawa *et al.* 2001) in which the instability of the vertical oscillations in a horizontal chain of dust particles (transverse mode) in the electrode sheath of a d.c. discharge was developed. The instability observed was attributed in this work to the charge delay mechanism considered in Section 11.4.1.

Laser excitation of waves was employed for an analysis of the longitudinal mode in a two–dimensional dust crystal (Homann *et al.* 1998). The experiments were performed at different pressures. With increasing pressure, the lattice pa-

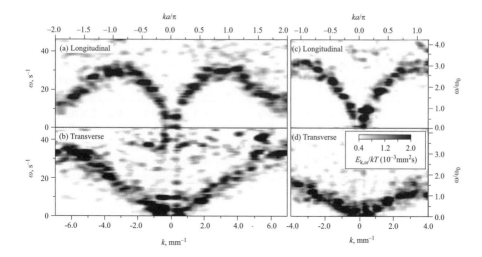

FIG. 11.19. Spectrum of fluctuation energy density in $(\mathbf{k},\omega)$ space in the absence of any intentional wave excitation (Nunomura *et al.* 2002b). Darker grays correspond to higher wave energy. Energy is concentrated along a curve corresponding to a dispersion relation. Figures (a) and (c) are for the longitudinal mode, while (b) and (d) are for the transverse mode. Vector $\mathbf{k}$ is directed parallel (a and b) and perpendicular (c and d) to a primitive translation lattice vector $\mathbf{a}$. Here $\omega_0 = 11.6$ s^{-1} is the plasma–dust frequency.

rameter κ estimated from the experiment slightly decreased. However, similar to most of the experiments in the sheath of an r.f. discharge, it was in the range $\kappa \sim 0.5$–2. Transverse modes in a two–dimensional dust crystal, excited by laser radiation, were analyzed by Nunomura *et al.* (2000). Finally, an analysis of laser–excited longitudinal and transverse waves in a two–dimensional crystal was made by Nunomura *et al.* (2002a). The experimental results are in qualitative agreement with the theory of Wang *et al.* (2001b), which allows us to use them for an estimation of Z_d and κ.

It is not necessary to excite the waves externally when studying the dispersion relations: this information is already contained in the random (thermal) motion of dust particles. Nunomura *et al.* (2002b) measured the dispersion relations for the longitudinal and transverse modes in the absence of any external excitation. Using a Fourier transform of the particle velocity, the wave amplitude $V_{\mathbf{k}\omega}$ and energy density $\sim V_{\mathbf{k}\omega}^2$ were computed. As follows from Fig. 11.19, the highest values of the energy density are concentrated in proximity to distinct curves in $(\mathbf{k},\omega)$–space; these curves are identified as dispersion curves. We note that in the experiment considered, the dust component was not in thermal equilibrium with the neutral component. The particle velocity distribution was close to Maxwellian, with the mean kinetic energy several times larger than the temperature of the neutrals. This indicates the existence of a source of (stochastic) energy for the dust particles, which is mainly dissipated in dust–neutral collisions.

Samsonov *et al.* (2002) investigated solitary waves in a hexagonal monolayer dust crystal. The crystal was made from monodispersed particles levitating in the electrode sheath of a capacitively coupled r.f. discharge. To excite the waves, a short potential pulse (-30 V, 100 ms) was applied to a thin filament situated below the crystal, which led to the appearance of a one–dimensional disturbance (compression) propagating perpendicular to the filament. It was found that the excited wave possessed the main features of solitons, in particular, the product of its amplitude and width squared remained constant during propagation. A theory developed by Samsonov *et al.* (2002) describes well the experiment and is based on the equation of motion for the one–dimensional chain of dust particles, the equation which takes into account dispersion, nonlinearity, and damping.

In conclusion, let us briefly discuss another effect associated with wave processes in dusty plasmas. The so–called Mach cones with an opening angle μ determined by the ratio of the wave propagation velocity (sound velocity) in a medium to a (supersonic) disturbance velocity, $\sin\mu = c_s/V$, can form in dusty plasmas due to acoustic dispersion and can be used for dusty plasma diagnostics (Havnes *et al.* 1995). In laboratory conditions they were observed by Samsonov *et al.* (1999, 2000) and their excitation was attributed to the presence of fast (supersonic) particles beneath the lattice of a two–dimensional dust crystal in the sheath of an r.f. discharge, whereas the origin of such fast particles was not fully understood. In work by Melzer *et al.* (2000), the focused spot of a laser beam propagating with a supersonic velocity through a plasma crystal was utilized as the supersonic object.

11.5   New directions in experimental research

In this section, some new directions in the laboratory investigation of dusty plasmas are considered. The term "new" is of course conditional here, because the field itself is very recent – active investigations began only ten years ago. Here we mainly focus on the following new directions of dusty plasma research: investigations under microgravity conditions, external perturbations, and the use of nonspherical particles. In conclusion, we briefly mention potential applications of dusty plasmas.

11.5.1   *Investigations of dusty plasmas under microgravity conditions*

In many cases, the force of gravity considerably limits the possibilities of laboratory experiments in ground–based conditions. For example, it prevents the formation of extended three–dimensional structures. To support particles against gravity, strong electric fields are required, which only exist in plasma sheaths or striations. These regions are characterized by a high degree of anisotropy and suprathermal ion flows. In these conditions, external forces are comparable with the interparticle forces. Hence, most dusty plasmas investigated on earth are essentially two–dimensional, strongly inhomogeneous, and anisotropic in the vertical direction. Most of these complications can be avoided by performing experiments under microgravity conditions (Thomas *et al.* 1994). Such experiments

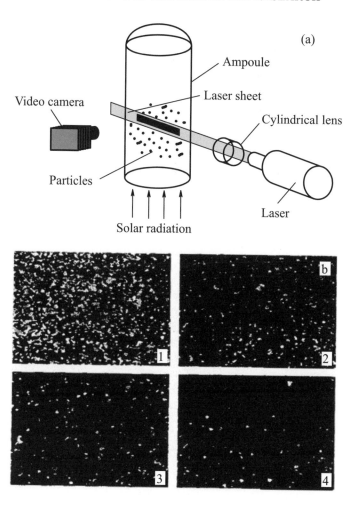

FIG. 11.20. (a) Schematic representation of the experimental setup for studying the behavior of dust particles charged by solar radiation under microgravity conditions. (b) Sequence of states of the dust particle system 2 s (1), 20 s (2), 50 s (3), and 110 s (4) after shaking the ampoule in experiments on the investigation of the behavior of dust particles charged by solar radiation in microgravity conditions.

were recently carried out in several types of plasmas. Below we briefly discuss each of them separately.

11.5.1.1 *Experiments on dusty plasma induced by UV radiation.* The behavior of an ensemble of particles charged by solar radiation was studied in a microgravity experiment on–board the "Mir" space station (Fortov *et al.* 1998). The main elements of the experimental setup (see Fig. 11.20(a)) comprised: glass ampoules with bronze particles of radius 25–50 µm coated with a monolayer of cesium; an illumination source – a semiconductor laser with an operating wavelength 0.67 μm and power 30 mW, and a recording system – a video camera and videomodule. Since in the initial state the particles were deposited on the ampoule walls, the experiment was performed employing the following scheme: dynamic effect (impact) on the system; relaxation for a period 3–4 s in order to reduce the random particle velocities acquired from the initial impulse (impact); illumination of the system by solar radiation for a period of several minutes with subsequent relaxation to the initial state – particle deposition on the walls. Figure 11.20(b) illustrates the evolution of the particle system in an ampoule at a buffer gas pressure of 40 torr. The particle charge was determined from a comparison of the experimentally established time dependence of dust density in the ampoule with numerical modeling or with results of particle trajectory analysis (Fortov *et al.* 1998; Vaulina *et al.* 2001). The charges determined in this way fell in the range $(5–10)\cdot10^4$ elementary charges. No strong correlation of interparticle distances was observed. The pair correlation functions obtained indicate that nonideal short–range–ordered structures were formed in this experiment.

11.5.1.2 *Experiments in a d.c. gas discharge.*

The experimental investigation of dusty plasmas in a d.c. gas discharge was performed on–board the "Mir" space station (Nefedov *et al.* 2002). The major difference between this experimental apparatus and a similar one used in ground–based experiments (see Fig. 11.12) was the presence of a two–grid electrode placed between the cathode and the anode. During the experiments, the electrode was at a floating potential and prevented negatively charged particles from escaping to the anode.

In the experiments, neon gas at a pressure $p = 1$ torr was utilized. The discharge current was varied in the range from 0.1 to 1 mA. Nonmonodispersed bronze spheres with a mean radius $\bar{a} \approx 65$ μm were used. The electron temperature and the plasma number density were estimated as $T_e \sim 3-7$ eV and $n_e \sim n_i \sim 10^8$–$3\cdot10^9$ cm^{-3}, respectively.

The experiment was performed using the following procedure. The particles were initially deposited on the tube walls. For this reason, the system was subjected to dynamic action after initiating a discharge with a given current I. In a plasma volume, the particles were charged by collecting electrons and ions and moved toward the anode. In the vicinity of the grid electrode a part of the particles trapped in this region formed a stationary three–dimensional structure (cloud). A typical image of such a structure recorded by a video camera is shown in Fig. 11.21. When the discharge was switched off, the particles went back to the tube walls. The experiment was repeated at a new discharge current value.

An analysis of video images of the stationary dust structure formed near the grid electrode allowed the measurement of the static (pair correlation function) and dynamic (diffusion coefficient) characteristics of the dust particle system. Comparison with numerical simulations of dissipative Debye–Hückel systems was employed for dusty plasma diagnostics (Fortov *et al.* 2003a, 2003b). In addition, an analysis of the particle's drift motion to the grid electrode at the initial stage

FIG. 11.21. Video image of typical structure of a dust cloud formed in the vicinity of a grid electrode in a d.c. discharge under microgravity conditions. The characteristic size of the cloud is about 2 cm in radial direction.

of the experiment was used to estimate the particle charge. Below we summarize the main experimental findings.

Measured pair correlation functions revealed the formation of ordered structures of dust particles of a liquid–like type (short–range order). This is in agreement with the estimate of the modified coupling parameter γ^* (made on the basis of measuring the diffusion coefficient) which decreases from ~ 75 to ~ 25 with an increase in the discharge current. The kinetic energy of chaotic particle motion was estimated as $T_\mathrm{d} \sim 10^5$ eV. Finally, the charge was estimated as $Z_\mathrm{d} \sim -2 \cdot 10^6$, which corresponds to the surface potential of about -40 V and is considerably larger than the magnitude predicted by the OML theory. Note that the physics of charging and interaction of large particles ($a \geq \lambda_{\mathrm{Di}}$, ℓ_i) has been relatively poorly studied, mainly because ground–based experiments with such large particles are typically impossible. For this reason, the above-described experiments in microgravity conditions are of obvious interest.

11.5.1.3 *Experiments in an r.f. discharge.* Dusty plasmas formed in an r.f. discharge have been intensively investigated under microgravity conditions. A typical scheme of an experimental setup is shown in Fig. 11.22. The results of the first rocket experiments (within ~ 6 min of microgravity) and their qualitative analysis were reported by Morfill *et al.* (1999b).

Currently, the "Plasma Crystal" laboratory, created within the framework

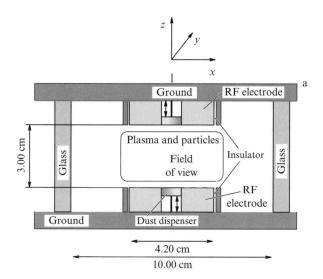

FIG. 11.22. Schematic of the experimental setup for investigating the dust structures in an r.f. gas discharge under microgravity conditions. The main dimensions are indicated in the figure.

of the Russian–German scientific cooperation program and operating on–board the International Space Station (ISS), is actively functioning. Its main tasks include investigations of dusty plasma crystals, phase transitions, wave phenomena, properties of boundaries between different plasma regions, etc. in a three–dimensional isotropic dusty plasma at the kinetic level. The first "basic" experiments designed to study the behavior of the dust component over a broad range of dusty plasma parameters were performed at the beginning of March 2001. Monodispersed systems and binary mixtures of particles of different size ($a = 1.7$ and $3.4\ \mu$m) were investigated at different argon pressures and r.f. powers. At present, most of the investigations are in the active phase or in the stage of data analysis. Thus, we first discuss several important phenomena which were observed in rocket experiments (Morfill *et al.* 1999b; Thomas and Morfill 2001), and then briefly mention some of the published results achieved on–board ISS.

The typical structure of the dust particle system in an r.f. discharge under microgravity conditions is shown in Fig. 11.23. As can be seen from the figure, the dust component does not fill the whole volume: the central part of the discharge (void) is free of dust particles, as are the regions adjacent to the electrodes and walls. The boundary between the dusty plasma and the usual electron–ion plasma is sharp both in the region of the void and at the periphery. At the periphery, far from the central axis of the discharge, the particles exhibit convective motions. Close to the axis, the structure is stable and does not support convective motion. Here, the dust cloud reveals the highest ordering. The particles form layers parallel to the electrode in the vicinity of the outer cloud boundary, but

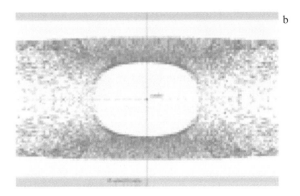

FIG. 11.23. Typical video image of the dust cloud in an rf discharge under microgravity conditions. The central part of the discharge is free of dust particles – the so called void is formed.

this symmetry is broken considerably toward the internal part of the cloud.

One of the main problems connected with the structure of dust clouds in an r.f. discharge under microgravity conditions is void formation. It is common to assume that the electric field in the discharge is directed from the center to the periphery, and hence the electrostatic force F_{el} acting on negatively charged dust particles is pointed to the center. The formation of a void implies the occurrence of some effect which not only compensates for the action of the electric force, but also leads to the repulsion of dust particles from the central region of the discharge. It is established that the void forms at a relatively small number of particles in the discharge too, i.e., its formation is not an essentially collective effect. This can correspond to an internal property of the discharge – the existence of some force directed from the center to the periphery and exceeding the electrostatic force. In the work by Morfill *et al.* (1999b), it was assumed that the void could be caused by the thermophoretic force. This assumption has not been confirmed so far. Moreover, the results of numerical simulation of the system considered (Akdim and Goedheer 2002) raise some doubts about this assumption. In work by Samsonov and Goree (1999); Goree *et al.* (1999); and Tsytovich *et al.* (2001b), it was suggested that the ion drag force F_i is responsible for the void formation. The necessary condition for the void formation is $|F_i/F_{el}| > 1$, at least in the vicinity of the discharge center. This condition was verified by Khrapak *et al.* (2002) using new results for the ion drag force acting on an individual particle in a collisionless regime (for ion scattering from a dust particle). Below, we present details of this estimation. In a weak electric field, $E \ll T_i/e\ell_i$, the ions drift with a subthermal velocity – this situation wittingly takes place in the vicinity of the center (the electric field is zero in the center). The ion drift velocity equals $u = \mu_i E$, where μ_i is the ion mobility defined as $\mu_i = e\ell_i v_{T_i}/T_i$. Using expressions (11.51) and (11.61) for F_{el} and F_i, and taking into account that both the forces are proportional to the electric field we get $|F_i/F_{el}| \approx \delta \ell_i/\lambda_D$, where δ is a weak function of dusty plasma parameters: $\delta \sim 0.3 - 0.5$ for typical experimental conditions. Therefore, in the collisionless regime when $\ell_i \gg \lambda_D$ at least should be satisfied, the condition $|F_i/F_{el}| > 1$ is always fulfilled. For this reason, it is quite natural to assume that the void formation is caused primarily by the ion drag force.

Reverting to the experiments performed on–board ISS we mention the work by Nefedov *et al.* (2003) describing basic experiments on dust component crystallization. Under certain conditions, the coexistence of domains of fcc, bcc, and hcp structures has been observed in the mostly ordered dusty plasma region – that is, the lower central part of the dust cloud. At the same time, string–like structures (when particles are aligned in vertical chains) which are typical for ground–based experiments in an r.f. discharge have not been observed. The results of the structure observation were compared with theoretical predictions of the Debye–Hückel model. A comparison of static (structure) and dynamic characteristics of the dust particle system obtained from r.f. and d.c. discharge research with those obtained from numerical modeling within the framework of Debye–Hückel systems was performed by Fortov *et al.* (2003a,b). This comparison can be used to estimate the coupling strength in dusty plasmas, which is mostly determined by the particle charge and plasma screening length. In (Khrapak *et al.* 2003c) the longitudinal oscillations excited by applying a low–frequency modulation voltage to the electrodes were analyzed. Comparison of the experimental dispersion relation with the theoretical prediction was used to estimate the dust particle charge. Excitation of shock waves in the dust component by a gas flow was described by Samsonov *et al.* (2003). Investigation of the boundaries between the conventional electron–ion plasma and dusty plasma, as well as between two different dusty plasmas composed of particles of different sizes, was reported by Annaratone *et al.* (2002). The "decharging" of dusty plasmas by switching off the r.f. voltage was studied by Ivlev *et al.* (2003b).

11.5.2   *External perturbations*

Investigation of different external perturbations of dusty plasma structures is of substantial interest for several reasons. First, the perturbations which affect the background plasma insignificantly can be employed as diagnostic tools. External perturbations can also be used to control the spatial location and ordering of dusty plasma structures, and to input additional energy in order to investigate their behavior in extreme conditions.

The action of laser radiation finds wide use. As discussed above, it is employed to manipulate particles when measuring the interparticle potential (see Section 11.2.4), excite the dust cluster rotation (see Section 11.3.3), excite vertical oscillations of individual particles (see Section 11.4.1) and low–frequency waves in dusty plasma structures (crystals) and form Mach cones (see Section 11.4). The application of laser radiation is still expanding. In this context, we note recent work where laser radiation was utilized for manipulating dust particles in a d.c. discharge in order to determine the particle charges (Fortov *et al.* 2001b) and forming a localized "point" source to investigate the waves emitted by this source in a two–dimensional dusty plasma crystal in an r.f. discharge (Piel *et al.* 2002).

"Electrostatic" action on dusty plasma structures is also employed as an external perturbation. Typically, an additional low–frequency voltage is applied either to the electrode(s) or to a small object (probe) situated near the dust particle structure. This method is widely used to excite waves in dusty plasma structures. In addition, Samsonov *et al.* (2001) used this method to analyze the long–range interaction between the dust particles levitating near the sheath edge of an r.f. discharge and a macroscopic object – a thin wire. It is shown that for a negative potential of the wire the nearest particles experience electrostatic repulsion, while at sufficiently long distances an attraction takes place. The ion drag force was taken to be a mechanism responsible for long–range attraction. Finally, the influence of a probe situated near the sheath edge of an r.f. discharge on the structure and the properties of dust crystals formed in this region was analyzed by Law *et al.* (1998).

A magnetic field can also serve as a source of external perturbation. As noted in Section 11.3.3, the vertical magnetic field can lead to rotation of dust crystals in a horizontal plane. The effect is associated with the tangential component of the ion drag force which appears due to distortion of ion trajectories in a magnetic field. This issue was discussed in detail by Konopka *et al.* (2000b).

Finally, a "gas–dynamic" action on a dusty plasma structure can also be employed. As an example, we consider the excitation of nonlinear waves in a d.c. discharge (Fortov *et al.* 2004). The main difference from the experimental setup shown in Fig. 11.12 was the following. Below the hollow cylindrical cathode, a plunger was set which could move with the aid of a permanent magnet. The plunger could be moved at a speed of about 30–40 $\mathrm{cm\,s}^{-1}$ (for a distance of 4–5 cm). A grid kept at a floating potential was inserted 7 cm above the cathode. When the plunger was moving downward, the particles were also moving downward to a region of stronger electric field, and the dust structure became unstable. It was possible to excite solitary waves as well as instabilities that appeared in the whole volume of the dusty plasma. The image of a solitary wave and profiles of the compression factors at different moments in time are shown in Fig. 11.24. As follows from the figure, the perturbation consists of two compression regions separated by a region of rarefaction. Both compression regions move downward with velocities of 1.5–3 $\mathrm{cm\,s}^{-1}$ (on the order of the dust–acoustic wave velocity), but the lower compression moves slightly faster. The minimum of the rarefaction region initially moves with the lower compression and then acquires the speed of the upper compression. In this way, an increase in the front steepness of the upper compression region is observed. In the rarefaction region, the upward motion of dust grains is observed with velocities up to 15 $\mathrm{cm\,s}^{-1}$, which is several times larger than the dust–acoustic wave velocity. The theoretical interpretation of the phenomenon observed constitutes an intriguing problem.

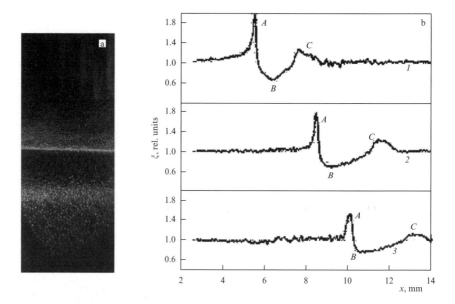

FIG. 11.24.   Video image of a solitary wave (a) and compressional factor ξ in the wave at different moments in time (b). Time interval between curves 1 and 2 is 120 ms, between curves 2 and 3 – 60 ms. The wave is excited by a gas–dynamic impact in the d.c. gas–discharge setup. The symbols A and C correspond to the compression region, and symbol B corresponds to the rarefaction region.

11.5.3   Dusty plasma of strongly asymmetric particles

Most of the experimental and theoretical works dealing with the investigation of dusty plasmas were performed with spherical particles. Only recently has dusty plasma with asymmetric particles been investigated experimentally and theoretically (Mohideen *et al.* 1998; Molotkov *et al.* 2000; Annaratone *et al.* 2001; Fortov *et al.* 2001c; Vladimirov and Nambu 2001; Shukla and Mamun 2001). Note that in (Mohideen *et al.* 1998) the geometrical aspect ratio was $\alpha \sim 3$, and the first experiments with strongly asymmetric particles, $\alpha = (40-80) \gg 1$, were carried out by Molotkov *et al.* (2000).

It is well known that colloidal solutions, which have much in common with dusty plasmas, show a much broader spectrum of possible states in the case of strongly asymmetric cylindrical or disk particles. In such solutions, liquid phase and several liquid–crystal and crystal phases with different degrees of orientational and positional ordering can be observed. It is also well known that the employment of cylindrical probes (in addition to spherical) considerably broaden the possibilities of low–temperature plasma diagnostics. It is obvious that the use of cylindrical (in addition to spherical) particles can considerably broaden the potentialities of noninvasive plasma diagnostics.

In work by Molotkov *et al.* (2000), in which an experimental setup analogous to that shown in Fig. 11.12 was adapted, nylon particles ($\rho = 1.1$ g cm^{-3}) of length 300 μm and diameters 7.5 and 15 μm, as well as particles of lengths 300 and 600 μm and diameter 10 μm, were introduced into the plasma of a d.c. discharge. The discharge was initiated in neon or a neon/hydrogen mixture at a pressure of 0.1–2 torr. The discharge current was varied from 0.1 to 10 mA. In this parameter range, standing striations were formed in the discharge, which made particle levitation possible. A neon/hydrogen mixture was used to levitate heavier particles of larger diameter (15 μm) or larger length (600 μm). In this case, the particles formed structures consisting of 3–4 horizontal layers. Lighter particles levitated in pure neon and formed much more extended structures in the vertical direction. In Fig. 11.25(a), a part of a horizontal section of an ordered structure levitating in a striation of a d.c. discharge excited in a neon/hydrogen mixture (1:1) at a pressure of 0.9 torr and discharge current of 3.8 mA is shown.

The observation of structures formed by microcylinders revealed clear ordering. All particles lay in the horizontal plane and were oriented in a certain direction. It could be expected that their orientation was determined by the cylindrical symmetry of the discharge. However, no correlation between the particle orientation and discharge tube symmetry was found. Nor could the preferential orientation of the particles be explained by the interparticle interaction, because individual particles were oriented in the same direction. Presumably, the preferential orientation was related to an insignificant asymmetry in the discharge tube. This was confirmed by the fact that the orientation could be changed by introducing an artificial perturbation into the discharge.

In later experiments (Fortov et al. 2001c), nylon particles of lengths 300 and 600 μm and diameter 10 μm coated by a thin layer of conducting polymer were utilized. In a d.c. discharge they formed structures identical to those formed by uncoated particles of the same size.

Levitation of cylindrical particles was also observed near the sheath edge of a capacitively coupled r.f. discharge (Annaratone et al. 2001). In this experiment, cylindrical particles of length 300 μm and diameters 7.5 and 15 μm were used, and a small fraction of very long particles (up to 800 μm) 7.5 μm in diameter was also present. A typical picture of a structure formed by these particles is shown in Fig. 11.25(b). Longer particles are oriented horizontally and mainly located in the central part limited by a ring placed on the electrode, while shorter particles are oriented vertically along the electric field. Levitation and ordering of the cylindrical particles occurred only for pressures higher than 5 Pa and a discharge power above 20 W. An increase in the discharge power did not significantly affect particle levitation. The average distance between vertically oriented particles varied from 1 to 0.3 mm. An increase in particle density leads to quasicrystalline structure degradation and an increase in particle kinetic energies. A further increase in density is impossible because the particles "fall down" from the structure. Levitation of particles coated by a conducting polymer was not observed in an r.f. discharge for the conditions at which the dielectric particles of the same size and mass could levitate. Instead, the conducting particles stuck

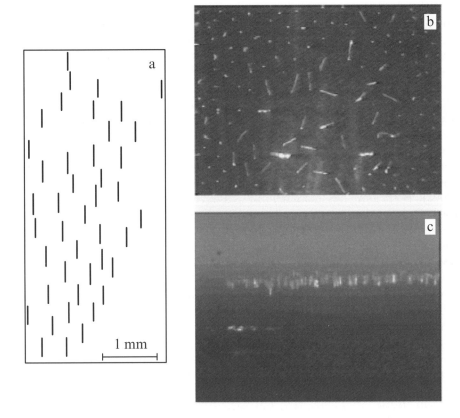

FIG. 11.25.   (a) Digitized image of a part of a horizontal section of a structure involving
        cylindrical macroparticles of length 300 μm and diameter 15 μm levitating in a striation of
        a d.c. discharge in a neon/hydrogen mixture at a pressure of 0.9 torr and discharge current
        3.8 mA. (b, c) Typical video images of structures formed by cylindrical particles levitating
        near the sheath edge of an r.f. discharge. The discharge was initiated in krypton at a
        pressure of 52 Pa and discharge power 80 W. Panel (b) shows a top view, dots correspond
        to vertically oriented particles; Panel (c) gives a side view.

to the electrode, preserving vertical orientation, and some stuck to each other
forming multiparticle fractal complexes with up to 10 particles.

    The preferential orientation of cylindrical particles is determined by the inter-
play between the interaction of nonuniform electric fields in striations or sheaths
with the particle charge and induced dipole and quadrupole moments. In a d.c.
discharge, the particle charge is typically larger than in an r.f. discharge, allowing particle levitation in weaker electric fields. In this case, the dipole moment,
which is proportional to the electric field strength squared, is much smaller than
in an r.f. discharge. This can explain the different orientations of similar–sized

particles: horizontal in a d.c. discharge, and vertical in an r.f. discharge. Theoretical investigations into the charging of horizontally and vertically oriented particles, their equilibrium positions (levitation height and angle), and the energy of electrostatic interaction between cylindrical particles depending on their orientation can be found in (Ivlev *et al.* 2003a).

11.5.4 *Dusty plasma at cryogenic temperatures*

The coupling in the dust component increases as the temperature of a gas discharge decreases. This is connected both with a direct fall in dust particle kinetic energy and a diminution in the screening length due to a decrease in the ion temperature. The first experiments with dusty plasmas in a cryogenic gas discharge under a liquid–nitrogen temperature of 77 K were performed by Fortov *et al.* (2002). Both glow d.c. discharge and capacitively coupled r.f. discharge were used. Ordered dust structures in the glow cryogenic discharge were similar to the structures observed at room temperatures. However, a much stronger influence of the thermophoretic force on the dynamics and stability of dust structures was observed. Very extensive (about 30 cm) ordered structures consisting of long chains and occupying practically the whole volume of the positive column of a d.c. gas discharge were observed for the first time in a d.c. discharge as well.

In experiments with an r.f. discharge it was found that at cryogenic temperatures the density of dust particles in the main volume of the ordered structures can increase considerably, while at the periphery it is similar to that in common r.f. discharges. In the lower part of dust structures, the propagation of travelling density waves was observed. The dust–acoustic wave velocity in cryogenic conditions was several times larger than in normal conditions. With decreasing pressure, instabilities led to structure separation into transverse layers with clear boundaries. Fortov *et al.* (2002) explain the formation of much more dense structures in cryogenic plasmas mainly by the decrease of the Debye radius, which leads to an exponential decrease in the interparticle interaction energy and allows the particles to be closer to each other.

11.5.5 *Possible applications of dusty plasmas*

Dusty plasmas have already been applied in industry for many decades, for example, in technologies such as precipitation of aerosol particles in combustion products of electric power stations, plasma spraying, and electrostatic painting, as well as in a number of other areas. In the beginning of the 1990s it was understood that a large part of the contamination on the surface of silicon wafers during manufacturing of semiconductor elements for electronics not come from insufficient cleaning of the production area of dust, but was an unavoidable consequence of plasma etching and deposition technologies. In widespread capacitively coupled r.f. discharge reactors, all particles are charged negatively and levitate close to one of the electrodes. After switching off the discharge they deposit on the surface of the silicon wafer. Submicron-sized particles deposited on the surface of processable wafers can reduce the working surface, cause dislocations and voids, and reduce adhesion of thin films. Great efforts focused on reduction of the number of undesirable dust particles in industrial plasma reactors have given positive results (Bouchoule 1999; Kersten *et al.* 2001; Selwyn *et al.* 1989), and this problem can be considered as practically solved.

For the power supply of space vehicles, automatic weather stations, antisubmarine buoys, etc., compact autonomous power–supply sources with a power about 1–10 kW and with working resources of several years are necessary. At present, solar power photoelectric converters, thermoelectric sources with thermoemissive elements on ^{90}Sr, ^{238}Pu, or ^{210}Po, and thermoemissive converters (TEC), where a nuclear reactor fueled by ^{235}U is used as the heat source, are widely applied. All these sources have disadvantages, in particular, very low efficiency. Moreover, a nuclear reactor is very complicated to produce.

Recently, Baranov *et al.* (2000) suggested the conversion of nuclear energy into electric energy by the photovoltaic effect in wide bandgap semiconductors based on diamond films obtained by precipitation of the carbon from the gas phase (CVD diamond) and boron nitride. Creation of such sources has become possible due to the investigation of diamond-film synthesis, resulting in the production of semiconducting structures, and the investigation of the physics of low–temperature dusty plasmas.

The principle of the action of the sources, which convert the energy of radioactive isotopes into electricity by the photovoltaic effect, is the following. Under the influence of ionizing radiation a specially selected gaseous mixture is excited and radiates in the UV range. The ultraviolet radiation due to the photovoltaic effect induces the electromotive force in the wide bandgap semiconductor. For this purpose, semiconductors based on diamond structures are the most convenient because they have high radiation resistance and high conversion efficiency (up to 70%). As a radioactive isotope it is possible to utilize β–active isotopes having a comparatively high half–life period (10–30 years), for example, ^{90}Sr or similar solid isotopes, e.g., α–active ^{238}Pu.

To use solid isotopes in the photovoltaic converters, it is necessary to have the isotope surface area as large as possible. A homogeneous mixture of gas and isotope dust, which has a maximum possible surface-to-volume ratio, is the most attractive. Excitation of the gas mixture is accomplished by β– or α–radiation from the radioactive dust. Estimates show that at a dust size of 1–20 μm and dust number density in the gas of 10^5–10^9 cm^{-3} it is possible to obtain a power density of ~ 1 W dm^{-3}. The gas pressure has to be on the order of 1–10 atm for effective energy conversion of β– or α–radiation into UV radiation.

The main physical problem arising during the creation of such a battery consists in the necessity of having the homogeneous gas–dust medium at pressures of several atmospheres. Results received recently in investigations of dusty plasmas and their condensation and crystallization (Pal' *et al.* 2001a,b) demonstrate such a possibility. Self–consistent processes in such plasmas result in the establishment of an ordered stationary state which is necessary for the transport of the radiation from the volume of the exited gas to photo–converters.

In recent years it has become obvious that the presence of dust in plasmas does not always result in undesirable consequences. Powders produced employing plasma technologies can have interesting and useful properties: very small sizes (from a nanometer to the micrometer range), monodispersity, and high chemical activity. The size, structure, and composition of the powder can be varied easily in compliance with the specific requirements of a certain technology. In this connection, two trends can be distinguished in applied research of dusty plasmas (Kersten *et al.* 2001). The first one represents a development of the well-established technologies of surface modification, except that in this case the surface of dust particles is the subject of treatment. With the aim of creating particles with specific properties, sputtering, surface activation, etching, or separation of clustered grains in plasmas can be adapted. The second important trend in applied dusty plasma research is the creation of new nanostructured materials, like thin films with an inclusion of nanometer–sized particles.

The typical size of the elements of integrated circuits in microelectronics is reduced every year and in the very near future it will likely reach 10 nm. Furthermore, there is a tendency to replace the capacitively coupled r.f. discharge by an inductively coupled one. These plasmas are characterized by higher electron density and lower sheath potential. Altogether, this results in a reduced particle trapping capacity and leads to the main part of the particles dropping on the surface of the silicon wafer during plasma processing. This implies that the solution introduced in the 1990s, which was mostly based on dust particle confinement in special traps, will not work anymore. This poses a serious problem to producers of integrated circuits of the next generation, which demands further applied research of the properties of dusty plasmas.

11.6   Conclusions

Despite a history spanning nearly a century, the investigation of dusty plasmas has acquired particular attention only during the last decade, after the experimental discovery of the crystallization of the dust component. Due to their unique properties dusty plasmas are successfully used in solving fundamental and applied problems. The simplicity of visualization permits measurements (of the dust component) at the kinetic level. In this case, a detailed analysis of the thermodynamic and kinetic properties of dislocations and other defects of the dust lattices becomes possible. The latter have much in common with the usual crystalline lattice of solid bodies. Much attention is attracted to studies of easily excited linear and nonlinear low–frequency waves and their instabilities. Investigation of phase transitions in systems of symmetric and asymmetric dust particles provides useful information about critical phenomena and self–organization processes, in particular, about the possibilities of natural formation of ordered dusty-plasma structures in the universe. The first space experiments under microgravity conditions were performed on–board the space stations "Mir" and ISS. A number of important and sometimes unexpected results have been obtained. The understanding of the observed effects is impossible without a detailed investigation of elementary processes in dusty plasmas, such as particle charging, interaction between the particles, the main forces acting on the particles, etc.

One of the most important application problems is the removal of dust particles when manufacturing computer chips by plasma–aided technologies. To solve this problem, a deep understanding of physical processes in a gas–discharge dusty plasma is necessary. Moreover, the unique possibility confining and controlling of physical and chemical properties of dust particles makes dusty plasma a perfect tool for creating powders with prescribed properties and for modifying them.

In this chapter, we have tried not only to discuss the most significant experimental and theoretical results obtained recently, but also to mention problems which are as yet unsolved. The field of dusty plasmas is one of the most rapidly growing fields in physics with more than one publication, on average, appearing every day. There is no doubt, therefore, that many intriguing and important results will be obtained in the future.

References

Akdim, M. R. and Goedheer, W. J. (2002). Modelling of voids in colloidal plasmas. *Phys. Rev. E*, **65**, 015401(R)/1–4.

Aleksandrov, A. F., Bogdankevich, L. S., and Rukhadze, A. A. (1984). *Principles of plasma electrodynamics*. Springer–Verlag, New York.

Allen, J. E. (1992). Probe Theory – The orbital motion approach. *Phys. Scripta*, **45**, 497–503.

Allen, J. E., Annaratone, B. M., and de Angelis, U. (2000). On the orbital motion limited theory for a small body at floating potential in a Maxwellian plasma. *J. Plasma Phys.*, **63**, 299–309.

Al'pert, Y. L., Gurevich, A. V., and Pitaevskii, L. P. (1965). *Space physics with artificial satellites*. Consultants Bureau, New York.

Amiranashvili, S. G., Gusein–zade, N. G., and Tsytovich, V. N. (2001). Spectral properties of small dusty clusters. *Phys. Rev. E*, **64**, 016407/1–6.

Annaratone, B. M., Khrapak, A. G., Ivlev, A. V., Soellner, G., Bryant, P., Suetterlin, R., Konopka, U., Yoshino, K., Zuzic, M., Thomas, H. M., and Morfill, G. E. (2001). Levitation of cylindrical particles in the sheath of r.f. plasma. *Phys. Rev. E*, **63**, 036406/1–6.

Annaratone, B. M., Khrapak, S. A., Bryant, P., Morfill, G. E., Rothermel, H., Thomas, H. M., Zuzic, M., Fortov, V. E., Molotkov, V. I., Nefedov, A. P., Krikalev, S., and Semenov, Yu. P. (2002). Complex–plasma boundaries. *Phys. Rev. E*, **66**, 056411/1–4.

Annou, R. (1998). Current–driven dust ion–acoustic instability in a collisional dusty plasma with a variable charge. *Phys. Plasmas*, **5**, 2813–2814.

Astrakharchik, G. E., Belousov, A. I., and Lozovik, Y. E. (1999a). Properties of two–dimensional dusty plasma clusters. *Phys. Lett. A*, **258**, 123–130.

Astrakharchik, G. E., Belousov, A. I., and Lozovik, Y. E. (1999b). Two–dimensional mesoscopic dusty plasma clusters: Structure and phase transitions. *JETP*, **89**, 696–703.

Balabanov, V. V., Vasilyak, L. M., Vetchinin, S. P., Nefedov, A. P., Polyakov, D. N., and Fortov, V. E. (2001). The effect of the gas temperature gradient on dust structures in a glow–discharge plasma. *JETP*, **92**, 86 92.

Baranov, V. Y., Belov, I. A., Dem'yanov, D. V., D. V., Ivanov, A. S., Mazalov, D. A., Pal', A. F., Petrushevich, Y. V., Pichugin, V. V., Starostin, A. N., Filippov, A. V., and Fortov, V. E. (2000). Radioactive isotopes as source of energy in photovoltanic nuclear battery on basis of plasma–dust structures. In *Isotopes: Properties, production, and applications (in Russian)*, V. Y. Baranov (ed.), pp. 626–641. IzdAT, Moscow.

Barkan, A., D'Angelo, N., and Merlino, R. L. (1994). Charging of dust grains in a plasma. *Phys. Rev. Lett.*, **73**, 3093–3096.

Barkan, A., Merlino, R. L., and D'Angelo, N. (1995). Laboratory observation of the dust–acoustic wave mode. *Phys. Plasmas*, **2**, 3563–3565.

Barnes, M. S., Keller, J. H., Forster, J. C., O'Neill, J. A., and Coultas, D. K. (1992). Transport of dust particles in glow–discharge plasmas. *Phys. Rev. Lett.*, **68**, 313–316.

Bhatt, J. R. (1997). Langmuir waves in a dusty plasma with variable grain charge. *Phys. Rev. E*, **55**, 1166–1169.

Bhatt, J. R. and Pandey, B. P. (1994). Self–consistent charge dynamics and collective modes in a dusty plasma. *Phys. Rev. E*, **50**, 3980–3983.

Bliokh, P., Sinitsin, V., and Yaroshenko, V. (1995). *Dusty and self–gravitational plasmas in space*. Kluwer Academic Publishing, Dordrecht.

Bolotovskii, B. M. and Stolyarov, S. N. (1992). Radiation from and energy–loss by charged–particles in moving media. *Phys. Uspekhi*, **35**, 143–150.

Bouchoule, A. (1999). Technological impacts of dusty plasmas. In *Dusty plasmas: Physics, chemistry and technological impacts in plasma processing*, Bouchoule, A. (ed.), pp. 305–396. Wiley, Chichester.

Candido, L., Rino, J.–P., Studart, N., and Peeters, F. M. (1998). The structure and spectrum of the anisotropically confined two–dimensional Yukawa system. *J. Phys.: Condens. Matter*, **10**, 11627–11644.

Chu, J. H. and I, L. (1994). Direct observation of Coulomb crystals and liquids in strongly coupled r.f. dusty plasmas. *Phys. Rev. Lett.*, **72**, 4009–4012.

Chu, J. H., Du, J. B., and I, L. (1994). Coulomb solids and low–frequency fluctuations in r.f. dusty plasmas. *J. Phys. D: Appl. Phys,*, **27**, 296–300.

Chung, P. M., Talbot, L., and Touryan, K. J. (1975). *Electric probes in stationary and flowing plasmas: Theory and application*. Springer–Verlag, New York.

Cui, C. and Goree, J. (1994). Fluctuations of the charge on a dust grain in a plasma. *IEEE Trans. Plasma Sci.*, **22**, 151–158.

D'Angelo, N. (1995). Coulomb solids and low–frequency fluctuations in r.f. dusty plasmas. *J. Phys. D: Appl. Phys.*, **28**, 1009–1010.

D'Angelo, N. (1997). Ionization instability in dusty plasmas. *Phys. Plasmas*, **4**, 3422–3426.

D'Angelo, N. (1998). Dusty plasma ionization instability with ion drag. *Phys. Plasmas*, **5**, 3155–3160.

D'Angelo, N. and Merlino, R. L. (1996). Current–driven dust–acoustic instability in a collisional plasma. *Planet. Space Sci.*, **44**, 1593–1598.

Daugherty, J. E. and Graves, D. B. (1993). Particulate temperature in radio frequency glow discharges. *J. Vac. Sci. Technol. A*, **11**, 1126–1134.

Daugherty, J. E., Porteous, R. K., Kilgore, M. D., and Graves, D. B. (1992). Sheath structure around particles in low–pressure discharges. *J. Appl. Phys.*, **72**, 3934–3942.

Daugherty, J. E., Porteus, R. K., and Graves, D. B. (1993). Electrostatic forces on small particles in low–pressure discharges. *J. Appl. Phys.*, **73**, 1617–1620.

Draine, B. T. and Salpeter, E. E. (1979). On the physics of dust grains in hot gas. *Astrophys. J.*, **231**, 77–94.

Dubin, D. H. E. (1990). First–order anharmonic correction to the free energy of Coulomb crystal in periodic boundary conditions. *Phys. Rev. A*, **42**, 4972–4982.

Dubin, D. H. E. (1996). Nonlinear Debye shielding in a dusty plasma. In *The physics of dusty plasmas*, Shukla, P. K., Mendis, D. A., and Chow V. W. (eds), pp. 15–21. Word Scientific, Singapore.

Dubin, D. H. E. and O'Neil, T. M. (1999). Trapped nonneutral plasmas, liquids, and crystals (the thermal equilibrium states). *Rev. Mod. Phys.*, **71**, 87–172.

Epstein, P. (1924). On the resistance experienced by spheres in their motion through gases. *Phys. Rev.*, **23**, 710–733.

Farouki, R. T. and Hamaguchi, S. (1993). Thermal energy of crystalline one–component plasma from dynamical simulation. *Phys. Rev. E*, **47**, 4330–4336.

Fortov, V. E., Filinov, V. S., Nefedov, A. P., Petrov, O. F., Samaryan, A. A., and Lipaev, A. M. (1997c). Creation of ordered structures in a classical thermal plasma containing macroparticles: Experiment and computer simulation. *JETP*, **84**, 489–496.

Fortov, V. E., Khrapak, A. G., Khrapak, S. A., Molotkov, V. I., Nefedov, A. P., Petrov, O. F., and Torchinsky, V. M. (2000). Mechanism of dust–acoustic instability in a current glow discharge plasma. *Phys. Plasmas*, **7**, 1374–1380.

Fortov, V. E., Khrapak, A. G., Molotkov, V. I., Nefedov, A. P., Poustylnik, M. Y., Torchinsky, V. M., and Yoshino, K. (2001c). Behavior of rod-like dust particles in striations. In *Proceedings of the XXV international conference on phenomena in ionized gases (ICPIG–2001)*, Nagoya, Japan. Contributed papers, Vol. 3, pp. 35–36.

Fortov, V. E., Molotkov, V. I., Nefedov, A. P., and Petrov, O. F. (1999a). Liquid– and crystallike structures in strongly coupled dusty plasmas. *Phys. Plasmas*, **6**, 1759–1768.

Fortov, V. E., Nefedov, A. P., Molotkov, V. I., Poustylnik, M. Y., and Torchinsky, V. M. (2001b). Dependence of the dust–particle charge on its size in a glow–discharge plasma. *Phys. Rev. Lett.*, **87**, 205002/1–4.

Fortov, V. E., Nefedov, A. P., Petrov, O. F., Samarian, A. A., and Chernyschev,

A. V. (1996c). Emission properties and structural ordering of strongly coupled dust particles in a thermal plasma. *Phys. Lett. A*, **219**, 89–94.

Fortov, V. E., Nefedov, A. P., Petrov, O. F., Samarian, A. A., and Chernyschev, A. V. (1996d). Particle ordered structures in a strongly coupled classical thermal plasma. *Phys. Rev. E*, **54**, R2236–R2239.

Fortov, V. E., Nefedov, A. P., Petrov, O. F., Samarian, A. A., Chernyschev, A. V., and Lipaev, A. M. (1996b). Experimental observation of Coulomb ordered structure in sprays of thermal dusty plasmas. *JETP Lett.*, **63**, 187–192.

Fortov, V. E., Nefedov, A. P., Torchinskii, V. M., Molotkov, V. I., Khrapak, A. G., Petrov, O. F., and Volykhin, K. F. (1996a). Crystallization of a dusty plasma in the positive column of a glow discharge, *JETP Lett.*, **64**, 92–98.

Fortov, V. E., Nefedov, A. P., Torchinsky, V. M., Molotkov, V. I., Petrov, O. F., Samarian, A. A., Lipaev, A. M., and Khrapak, A. G. (1997a). Crystalline structures of strongly coupled dusty plasmas in d.c. glow discharge strata. *Phys. Lett. A*, **229**, 317–322.

Fortov, V. E., Nefedov, A. P., Petrov, O. F., Samarian, A. A., and Chernyschev, A. V. (1997b). Highly nonideal classical thermal plasmas: experimental study of ordered macroparticle structures. *JETP*, **84**, 256–261.

Fortov, V. E., Nefedov, A. P., Vaulina, O. S., Lipaev, A. M., Molotkov, V. I., Samaryan, A. A., Nikitskii, V. P., Ivanov, A. I., Savin, S. F., Kalmykov, A. V., Solov'ev, A. Ya., and Vinogradov, P. V. (1998). Dusty plasma induced by solar radiation under microgravitational conditions: an experiment on board the Mir orbiting space station. *JETP*, **87**, 1087–1097.

Fortov, V. E., Nefedov, A. P., Vladimirov, V. I., Deputatova, L. V., Budnik, A. P., Khudyakov, A. V., and Rykov, V. A. (2001a). Dust grain charging in the nuclear–induced plasma. *Phys. Lett. A*, **284**, 118–123.

Fortov, V. E., Nefedov, A. P., Vladimirov, V. I., Deputatova, L. V., Molotkov, V. I., Rykov, V. A., and Khudyakov, A. V. (1999b). Dust particles in a nuclear–induced plasma. *Phys. Lett. A*, **258**, 305–311.

Fortov, V. E., Petrov, O. F., Molotkov, V. I., Poustylnik, M. Y., Torchinsky, V. M., Khrapak, A. G., and Chernyshev, A. V. (2004). Large–amplitude dust waves excited by the gas–dynamic impact in a d.c. glow discharge plasma. *Phys. Rev. E*, **69**, 016402/1–5.

Fortov, V. E., Vaulina, O. S., Petrov, O. F., Molotkov, V. I., Chernyshev, A. V., Lipaev, A. M., Morfill, G., Thomas, H., Rotermell, H., Khrapak, S. A., Semenov, Y. P., Ivanov, A. I., Krikalev, S. K., and Gidzenko, Y. P. (2003b). Dynamics of macroparticles in a dusty plasma under microgravity conditions (First experiments on board the ISS). *JETP*, **96**, 704–718.

Fortov, V. E., Vaulina, O. S., Petrov, O. F., Molotkov, V. I., Lipaev, A. M., Torchinsky, V. M., Thomas, H. M., Morfill, G. E., Khrapak, S. A., Semenov, Y. P., Ivanov, A. I., Krikalev, S. K., Kalery, A. Y., Zaletin, S. V., and Gidzenko, Y. P. (2003a). Transport of microparticles in weakly ionized gas–discharge plasmas under microgravity conditions. *Phys. Rev. Lett.*, **90**, 245005/1–4.

Fortov, V. E., Vladimirov, V.I., Deputatova, L. V., Molotkov, V. I., Nefedov, A.

P., Rykov, V. A., Torchinskii, V. M. and Khudyakov, A. V. (1999c). Ordered dusty structures in plasma produced by nuclear particles. *Dokl. Phys.*, **44**, 279–282.

Fortov, V. E., Vladimirov, V. I., Deputatova, L. V., Nefedov, A. P., Rykov, V. A., and Khudyakov, A. V. (2002). Removal of dust particles from technological plants, *Dokl. Phys.*, **47**, 367–369.

Fried, B. D. and Conte, S. D. (1961). *The plasma dispersion function; the Hilbert transform of the Gaussian.* Academic Press, New York.

Gilbert, S. L., Bollinger, J. J., and Wineland, D. J. (1988). Shell–structure phase of magnetically confined strongly coupled plasmas. *Phys. Rev. Lett.*, **60**, 2022–2025.

Ginzburg, V. L. (1996). Radiation by uniformly moving sources (Vavilov–Cherenkov effect, transition radiation, and other phenomena). *Phys. Usp.*, **39**, 973–982.

Goertz, C. K. (1989). Dusty plasmas in the solar system. *Rev. Geophys.*, **27**, 271–292.

Golubovskii, Y. B., and Nisimov, S. U. (1995). Two–dimensional character of striations in a low–pressure discharge in inert gases. II. *Tech. Phys.*, **40**, 24–28.

Golubovskii, Y. B., and Nisimov, S. U. (1996). Ionization transport waves in a neon discharge. *Tech. Phys.*, **41**, 645–651.

Golubovskii, Y. B., Nisimov, S. U., and Suleimenov, I. E. (1994). Instability–triggered phase transition to a dusty–plasma condensate. *Tech. Phys.*, **39**, 1005–1008.

Goree, J. (1992). Ion trapping by a charged dust grain in a plasma. *Phys. Rev. Lett.*, **69**, 277–280.

Goree, J. (1994). Charging of particles in a plasma. *Plasma Sources Sci. Technol.*, **3**, 400–406.

Goree J., Morfill, G. E., Tsytovich, V. N., and Vladimirov, S. V. (1999). Theory of dust voids in plasmas. *Phys. Rev. E*, **59**, 7055–7067.

Grier, D. G. and Murray C. A. (1994). The microscopic dynamics of freezing in supercooled colloidal fluids. *J. Chem. Phys.*, **100**, 9088–9095.

Hahn, H.–S., Mason, E. A., and Smith, F. J. (1971). Quantum transport cross–sections for ionized gases. *Phys. Fluids*, **14**, 278–287.

Hamaguchi, S. (1997). Electrostatic forces of plasma dust grains with position–dependent charge. *Comments Plasma Phys. Control. Fusion*, **18**, 95–102.

Hamaguchi, S. and Farouki, R. T. (1994a). Polarization force on a charged particulate in a nonuniform plasma. *Phys Rev. E*, **49**, 4430–4441.

Hamaguchi, S. and Farouki, R. T. (1994b). Plasma–particulate interactions in nonuniform plasmas with finite flows. *Phys. Plasmas*, **1**, 2110–2118.

Hamaguchi, S., Farouki, R. T., and Dubin, D. H. E. (1997). Triple point of Yukawa systems, *Phys. Rev. E*, **56**, 4671–4682.

Hansen, J. P. and Verlet, L. (1969). Phase transitions of the Lennard–Jones system. *Phys. Rev.*, **184**, 151–161.

Havnes, O., Aslaksen, T., Hartquist, T. W., Li, F., Melandsø, F., Morfill, G.

E., and Nitter, T. (1995). Probing the properties of planetary ring dust by the observation of Mach cones. *J. Geophys. Res.*, **100**, 1731–1734.

Havnes, O., Goertz, C. K., Morfill, G. E., Grün, E., and Ip, W. (1987). Dust charges, cloud potential, and instabilities in a dust cloud embedded in a plasma. *J. Geophys. Res. A*, **92**, 2281–2293.

Havnes, O., Nitter, T., Tsytovich, V., Morfill, G. E., and Hartquist, T. (1994). On the thermophoretic force close to walls in dusty plasma experiments. *Plasma Sources Sci. Technol.*, **3**, 448–457.

Hayashi, Y. (1999). Structure of a three–dimensional Coulomb crystal in a fine–particle plasma. *Phys. Rev. Lett.*, **83**, 4764–4767.

Hayashi, Y. and Tachibana, S. (1994). Observation of coulomb–crystal formation from carbon particles grown in a methane plasma. *Jpn. J. Appl. Phys.*, **33**, L804–L806.

Homann, A., Melzer, A., Peters, S., Madani, R., and Piel, A. (1998). Laser–excited dust lattice waves in plasma crystals. *Phys. Lett. A*, **242**, 173–180.

Homann, A., Melzer, A., Peters, S., and Piel, A. (1997). Determination of the dust screening length by laser–excited lattice waves. *Phys. Rev. E*, **56**, 7138–7141.

Homann, A., Melzer, A., and Piel, A. (1999). Measuring the charge on single particles by laser–excited resonances in plasma crystals. *Phys. Rev. E*, **59**, R3835–R3838.

Hou, L.–J., Wang, Y.–N., and Miskovic, Z. L. (2001). Interaction potential among dust grains in a plasma with ion flow. *Phys. Rev. E*, **64**, 046406/1–7.

Hou, L.–J., Wang, Y.–N., and Miskovic, Z. L. (2003). Induced potential of a dust particle in a collisional radio–frequency sheath. *Phys. Rev. E*, **68**, 016410/1–7.

Ichimaru, S. (1982). Strongly coupled plasmas: High–density classical plasmas and degenerate electron fluids. *Rev. Mod. Phys.*, **54**, 1017–1059.

Ichimaru, S. (1992). *Statistical plasma physics Vol. 1 Basic principles*. Addison–Wesley, Redwood City.

Ignatov, A. M. (1996). Lesage gravity in dusty plasmas. *Plasma Phys. Rep.*, **22**, 585–589.

Ikezi, H. (1986). Coulomb solid of small particles in plasmas. *Phys. Fluids*, **29**, 1764–1766.

Ishihara, O. and Vladimirov, S. V. (1997). Wake potential of a dust grain in a plasma with ion flow. *Phys. Plasmas*, **4**, 69–74.

Ishihara, O., Vladimirov, S. V., and Cramer, N. F. (2000). Effect of dipole moment on the wake potential of a dust grain. *Phys. Rev. E*, **61**, 7246–7248.

Ishihara, O., Kamimura, T., Hirose, K. I., and Sato, N. (2002). Rotation of a two–dimensional Coulomb cluster in a magnetic field. *Phys. Rev. E*, **66**, 046406/1–6.

Ivlev, A. V. and Morfill, G. (2000). Acoustic modes in a collisional dusty plasma: Effect of the charge variation. *Phys. Plasmas*, **7**, 1094–1102.

Ivlev, A. V. and Morfill, G. (2001). Anisotropic dust lattice modes. *Phys. Rev. E*, **63**, 016409/1–3.

Ivlev, A. V., Khrapak, A. G., Khrapak, S. A., Annaratone, B. M., Morfill, G., and Yoshino, K. (2003a). Rodlike particles in gas discharge plasmas: Theoretical model, *Phys. Rev. E*, **68**, 026403/1–10.

Ivlev, A. V., Khrapak, S. A., Zhdanov, S. K., Morfill G. E., Joyce, G. (2004). Force on a charged test particle in a collisional flowing plasma. *Phys. Rev. Lett.*, **92**, 205007/1–4.

Ivlev, A. V., Konopka, U., and Morfill, G. (2000a). Influence of charge variations on particle oscillations in the plasma sheath. *Phys. Rev. E*, **62**, 2739–2744.

Ivlev, A. V., Kretschmer, M., Zuzic, M., Morfill, G. E., Rothermel, H., Thomas, H. M., Fortov, V. E., Molotkov, V. I., Nefedov, A. P., Lipaev, A. M., Petrov, O. F., Baturin, Y. M., Ivanov, A. I., and Goree, J. (2003b). Decharging of complex plasmas: First kinetic observations. *Phys. Rev. Lett.*, **90**, 055003/1–4.

Ivlev, A. V., Samsonov, D., Goree, J., Morfill, G., and Fortov, V. E. (1999). Acoustic modes in a collisional dusty plasma. *Phys. Plasmas*, **6**, 741–750.

Ivlev, A. V., Sutterlin, R., Steinberg, V., Zuzic, M., and Morfill, G. (2000b). Nonlinear vertical oscillations of a particle in a sheath of r.f. discharge. *Phys. Rev. Lett.*, **85**, 4060–4063.

Ivlev, A. V., Zhdanov, S. K., Khrapak, S. A., Morfill, G. E. (2005). Kinetic approach for the ion drag force in a collisional plasma. *Phys. Rev. E*, **71**, 016405/1–7.

Jana, M. R., Sen, A., and Kaw, P. K. (1993). Collective effects due to charge–fluctuations dynamics in a dusty plasma. *Phys. Rev. E*, **48**, 3930–3933.

Jellum, G. M., Daugherty, J. E., and Graves, D. B. (1991). Particle thermophoresis in low pressure glow discharges. *J. Appl. Phys.*, **69**, 6923–6934.

Joyce, G., Lampe, M., and Ganguli, G. (2002). Instability–triggered phase transition to a dusty–plasma condensate. *Phys. Rev. Lett.*, **88**, 095006/1–4.

Juan, W.-T., Huang, Z.-H., Hsu, J.-W., Lai, Y.-J., and I, L. (1998). Observation of dust Coulomb clusters in a plasma trap. *Phys. Rev. E*, **58**, R6947–R6950.

Kalman, G. and Golden, K. I. (1990). Response function and plasmon dispersion for strongly coupled Coulomb liquids. *Phys. Rev. A*, **41**, 5516–5527.

Kalman, G., Rosenberg, M., and DeWitt, H. E. (2000). Collective modes in strongly correlated Yukawa liquids: Waves in dusty plasmas. *Phys. Rev. Lett.*, **84**, 6030–6033.

Kaw, P. K. (2001). Collective modes in a strongly coupled dusty plasma. *Phys. Plasmas*, **8**, 1870–1878.

Kaw, P. and Singh, R. (1997). Collisional instabilities in a dusty plasma with recombination and ion–drift effects. *Phys. Rev. Lett.*, **79**, 423–426.

Kaw, P. K. and Sen, A. (1998). Low frequency modes in strongly coupled dusty plasmas. *Phys. Plasmas*, **5**, 3552–3559.

Kaw, P. K., Nishikawa, K., and Sato, N. (2002). Rotation in collisional strongly coupled dusty plasmas in a magnetic field. *Phys. Plasmas*, **9**, 387–390.

Kersten, H., Deutsch, H., Stoffels, E., Stoffels, W. W., Kroesen, G. M. W., and Hippler, R. (2001). Micro–disperse particles in plasmas: From disturbing side effects to new applications. *Contrib. Plasma Phys.*, **41**, 598–609.

Khodataev, Y. K., Khrapak, S. A., Nefedov, A. P., and Petrov, O. F. (1998). Dynamics of the ordered structure formation in a thermal dusty plasma. *Phys. Rev. E*, **57**, 7086–7092.

Khrapak, S. A. and Morfill, G. (2001). Waves in two component electron–dust plasma. *Phys. Plasmas*, **8**, 2629–2634.

Khrapak, S. A. and Morfill, G. E. (2002). Dust diffusion across a magnetic field due to random charge fluctuations. *Phys. Plasmas*, **9**, 619–623.

Khrapak, S. A. and Morfill, G. E. (2004). Dusty plasmas in a constant electric field: Role of the electron drag force. *Phys. Rev. E*, **69**, 066411/1–5.

Khrapak, S. A. and Yaroshenko, V. V. (2003). Low–frequency waves in collisional complex plasmas with an ion drift. *Phys. Plasmas*, **10**, 4616–4621.

Khrapak, S. A., Ivlev, A. V., and Morfill, G. (2001). Interaction potential of microparticles in a plasma: Role of collisions with plasma particles. *Phys. Rev. E*, **64**, 046403/1–7.

Khrapak, S. A., Ivlev, A. V., Morfill, G. E., and Thomas, H. M. (2002). Ion drag force in complex plasmas. *Phys. Rev. E*, **66**, 046414/1–4.

Khrapak, S., Samsonov, D., Morfill, G., Thomas, H., Yaroshenko, V., Rothermel, H., Hagl, T., Fortov, V., Nefedov, A., Molotkov, V., Petrov, O., Lipaev, A., Ivanov, A., and Baturin Y. (2003c). Compressional waves in complex (dusty) plasmas under microgravity conditions. *Phys. Plasmas*, **10**, 1–4.

Khrapak, S. A., Ivlev, A. V., Morfill, G. E., Thomas, H. M., Zhdanov, S. K., Konopka, U., Thoma, M. H., and Quinn, R. A. (2003b). Comment on "Measurement of the ion drag force on falling dust particles and its relation to the void formation in complex (dusty) plasmas" [*Phys. Plasmas* **10**, 1278 (2003)]. *Phys. Plasmas*, **10**, 4579–4581.

Khrapak, S. A., Ivlev, A. V., Morfill, G. E., and Zhdanov, S. K. (2003a). Scattering in the attractive Yukawa potential in the limit of strong interaction. *Phys. Rev. Lett.*, **90**, 225002/1–4.

Khrapak, S. A., Ivlev, A. V., Morfill, G. E., Zhdanov, S. K., and Thomas, H. M. (2004). Scattering in the attractive Yukawa potential: Application to the ion–drag force in complex plasmas. *IEEE Trans. Plasma Sci.*, **32**, 555–560.

Khrapak, S. A., Nefedov, A. P., Petrov, O. F., and Vaulina, O. S. (1999). Dynamical properties of random charge fluctuations in a dusty plasma with different charging mechanisms. *Phys. Rev. E*, **59**, 6017–6022, and Erratum. *Phys. Rev. E*, **60**, 3450.

Kilgore, M. D., Daugherty, J. E., Porteous, R. K., and Graves, D. B. (1993). Ion drag on an isolated particulate in low–pressure discharge. *J. Appl. Phys.*, **73**, 7195–7202.

Kilgore, M. D., Daugherty, J. E., Porteous, R. K., and Graves, D. B. (1994). Transport and heating of small particles in high density plasma sources. *J. Vac. Sci. Technol. B*, **12**, 486–493.

Klindworth, M., Melzer, A., and Piel, A. (2000). Laser–excited intershell rotation of finite Coulomb clusters in a dusty plasma. *Phys. Rev. B*, **61**, 8404–8415.

Konopka, U., Ratke, L., and Thomas, H. M. (1997). Central collisions of charged dust particles in a plasma. *Phys. Rev. Lett.*, **79**, 1269–1272.

Konopka, U., Morfill, G. E., and Ratke, L. (2000a). Measurement of the interaction potential of microspheres in the sheath of a r.f. discharge. *Phys. Rev. Lett.*, **84**, 891–894.

Konopka, U., Samsonov, D., Ivlev, A. V., Goree, J., Steinberg, V., and Morfill, G. E. (2000b). Rigid and differential plasma crystal rotation induced by magnetic fields. *Phys. Rev. E*, **61**, 1890–1898.

Kremer, K., Robbins, M. O., and Grest, G. S. (1986). Phase diagram of Yukawa systems: Model for charge–stabilized colloids. *Phys. Rev. Lett.*, **57**, 2694–2697.

Kremer, K. Grest, G. S., and Robbins, M. O. (1987). Dynamics of supercooled liquids interacting with a repulsive Yukawa potential. *J. Phys. A: Math. Gen.*, **20**, L181–L187.

Lai, Y.–J., I, L. (1999). Packings and defects of strongly coupled two–dimensional Coulomb clusters: Numerical simulation. *Phys. Rev. E*, **60**, 4743–4753.

Lampe, M., Joyce, G., Ganguli, G., and Gavrishchaka, V. (2000). Interactions between dust grains in a dusty plasma. *Phys. Plasmas*, **7**, 3851–3861.

Lampe, M., Gavrishchaka, V., Ganguli, G., and Joyce, G. (2001a). Effect of trapped ions on shielding of a charged spherical object in a plasma. *Phys. Rev. Lett.*, **86**, 5278–5281.

Lampe, M., Joyce, G., and Ganguli, G. (2001b). Analytic and simulation studies of dust grain interaction and structuring. *Phys. Scripta*, **T89**, 106–111.

Landau, L. D. and Lifshitz, E. M. (1976). *Mechanics.* Pergamon, Oxford.

Landau, L. D. and Lifshitz, E. M. (1987). *Fluid mechanics.* Pergamon Press, Oxford.

Langmuir, I., Found, G., and Dittmer, A. F. (1924). A new type of electric discharge: The streamer discharge. *Science*, **60**, 392–394.

Lapenta, G. (1999). Simulation of charging and shielding of dust particles in drifting plasmas. *Phys. Plasmas*, **6**, 1442–1447.

Lapenta, G. (2000). Linear theory of plasma wakes. *Phys. Rev. E*, **62**, 1175–1181.

Lapenta, G. (2002). Nature of the force field in plasma wakes. *Phys. Rev. E*, **66**, 026409/1–6.

Law, D. A., Steel, W. H., Annaratone, B. M., and Allen, J. E. (1998). Probe–induced particle circulation in a plasma crystal. *Phys. Rev. Lett.*, **80**, 4189–4192.

Lemons, D. S., Murillo, M. S., Daughton, W., and Winske, D. (2000). Two–dimensional wake potentials in sub– and supersonic dusty plasmas. *Phys. Plasmas*, **7**, 2306–2313.

Liboff, R. L. (1959). Transport coefficients determined using the shielded Coulomb potential. *Phys. Fluids*, **2**, 40–46.

Lifshitz, E. M. and Pitaevskii, L. P. (1981). *Physical kinetics.* Pergamon Press, Oxford.

Lindemann, F. A. (1910). The calculation of molecular vibration frequencies. *Z. Phys.*, **11**, 609.

Lipaev, A. M., Molotkov, V. I., Nefedov, A. P., Petrov, O. F., Torchinskii, V. M., Fortov, V. E., Khrapak, A. G., and Khrapak, S. A. (1997). Ordered structures in a nonideal dusty glow–discharge plasma. *JETP*, **85**, 1110–1118.

Löwen, H. (1996). Dynamical criterion for two–dimension freezing. *Phys. Rev. E*, **53**, R29–R32.

Löwen, H., Palberg, T., and Simon, R. (1993). Dynamical criterion for freezing of colloidal liquids. *Phys. Rev. Lett.*, **70**, 1557–1560.

Maiorov, S. A., Vladimirov, S. V., and Cramer, N. F. (2001). Plasma kinetics around a dust grain in an ion flow. *Phys. Rev. E*, **63**, 017401/1–4.

Mamun, A. A. and Shukla, P. K. (2000). Streaming instabilities in a collisional dusty plasma. *Phys. Plasmas*, **7**, 4412–4417.

Matsoukas, T. and Russell, M. (1995). Particle charging in low–pressure plasmas. *J. Appl. Phys.*, **77**, 4285–4292.

Matsoukas, T., Russel, M., and Smith, M. (1996). Stochastic charge fluctuations in dusty plasmas. *J. Vac. Sci. Technol.*, **14**, 624–630.

Matsoukas, T. and Russell, M. (1997). Fokker–Planck description of particle charging in ionized gases. *Phys. Rev. E*, **55**, 991–994.

Meijer, E. J. and Frenkel, D. (1991). Melting line of Yukawa system by computer simulation. *J. Chem. Phys.*, **94**, 2269–2271.

Melandsø, F. (1996). Lattice waves in dust plasma crystals. *Phys. Plasmas*, **3**, 3890–3901.

Melandsø, F. and Goree, J. (1995). Polarized supersonic plasma flow simulation for charged bodies such as dust particles and spacecraft. *Phys. Rev. E*, **52**, 5312–5326.

Melandsø, F., Aslaksen, T., and Havnes, O. (1993). A new damping effect for the dust–acoustic wave. *Planet. Space Sci.*, **41**, 321–325.

Melzer, A., Trottenberg, T., and Piel, A. (1994). Experimental determination of the charge on dust particles forming Coulomb lattices. *Phys. Lett. A*, **191**, 301–307.

Melzer, A., Homann, A., and Piel, A. (1996). Experimental investigation of the melting transition of the plasma crystal. *Phys. Rev. E*, **53**, 2757–2766.

Melzer, A. , Schweigert, V. A., and Piel, A. (1999). Transition from attractive to repulsive forces between dust molecules in a plasma sheath. *Phys. Rev. Lett.*, **83**, 3194–3197.

Melzer, A., Nunomura, S., Samsonov, D., Ma, Z. W., and Goree, J. (2000). Laser–excited Mach cones in a dusty plasma crystal. *Phys. Rev. E*, **62**, 4162–4168.

Melzer, A., Klindworth, M., and Piel, A. (2001). Normal modes of 2D finite clusters in complex plasmas. *Phys. Rev. Lett.*, **87**, 115002/1–4.

Mendis, D. A. (2002). Progress in the study of dusty plasmas. *Plasma Sources Sci. Technol.*, **11**, A219–A228.

Merlino, R. L., Barkan, A., Thompson, C., and D'Angelo, N. (1998). Laboratory studies of waves and instabilities in dusty plasmas. *Phys. Plasmas*, **5**, 1607–1614.

Misawa, T., Ohno, N., Asano, K., Sawai, M., Takamura, S., and Kaw, P. K. (2001). Experimental observation of vertically polarized transverse dust–lattice wave propagating in a one–dimensional strongly coupled dust chain. *Phys. Rev. Lett.*, **86**, 1219–1222.

Mohideen, U., Rahman, H. U., Smith, M. A., Rosenberg, M., and Mendis, D. A. (1998). Intergrain coupling in dusty–plasma coulomb crystals. *Phys. Rev. Lett.*, **81**, 349–352.

Molotkov, V. I., Nefedov, A. P., Pustylnik, M. Yu., Torchinsky, V. M., Fortov, V. E., Khrapak, A. G., and Yoshino, K. (2000). Liquid plasma crystal: Coulomb crystallization of cylindrical macroscopic grains in a gas–discharge plasma. *JETP Lett.*, **71**, 102–105.

Molotkov, V. I., Nefedov, A. P., Torchinskii, V. M., Fortov, V. E., and Khrapak, A. G. (1999). Dust acoustic waves in a d.c. glow–discharge plasma. *JETP*, **89**, 477–480.

Morfill, G. E. and Grün, E. (1979). The motion of charged dust particles in interplanetary space. I – The zodiacal dust cloud. II – Interstellar grains. *Planet. Space Sci.*, **27**, 1269–1292.

Morfill, G. E. and Thomas, H. (1996). Plasma crystal. *J. Vac. Sci. Technol. A*, **14**, 490–495.

Morfill, G., Ivlev, A. V., and Jokipii, J. R. (1999a). Charge fluctuation instability of the dust lattice wave. *Phys. Rev. Lett.*, **83**, 971–974.

Morfill, G. E., Thomas, H. M., Konopka, U., Rothermel, H., Zuzic, M., Ivlev, A., and Goree, J. (1999b). Condensed plasmas under microgravity. *Phys. Rev. Lett.*, **83**, 1598–1602.

Morfill, G. E., Thomas, H. M., Konopka, U., and Zuzic, M. (1999c). The plasma condensation: Liquid and crystalline plasmas. *Phys. Plasmas*, **6**, 1769–1780.

Morfill, G. E., Tsytovich, V. N., and Thomas H. (2003). Complex plasmas: II. Elementary processes in complex plasmas. *Plasma Phys. Rep.*, **29**, 1–30.

Murillo, M. S. (1998). Static local field correction description of acoustic waves in strongly coupling dusty plasmas. *Phys. Plasmas*, **5**, 3116–3121.

Murillo, M. S. (2000a). Longitudinal collective modes of strongly coupled dusty plasmas at finite frequencies and wavevectors. *Phys. Plasmas*, **7**, 33–38.

Murillo, M. S. (2000b). Critical wavevectors for transverse modes in strongly coupled dusty plasmas. *Phys. Rev. Lett.*, **85**, 2514–2517.

Nambu, M., Vladimirov, S. V., and Shukla, P. K. (1995). Attractive forces between charged particulates in plasmas. *Phys. Lett. A*, **203**, 40–42.

Nefedov, A. P., Petrov, O. F., and Fortov, V. E. (1997). Quasicrystalline structures in strongly coupled dusty plasma. *Phys. Usp.*, **40**, 1163–1173.

Nefedov, A. P., Petrov, O. F., and Khrapak, S. A. (1998). Potential of electrostatic interaction in a thermal dusty plasma. *Plasma Phys. Rep.*, **24**, 1037–1040.

Nefedov, A. P., Petrov, O. F., Khodataev, Ya. K., and Khrapak, S. A. (1999). Dynamics of formation of ordered structures in a thermal plasma with macroparticles. *JETP*, **88**, 460–464.

Nefedov, A. P., Petrov, O. F., Molotkov, V. I., and Fortov, V. E. (2000). Formation of liquidlike and crystalline structures in dusty plasmas. *JETP Lett.*, **72**, 218–226.

Nefedov, A. P., Vaulina, O. S., Petrov, O. F., Molotkov, V. I., Torchinskii, V. M., Fortov, V. E., Chernyshev, A. V., Lipaev, A. M., Ivanov, A. I., Kaleri, A. Y., Semenov, Y. P., and Zaletin, S. V. (2002). The dynamics of macroparticles in a direct current glow discharge plasma under microgravitation conditions. *JETP*, **95**, 673–681.

Nefedov, A. P., Morfill, G. E., Fortov, V. E., Thomas, H. M., Rotherme, H., Hag, T., Ivlev, A. V., Zuzic, M., Klumov, B. A., Lipaev, A. M., Molotkov, V. I., Petrov, O. F., Gidzenko, Y. P., Krikalev, S. K., Shepherd, W., Ivanov, A. I., Roth, M., Binnenbruck, H., Goree, J. A., and Semenov, Y. P. (2003). PKE–Nefedov: Plasma crystal experiments on the International Space Station. *New J. Phys.*, **5**, 33.1–33.10.

Nitter, T. (1996). Levitation of dust in r.f. and d.c. glow discharges. *Plasma Sources Sci. Technol.*, **5**, 93–99.

Northrop, T. G. (1992). Dusty plasmas. *Phys. Scr.*, **45**, 475–490.

Northrop, T. G. and Birmingham, T. J. (1990). Plasma drag on a dust grain due to Coulomb collisions. *Planet. Space Sci.*, **38**, 319–326.

Nunomura, S., Misawa, T., Ohno, N., and Takamura, S. (1999). Instability of dust particles in a Coulomb crystal due to delayed charging. *Phys. Rev. Lett.*, **83**, 1970–1973.

Nunomura, S., Samsonov, D., and Goree, J. (2000). Transverse waves in a two–dimensional screened–Coulomb crystal (dusty plasma). *Phys. Rev. Lett.*, **84**, 5141–5144.

Nunomura, S., Goree, J., Hu, S., Wang, X., and Bhattacharjee, A. (2002a). Dispersion relations of longitudinal and transverse waves in two–dimensional screened Coulomb crystals. *Phys. Rev. E*, **65**, 066402/1–11.

Nunomura, S., Goree, J., Hu, S., Wang, X., Bhattacharjee, A., and Avinash, K. (2002b). Phonon spectrum in a plasma crystal. *Phys. Rev. Lett.*, **89**, 035001/1–4.

Ohta, H. and Hamaguchi, S. (2000a). Molecular dynamics evaluation of self–diffusion in Yukawa systems. *Phys. Plasmas*, **7**, 4506–4514.

Ohta, H. and Hamaguchi, S. (2000b). Wave dispersion relations in Yukawa fluids. *Phys. Rev. Lett.*, **84**, 6026–6029.

Ostrikov, K. N., Yu, M. Y., and Stenflo, L. (2000a). Surface waves in strongly irradiated dusty plasmas. *Phys. Rev. E*, **61**, 4314–4321.

Ostrikov, K. N., Vladimirov, S. V., Yu, M. Y., and Morfill, G. E. (2000b). Low–frequency dispersion properties of plasmas with variable–charge impurities. *Phys. Plasmas*, **7**, 461–465.

Pal', A. F., Starostin, A. N., and Filippov, A. V. (2001a). Charging of dust grains in a nuclear–induced plasma at high pressures, *Plasma Phys. Rep.*, **27**, 143–152.

Pal', A. F., Serov, A. O., Starostin, A. N., Filippov, A. V., and Fortov, V. E. (2001b). Non–self–sustained discharge in nitrogen with a condensed dispersed phase, *JETP*, **92**, 235–245.

Pal', A. F., Sivokhin, D. V., Starostin, A. N., Filippov, A. V., and Fortov, V. E. (2002). Potential of a dust grain in a nitrogen plasma with a condensed disperse phase at room and cryogenic temperatures. *Plasma Phys. Rep.*, **28**, 28–39.

Paul, W. and Raether, M. (1955). Das Elektrische Massenfilter. *Z. Phys.*, **140**, 262–273.

Perrin, J. and Hollenstein, Ch. (1999). Sources and growth of particles. In *Dusty plasmas: Physics, chemistry, and technological impacts in plasma processing*, A. Bouchoule (ed.), pp. 77–180. Wiley, Chichester UK.

Peters, S., Homann, A., Melzer, A., and Piel, A. (1996). Measurement of dust particle shielding in a plasma from oscillations of a linear chain. *Phys. Lett. A*, **223**, 389–393.

Piel, A. and Melzer, A. (2002). Dynamical processes in complex plasmas. *Plasma Phys. Control. Fusion*, **44**, R1–R26.

Piel, A., Nosenko, V., and Goree, J. (2002). Experiments and molecular–dynamics simulation of elastic waves in a plasma crystal radiated from a small dipole source. *Phys. Rev. Lett.*, **89**, 085004/1–4.

Piel, A., Homann, A., Klindworth, M., Melzer, A., Zafiu, C., Nosenko, V., and Goree, J. (2003). Waves and oscillations in plasma crystals. *J. Phys. B: At. Mol. Opt. Phys.*, **36**, 533–543.

Pieper, J. B. and Goree, J. (1996). Dispersion of plasma dust acoustic waves in the strong–coupling regime. *Phys. Rev. Lett.*, **77**, 3137–3140.

Pilipp, W., Hartquist, T. W., Havnes, O., and Morfill, G. E. (1987). The effect of dust on the propagation and dissipation of Alfven waves in interstellar clouds. *The Astronomical J.*, **314**, 341–351.

Praburam, G. and Goree, J. (1996). Experimental observation of very low–frequency macroscopic modes in a dusty plasma. *Phys. Plasmas*, **3**, 1212–1219.

Quinn, R. A. and Goree, J. (2000). Single–particle Langevin model of particle temperature in dusty plasmas. *Phys. Rev. E*, **61**, 3033–3041.

Quinn, R. A., Cui, C., Goree, J., Pieper, J. B., Thomas, H., and Morfill, G. E. (1996). Structural analysis of a Coulomb lattice in a dusty plasma. *Phys. Rev. E*, **53**, 2049–2052.

Raizer, Y. P. (1991). *Gas discharge physics*. Springer–Verlag, Berlin.

Rao, N. N., Shukla, P. K., and Yu, M. Y. (1990). Dust–acoustic waves in dusty plasmas. *Planet. Space Sci.*, **38**, 543–546.

Ratynskaia, S., Khrapak, S., Zobnin, A., Thoma, M. H., Kretschmer, M., Usachev, A., Yaroshenko, V., Quinn, R. A., Morfill, G. E., Petrov, O., Fortov, V. (2004). Experimental determination of dust–particle charge in a discharge plasma at elevated pressures. *Phys. Rev. Lett.*, **93**, 085001/1–4.

Robbins, M. O., Kremer, K., and Grest, G. S. (1988). Phase diagram and dynamics of Yukawa systems. *J. Chem. Phys.*, **88**, 3286–3312.

Robinson, P. A. and Coakley, P. (1992). Spacecraft charging–progress in the study of dielectrics and plasmas. *IEEE Trans. Electr. Insul.*, **27**, 944–960.

Rosenberg, M. (1993). Ion– and dust–acoustic instabilities in dusty plasmas. *Planet. Space Sci.*, **41**, 229–233.

Rosenberg, M. (1996). Ion–dust streaming instability in processing plasmas. *J. Vac. Sci. Technol. A*, **14**, 631–633.

Rosenberg M. (2002). A note on ion–dust streaming instability in a collisional dusty plasma. *J. Plasma Phys.*, **67**, 235–242.

Rosenberg, M. and Mendis, D. A. (1995). UV–induced Coulomb crystallization in a dusty gas. *IEEE Trans. Plasma Sci.*, **23**, 177–179.

Rosenberg M. and Kalman, G. (1997). Dust acoustic waves in strongly coupled dusty plasmas. *Phys. Rev. E*, **56**, 7166–7173.

Rosenberg, M., Mendis, D. A., and Sheehan, D. P. (1996). UV–induced Coulomb crystallization of dust grains in high–pressure gas. *IEEE Trans. Plasma Sci.*, **24**, 1422–1430.

Rosenberg, M., Mendis, D. A., and Sheehan, D. P. (1999). Positively charged dust crystals induced by radiative heating. *IEEE Trans. Plasma Sci.*, **27**, 239–242.

Rosenberg, R. O. and Thirumalai, D. (1986). Structure and dynamics of screened–Coulomb colloidal liquids. *Phys. Rev. A*, **33**, 4473–4476.

Rothermel, H., Hagl, T., Morfill, G. E., Thoma, M. H., and Thomas, H. M. (2002). Gravity compensation in complex plasmas by application of a temperature gradient. *Phys. Rev. Lett.*, **89**, 175001/1–4.

Samarian, A. A. and Vladimirov, S. V. (2002). Comment on "Dependence of the dust–particle charge on its size in a glow–discharge plasma". *Phys. Rev. Lett.*, **89**, 229501.

Samaryan, A. A., Chernyshev, A. V., Nefedov, A. P., Petrov, O. F., Mikhailov, Y. M., Mintsev, V. B., and Fortov, V. E. (2000). Structures of the particles of the condensed dispersed phase in solid fuel combustion products plasma. *JETP*, **90**, 817–822.

Samarian, A. A., James, B. W., Vladimirov, S. V., and Cramer, N. F. (2001). Self–excited vertical oscillations in an r.f.–discharge dusty plasma. *Phys. Rev. E*, **64**, 025402(R)/1–4.

Samsonov, D. and Goree, J. (1999). Instabilities in a dusty plasma with ion drag and ionization. *Phys. Rev. E*, **59**, 1047–1058.

Samsonov, D., Goree, J., Ma, Z. W., Bhattacharjee, A., Thomas, H. M., and Morfill, G. E. (1999). Mach cones in a Coulomb lattice and a dusty plasma. *Phys. Rev. Lett.*, **83**, 3649–3652.

Samsonov, D., Goree, J., Thomas, H. M., and Morfill, G. E. (2000). Mach cone shocks in a two–dimensional Yukawa solid using a complex plasma. *Phys. Rev. E*, **61**, 5557–72.

Samsonov, D., Ivlev, A. V., Morfill, G. E., and Goree, J. (2001). Long–range attractive and repulsive forces in a two–dimensional complex (dusty) plasma. *Phys. Rev. E*, **63**, 025401(R)/1–4.

Samsonov, D., Ivlev, A. V., Quinn, R. A., Morfill, G., and Zhdanov, S. (2002). Dissipative longitudinal solitons in a two–dimensional strongly coupled complex (dusty) plasma. *Phys. Rev. Lett.*, **88**, 095004/1–4.

Samsonov, D., Morfill, G., Thomas, H., Hagl, T., Rothermel, H., Fortov, V., Lipaev, A., Molotkov, V., Nefedov, A., Petrov, O., Ivanov, A., and Krikalev, S. (2003). Kinetic measurements of shock wave propagation in a three–dimensional complex (dusty) plasma. *Phys. Rev. E*, **67**, 036404/1–5.

Schweigert, V. A., Schweigert, I. V., Melzer, A., Homann, A., and Piel, A. (1998). Plasma crystal melting: a nonequilibrium phase transition. *Phys. Rev. Lett.*, **80**, 5345–5348.

Selwyn, G. S. (1996). Electrode engineering: A new technology for control of particle contamination and process uniformity in plasma processing. In *The physics of dusty plasmas*, Shukla, P. K., Mendis, D. A., and Chow V. W. (eds), pp. 177–198. Word Scientific, Singapore.

Selwyn, G. S., Singh, J.,and Bennett, R. S. (1989). In situ laser diagnostic studies of plasma–generated particulate contamination, *J. Vac. Sci. Technol. A*, **7**, 2758–2765.

Selwyn, G. S., Haller, K. L., and Patterson, E. F. (1993). Trapping and behavior of particulates in a radio frequency magnetron plasma etching tool. *J. Vac. Sci. Technol. A*, **11**, 1132–1135.

Shikin, V. B. (1989). Wigner crystallization on the surface liquid helium. *Sov. Phys. Usp.*, **32**, 452–455.

Shukla, P. K. and Silin, V. P. (1992). Dust ion–acoustic wave. *Phys. Scripta*, **45**, 508.

Shukla, P. K. and Mamun, A. A. (2001). *Introduction to dusty plasma physics.* IOP Publ., Bristol.

Sickafoose, A. A., Colwell, J. E., Horanyi, M., and Robertson, S. (2000). Photoelectric charging of dust particles in vacuum. *Phys. Rev. Lett.*, **84**, 6034–6037.

Smirnov, B. M. (2000). *Clusters and small particles.* Springer, New York.

Sodha, M. S. and Guha, S. (1971). Physics of colloidal plasmas. *Adv. Plasma Phys.*, **4**, 219–309.

Soo, S. L. (1990). *Multiphase fluid dynamics.* Gower Technical, Brookfield.

Sorasio, G., Resendes, D. P., and Shukla, P. K. (2002). Induced oscillations of dust grains in a plasma sheath under low pressures. *Phys. Lett. A*, **293**, 67–73.

Stevens, M. J. and Robbins, M. O. (1993). Melting of Yukawa systems: A test of phenomenological melting criteria. *J. Chem. Phys.*, **98**, 2319–2324.

Su, C. H. and Lam, S. H. (1963). Continuum theory of spherical electrostatic probes. *Phys. Fluids*, **6**, 1479–1491.

Takahashi, K., Oishi, T., Shimonai, K. I., Hayashi, Y., and Nishino, S. (1998). Analyses of attractive forces between particles in Coulomb crystal of dusty plasmas by optical manipulations. *Phys. Rev. E*, **58**, 7805–7811.

Talbot, L., Cheng, R. K., Schefer, R. W., and Willis, D. R. (1980). Thermophoresis of particles in a heated boundary layer. *J. Fluid Mech.*, **101**, 737–758.

Thomas, H. M. and Morfill, G. E. (1996a). Melting dynamics of a plasma crystal. *Nature (London)*, **379**, 806–809.

Thomas, H. M. and Morfill, G. E. (1996b). Solid/liquid/gaseous phase transitions in plasma crystals. *J. Vac. Sci. Technol. A*, **14**, 501–505.

Thomas, H. M. and Morfill, G. E. (2001). Complex plasmas – An introduction to a new field of research. *Contrib. Plasma Phys.*, **41**, 255–258.

Thomas, H., Morfill, G. E. Demmel, V., Goree, J., Feuerbacher, B., and Mohlmann, D. (1994). Plasma crystal: Coulomb crystallization in a dusty plasma. *Phys. Rev. Lett.*, **73**, 652–655.

Thompson, C., Barkan, A., D'Angelo, N., and Merlino, R. L. (1997). Dust acoustic waves in a direct current glow discharge. *Phys. Plasmas*, **4**, 2331–2335.

Tomme, E. B., Law, D. A., Annaratone, B. M., Allen, J. E. (2000). Parabolic plasma sheath potentials and their implications for the charge on levitated dust particles. *Phys. Rev. Lett.*, **85**, 2518–2521.

Totsuji, H. (2001). Structure and melting of two–dimensional dust crystals. *Phys. Plasmas*, **8**, 1856–1862.

Totsuji, H., Totsuji, C., and Tsuruta, K. (2001). Structure of finite two–dimensional Yukawa lattices: Dust crystals. *Phys. Rev. E*, **64**, 066402/1–7.

Trigger, S. A. (2003a). Fokker–Planck equation for Boltzmann–type and active particles: Transfer probability approach. *Phys. Rev. E*, **67**, 046403/1–10.

Trigger, S. A., Ebeling, W., Ignatov, A. M., and Tkachenko, I. M. (2003b). Fokker–Planck equation with velocity–dependent coefficients: Application to dusty plasmas and active particles. Contrib. *Plasma Phys.*, **43**, 377–380.

Trottenberg, T., Melzer, A., and Piel, A. (1995). Measurement of the electric charge on particulates forming Coulomb crystals in the sheath of a radiofrequency plasma. *Plasma Sources Sci. Technol.*, **4**, 450–458.

Tsytovich, V. N. (1994). A possibility of attraction negative charges and formation of a new state of matter in plasma. *Comments Plasma Phys. Control. Fusion*, **15**, 349–358.

Tsytovich, V. N. (1997). Dust plasma crystals, drops, and clouds. *Physics–Uspekhi*, **40**, 53–94.

Tsytovich, V. (2001). Evolution of voids in dusty plasmas. *Phys. Scripta*, **T89**, 89–94.

Tsytovich, V. N. and de Angelis, U. (1999). Kinetic theory of dusty plasmas. General approach. *Phys. Plasmas*, **6**, 1093–1106.

Tsytovich, V. N. and de Angelis, U. (2000). Kinetic theory of dusty plasmas II. Dust–plasma particle collision integrals. *Phys. Plasmas*, **7**, 554–563.

Tsytovich, V. N. and de Angelis, U. (2001). Kinetic theory of dusty plasmas. III. Dust–dust collision integrals. *Phys. Plasmas*, **8**, 1141–1153.

Tsytovich, V. N. and de Angelis, U. (2002). Kinetic theory of dusty plasmas. IV. Distribution and fluctuations of dust charges. *Phys. Plasmas*, **9**, 2497–2506.

Tsytovich, V. N. and Winter, J. (1998). On the role of dust in fusion devices. *Phys. Usp.*, **41**, 815–822.

Tsytovich, V. N., de Angelis, U., and Bingham, R. (2001a). Low–frequency responses and wave dispersion in dusty plasmas. *Phys. Rev. Lett.*, **87**, 185003/1–4.

Tsytovich, V. N., Khodataev, Ya. K., and Bingham, R. (1996). Formation of a dust molecule in plasmas as a first step to supper–chemistry. *Comments Plasma Phys. Control. Fusion*, **17**, 249–265.

Tsytovich, V. N., Khodataev, Ya. K., Morfill, G. E., Bingham, R., and Winter, D. J. (1998). Radiative dust cooling and dust agglomeration in plasmas. *Comments Plasma Phys. Control. Fusion*, **18**, 281–291.

Tsytovich, V. N., Morfill, G. E., and Thomas H. (2002). Complex plasmas: I. Complex plasmas as unusual state of matter. *Plasma Phys. Rep.*, **28**, 623–651.

Tsytovich, V. N., Vladimirov, S. V., Morfill, G. E., and Goree, J. (2001b). Theory of collision–dominated dust voids in plasmas. *Phys. Rev. E*, **63**, 056609/1–11.

Uglov, A. A. and Gnedovets, A. G. (1991). Effect of particle charging on momentum and heat–transfer from rarefied plasma–flow. *Plasma Chem. Plasma Proc.*, **11**, 251–267.

Uhlenbeck, G. E. and Ornstein, L. S. (1930). On the theory of the Brownian motion. *Phys. Rev.*, **36**, 823–841.

Varma, R. K., Shukla, P. K., and Krishan, V. (1993). Electrostatic oscillations in the presence of grain–charge perturbations in dusty plasmas. *Phys. Rev. E*, **47**, 3612–3616.

Vaulina, O. S. and Khrapak, S. A. (2000). Scaling law for the fluid–solid phase transition in Yukawa systems (dusty plasmas). *JETP*, **90**, 287–289.

Vaulina, O. S. and Khrapak, S. A. (2001). Simulation of the dynamics of strongly interacting macroparticles in a weakly ionized plasma. *JETP*, **92**, 228–234.

Vaulina, O., Khrapak, S., and Morfill, G. (2002). Universal scaling in complex (dusty) plasmas. *Phys. Rev. E*, **66**, 016404/1–5.

Vaulina, O. S., Khrapak, S. A., Nefedov, A. P., and Petrov, O. F. (1999b). Charge fluctuations induced heating of dust particles in a plasma. *Phys. Rev. E*, **60**, 5959–5964.

Vaulina, O. S., Khrapak, S. A., Samarian, A. A., and Petrov, O. F. (2000). Effect of stochastic grain charge fluctuations on the kinetic energy of the particles in dusty plasma. *Phys. Scripta* **T84**, 229–231.

Vaulina, O. S., Nefedov, A. P., Petrov, O. F., and Fortov, V. E. (2001). Transport properties of macroparticles in dust plasma induced by solar radiation. *JETP*, **92**, 979–985.

Vaulina, O. S., Nefedov, A. P., Petrov, O. F., and Khrapak, S. A. (1999a). Role of stochastic fluctuations in the charge on macroscopic particles in dusty plasmas. *JETP*, **88**, 1130–1136

Vaulina, O. S. and Vladimirov, S. V. (2002). Diffusion and dynamics of macro–particles in a complex plasma. *Phys. Plasmas*, **9**, 835–840.

Vladimirov, S. V. (1994). Propagation of waves in dusty plasmas with variable charges on dust particles. *Phys. Plasmas*, **1**, 2762–2767.

Vladimirov, S. V. and Ishihara, O. (1996). On plasma crystal formation. *Phys. Plasmas*, **3**, 444–446.

Vladimirov, S. V. and Nambu, M. (1995). Attraction of charged particulates in plasmas with finite Iows. *Phys. Rev. E*, **52**, R2172–R2174.

Vladimirov, S. V. and Nambu, M. (2001). Interaction of a rod–like charged macroparticle with a flowing plasma, *Phys. Rev. E*, **64**, 026403/1–7.

Vladimirov, S. V., Maiorov, S. A., and Ishihara, O. (2003). Molecular dynamics simulation of plasma flow around two stationary dust grains. *Phys. Plasmas*, **10**, 3867–3873.

Vladimirov, S. V., Ostrikov, K. N., Yu, M. Y., and Stenflo, L. (1998). Evolution of Langmuir waves in a plasma contaminated by variable–charge impurities. *Phys. Rev. E*, **58**, 8046–8048.

Vladimirov, S. V., Shevchenko, P. V., and Cramer, N. F. (1997). Vibrational modes in the dust–plasma crystal. *Phys. Rev. E*, **56**, R74–R76.

Vladimirov, V. I., Deputatova, L. V., Molotkov, V. I., Nefedov, A. P., Rykov, V. A., Filinov, V. S., Fortov, V. E., and Khudyakov. A. V. (2001a). Ordered dusty structures in nuclear–track neon and argon plasmas. *Plasma Phys. Reports*, **27**, 36–43.

Vladimirov, V. I., Deputatova, L. V., Nefedov, A. P., Fortov, V. E., Rykov, V. A., and Khudyakov. A. V. (2001b). Dust vortices, clouds, and jets in nuclear–induced plasmas. *JETP*, **93**, 313–323.

Walch, B., Horanyi, M., and Robertson, S. (1995). Charging of dust grains in plasma with energetic electrons. *Phys. Rev. Lett.*, **75**, 838–840.

Wang, X., Bhattacharjee, A., Gou, S. K., and Goree, J. (2001a). Ionization instabilities and resonant acoustic modes. *Phys. Plasmas*, **8**, 5018–5024.

Wang, X., Bhattacharjee, A., and Hu, S. (2001b). Longitudinal and transverse waves in Yukawa crystals. *Phys. Rev. Lett.*, **86**, 2569–2572.

Wang, Y.–N., Hou, L.–J., and Wang, X. (2002). Self–consistent nonlinear resonance and hysteresis of a charged microparticle in a r.f. sheath. *Phys. Rev. Lett.*, **89**, 155001/1–4.

Whipple, E. C. (1981). Potentials of surfaces in space. *Rep. Prog. Phys.*, **44**, 1197–1250.

Winske, D. (2001). Nonlinear wake potential in a dusty plasma. *IEEE Trans. Plasma Sci.*, **29**, 191–197.

Winske, D., Murillo, M. S., and Rosenberg, M. (1999). Numerical simulation of dust–acoustic waves. *Phys. Rev. E*, **59**, 2263–2272.

Winter, J. (2000). Dust: A new challenge in nuclear fusion research? *Phys. Plasmas*, **7**, 3862–3866.

Winter, J. and Gebauer, G. (1999). Dust in magnetic confinement fusion devices and its impact on plasma operation. *J. Nucl. Mater.*, **266**, 228–233.

Wuerker, R. F., Shelton, H., and Langmuir, R. V. (1959). Electrodynamic containment of charged particles. *J. Appl. Phys.*, **30**, 342–349.

Xie, B. S. and Yu, M. Y. (2000). Dust acoustic waves in strongly coupled dissipative plasmas. *Phys. Rev. E*, **62**, 8501–8507.

Xie, B., He, K., and Huang, Z. (1999). Attractive potential in weak ion flow coupling with dust–acoustic waves. *Phys. Lett. A*, **253**, 83–87.

Yakubov, I. T. and Khrapak, A. G. (1989). Thermophysical and electrophysical properties of low temperature plasma with condensed disperse phase. *Sov. Tech. Rev. B. Therm. Phys.*, **2**, 269–337.

Zafiu, C., Melzer, A., and Piel, A. (2001). Nonlinear resonances of particles in a dusty plasma sheath. *Phys. Rev. E*, **63**, 066403/1–8.

Zafiu, C., Melzer, A., and Piel, A. (2002). Ion drag and thermophoretic forces acting on free falling charged particles in an r.f.–driven complex plasma. *Phys. Plasmas*, **9**, 4794–4803.

Zafiu, C., Melzer, A., and Piel, A. (2003). Measurement of the ion drag force on falling dust particles and its relation to the void formation in complex (dusty) plasmas. *Phys. Plasmas*, **10**, 1278–1282.

Zamalin, V. M., Norman, G. E., and Filinov, V. S. (1977). *Monte Carlo method in statistical thermodynamics* (in Russian). Nauka, Moscow.

Zhakhovskii, V. V., Molotkov, V. I., Nefedov, A. P., Torchinskii, V. M., Khrapak, A. G., and Fortov, V. E. (1997). Anomalous heating of a system of dust particles in a gas–discharge plasma. *JETP Lett.*, **66**, 419–425.

Zhukhovitskii, D. I., Khrapak, A. G., and Yakubov, I. T. (1984). Ionization equilibrium in plasma with condensed disperse phase (in Russian), In *Khimiya Plazmy (Plasma Chemistry). Vol. 11*, B.M. Smirnov (ed.), pp. 130–170. Energoatomizdat, Moscow.

Zobnin, A. V., Nefedov, A. P., Sinel'shchikov, V. A., and Fortov, V. E. (2000). On the charge of dust particles in a low–pressure gas discharge plasma. *JETP*, **91**, 483–487.

Zobnin, A. V., Usachev, A. D., Petrov, O. F., and Fortov, V. E. (2002). Dust–acoustic instability in an inductive gas–discharge plasma. *JETP*, **95**, 429–439.

Zuzic, M., Thomas, H. M., and Morfill, G. E. (1996). Wave propagation and damping in plasma crystals. *J. Vac. Sci. Technol. A*, **14**, 496–500.

Zuzic, M., Ivlev, A. V., Goree, J., Morfill, G. E., Thomas, H. M., Rothermel, H., Konopka, U., Sutterlin, R., and Goldbeck, D. D. (2000). Three–dimensional strongly coupled plasma crystal under gravity conditions. *Phys. Rev. Lett.*, **85**, 4064–4067.

APPENDIX A

CRITICAL PARAMETERS OF METALS

Element	T_c, K	p_c, 10^8 Pa	ρ_c, g cm^{-3}	S_c, cal mol^{-1}K^{-1}
Zr	16250	7.52	1.79	40.45
Zn	3196	2.63	2.29	31.27
Y	10800	3.74	1.30	38.92
Yb	4280	1.38	2.36	36.92
V	12500	10.78	1.86	37.26
U	11630	6.11	5.30	43.07
W	21010	15.83	5.87	43.34
Ti	11790	7.63	1.31	37.92
Sn	8200	3.35	2.05	39.44
Th	14950	4.88	3.21	43.67
Tu	5910	2.65	3.22	38.52
Tl	4470	1.63	3.16	38.67
Tb	8060	3.08	2.57	42.34
Te	1850	0.75	2.21	35.22
Ta	20570	13.5	5.04	43.07
Sr	3860	0.90	0.86	35.23
Ag	7010	4.50	2.93	36.86
Se	8350	4.08	0.93	36.82
Sm	5340	2.10	2.51	40.62
Ru	15500	13.74	3.79	37.14
Rh	13510	11.23	3.62	39.66
Re	19600	15.7	6.32	43.42
Ra	3830	0.77	1.93	38.32
Pr	9160	2.85	1.86	41.39
Tc	15930	11.81	3.09	41.00
Pa	12650	4.82	3.72	43.34
Po	2050	0.62	2.67	37.52
Pt	14330	8.70	5.02	41.8
Pd	10760	7.64	3.20	36.64
Os	17110	14.49	6.83	42.60
Nb	19040	12.52	2.59	41.65
Ni	10330	9.12	2.19	36.50
Nd	7920	2.65	2.05	41.33
Mo	16140	12.63	3.18	41.21

Element	T_c, K	p_c, 10^8 Pa	ρ_c, $\mathrm{g\,cm}^{-3}$	S_c, $\mathrm{cal\,mol}^{-1}\mathrm{K}^{-1}$
Lu	7060	2.84	2.97	39.41
Fr	1810	0.12	0.65	39.50
Bi	4200	1.26	2.66	40.2
Hg	1763 ± 15	1.51 ± 0.025	4.2 ± 0.4	33.790
Hg	1753 ± 10	1.52 ± 0.01	5.7 ± 0.2	33.241
Na	2573	0.275	0.206	31.683
K	2223	0.152	0.194	33.931
Rb	2093	0.159	0.346	36.485
Cs	2057 ± 40	0.144	0.428 ± 0.012	38.111
Li	3223	0.689	0.105	27.666
Mn	5940	6.28	2.46	35.68
Mg	3590	1.98	0.56	29.48
Pb	4980	1.84	3.25	39.81
La	11060	3.35	1.78	40.58
Fe	9600	8.25	2.03	37.20
Ir	15380	12.78	6.77	41.91
In	6120	2.43	1.84	37.95
Ho	7240	2.94	2.84	40.63
Hf	18270	9.38	3.88	42.59
Au	8970	6.10	5.68	39.40
Ge	9170	4.90	1.64	37.77
Ga	7210	4.31	1.77	35.91
Gd	8670	3.25	2.50	42.79
Er	8250	3.34	2.86	40.51
Eu	4680	1.21	1.67	41.36
Dy	7240	2.91	2.76	41.05
Cu	8390	7.46	2.39	35.30
Co	10460	9.23	2.20	37.12
Cr	9620	9.68	2.22	37.35
Ce	8860	3.03	2.03	40.51
Ca	4180	1.21	0.49	32.70
Cd	2790	1.60	2.74	33.20
Be	8080	11.70	0.55	27.02
Ba	4100	0.81	1.15	37.06
Al	8000	4.47	0.64	33.52
Sb	2570	1.30	2.61	36.1
B	8200	9.57	0.69	28.9
Se	1010	0.37	1.60	31.8
S	730	0.30	0.72	27.7

APPENDIX B

TRANSPORT CROSS–SECTIONS OF ELECTRON SCATTERING BY RARE GAS ATOMS

E, eV	0	0.01	0.02	0.03
He	–	5.21 (–16)	5.35 (–16)	5.46 (–16)
Ne	–	–	–	4.69 (–17)
Ar	8.05 (–16)*	6.1 (–16)	3.74 (–16)	2.08 (–16)
Kr	3.07 (–15)	2.6 (–16)	1.97 (–15)	1.6 (–15)
Xe	1.78 (–14)	1.16 (–14)	8.0 (–15)	6.13 (–15)

E, eV	0.05	0.07	0.1	0.15
He	5.62 (–16)	5.74 (–16)	5.86 (–16)	6.04 (–16)
Ne	5.36 (–17)	6.01 (–17)	7.01 (–17)	8.28 (–17)
Ar	1.84 (–16)	1.14 (–16)	4.5 (–17)	2.2 (–17)
Kr	1.14 (–15)	9.0 (–16)	6.8 (–16)	4.4 (–16)
Xe	3.9 (–15)	2.9 (–15)	2.1 (–15)	1.29 (–15)

E, eV	0.25	0.3	0.5	1.0
He	6.26 (–16)	6.35 (–16)	6.59 (–16)	6.85 (–16)
Ne	1.02 (–16)	1.09 (–16)	1.32 (–16)	1.62 (–16)
Ar	1.56 (–16)	1.09 (–16)	2.83 (–17)	1.05 (–16)
Kr	1.5 (–16)	1.0 (–16)	5.3 (–17)	6.17 (–17)
Xe	5.35 (–16)	3.15 (–16)	1.38 (–17)	2.47 (–16)

E, eV	2.0	3.0	4.0	5.0
He	6.99 (–16)	6.89 (–16)	5.6 (–16)	6.26 (–16)
Ne	1.81 (–16)	1.9 (–16)	1.98 (–16)	2.07 (–16)
Ar	2.48 (–16)	4.07 (–16)	5.8 (–16)	–
Kr	2.1 (–16)	4.84 (–16)	–	1.02 (–15)
Xe	8.25 (–16)	1.7 (–15)	2.48 (–15)	3.08 (–15)

E, eV	6.0	7.0	8.0	10.0
He	6.01 (–16)	–	–	–
Ne	2.14 (–16)	2.21 (–16)	–	–
Ar	8.7 (–16)	–	1.17 (–15)	1.38 (–15)
Kr	–	1.5 (–15)	–	1.93 (–15)
Xe	–	–	3.37 (–15)	3.2 (–15)

* In the table following notation is adopted: $1.6\,(-16) = 1.6 \cdot 10^{-16}$

APPENDIX   C

THE ELECTRON WORK FUNCTION FOR METALS

Metal	W, eV	Metal	W, eV	Metal	W, eV	Metal	W, eV
Li	2.38	Be	3.92	Cu	4.4	Pb	4.0
Na	2.35	Mg	3.64	Zn	4.24	Mo	4.3
K	2.22	Ca	2.80	Cd	4.1	W	4.5
Rb	2.16	Sr	2.35	Hg	4.25	U	3.3
Cs	1.81	Ba	2.49	Sn	4.38		

INDEX